AF595059

Design of Micro- and Nanoparticles: Self-Assembly and Application

Design of Micro- and Nanoparticles: Self-Assembly and Application

Editors

Pavel Padnya
Ivan Stoikov

MDPI • Basel • Beijing • Wuhan • Barcelona • Belgrade • Manchester • Tokyo • Cluj • Tianjin

Editors

Pavel Padnya
A.M. Butlerov Chemistry
Institute
Kazan Federal University
Kazan
Russia

Ivan Stoikov
A.M. Butlerov Chemistry
Institute
Kazan Federal University
Kazan
Russia

Editorial Office
MDPI
St. Alban-Anlage 66
4052 Basel, Switzerland

This is a reprint of articles from the Special Issue published online in the open access journal *Nanomaterials* (ISSN 2079-4991) (available at: www.mdpi.com/journal/nanomaterials/special_issues/Design_Nanoparticles).

For citation purposes, cite each article independently as indicated on the article page online and as indicated below:

LastName, A.A.; LastName, B.B.; LastName, C.C. Article Title. *Journal Name* **Year**, *Volume Number*, Page Range.

ISBN 978-3-0365-4764-0 (Hbk)
ISBN 978-3-0365-4763-3 (PDF)

Contents

About the Editors

Pavel Padnya

Dr. Pavel Padnya graduated from the Chemistry Faculty of Kazan Federal University in 2010 and defended his PhD thesis in 2015 at Kazan Federal University. After that, he worked as an engineer and researcher at the Organic Chemistry Department of Kazan Federal University, Russia. Since 2019, he has been a senior researcher in the above department. He is the author of 59 publications in a peer-reviewed international journal (h-index of 14, Scopus). He is a topical advisory panel member of {Nanomaterials} (MDPI) and {Journal of Functional Biomaterials} (MDPI). His research activity covers the design, synthesis and study of applications of macrocyclic compounds and nanomaterials.

Ivan Stoikov

Prof. Dr. Ivan Stoikov is the head of the Department of Organic and Medicinal Chemistry of the Kazan Federal University (Russia). He received his senior doctor's degree in chemistry in 2008. His research interests cover areas of organic, bioorganic, and supramolecular chemistry, molecular recognition, polymer chemistry, and nanoscience, including the synthesis of macrocycles (calix[4]arenes, thiacalix[4]arenes, pillar[5]arene), artificial ion channels, and membrane extraction. The current research interest of his group involves supramolecular biomedical materials. He is the author of more than 260 articles indexed by Scopus and Web of Science and has been cited more than 2800 times with an H-index of 25.

Preface to "Design of Micro- and Nanoparticles: Self-Assembly and Application"

Micro-/nanoparticles (M&NPs) have attracted researchers'interest due to their unique chemical and physical properties. Recent advances in the synthesis of M&NPs (metallic, metal oxide, silica, polymeric, lipid-based, supramolecular, colloidal, and carbon-based M&NPs) have offered exciting opportunities for many applications, such as catalysis, plasmonics, sensors, magnetism, drug delivery, and nanomedicine. The use of self-assembly for the synthesis or modification of M&NPs provides unlimited possibilities for designing nanomaterials with desired properties. This reprint presents the most recent developments in the synthesis, modification, properties, and applications of M&NPs. This book will be useful to scientists of diverse backgrounds, i.e., material science, organic and inorganic chemistry, biochemistry, and biology.

Pavel Padnya and Ivan Stoikov
Editors

 *nanomaterials* MDPI

Editorial

Design of Micro- and Nanoparticles: Self-Assembly and Application

Pavel Padnya * and Ivan Stoikov *

A.M. Butlerov Chemistry Institute, Kazan Federal University, 420008 Kazan, Russia
* Correspondence: padnya.ksu@gmail.com (P.P.); ivan.stoikov@mail.ru (I.S.); Tel.: +7-843-233-7241 (I.S.)

The modern world throws down an increasing number of challenges to humanity. The development of nanotechnology and the creation of nanomaterials of the future are possible solution to the emerging threats. Today, nanomaterials are in demand in the most pressing research areas starting from solving problems with diagnosing COVID-19 and developing antiviral drugs and finishing with new energy sources and environmental challenges.

The use of self-assembly of organic or/and inorganic components to create the nanomaterials of the future is a promising area of research for many scientists due to the limitless possibilities of designing various architectures for many tasks of modern technologies [1]. This Special Issue covers a wide range of research topics related to the design and application of nanomaterials, including micro- and nanoparticles based on self-assembly.

The development of new approaches to the synthesis of such nanomaterials is urgent due to the promising prospects of the resulting nanostructures with a broad application potential. The use of nanoparticles in biomedicine offers new opportunities for the development of drug delivery systems [2]. Nakamura et al. has proposed a novel protein carrier system capable of delivering various proteins to cells in vitro and in vivo [3]. The uniqueness of this system is in its easy two-step method of self-assembling of protamine, low molecular heparin, and cell penetrating peptides. The obtained system has great potential for use, especially in the field of animal biotechnology and biomedicine.

The authors of the article [4] have developed a pH-sensitive supramolecular nanosystem based on decasubstituted pillar[5]arenes and tetrazole-containing polymers. A series of pillar[5]arene derivatives containing tosylate, phthalimide, and primary and tertiary amino groups was synthesized to create novel pH-sensitive nanosystems. It was shown that nanoparticles, stable at neutral pH and degraded in acidic medium (pH = 5), were formed by the interaction of the synthesized pillar[5]arenes with tetrazole-containing polymers. The results obtained are a step toward the creation of new universal stimulus-sensitive drug delivery systems.

Due to the active influence of viral diseases, it is important to search for new antiviral agents, as well as to establish the mechanism and potential of existing medicines. Severe acute respiratory syndrome (SARS), Middle East respiratory syndrome (MERS), and coronavirus disease (COVID-19) are the most well-known diseases affecting the life quality of the population. The prospect of using nanomaterials and nanoparticles (silver, gold, quantum dots, organic nanoparticles, liposomes, dendrimers, and polymers) against coronaviruses was discussed in the review [5]. The results obtained during the discussion can be actively used for the development of relevant antiviral drugs against coronaviruses and other viral diseases.

The use of various biomolecules in modern bioengineering and nanotechnology makes it possible to create new nanomaterials with attractive properties such as biocompatibility, biodegradability, the ability to reproduce natural cellular microenvironment, and minimal toxicity. A recent review [6] brings together publications on the design, construction, and in vitro and in vivo research of multilayer capsules and nanoparticles based on biopolymers mimicked on non-porous and porous particles and crystalline drugs. Based on the analysis

Citation: Padnya, P.; Stoikov, I. Design of Micro- and Nanoparticles: Self-Assembly and Application. *Nanomaterials* **2022**, *12*, 430. https://doi.org/10.3390/nano12030430

Received: 16 December 2021
Accepted: 18 January 2022
Published: 27 January 2022

and summary of the results obtained for many years, the authors conclude about the prospects and problems of this topic.

Environmental pollution by industry and non-ecological modes of transport is another problem of the 21st century. The transfer to electric vehicles can solve it. However, the problem of creating modern storage batteries has appeared in this way. $LiFePO_4$, often used as a cathode, is one of the promising materials in lithium-ion batteries. Ruiz-Jorge et al. [7] investigated the effect of the heating rate on the hydrothermal synthesis of $LiFePO_4$ particles. It turned out that it is important to provide low heating rates in the temperature range of 130–150 °C for the formation of large amounts of $LiFePO_4$ particles with good levels of crystallinity. The results obtained open up possibilities for the rapid industrial synthesis of the materials required for storage batteries.

When creating micro- and nanoparticles, it is important to learn how to control the synthesis process, not only for the reproducibility of the results obtained but also in order to search for the possibility of predicting the properties of the target nanostructures [8]. Prof. Duguet and co-workers [9] obtained submicron polystyrene particles with several 100 nm silica or magnetic silica patches emerging at their surface by self-assembly under the solvent action. Thus, the authors were able to develop a method for obtaining and controlling the morphology of polymeric multipatch particles. Further use of the obtained microparticles with inorganic patches was expected in various applications as nanomotors, nanostirrers, nanoswimmers, e-paper pixels, targeted drug delivery, and water treatment.

Prof. Toprak's scientific group suggested another example of the controlled synthesis of the nanoparticles [10]. The effect of the addition of a series of salts and surfactant additives (poly(vinyl pyrrolidone), sodium acetate, sodium citrate, hexadecyltrimethylammonium bromide, hexadecyltrimethylammonium chloride, and potassium bromide) on the morphology of Rh nanoparticles was investigated. The correlation between the use of the studied additives and the shape of rhodium nanoparticles (trigonal, cubic, or spherical ones) was shown. The authors studied the cytotoxicity of the obtained Rh nanoparticles and assessed the cell lines of macrophages and ovarian cancer. The obtained results are important and can be useful for the design of the Rh-nanoparticles-based contrast agents for X-ray fluorescence computed tomography bio-imaging.

Supramolecular self-assembly is a promising way to obtain functional nanosystems [11]. The use of synthetic macrocyclic compounds such as crown ethers, cyclodextrins, cucurbiturils, calixarenes, porphyrins, and pillararenes opens up wide possibilities for creating such systems [12–15]. New amphiphilic calix[4]arene derivatives containing *N*-alkyl/aryl imidazolium/benzimidazolium fragments capable of self-assembly into the submicron nanoparticles in water were synthesized [16]. The addition of the obtained compounds to multilayer 1,2-dipalmitoyl-sn-glycero-3-phosphocholine (DPPS) vesicles resulted in the formation of single-layer calixarene/DPPS vesicles, which are capable of enhancing the catalytic activity in Suzuki–Miyaura coupling.

Organic dyes and their nanosized polymeric forms are widely used in the design of electrochemical sensors based on the self-assembly of organic components [17]. A novel DNA sensor for the determination of doxorubicin has been developed by the consecutive electropolymerization of an equimolar mixture of organic dyes (azure B and proflavine) and the adsorption of the DNA from salmon sperm on the polymer film [18]. The developed electrochemical biosensor can detect the anticancer drug doxorubicin in the concentration range from 0.03 to 10 nM with the limit of detection of 0.01 nM. The DNA sensor was tested on doxorubicin preparations and on spiked samples of artificial blood serum. It can be used for the determination of drug residues in blood and for pharmacokinetic studies.

A cerium oxide (CeO_2) nanostructures possess unique catalytic and redox properties, comprehensively described in the review [19]. Ceria's nanomorphology is an important factor affecting its reactivity. Therefore, the development of new methods for obtaining these nanomaterials with different shapes based on various biotemplates is actual. The authors analyzed the relationship between the nanomorphology of nanocerium and its reactivity. They described biotemplates used in recent years for the synthesis of cerium nanostructures and their (bio)applications in many areas such as catalysis for energy consumption, environmental preservation, and remediation.

Due to the active development of new synthetic methods for nanomaterials, the influence of impurities formed in the synthesis are challenges to the quality of the obtained nanostructures. The search for the methods of their purification is an inalienable part of the nanoparticles' design. Prof. Prior and co-workers, in their review article [20], summarized the latest advances in silver nanoparticle purification techniques. The authors identified several commonly used approaches, such as magnetic and hydrodynamic force-based methods, chromatography, density gradient centrifugation, electrophoresis, selective precipitation, membrane filtration, and liquid extraction. Due to the possible toxicity of organic and/or inorganic impurities, this review is relevant for many researchers planning their work in the nanotechnology field, especially in case of biological applications of the obtained nanomaterials.

Two-dimensional nanomaterials are another interesting type of nanomaterials. There are two key approaches of obtaining 2D nanomaterials, namely, top-down and bottom-up technologies [21]. Each of them has its own advantages and disadvantages [22]. The production of nanolayers using top-down approaches, such as the layering of bulk layered oxides, is one of the most promising directions in the field of 2D nanomaterials. Using the top-down approach, Silyukov and co-workers [23] showed the formation of stable suspensions of perovskite nanolayers and their coatings. They were obtained by exfoliation of the protonated bismuth titanate and its *N*-alkylamine derivatives in an aqueous solution of tetrabutylammonium hydroxide. The resulting nanolayers and coatings can be used as functional materials for catalysis and electronics.

Self-assembling in solid state 2D nanosheets based on thiacalix[4]arene derivatives functionalized by geranyl fragments at the lower rim in *cone* and *1,3-alternate* conformations were obtained by the bottom-up approach [24]. The choice of terpenoid macrocyclic derivatives was explained by their ability to polymorphism and crystalline modifications, as well as non-toxicity. The obtained 2D nanomaterials can be used for the creation of high-energy materials and polymorphic structures for pharmaceuticals.

Overall, this Special Issue includes 13 excellent papers of the contributors from 10 countries in the design and application of nanomaterials, including micro- and nanoparticles, and covers various fields of nanotechnology, chemistry, biology, and material science. As guest editors of the Special Issue "Design of Micro- and Nanoparticles: Self-Assembly and Application", we thank all authors and reviewers for their valuable contributions. We hope that the publications presented in this Special Issue will attract even greater interest from scientists around the world.

Author Contributions: Conceptualization, P.P. and I.S.; writing—original draft preparation, P.P.; writing—review and editing, I.S.; supervision, P.P. and I.S.; project administration, P.P. and I.S.; funding acquisition, I.S. All authors have read and agreed to the published version of the manuscript.

Funding: This research was funded by the grant of the President of the Russian Federation for state support of leading scientific schools of the Russian Federation (NSh-2499.2020.3).

Conflicts of Interest: The authors declare no conflict of interest. The funders had no role in the design of the study; in the collection, analyses, or interpretation of data; in the writing of the manuscript, or in the decision to publish the results.

References

1. Ariga, K.; Jia, X.; Song, J.; Hill, J.P.; Leong, D.T.; Jia, Y.; Li, J. Nanoarchitectonics beyond Self-Assembly: Challenges to Create Bio-Like Hierarchic Organization. *Angew. Chem. Int. Ed.* **2020**, *59*, 15424–15446. [CrossRef] [PubMed]
2. Kashapov, R.; Gaynanova, G.; Gabdrakhmanov, D.; Kuznetsov, D.; Pavlov, R.; Petrov, K.; Zakharova, L.; Sinyashin, O. Self-Assembly of Amphiphilic Compounds as a Versatile Tool for Construction of Nanoscale Drug Carriers. *IJMS* **2020**, *21*, 6961. [CrossRef] [PubMed]
3. Nakamura, S.; Ando, N.; Ishihara, M.; Sato, M. Development of Novel Heparin/Protamine Nanoparticles Useful for Delivery of Exogenous Proteins In Vitro and In Vivo. *Nanomaterials* **2020**, *10*, 1584. [CrossRef] [PubMed]

4. Shurpik, D.N.; Makhmutova, L.I.; Usachev, K.S.; Islamov, D.R.; Mostovaya, O.A.; Nazarova, A.A.; Kizhnyaev, V.N.; Stoikov, I.I. Towards Universal Stimuli-Responsive Drug Delivery Systems: Pillar[5]Arenes Synthesis and Self-Assembly into Nanocontainers with Tetrazole Polymers. *Nanomaterials* **2021**, *11*, 947. [CrossRef]
5. Gurunathan, S.; Qasim, M.; Choi, Y.; Do, J.T.; Park, C.; Hong, K.; Kim, J.-H.; Song, H. Antiviral Potential of Nanoparticles—Can Nanoparticles Fight Against Coronaviruses? *Nanomaterials* **2020**, *10*, 1645. [CrossRef]
6. Vikulina, A.S.; Campbell, J. Biopolymer-Based Multilayer Capsules and Beads Made via Templating: Advantages, Hurdles and Perspectives. *Nanomaterials* **2021**, *11*, 2502. [CrossRef]
7. Ruiz-Jorge, F.; Benítez, A.; García-Jarana, M.B.; Sánchez-Oneto, J.; Portela, J.R.; Martínez de la Ossa, E.J. Effect of the Heating Rate to Prevent the Generation of Iron Oxides during the Hydrothermal Synthesis of LiFePO4. *Nanomaterials* **2021**, *11*, 2412. [CrossRef]
8. Wang, L.; Gong, C.; Yuan, X.; Wei, G. Controlling the Self-Assembly of Biomolecules into Functional Nanomaterials through Internal Interactions and External Stimulations: A Review. *Nanomaterials* **2019**, *9*, 285. [CrossRef]
9. Yammine, E.; Adumeau, L.; Abboud, M.; Mornet, S.; Nakhl, M.; Duguet, E. Towards Polymeric Nanoparticles with Multiple Magnetic Patches. *Nanomaterials* **2021**, *11*, 147. [CrossRef]
10. Li, Y.; Saladino, G.M.; Shaker, K.; Svenda, M.; Vogt, C.; Brodin, B.; Hertz, H.M.; Toprak, M.S. Synthesis, Physicochemical Characterization, and Cytotoxicity Assessment of Rh Nanoparticles with Different Morphologies-as Potential XFCT Nanoprobes. *Nanomaterials* **2020**, *10*, 2129. [CrossRef]
11. Antipin, I.S.; Alfimov, M.V.; Arslanov, V.V.; Burilov, V.A.; Vatsadze, S.Z.; Voloshin, Y.Z.; Volcho, K.P.; Gorbatchuk, V.V.; Gorbunova, Y.G.; Gromov, S.P.; et al. Functional Supramolecular Systems: Design and Applications. *Russ. Chem. Rev.* **2021**, *90*, 895–1107. [CrossRef]
12. Nazarova, A.A.; Padnya, P.L.; Gilyazeva, A.I.; Khannanov, A.A.; Evtugyn, V.G.; Kutyreva, M.P.; Klochkov, V.V.; Stoikov, I.I. Supramolecular Motifs for the Self-Assembly of Monosubstituted Pillar[5]Arenes with an Amide Fragment: From Nanoparticles to Supramolecular Polymers. *New J. Chem.* **2018**, *42*, 19853–19863. [CrossRef]
13. Braga, S.S. Cyclodextrins: Emerging Medicines of the New Millennium. *Biomolecules* **2019**, *9*, 801. [CrossRef]
14. Padnya, P.; Gorbachuk, V.; Stoikov, I. The Role of Calix[n]Arenes and Pillar[n]Arenes in the Design of Silver Nanoparticles: Self-Assembly and Application. *IJMS* **2020**, *21*, 1425. [CrossRef]
15. Shurpik, D.N.; Sevastyanov, D.A.; Zelenikhin, P.V.; Padnya, P.L.; Evtugyn, V.G.; Osin, Y.N.; Stoikov, I.I. Nanoparticles Based on the Zwitterionic Pillar[5]Arene and Ag+: Synthesis, Self-Assembly and Cytotoxicity in the Human Lung Cancer Cell Line A549. *Beilstein J. Nanotechnol.* **2020**, *11*, 421–431. [CrossRef] [PubMed]
16. Burilov, V.; Garipova, R.; Sultanova, E.; Mironova, D.; Grigoryev, I.; Solovieva, S.; Antipin, I. New Amphiphilic Imidazolium/Benzimidazolium Calix[4]Arene Derivatives: Synthesis, Aggregation Behavior and Decoration of DPPC Vesicles for Suzuki Coupling in Aqueous Media. *Nanomaterials* **2020**, *10*, 1143. [CrossRef]
17. Evtugyn, G.; Hianik, T. Electrochemical DNA Sensors and Aptasensors Based on Electropolymerized Materials and Polyelectrolyte Complexes. *TrAC Trends Anal. Chem.* **2016**, *79*, 168–178. [CrossRef]
18. Porfireva, A.; Evtugyn, G. Electrochemical DNA Sensor Based on the Copolymer of Proflavine and Azure B for Doxorubicin Determination. *Nanomaterials* **2020**, *10*, 924. [CrossRef]
19. Rozhin, P.; Melchionna, M.; Fornasiero, P.; Marchesan, S. From Impure to Purified Silver Nanoparticles: Advances and Timeline in Separation Methods. *Nanomaterials* **2021**, *11*, 2259. [CrossRef]
20. Martins, C.S.M.; Sousa, H.B.A.; Prior, J.A.V. Nanostructured Ceria: Biomolecular Templates and (Bio)Applications. *Nanomaterials* **2021**, *11*, 3407. [CrossRef]
21. Fang, Z.; Xing, Q.; Fernandez, D.; Zhang, X.; Yu, G. A Mini Review on Two-Dimensional Nanomaterial Assembly. *Nano Res.* **2019**, *13*, 1179–1190. [CrossRef]
22. Lombardo, D.; Calandra, P.; Pasqua, L.; Magazù, S. Self-Assembly of Organic Nanomaterials and Biomaterials: The Bottom-Up Approach for Functional Nanostructures Formation and Advanced Applications. *Materials* **2020**, *13*, 1048. [CrossRef] [PubMed]
23. Minich, I.A.; Silyukov, O.I.; Kurnosenko, S.A.; Gak, V.V.; Kalganov, V.D.; Kolonitskiy, P.D.; Zvereva, I.A. Physical–Chemical Exfoliation of n-Alkylamine Derivatives of Layered Perovskite-like Oxide H2K0.5Bi2.5Ti4O13 into Nanosheets. *Nanomaterials* **2021**, *11*, 2708. [CrossRef] [PubMed]
24. Vavilova, A.; Padnya, P.; Mukhametzyanov, T.; Buzyurov, A.; Usachev, K.; Islamov, D.; Ziganshin, M.; Boldyrev, A.; Stoikov, I. 2D Monomolecular Nanosheets Based on Thiacalixarene Derivatives: Synthesis, Solid State Self-Assembly and Crystal Polymorphism. *Nanomaterials* **2020**, *10*, 2505. [CrossRef] [PubMed]

Communication

Development of Novel Heparin/Protamine Nanoparticles Useful for Delivery of Exogenous Proteins In Vitro and In Vivo

Shingo Nakamura [1,*], Naoko Ando [1], Masayuki Ishihara [1] and Masahiro Sato [2]

[1] Division of Biomedical Engineering, National Defense Medical College Research Institute, Saitama 359-8513, Japan; naoandokoro@gmail.com (N.A.); ishihara@ndmc.ac.jp (M.I.)

[2] Section of Gene Expression Regulation, Frontier Science Research Center, Kagoshima University, Kagoshima 890-8544, Japan; masasato@m.kufm.kagoshima-u.ac.jp

* Correspondence: snaka@ndmc.ac.jp; Tel.: +81-4-2995-1211

Received: 1 July 2020; Accepted: 29 July 2020; Published: 12 August 2020

Abstract: We previously reported that heparin/protamine particles (LHPPs) produced as nanoparticles through simple mixing of raw materials exhibit sustained protein release and can be retained in cells. In the present study, we modified LHPPs without employing any organic synthetic approach. The resulting LHPPs were re-named as improved LHPPs (*i*-LHPPs) and have the ability to retain cell-penetrating peptides (GRKKRRQRRRPPQ) based on electrostatic interactions. We examined whether *i*-LHPPs can introduce exogenous proteins (i.e., lacZ protein encoding bacterial β-galactosidase) into cultured cells in vitro, or into murine hepatocytes in vivo through intravenous injection to anesthetized mice. We found an accumulation of the transferred protein in both in vitro cultured cells and in vivo hepatocytes. To the best of our knowledge, reports of successful in vivo delivery to hepatocytes are rare. The *i*-LHPP-based protein delivery technique will be useful for in vivo functional genetic modification of mouse hepatocytes using Cas9 protein-mediated genome editing targeting specific genes, leading to the creation of hepatic disease animal models for research that aims to treat liver diseases.

Keywords: heparin/protamine particles; hepatocyte; intravenous injection; lacZ protein; nanoparticles; protein delivery; self-assembling

---

1. Introduction

Introduction of genetic material is one of the representative techniques used to modify cell function. A foreign gene introduced into a cell can produce functional protein, via several processes, namely transcription (mRNA synthesis from the foreign gene), translation of mRNA into protein, and post-translational modifications. This cellular event is termed protein biosynthesis. In contrast, direct introduction of in vitro synthesized protein into a cell does not require protein biosynthesis. As a result, a protein introduced into a cell can rapidly exert its function. Furthermore, its function can be readily controlled by adjusting the protein dose [1]. However, in the case of protein delivery to a cell, maintaining the three-dimensional (3D) structure of the delivered protein itself is very important. If the 3D protein structure is destroyed and metamorphosed upon delivery into a cell, the protein will rapidly lose its function.

Generally, naked proteins introduced in vivo are rapidly degraded by proteolytic enzymes ubiquitously present in a given organism. Furthermore, naked proteins administered into the bloodstream may be difficult to be successfully delivered to target organs or tissues because they do not have specificity towards specific organs (or tissues) [1]. The use of a protein delivery carrier enabling protein encapsulation and subsequent release would overcome these limitations [2].

Small particles in the size range between 1 and 1000 nm, or sheets of that thickness, are generally designated as nanomaterials (or nanoparticles) and may be potentially harmful or beneficial to cells [3]. Generally, nanoparticles are potentially toxic to the human body; therefore, nanoparticles must undergo various removal reactions in vivo, depending on their size. For instance, nanomaterials smaller than 5.5 nm are subject to renal excretion [4]. Therefore, for in vivo applications, particularly when they are used as delivery carriers for intravenous administration, nanomaterials ranging between 100 and 200 nm would be desirable [3].

The liver is the largest organ in the body, and plays many roles in metabolism, excretion, detoxification, body fluid homeostasis, and digestion [5]. Therefore, hepatic diseases are highly likely to be fatal, and studies targeting the liver have been actively conducted [6]. In addition to basic research aimed at developing animal models of liver disease [7] and exploring liver disease treatments [8], the development of efficient methods to introduce foreign genes into mouse hepatocytes in vivo is important. Hydrodynamics-based gene delivery (HGD) is an efficient method that enables in vivo transfer of nucleic acids (NAs) into mouse hepatocytes via the caudal vein [9,10]. Unfortunately, few technical reports are available concerning protein introduction in vivo and in vitro [11–13]. For example, Rouet et al. demonstrated that protein introduction proceeds using protein delivery carriers capable of binding to the asialoglycoprotein receptor (ASGP-R) with a high degree of specificity [14]. ASGP-R is a hepatocyte-specific receptor, to which galactose specifically binds. Thus, such protein delivery carriers are modified to have galactose residues [15]. To create such protein delivery carriers requires technical skills and knowledge regarding organic synthesis, as exemplified by the introduction of modifying groups.

Here, we developed a novel biomaterial referred to as low-molecular-weight heparin/protamine particles (LHPPs) capable of interacting with NAs, proteins, and cells to protect them from proteolytic enzymes and thermal damage [16]. These materials are proven to exhibit sustained release of proteins [17–19] and are useful for retaining cells, which allows them to adapt to cell culture environments [20]. LHPPs can be synthesized by simply mixing two types of polymer chains (heparin and protamine) bearing opposite charges as monodisperse fine particles (also called micro- and nanoparticles). Their synthesis does not depend on organic reactions, and is performed via electrostatic interactions.

Since LHPPs can interact with proteins, we speculated that it might possible to deliver a target protein to hepatocytes in vivo if LHPPs are used as protein carriers. In this case, HGD, which is known as a technique allowing liver-directed gene introduction, can be used for targeted delivery into hepatocytes. Furthermore, we hypothesized that it might be possible to modify the surface of LHPPs through electrostatic interactions, because LHPPs have surface charge. For instance, it is possible to bind cell penetrating peptides (CPPs) [21], which can increase the efficiency of transmembrane transport by directly passing through a lipid membrane to the surface of LHPPs. This new type of LHPPs is referred to as improved LHPPs (*i*-LHPPs).

In this study, we examined whether *i*-LHPPs could be produced without organic synthesis, and the resulting *i*-LHPPs were used as protein carriers for delivery of a target protein to mouse hepatocytes via HGD.

2. Materials and Methods

2.1. Preparation of LHPPs and i-LHPPs

LHPPs were prepared as previously described [17]. In brief, 14 mL of a low-molecular-weight heparin solution containing 10-fold diluted dalteparin (Fragmin: 6.4 mg/mL; Kissei Pharmaceutical Co., Tokyo, Japan) in phosphate-buffered saline without $Ca^{2+}$and Mg^{2+}, pH 7.2 [PBS(-)], was mixed with 6 mL of protamine solution containing 10-fold diluted protamine (10 mg/mL; Mochida Pharmaceutical Co., Tokyo, Japan) in PBS(-), and vortexed for 2 min. The final concentration of LHPPs was estimated to be 1.57 mg/mL [22]. The resulting LHPPs are referred to as "LHPPs 1.57 mg/mL" and can be stored at 4 °C for at least six months. For CPP, we used the human immunodeficiency virus-1 (HIV-1) viral protein Tat (trans-activator of transcription) (comprised of GRKKRRQRRRPPQ), which is known to have the ability to transport proteins into cells [21]. We hereinafter term this peptide 'TAT', and its synthesis was outsourced from Greiner Bio-One International (Tokyo, Japan). TAT was dissolved in PBS(-) at a concentration of 1 mg/mL and used as a stock solution. One milliliter of LHPPs 1.57 mg/mL was added with designated amounts of TAT, and the mixtures were kept on ice before measuring the particle size and zeta charge (electric charge on the nanoparticle surface) of the *i*-LHPPs using a particle size measurement system (ELSZ-1000; Otsuka Electronics Co. Ltd., Osaka, Japan). The resulting *i*-LHPPs containing TAT are referred to as "*i*-LHPPs 1.57 mg/mL" and can be stored at 4 °C for at least six months almost no change in charge or particle size. This experiment used the materials from three different batches, and a total of three experiments for each batch were performed.

2.2. Protein Delivery into Cultured Cells in Vitro

NIH3T3 cell [23] using in this study was obtained from ATCC (#CRL-1658) [24]. The cells were first seeded onto a well of a six-well gelatin-coated plate (#4810-020; Iwaki Glass Co., Tokyo, Japan) at a density of 10^6 cells/well one day before introducing a protein and grown in 2 mL of Dulbecco 's modified Eagle's medium (DMEM) (#11995065; Thermo Fisher Scientific K.K., Tokyo, Japan) supplemented with 10% fetal bovine serum (FBS) (Invitrogen Co., Carlsbad, CA, USA) at 37 °C in an atmosphere of 10% CO_2.

Before introducing of *i*-LHPPs 1.57 mg/mL complexed with a protein, 1 μg of bacterial β-galactosidase (β-gal) (#072-04141; FUJIFILM Wako Pure Chemical Corp., Osaka, Japan) or RNase T1 (#109193; F. Hoffmann-La Roche Ltd., Basel, Switzerland) was added to 1 mL of *i*-LHPPs 1.57 mg/mL, and placed on ice for 30 min. Then, each mixture was added to the cell culture medium in a well. Notably, as reported in our previous study and in other studies, ~5 μg of protein can be immobilized on 1 mg of LHPPs [25–27]. One day after the addition of the *i*-LHPPs-protein mixture, cells were subjected to inspection for the presence of exogenously-added protein.

To detect the presence of β-gal in a cell, cells were subjected to cytochemical staining with X-gal, a substrate for β-gal, as described previously [28]. In brief, after washing with PBS(-), cells were fixed in 4% paraformaldehyde (PFA) in PBS(-) at 4 °C for 4 h, then, a solution of X-gal staining kit (GX10003; OZ Biosciences SAS, Marseille, France) was added and incubated at 30 °C for 24 h.

In the case of addition of RNase T1, cells were subjected to cell counting using a Cell Counting Kit (#343-07623; FUJIFILM Wako Pure Chemical Corp.), as described previously [20,29]. In brief, after washing with PBS(-), cells were incubated in 100 μL of fresh medium containing 10 μL of WST-1 reagent (containing tetrazolium salts) at 37 °C for 1 h. The optical density (OD) of each well was read at 450 nm using a microplate spectrophotometer (Multiskan FC; Thermo Fisher Scientific K.K.).

2.3. Protein Delivery into Murine Hepatocytes via HGD

This experiment involved HGD-based in vivo delivery of proteins using mice. Institute of Cancer Research (ICR) male mice (7-weeks old) were purchased from Japan SLC Inc. (Hamamatsu, Japan), maintained on a 12 h light/dark schedule (lights on from 07:00 to 19:00), and were allowed food and water ad libitum. All animal experiments were performed at the National Defense Medical College (Saitama, Japan) in accordance with the guidelines of the *National Defense Medical College Committee*, and approved by the *Care and Use of Laboratory Animals* (permission no. 17064; valid from 27 July 2017 to 31 March, 2020). All efforts were made to reduce the number of animals used and to minimize their suffering (n = 3 in each group).

First, 1 mL of a stock solution containing 10 μg of β-gal and 100 μg of TAT in *i*-LHPPs 1.57 mg/mL was prepared. Prior to injection, mice were subjected to sufficient anesthesia by intraperitoneal (IP) injection of three combined anesthetics [medetomidine (0.75 mg/kg; Nippon Zenyaku Kogyo Co. Ltd., Fukushima, Japan), midazolam (4 mg/kg; Sandoz K.K., Tokyo, Japan), and butorphanol (5 mg/kg; Meiji Seika Pharma Co., Ltd., Tokyo, Japan)]. HGD was performed according to methods described previously [30–32]. Briefly, anesthetized mice were intravenously injected with PBS(-) containing 10 μL of stock solution using a syringe (3 mL Luer lock type; Nipro, Inc., Osaka, Japan) fitted with a 30-gauge needle (Dentronics Co., Ltd., Tokyo, Japan). The volume (mL) used for HGD is calculated as one-tenth of the body weight (g). For example, for 30 g of a mouse it requires 3 mL of *i*-LHPPs-containing solution. Injections were performed at a constant injection speed via the tail vein and completed within 10 s by the same researcher in order to avoid artefactual effects in each experiment. After intravenous injection, anesthetized mice were recovered by subcutaneous injection of atipamezole (3.75 mg/kg; Nippon Zenyaku Kogyo Co. Ltd.), a medetomidine antagonist, and then maintained on an electric plate warmer for recovery from anesthesia.

Two days after injection, the mice were euthanized and subjected to perfusion fixation, according to Nakamura et al. [33]. Liver samples were dissected out and fixed in 4% PFA in PBS(-) at 4 °C for one day and then subjected to standard immunohistochemical staining using rabbit anti-β-gal antibody with the aid of a professional company (KYODO BYORI. Inc., Kobe, Japan). In brief, cryostat specimens were incubated with rabbit anti-β-gal polyclonal antibody (diluted 1/200; #100-4136; Rockland Immunochemicals, Inc., Limerick, PA, USA) (as the primary antibody) at 4 °C for one day, and subsequently with peroxidase-labeled anti-rabbit IgG polyclonal antibody (#424144; NICHIREI BIOSCIENCE INC, Tokyo, Japan) (as the secondary antibody) at 4 °C for 3 h. The specimens were then color-developed using 3,3′-diamino benzidine (DAB) solution. After counterstaining with 4′,6-diamidino-2-phenylindole (DAPI), the immunostained specimens were inspected using a phase-contrast microscope (#BZ-8000; Keyence Co., Osaka, Japan).

2.4. Statistical Analysis

The zeta potential [the electric charge that develops at the interface between a solid surface (nanoparticle) and its liquid medium], particle size, and number of cells are presented as means ± standard deviation. Statistical analysis was performed using unpaired Student's *t*-test, and one-way factorial analysis of variance. Scheffe's post-hoc test was used for multiple comparisons. The *p*-values were calculated using JMP14 for Windows software (SAS Institute Japan Ltd., Tokyo, Japan). A *p*-value of <0.05 was considered to indicate statistical significance. Values of $p < 0.01$ are marked with a double asterisk; <0.05 with a single asterisk.

3. Results

3.1. Preparation of *i*-LHPPs

We added TAT (comprised of GRKKRRQRRRPPQ) to LHPPs to create *i*-LHPPs. Since the TAT protein used contains many arginine residues, the peptide itself is thought to be cationic. Thus, it is likely that the addition of TAT to LHPPs will result in a charge shift towards the cationic state. Measurement of the zeta potential in the TAT-containing LHPPs demonstrated that as the dosage of TAT increased, the surface charge of LHPPs shifted from negative (approximately between −25 mV and −30 mV) to the neutral charge side, stabilizing at −5 mV (Figure 1). The particle size remained stable, although a slight fluctuation in the particle size (120.27 ± 4.46 nm, 123.47 ± 4.41 nm, and 196.7 ± 6.75 nm) was observed during the addition of 0, 10, and 100 μg of TAT, respectively. However, the error range of particle size (256.17 ± 33.89 nm to 751.13 ± 86.94 nm) increased, along with an increase in the amount of TAT from 300 to 800 μg. These findings suggested an interaction between TAT and LHPPs. Furthermore, particle stability is guaranteed when 100 μg of TAT is added to a solution containing 1.57 mg/mL of LHHPs. Thus, hereinafter we decided to use *i*-LHPPs at a ratio of 100 μg of TAT to 1.57 mg/mL of LHHPs.

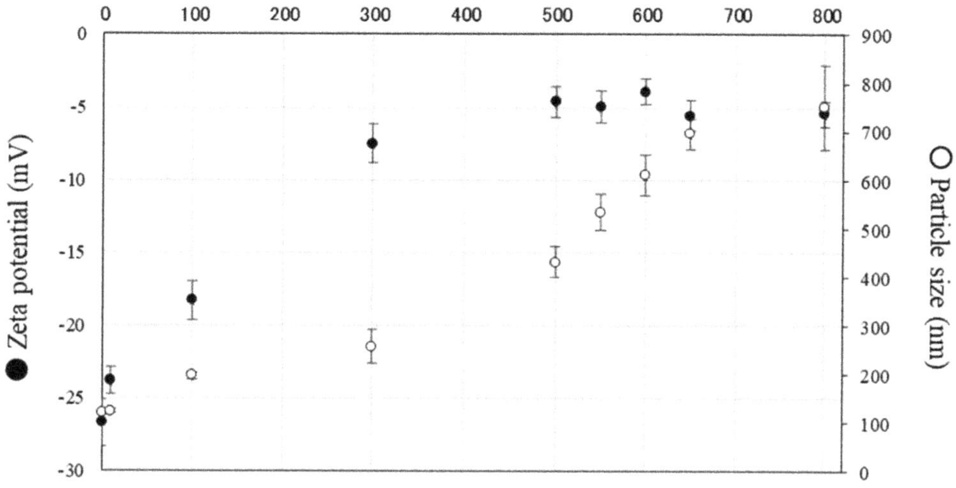

Figure 1. Relationship between LHPPs particle size and zeta potential after addition of TAT as CPP. Black circles in the graph indicate zeta potential (mV) (shown at left vertical axis); white circles indicate particle size (nm) (shown at the right vertical axis). The horizontal axis indicates the amounts (μg) of TAT added. As TAT (up to 300 μg) was added to LHPPs at 1.57 mg/mL, both zeta potential and particle size increased. However, although the zeta potential error remained unaltered, the addition of >300 μg of TAT caused increased particle size error, suggesting particle instability due to size heterogeneity.

3.2. Incorporation of Proteins Using i-LHPPs into NIH 3T3 Cells

To test the possibility that *i*-LHPPs enabled the incorporation of a protein into cultured cells in vitro, NIH3T3 cells were exposed to *i*-LHPPs coupled with β-gal for 24 h at 37 °C. After that, the cells were fixed and then subjected to cytochemical staining in the presence of x-gal, a β-gal substrate. As a result, numerous cells exhibited blue cytoplasmic deposits when the *i*-LHPPs/β-gal complex was employed (Figure 2A-e). Notably, in the groups where β-gal alone was used, or when the LHPPs/β-gal complexes (without TAT) were used, x-gal-positive cells were observed (Figure 2A-c,d). However, the rate of x-gal-positive cells in the latter two groups was significantly lower than that in the *i*-LHPPs/β-gal complex-treated groups (c,d, vs. e in Figure 2A). Only marginal staining was seen in cells treated with *i*-LHPPs alone (Figure 2A-b), and in those without any treatment (intact) (Figure 2A-a), suggesting residual endogenous β-gal activity present in NIH3T3 cells.

Next, to confirm that exogenous protein delivered via *i*-LHPPs is indeed internalized inside cells, RNase T1 was used in this *i*-LHPPs-based protein delivery system. RNase T1 is known to elicit cell death caused by inhibition of RNA synthesis when delivered inside a cell [34]. Incubation of NIH3T3 cells in the presence of the *i*-LHPPs/RNase T1 complex for up to five days at 37 °C resulted in retardation of cell growth, which can be assessed by staining cells with tetrazolium salt-containing solution (black circles in Figure 2B-a). This mode of cell growth in the group treated with the *i*-LHPPs/RNase T1 complex was significantly different from that in the other groups (treatment with RNase T1 alone, LHPPs/RNase T1 complex, or none) (Figure 2B-a). If the survival rate of cells in the intact group at each day of culture was defined as 100%, cells treated with the *i*-LHPPs/RNase T1 complex reached approximately 50% survival after 3 days in culture (Figure 2B-b). Notably, cells treated with RNase T1 alone exhibited transient reductions in cell numbers (Figure 2B-b). These findings clearly suggested that exogenous protein is incorporated by a cell when delivered in the form of an *i*-LHPPs/protein complex.

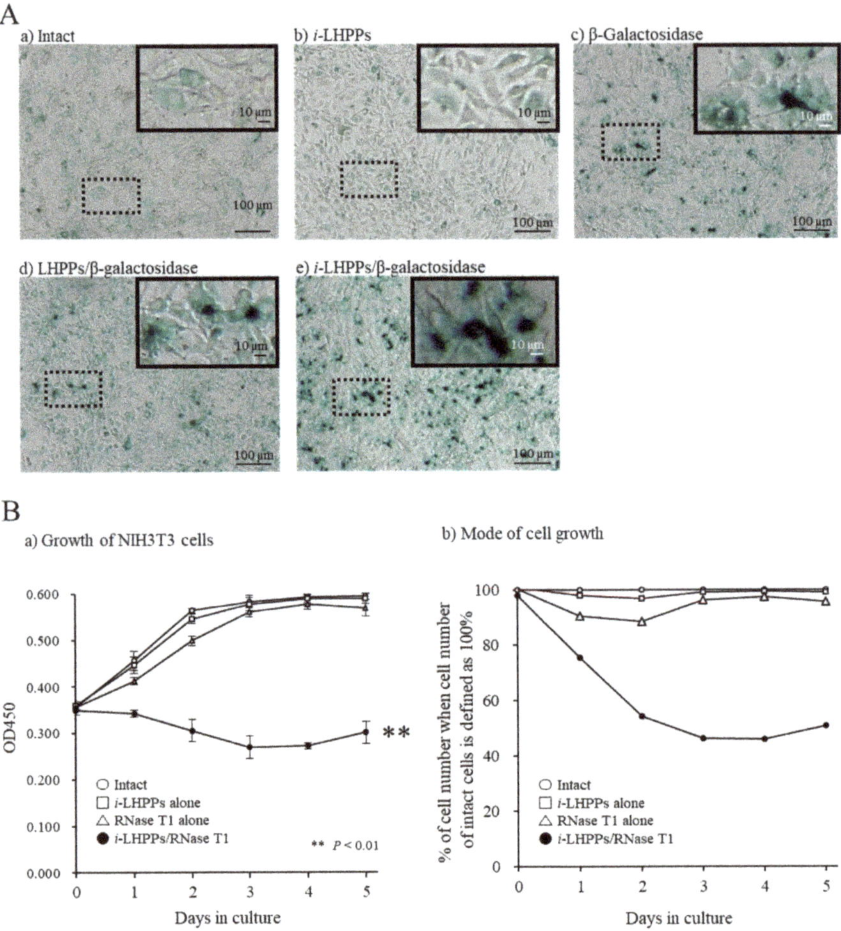

Figure 2. *i*-LHPPs-mediated protein delivery in vitro. (**A**) NIH3T3 cells stained in the presence of x-gal for β-galactosidase (β-gal) activity after in vitro protein delivery. In each microphotograph, the upper right image represents the magnified image of the dotted square area. Notably, in untreated (intact) cells (**A-a**), or cells treated with *i*-LHPPs alone (**A-b**), nonspecific positive reactions were observed, probably due to endogenous β-gal activity. A slight increase in blue deposition was noted in cells treated with β-gal alone (**A-c**), or in those treated with LHPPs/β-gal complexes (**A-d**). However, cells exposed to *i*-LHPPs/β-gal complexes exhibited distinct blue cytoplasmic deposits (**A-e**). (**B**) Survivability of NIH3T3 cells after in vitro protein (RNase T1) delivery. (**B-a**) Cell growth in groups of intact cells (shown by white circles), cells treated with *i*-LHPPs alone (shown by squares), cells treated with RNase T1 alone (shown by triangles), or cells treated with *i*-LHPPs/RNase T1 complex (black circles). Cell growth was assessed by measuring optical absorbance at 450 nm, and represented by OD values on the longitudinal axis. Data are shown as means ± SD. ** indicates $p < 0.01$ when OD 450 nm values were evaluated between cells treated with *i*-LHPPs/RNase T1 complex and those treated with none (intact), *i*-LHPPs alone, or RNase T1 only. (**B-b**) Cell survivability for each treatment group when the survival rate of cells in the intact group on each culture day was defined as 100%. Notably, the number of cells after treatment with *i*-LHPPs/RNase T1 complexes (black circle) was greatly reduced after up to three days in culture, suggesting that exogenous protein is indeed incorporated by cells when delivered in the form of *i*-LHPPs/protein complexes. Measurements were repeated three times on different days.

3.3. Introduction of Protein to Mouse Hepatocytes via the Tail Vein Using i-LHPPs

To examine whether *i*-LHPP-based protein delivery is also possible in vivo, we introduced the *i*-LHPPs/β-gal complex intravenously using the HGD method, which is known to be effective for targeted delivery of NAs to hepatocytes [9,10]. Two days after protein delivery, HGD-treated mice were subjected to perfusion fixation, liver dissection, and subsequent post-fixation prior to histochemical staining for the presence of the introduced β-gal using anti-β-gal antibody. If β-gal is successfully delivered to hepatocytes and taken up, liver specimens should exhibit brown cytoplasmic deposits [products generated after enzymatic reaction involving DAB and peroxidase (to which the secondary antibody had been conjugated)]. Consequently, extensive and substantial brown deposits were recognizable throughout the liver specimens obtained after delivery of *i*-LHPPs/β-gal complexes (Figure 3e). Minor deposition of β-gal-reacted products was also observed in the specimens derived from delivery of β-gal alone (Figure 3c), or in those from LHPPs/β-gal complexes (without using TAT) (Figure 3d), whereas the extent was less than that seen in specimens derived from delivery of the *i*-LHPPs/β-gal complexes (Figure 3c,d vs. Figure 3e). No such deposits were noted in untreated (intact) specimens (Figure 3a), or in those that had been given *i*-LHPPs alone (Figure 3b). From these findings, it was found that it is possible to deliver proteins into hepatocytes without loss of integrity when *i*-LHPPs are used as protein carriers.

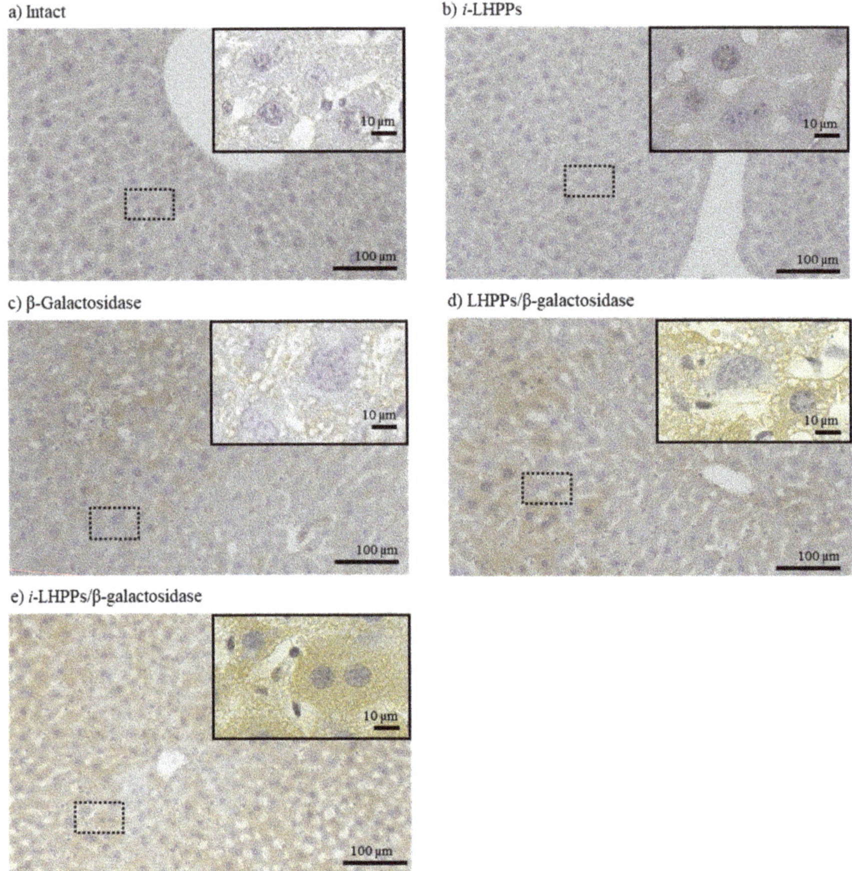

Figure 3. Immunocytochemical staining of mouse liver two days after HGD-based protein (β-galactosidase, β-gal) delivery in vivo. After staining with antibodies, specimens were subjected to 3,3′-diamino benzidine (DAB)-based coloration reactions, which cause generation of brown deposits if the specimens have cytoplasmic β-gal. Blue deposits indicate cell nuclei revealed by 4′,6-diamidino-2-phenylindole (DAPI) staining. In each microphotograph, the upper right image represents the magnified image of the dotted square area. Extensive and substantial brown deposits are discernible when *i*-LHPPs/β-gal complexes were used for HGD (**e**). In contrast, in specimens where no protein was introduced (**a**), or in those where *i*-LHPPs alone were introduced (**b**), poor coloration of immunoreactive substances was observed (which may reflect endogenous nonspecific reactivity against the antibody). Introduction of β-gal alone (**c**) or LHPPs/β-gal complexes (**d**) resulted in generation of minor amounts of immunoreactive substances in hepatocytes.

4. Discussion

LHPPs are unique materials that are developed and produced using our methodology, which is based on the simple and clear hypothesis that mixing a polymer with positive charge and that with negative charge might form complexes via electrostatic interactions. In our previous study, we reported the usefulness of LHPPs complexed with growth-stimulating proteins to enhance the biological function of certain specific cells through interaction between the cells and LHPPs. For example, we reported that local administration of LHPPs carrying recombinant fibroblast growth factor-2 (FGF-2), known as an angiogenic factor, resulted in the controlled release of FGF-2, finally causing angiogenesis at the injection site of a mouse [17,18]. In this case, it was impossible to introduce a protein of interest into a cell. If this process was successful, the utility of LHPPs with other proteins could be greatly expanded. In this study, we realized the aforementioned idea by developing a new protein carrier (*i*-LHPP), which was easily produced by mixing CPP with LHPPs. This is a novel achievement that is different from the results of our previous work. The resulting complexes are generally referred to as "polyelectrolyte complexes (PECs)" [16,35,36], with both positive and negative surface charges. Accordingly, LHPPs appear to have similar surface conditions. Analysis of the properties of LHPPs using a particle size measurement system suggested that the surface charge of LHPPs remained stable, with a charge of approximately –30 mV (see Figure 1), which indicates that LHPPs are predominantly occupied by negative charges. This finding, in turn, prompted us to suppose that a peptide with net positive charge could be easily introduced into LHPPs through electrostatic interactions with modification groups. In the present study, we decided to use TAT as a modification group of CPPs [21] in order to enhance the cell permeability of LHPPs. When we added TAT (100 μg) to LHPPs (1.57 mg/mL), the zeta potential value was found to change from negative to neutral (Figure 1). This phenomenon suggested molecular interactions involving TATs and LHPPs.

To test whether the resulting LHPPs mixed with TAT (defined as *i*-LHPPs) have the ability to transport a protein into a cell, NIH3T3 cells were co-incubated with medium containing *i*-LHPPs/β-gal complex for 24 h. As expected, blue deposits were more frequently discernible on the cells treated with *i*-LHPPs/β-gal complexes than on those treated with β-gal alone, or with LHPPs/β-gal complexes (see Figure 2A). It may be claimed that the extensive accumulation of blue deposits on cells treated with *i*-LHPPs/β-gal complexes could be ascribed to simple attachment of these complexes on the surface of NIH3T3 cells, but not to the internalization of β-gal inside a cell. To eliminate this possibility, we next introduced *i*-LHPPs/RNase T1 complexes into NIH3T3 cells. It is well known that intracellular uptake of RNase T1, which interferes with cellular RNA synthesis, causes apoptosis-based cell death [34]. Incubation of NIH3T3 cells in a medium containing *i*-LHPPs/RNase T1 complexes led to reduced cell survival, as shown in Figure 2B. In contrast, treatment of cells with other reagents (*i*-LHPPs, RNase T1 alone, or LHPPs/RNase T1 complexes) did not affect NIH3T3 cell survival (Figure 2B). From these results, it can be concluded that *i*-LHPPs enable the internalization of a target protein inside a cell. Similar to other reagents used with CPP for protein delivery [37], the mode of protein internalization mediated by *i*-LHPPs may be as follows: (1) attachment of protein (β-gal, or RNase T1 in this study) to the cell surface, and (2) internalization of the protein by endocytosis (micropinocytosis), a general process by which cells absorb external material by engulfing it within the cell membrane [38]. Since the size of *i*-LHPPs used in this study was 100–200 nm (see Figure 1), it is reasonable to consider that these molecules can be internalized via micropinocytosis, namely via vesicle-mediated endocytosis, as shown in the case of another CPP fusion protein. The *i*-LHPPs that are uptaken by a cell through micropinocytosis will then be degraded within intracellular lysosomes, as suggested by Commisso et al. [39].

Our *i*-LHPPs are useful for targeted protein delivery to hepatocytes in vivo, and were also confirmed to be effective using HGD, a well-known technology enabling efficient gene delivery and targeting mouse hepatocytes [9,10]. HGD is performed by intravenous injection of a large amount of solution at once to rapidly increase intravenous pressure to pierce the pores (approximately 100 nm in mice) of hepatocytes and introduce NAs administered into the blood into hepatocytes [40]. Since the particle sizes of LHPPs and *i*-LHPPs are expected to be > 100 nm, it is unlikely that these materials are directly introduced into hepatocytes by passing through cell pores. The probability of this event is that *i*-LHPPs/protein (β-gal) complexes provided by HGD accumulates in the murine liver and adheres to the pore surface of hepatocytes by the action of TAT as CPP, and is then taken up by hepatocytes through the mechanism of cellular phagocytosis. Notably, successful HGD-based delivery of NAs into mouse hepatocytes has been identified by many laboratories; however, to our knowledge, reports of the delivery of proteins by the HGD technique appear to be few, probably due to instability in the blood and poor membrane permeability [1]. In this context, the *i*-LHPPs/HGD-based protein delivery method is important for delivering polypeptides of interest to the liver in vivo. Furthermore, this technique will provide a novel approach to create animal disease models, or to develop techniques to cure diseases in such models. It may also be possible to apply this technology to the field of in vivo genome editing, which has been extensively used in recent years for research and disease treatment. For instance, tail vein-mediated injection of *i*-LHPPs coupled with Cas9 protein (now commercially available) and single guide RNA, which is a synthetic RNA comprised of CRISPR RNA (crRNA) and trans-activating crRNA (tracrRNA) and can be easily obtained by outside sourcing, will be possible to realize this idea.

The most remarkable advantage of using *i*-LHPPs is that it does not require any organic synthetic process, which is laborious and time-consuming. *i*-LHPPs are produced through "simple mixing of materials", which makes it easier to perform in vivo protein transfer experiments. In this study, we used TAT as CPP to enhance the incorporation rate of a target protein into cells. Unfortunately, CPP does not exhibit affinity for any specific cells or tissues. In this case, instead of CPP, the use of a tissue-specific protein, as exemplified by ASGP-R [14], appears to be of interest. LHPPs incorporating ASGP-R would show preferential affinity for the liver when they are provided intravenously. We believe that HGD-based protein delivery using LHPPs coupled with ASGP-R could be more effective than HGD-based protein delivery using *i*-LHPPs. Notably, organ- or cancer cell-specific receptors exist [10]. If we use such receptors as guide proteins for LHPPs, it might be possible to produce *i*-LHPPs with tissue or organ specificity. Alternatively, the use of mannose as a modifying group that can bind specifically to mannose receptors expressed in the lung [41] would be of interest. In this case, *i*-LHPPs would be used as "inhaled drug delivery material", which does not require bloodstream-mediated introduction of drugs such as HGD.

We used RNase T1 (MW = 11 kDa) and β-gal (MW = 54 kDa) for protein delivery with our *i*-LHPPs. In our previous study, we had succeeded in the controlled release of recombinant FGF-2 (MW = 17 kDa) from the LHPPs/FGF-2 complex that was injected in vivo [17,18]. We believe that it is possible to deliver up to 5 μg of protein per 1 mg of LHPPs with a one-shot injection [25–27]. This appears to be solely due to the nature of LHPPs, which mainly comprise heparin and easily interact with cationic proteins, as exemplified by heparin-binding protein. Notably, since the surface of LHPPs has positive and negative charges, it is highly likely that LHPPs can potentially retain anionic proteins. Thus, it is conceivable that *i*-LHPPs have a similar property to that of LHPPs. In this context, it would be worthwhile to examine the types of proteins trapped by *i*-LHPPs, in view of increasing their utility.

HGD is widely accepted as an efficient method that enables the in vivo transfer of NAs into mouse hepatocytes via the caudal vein [42,43]. Gene delivery to the kidneys and lungs is also possible with this method, although the transfection efficiency is low and the number of transferred NAs appears to be low, in the case of a successful transfection [9]. Similarly, it is reasonable to consider that protein delivery to these organs can be successfully achieved by using *i*-LHPPs. Unfortunately, we failed to detect the presence of the introduced protein (β-gal) in both the kidneys and the lungs (unpublished data). This failure appears to be closely associated with the presence of TAT as CCP included in *i*-LHPPs, because it is known that the composition and organization of the extracellular matrix can affect the tissue localization pattern of substances that are introduced in vivo [44,45].

Notably, HGD itself appears to be infeasible in humans because it requires the rapid administration of a large amount of solution via the tail vein. Suda and Liu [46] reported that fluid overload in systemic circulation induces irregular cardiac function, leading to transient heart failure. However, the application of this technology to large animals is also being attempted. For instance, the feasibility of local administration with the aid of a catheter, instead of systemic administration, is now considered a safer approach to achieve gene delivery to individuals [47]. This way, catheter-based local administration of *i*-LHPPs in large animals would be valuable, in view of increasing the utility of *i*-LHPPs.

Determination of the optimal dose of *i*-LHPPs is important for safe in vivo administration. Since LHPPs mainly comprise heparin, there is a concern regarding occasional bleeding when these substances are intravenously administered to mice. We already checked this possibility and found almost no sign of occasional bleeding after administration of LHPPs [17]. Furthermore, we did not observe any bleeding in the liver specimens that are thought to be mostly affected by the HGD treatment (Figure 3,b,d,e). Although there is a still room to take steps to avoid possible bleeding elicit by LHPPs, we believe that as long as *i*-LHPPs (which are constructed at a heparin/protamine ratio determined in this study) are used, blood leakage is not possible.

Regarding possible changes in body weight and behavior during the experimental period of two days after the administration of one shot of *i*-LHPPs, we did not observe any noticeable changes in the mice. However, repeated administration of *i*-LHPPs may elicit unwanted immunological issues. In this context, we plan to evaluate this possibility through the periodic collection of mouse serum.

5. Conclusions

We developed a novel protein carrier system, termed *i*-LHPPs, allowing delivery of a target protein into cells both in vitro and in vivo. This system can be distinguished from other existing systems based on organic synthesis for creating biomaterials, which always require skilled techniques and advanced expertise. *i*-LHPPs can be created via an extremely easy method: it is sufficient to mix raw materials in two steps; mixing protamine and low-molecular-weight heparin to produce LHPPs in a first step, and mixing LHPPs and CPP in a second step. With HGD-based delivery of *i*-LHPPs, we introduced a protein (β-gal) into mouse hepatocytes in vivo. By replacing CPP with other positively charged short peptides, such as receptor proteins in *i*-LHPPs, it would be possible to deliver a target protein into tissues or organs other than the liver. Thus, our protein carrier system, which does not rely on organic synthesis, has great potential for use in various research fields, including animal biotechnology and biomedicine. On the other hand, there are some subjects to be overcome in future, which include physicochemical characterization of *i*-LHPPs, as exemplified by protein loading efficiency and stability in vitro and in vivo. These data will be especially important when *i*-LHPPs produced by a simple synthetic method are applied to various research fields.

Author Contributions: Conceptualization, S.N.; methodology, S.N., M.I., and M.S.; software, N.A.; validation, M.I., and M.S.; formal analysis, S.N., and N.A.; investigation, S.N., and N.A.; resources, S.N.; data curation, S.N., and N.A.; writing—original draft preparation, S.N.; writing—review and editing, S.N., and M.S.; visualization, N.A.; supervision, S.N.; project administration, S.N.; funding acquisition, S.N., and M.I.; All authors have read and agreed to the published version of the manuscript.

Funding: This study was partly supported by JSPS KAKENHI (no. 16K15063 for S.N.).

Conflicts of Interest: The founding sponsors had no role in the design of the paper.

References

1. Lee, M.S.; Kim, N.W.; Lee, J.E.; Kim, M.G.; Yin, Y.; Kim, S.Y.; Ko, B.S.; Kim, A.; Lee, J.H.; Lim, S.Y.; et al. Targeted cellular delivery of robust enzyme nanoparticles for the treatment of drug-induced hepatotoxicity and liver injury. *Acta Biomater.* **2018**, *81*, 231–241. [CrossRef] [PubMed]
2. Qin, X.; Yu, C.; Wei, J.; Li, L.; Zhang, C.; Wu, Q.; Liu, J.; Yao, S.Q.; Huang, W. Rational Design of Nanocarriers for Intracellular Protein Delivery. *Adv. Mater.* **2019**, *31*, e1902791. [CrossRef] [PubMed]
3. Buzea, C.; Pacheco, I.I.; Robbie, K. Nanomaterials and nanoparticles: Sources and toxicity. *Biointerphases* **2007**, *2*, MR17–MR71. [CrossRef] [PubMed]
4. Choi, H.S.; Liu, W.; Misra, P.; Tanaka, E.; Zimmer, J.P.; Itty Ipe, B.; Bawendi, M.G.; Frangioni, J.V. Renal clearance of quantum dots. *Nat. Biotechnol.* **2007**, *25*, 1165–1170. [CrossRef] [PubMed]
5. Miyajima, A.; Tanaka, M.; Itoh, T. Stem/progenitor cells in liver development, homeostasis, regeneration, and reprogramming. *Cell Stem Cell* **2014**, *14*, 561–574. [CrossRef]
6. Beckwitt, C.H.; Clark, A.M.; Wheeler, S.; Taylor, D.L.; Stolz, D.B.; Griffith, L.; Wells, A. Liver 'organ on a chip'. *Exp. Cell Res.* **2018**, *363*, 15–25. [CrossRef]
7. Lamas-Paz, A.; Hao, F.; Nelson, L.J.; Vazquez, M.T.; Canals, S.; Gomez Del Moral, M.; Martinez-Naves, E.; Nevzorova, Y.A.; Cubero, F.J. Alcoholic liver disease: Utility of animal models. *World J. Gastroenterol.* **2018**, *24*, 5063–5075. [CrossRef]

8. Mitra, A.; Ahn, J. Liver Disease in Patients on Total Parenteral Nutrition. *Clin. Liver Dis.* **2017**, *21*, 687–695. [CrossRef]
9. Liu, F.; Song, Y.; Liu, D. Hydrodynamics-based transfection in animals by systemic administration of plasmid DNA. *Gene Ther.* **1999**, *6*, 1258–1266. [CrossRef]
10. Budker, V.; Budker, T.; Zhang, G.; Subbotin, V.; Loomis, A.; Wolff, J.A. Hypothesis: Naked plasmid DNA is taken up by cells in vivo by a receptor-mediated process. *J. Gene Med.* **2000**, *2*, 76–88. [CrossRef]
11. Lee, E.J.; Lee, N.K.; Kim, I.S. Bioengineered protein-based nanocage for drug delivery. *Adv. Drug Deliv. Rev.* **2016**, *106*, 157–171. [CrossRef]
12. Bolhassani, A.; Jafarzade, B.S.; Mardani, G. In vitro and in vivo delivery of therapeutic proteins using cell penetrating peptides. *Peptides* **2017**, *87*, 50–63. [CrossRef]
13. Hettiaratchi, M.H.; Shoichet, M.S. Modulated Protein Delivery to Engineer Tissue Repair. *Tissue Eng. Part A* **2019**, *25*, 925–930. [CrossRef]
14. Rouet, R.; Christ, D. Efficient Intracellular Delivery of CRISPR-Cas Ribonucleoproteins through Receptor Mediated Endocytosis. *ACS Chem. Biol.* **2019**, *14*, 554–561. [CrossRef]
15. Miller, C.M.; Tanowitz, M.; Donner, A.J.; Prakash, T.P.; Swayze, E.E.; Harris, E.N.; Seth, P.P. Receptor-Mediated Uptake of Phosphorothioate Antisense Oligonucleotides in Different Cell Types of the Liver. *Nucleic Acid Ther.* **2018**, *28*, 119–127. [CrossRef]
16. Ishihara, M.; Kishimoto, S.; Nakamura, S.; Sato, Y.; Hattori, H. Polyelectrolyte Complexes of Natural Polymers and Their Biomedical Applications. *Polymers* **2019**, *11*, 672. [CrossRef]
17. Nakamura, S.; Kanatani, Y.; Kishimoto, S.; Nakamura, S.; Ohno, C.; Horio, T.; Masanori, F.; Hattori, H.; Tanaka, Y.; Kiyosawa, T.; et al. Controlled release of FGF-2 using fragmin/protamine microparticles and effect on neovascularization. *J. Biomed. Mater. Res. A* **2009**, *91*, 814–823. [CrossRef]
18. Nakamura, S.; Ishihara, M.; Takikawa, M.; Kishimoto, S.; Isoda, S.; Fujita, M.; Sato, M.; Maehara, T. Attenuation of limb loss in an experimentally induced hindlimb ischemic model by fibroblast growth factor-2/fragmin/protamine microparticles as a delivery system. *Tissue Eng. Part A* **2012**, *18*, 2239–2247. [CrossRef]
19. Nakamura, S.; Takikawa, M.; Ishihara, M.; Nakayama, T.; Kishimoto, S.; Isoda, S.; Ozeki, Y.; Sato, M.; Maehara, T. Delivery system for autologous growth factors fabricated with low-molecular-weight heparin and protamine to attenuate ischemic hind-limb loss in a mouse model. *J. Artif. Organs* **2012**, *15*, 375–385. [CrossRef]
20. Nakamura, S.; Kishimoto, S.; Nakamura, S.; Nambu, M.; Fujita, M.; Tanaka, Y.; Mori, Y.; Tagawa, M.; Maehara, T.; Ishihara, M. Fragmin/protamine microparticles as cell carriers to enhance viability of adipose-derived stromal cells and their subsequent effect on in vivo neovascularization. *J. Biomed. Mater. Res. A* **2010**, *92*, 1614–1622. [CrossRef]
21. Gagat, M.; Zielinska, W.; Grzanka, A. Cell-penetrating peptides and their utility in genome function modifications. *Int. J. Mol. Med.* **2017**, *40*, 1615–1623. [CrossRef]
22. Mori, Y.; Nakamura, S.; Kishimoto, S.; Kawakami, M.; Suzuki, S.; Matsui, T.; Ishihara, M. Preparation and characterization of low-molecular-weight heparin/protamine nanoparticles (LMW-H/P NPs) as FGF-2 carrier. *Int. J. Nanomed.* **2010**, *5*, 147–155. [CrossRef]
23. Jainchill, J.L.; Aaronson, S.A.; Todaro, G.J. Murine sarcoma and leukemia viruses: Assay using clonal lines of contact-inhibited mouse cells. *J. Virol.* **1969**, *4*, 549–553. [CrossRef]
24. Nakamura, S.; Watanabe, S.; Ohtsuka, M.; Maehara, T.; Ishihara, M.; Yokomine, T.; Sato, M. Cre-loxP system as a versatile tool for conferring increased levels of tissue-specific gene expression from a weak promoter. *Mol. Reprod. Dev.* **2008**, *75*, 1085–1093. [CrossRef]
25. Pan, M.; Suarez de Lezo, J.; Medina, A.; Romero, M.; Hernandez, E.; Segura, J.; Melian, F.; Wanguemert, F.; Landin, M.; Benitez, F.; et al. In-laboratory removal of femoral sheath following protamine administration in patients having intracoronary stent implantation. *Am. J. Cardiol.* **1997**, *80*, 1336–1338. [CrossRef]
26. Houska, M.; Brynda, E.; Bohata, K. The effect of polyelectrolyte chain length on layer-by-layer protein/polyelectrolyte assembly–an experimental study. *J. Colloid Interface Sci.* **2004**, *273*, 140–147. [CrossRef]
27. Takabayashi, Y.; Nambu, M.; Ishihara, M.; Kuwabara, M.; Fukuda, K.; Nakamura, S.; Hattori, H.; Kiyosawa, T. Enhanced effect of fibroblast growth factor-2-containing dalteparin/protamine nanoparticles on hair growth. *Clin. Cosmet. Investig. Dermatol.* **2016**, *9*, 127–134. [CrossRef]

28. Kikuchi, N.; Nakamura, S.; Ohtsuka, M.; Kimura, M.; Sato, M. Possible mechanism of gene transfer into early to mid-gestational mouse fetuses by tail vein injection. *Gene Ther.* **2002**, *9*, 1529–1541. [CrossRef]
29. Nakamura, S.; Nambu, M.; Ishizuka, T.; Hattori, H.; Kanatani, Y.; Takase, B.; Kishimoto, S.; Amano, Y.; Aoki, H.; Kiyosawa, T.; et al. Effect of controlled release of fibroblast growth factor-2 from chitosan/fucoidan micro complex-hydrogel on in vitro and in vivo vascularization. *J. Biomed. Mater. Res. A* **2008**, *85*, 619–627. [CrossRef]
30. Nakamura, S.; Maehara, T.; Watanabe, S.; Ishihara, M.; Sato, M. Improvement of hydrodynamics-based gene transfer of nonviral DNA targeted to murine hepatocytes. *Biomed. Res. Int.* **2013**, *2013*, 928790. [CrossRef]
31. Nakamura, S.; Maehara, T.; Watanabe, S.; Ishihara, M.; Sato, M. Liver lobe and strain difference in gene expression after hydrodynamics-based gene delivery in mice. *Anim. Biotechnol.* **2015**, *26*, 51–57. [CrossRef] [PubMed]
32. Nakamura, S.; Ishihara, M.; Watanabe, S.; Ando, N.; Ohtsuka, M.; Sato, M. Intravenous Delivery of piggyBac Transposons as a Useful Tool for Liver-Specific Gene-Switching. *Int. J. Mol. Sci.* **2018**, *19*, 3452. [CrossRef] [PubMed]
33. Nakamura, S.; Terashima, M.; Kikuchi, N.; Kimura, M.; Maehara, T.; Saito, A.; Sato, M. A new mouse model for renal lesions produced by intravenous injection of diphtheria toxin A-chain expression plasmid. *BMC Nephrol.* **2004**, *5*, 4. [CrossRef] [PubMed]
34. Yuki, S.; Kondo, Y.; Kato, F.; Kato, M.; Matsuo, N. Noncytotoxic ribonuclease, RNase T1, induces tumor cell death via hemagglutinating virus of Japan envelope vector. *Eur. J. Biochem.* **2004**, *271*, 3567–3572. [CrossRef]
35. Dautzenberg, H. Polyelectrolyte Complex Formation in Highly Aggregating Systems. 1. Effect of Salt: Polyelectrolyte Complex Formation in the Presence of NaCl. *Macromolecules* **1997**, *30*, 7810–7815. [CrossRef]
36. Ishihara, M.; Nakamura, S.; Sato, Y.; Takayama, T.; Fukuda, K.; Fujita, M.; Murakami, K.; Yokoe, H. Heparinoid Complex-Based Heparin-Binding Cytokines and Cell Delivery Carriers. *Molecules* **2019**, *24*. [CrossRef]
37. Futaki, S. Oligoarginine vectors for intracellular delivery: Design and cellular-uptake mechanisms. *Biopolymers* **2006**, *84*, 241–249. [CrossRef]
38. Conner, S.D.; Schmid, S.L. Regulated portals of entry into the cell. *Nature* **2003**, *422*, 37–44. [CrossRef]
39. Commisso, C.; Davidson, S.M.; Soydaner-Azeloglu, R.G.; Parker, S.J.; Kamphorst, J.J.; Hackett, S.; Grabocka, E.; Nofal, M.; Drebin, J.A.; Thompson, C.B.; et al. Macropinocytosis of protein is an amino acid supply route in Ras-transformed cells. *Nature* **2013**, *497*, 633–637. [CrossRef]
40. Suda, T.; Liu, D. Hydrodynamic delivery. *Adv. Genet.* **2015**, *89*, 89–111. [CrossRef]
41. Ezekowitz, R.A.; Stahl, P.D. The structure and function of vertebrate mannose lectin-like proteins. *J. Cell Sci. Suppl.* **1988**, *9*, 121–133. [CrossRef] [PubMed]
42. Yokoo, T.; Kamimura, K.; Abe, H.; Kobayashi, Y.; Kanefuji, T.; Ogawa, K.; Goto, R.; Oda, M.; Suda, T.; Terai, S. Liver-targeted hydrodynamic gene therapy: Recent advances in the technique. *World J. Gastroenterol.* **2016**, *22*, 8862–8868. [CrossRef] [PubMed]
43. Huang, M.; Sun, R.; Huang, Q.; Tian, Z. Technical Improvement and Application of Hydrodynamic Gene Delivery in Study of Liver Diseases. *Front. Pharmacol.* **2017**, *8*, 591. [CrossRef] [PubMed]
44. Davies Cde, L.; Berk, D.A.; Pluen, A.; Jain, R.K. Comparison of IgG diffusion and extracellular matrix composition in rhabdomyosarcomas grown in mice versus in vitro as spheroids reveals the role of host stromal cells. *Br. J. Cancer* **2002**, *86*, 1639–1644. [CrossRef]
45. Markman, J.L.; Rekechenetskiy, A.; Holler, E.; Ljubimova, J.Y. Nanomedicine therapeutic approaches to overcome cancer drug resistance. *Adv. Drug Deliv. Rev.* **2013**, *65*, 1866–1879. [CrossRef]
46. Suda, T.; Liu, D. Hydrodynamic gene delivery: Its principles and applications. *Mol. Ther.* **2007**, *15*, 2063–2069. [CrossRef]
47. Kamimura, K.; Zhang, G.; Liu, D. Image-guided, intravascular hydrodynamic gene delivery to skeletal muscle in pigs. *Mol. Ther.* **2010**, *18*, 93–100. [CrossRef]

Review

Antiviral Potential of Nanoparticles—Can Nanoparticles Fight against Coronaviruses?

Sangiliyandi Gurunathan [1], Muhammad Qasim [2], Youngsok Choi [1], Jeong Tae Do [1], Chankyu Park [1], Kwonho Hong [1], Jin-Hoi Kim [1] and Hyuk Song [1,*]

[1] Department of Stem Cell and Regenerative Biotechnology, Konkuk University, Seoul 05029, Korea; sangiliyandi@konkuk.ac.kr (S.G.); choiys3969@konkuk.ac.kr (Y.C.); dojt@konkuk.ac.kr (J.T.D.); chankyu@konkuk.ac.kr (C.P.); hongk@konkuk.ac.kr (K.H.); jhkim541@konkuk.ac.kr (J.-H.K.)

[2] Center of Bioengineering and Nanomedicine, Department of Food Science, University of Otago, Dunedin 9054, New Zealand; muhammadqasim.qasim@otago.ac.nz

* Correspondence: songh@konkuk.ac.kr; Tel.: +82-2-450-0050

Received: 10 July 2020; Accepted: 18 August 2020; Published: 21 August 2020

Abstract: Infectious diseases account for more than 20% of global mortality and viruses are responsible for about one-third of these deaths. Highly infectious viral diseases such as severe acute respiratory (SARS), Middle East respiratory syndrome (MERS) and coronavirus disease (COVID-19) are emerging more frequently and their worldwide spread poses a serious threat to human health and the global economy. The current COVID-19 pandemic, caused by severe acute respiratory syndrome coronavirus 2 (SARS-CoV-2). As of 27 July 2020, SARS-CoV-2 has infected over 16 million people and led to the death of more than 652,434 individuals as on 27 July 2020 while also causing significant economic losses. To date, there are no vaccines or specific antiviral drugs to prevent or treat COVID-19. Hence, it is necessary to accelerate the development of antiviral drugs and vaccines to help mitigate this pandemic. Non-Conventional antiviral agents must also be considered and exploited. In this regard, nanoparticles can be used as antiviral agents for the treatment of various viral infections. The use of nanoparticles provides an interesting opportunity for the development of novel antiviral therapies with a low probability of developing drug resistance compared to conventional chemical-based antiviral therapies. In this review, we first discuss viral mechanisms of entry into host cells and then we detail the major and important types of nanomaterials that could be used as antiviral agents. These nanomaterials include silver, gold, quantum dots, organic nanoparticles, liposomes, dendrimers and polymers. Further, we consider antiviral mechanisms, the effects of nanoparticles on coronaviruses and therapeutic approaches of nanoparticles. Finally, we provide our perspective on the future of nanoparticles in the fight against viral infections.

Keywords: antiviral agent; nanoparticle; coronavirus; viral mechanism of entry; antiviral mechanism; therapeutic approaches; SARS-CoV-2; COVID-19

1. Introduction

Infectious diseases are caused by pathogenic microorganisms that spread directly or indirectly from one person to another [1]. Zoonotic diseases are infectious diseases of animals that can cause disease in humans when the causal agent is transmitted from animal to human; the diseases can account for hundreds of thousands of deaths worldwide. Infectious diseases pose a significant threat to both human health and the global economy; to date, we know of about 200 infectious diseases. Fortunately, only a handful of these diseases are responsible for significant morbidity and mortality [2,3]. Among them, human immunodeficiency virus/acquired immune deficiency syndrome (HIV/AIDS), tuberculosis and malaria are the most pronounced. In addition, several outbreaks of infectious diseases have occurred recently including Ebola, Zika and avian influenza as well as the coronavirus (CoV) diseases severe acute respiratory syndrome (SARS; caused by SARS-CoV), Middle East Respiratory Syndrome (MERS; caused by MERS-coronavirus (MERS-CoV)) and COVID-19 (caused by SARS coronavirus 2 (SARS-CoV-2). These diseases originated in West Africa, South America, Asia, Asia, the Middle East and Asia, respectively, before spreading to other parts of the world [4–8]. In addition several other viral diseases also found to be widely spread such as Hantavirus which is transmitted by rodents, Chikungunya and Rift Valley fever virus (RVFV) virus by mosquitos, paramyxoviruses such as rubulaviruses, Nipah and Hendra viruses from bats and so forth. Infectious diseases account for ~20% of global mortality and viruses are responsible for about one-third of these deaths [1]. For example, SARS-CoV-2, the causal virus of coronavirus disease (COVID-19), is transmitted directly from one human to another. The outbreak of COVID-19 began in late 2019 and as of 27 July 2020, SARS-CoV-2 has infected more than 16,430,566 individuals and led to the death of more than 652,434 individuals in 215 countries around the world (https://www.worldometers.info/coronavirus/accessed on 27 July 2020). The number of infected cases and deaths are still seriously increasing every day, affecting essentially every country worldwide.

Coronaviruses (CoVs) are a large family of RNA viruses. They are the major pathogen of emerging respiratory disease outbreaks and can cause a variety of diseases in mammals and birds, from which they can be isolated [9]. CoVs primarily infect the upper respiratory and gastrointestinal tract of mammals and birds. The major symptoms of CoV infection of the upper respiratory tract in humans include cough, fever, and, in more severe cases, difficulty in breathing has been reported with potential fatality from SARS-CoV, MERS-CoV and SARS-CoV-2. The clinical symptoms of COVID-19, including acute respiratory disorder induced by either highly homogenous SARS-CoV-2 or other secondary pathogens, suggest that excessive inflammation, oxidation and an exaggerated immune response very likely contribute to COVID-19 pathology. This initial response can lead to a cytokine storm and subsequent progression to acute lung injury/acute respiratory distress syndrome and often death [10]. SARS-CoV-2 binds to human angiotensin-converting enzyme-2 (ACE2) receptors and not only induces pneumonia but also it induces multisystem illness with involvement of different organs and potential for systemic complications [11]. Increasing evidence suggest that SARS-CoV-2 induce significant abnormalities compatible with hypercoagulability with hyperfibrinogenemia and clinically a high prevalence of thromboembolic events. Further it causes potential for large vessel thrombosis and major thromboembolic sequelae including pulmonary embolism (PE) deep vein thrombosis (DVT) and thrombosis in extracorporeal circuits and arterial thrombosis [12–14]. COVID-19 is closely related to SARS, which swept the world in 2002 and 2003. SARS-CoV infected about 8000 people and killed about 800. MERS outbreaks have occurred sporadically since 2012, infecting about 2500 people and resulting in nearly 900 deaths. COVID-19 is different from other two CoVs resulting in more severe effects and spreading faster in humans. COVID-19 has become a pandemic with millions of infected patients over a period of less than one year since its suspected outbreak. These alarming disease statistics serve to emphasize the global concern over infectious diseases and the enormous influence on the global socio-economic and health-care sectors. Although we may develop a drug for a particular viral infection, viral mutations may lead to drug resistance, rendering it ineffective. This was observed in the case of HIV and influenza treatment. However, mixture of antiviral agents such as HIV2 nucleoside reverse transcriptase inhibitors (NRTIs) plus an integrase strand transfer inhibitor (InSTI) and also other effective regimens include nonnucleoside reverse transcriptase inhibitors or boosted protease inhibitors with 2 NRTIs showed promising effect on HIV patients. These antiretroviral drugs (ARVs) can sustain HIV suppression and can prevent new HIV infection. Hence, drug resistance is a public health threat, eventually increasing morbidity and mortality [15–17].

The best approach to preventing viral infections is vaccination; however, the development of vaccines is time consuming, expensive and requires sophisticated equipment and lengthy protocols. A limited number of vaccines are available for infectious diseases but the ones that exist are not always equally available worldwide. Some currently available broad-spectrum antivirals including lopinavir/ritonavir, neuraminidase inhibitors, EK1 peptide, RNA synthesis inhibitors, nucleoside analogs and HIV-protease inhibitors could be effective, alternative medicines for COVID-19 [18]. Further, interferons (IFNs) seem to be partially effective against CoVs; a combination of IFNs and ribavirin exhibited increased inhibitory activity in vitro when compared to IFNs alone against some CoVs [19,20]. The drug EIDD-2801, used for pandemic influenza viral infections, is another alternative worthy of consideration for the treatment of COVID-19 [21].

An in vitro study suggested that the antiviral drug remdesivir and the anti-malaria drug chloroquine can potentially control COVID-19. Chloroquine has been used to treat malaria for many years; however, the mechanism of action of chloroquine against viral infections remains elusive [22]. Liu et al. performed the comparative analysis between hydroxychloroquine and chloroquine against SARS-CoV-2 infected patients [23]. The results shows that hydroxychloroquine is less toxic and more effective to inhibit SARS-CoV-2 infection. Corticosteroid ciclesonide blocks coronavirus RNA replication by targeting viral NSP15. Conversely, corticosteroid did not suppress replication of respiratory syncytial virus or influenza virus [24]. Bioinformatics analysis revealed that both esterified and non-esterified derivatives of ciclesonide had the capacity to interact with NSP-15, thereby possessing the capacity to inhibit replication of the SARS-CoV-2 viral genome [25]. Ivermectin, an FDA-approved anti-parasitic drug inhibits RNA replication and also decrease many fold RNA virus load [26]. Ivermectin plays significant role in several biological mechanisms and it could be potential antiviral agents against various type of viruses including SARS-CoV-2 [27]. Ianevski et al. (2020) adopted a screening strategy to find suitable drug against SARS-C0V-2 by neutralization assay using sera from various SARS-CoV-2 infected patients [28]. They found that the most potent sera from recovered patients for the treatment of SARS-CoV-2-infected patients. They found that a combination of orally available virus directed nelfinavir and host-directed amodiaquine exhibited the highest synergy against SARS-CoV-2. Wang et al. reported that anti-influenza drug called arbidol efficiently inhibited SARS-CoV-2 infection by the mechanism of blocking of entry of virus into the host cells by impeding viral attachment [29]. Remdesivir (GS-5734) shows broad spectrum antiviral activity against several RNA viruses by interfering with RNA-dependent RNA polymerase (RdRp, also called NSP12 polymerase), even in the presence of an exoribonuclease (ExoN) with proof-reading activity, as demonstrated in an in vitro cell line and mouse model [30]. Remdesivir also showed positive results when tested in a rhesus macaque model of MERS-CoV infection [31,32]. Therefore, the drug could be effective for both the prevention and treatment of human CoV (HCoV) infections. Holshue et al. (2020) reported the first case of a COVID-19 patient being treated with remdesivir in the United States [33].

Protease inhibitors such as lopinavir and ritonavir have been successfully used to treat HIV infection and they have improved the outcomes of MERS-CoV and SARS-CoV patients [34–36]. Recent studies reported that administration of lopinavir/ritonavir (Kaletra®, AbbVie, North Chicago, IL, USA) significantly reduced β-CoV titers of a COVID-19 patient in Korea [37]. The combination of Chinese and Western medicine treatments including lopinavir/ritonavir (Kaletra®), arbidol and Shufeng Jiedu Capsule (a traditional Chinese medicine) which is composed of mixture of flavonoids such as resveratrol and quercetin significantly improved pneumonia-associated symptoms in patients in the Shanghai Public Health Clinical Center, China [38,39]. Recently, siRNA-induced RNA interference (RNAi) play significant role and one of the fascinating techniques to explore the possible novel approach for emerging novel pathogenic viruses. For instance, siRNAs directed against Spike sequences and the 3′-UTR can inhibit the replication of SARS-CoV in Vero-E6 cells [40]. Abbott et al. demonstrated a CRISPR-Cas13-based strategy to inhibit SARS-CoV-2 inhibition by designing and screening of CRISPR RNAs (crRNAs) targeting conserved viral regions. The results revealed that they identified functional crRNAs targeting SARS-CoV-2 [41]. This technique could contribute immense level to inhibit spreading and infection of SARS-CoV-2. Taking all of this information into account, the increasing number of outbreaks and severity of viral infections call for novel, multidirectional, safe, biocompatible, cost-effective, target specific and tunable-based alternative approaches to prevent and treat the diseases caused by infectious viruses.

One such alternative approach involves nanomaterials and their fascinating properties, which include optimal size, shape, tunable surface charge, superparamagnetism, high surface plasmon resonance, luminescence, photon upconversion, bioavailability, biocompatibility, immunocompatibility/tolerability and biodegradability. Furthermore, the versatility of nanomaterials are can be easily decorated/anchored/conjugated with one or more type of functional groups, linkers and various bioactive molecules and some of the nanomaterials are being capable of simultaneous therapy and diagnosis [42–45].

In addition, the major requirements including cellular entry through the blood-brain barrier and blood-air barrier, tenability and targeted control discharge are feasible with nanomaterials, thus qualifying these materials as potential novel candidates for use in biomedical therapy [28–30] Nanoparticles have been widely used in antiviral therapy over the last few decades, owing to the development of surface functionalization strategies [45]. For example, Ag [46], Au [47], TiO_2 [48], SiO_2 [49], CeO_2 [50] and $CuCl_2$ [51] nanoparticles have been employed against different viruses including hepatitis B virus (HBV) [46], H3N2 and H1N1 [52], HIV-1 [53], herpes simplex virus (HSV) [54], vesicular stomatitis [55], foot-and-mouth disease [47] and dengue virus type-2 [51]. Recently, Sportelli et al. (2020) stated that researchers need to focus on the development of nanomaterial-based technological solutions to fight COVID-19 [56]. Therefore, several articles are expected to provide the basic knowledge regarding nanomaterials and describe how to use these materials for the development of antiviral therapies. Considering the seriousness of infectious disease transmission and the potential of nanomaterials in treating these diseases, our review first focuses on the mechanism of entry of viruses into host cells and then on the use of major and important types of nanomaterials such as silver, gold, quantum dots, organic nanoparticles, liposomes, dendrimers and polymers against various types of viral infections. Further, we discuss antiviral mechanisms, therapeutic approaches of nanoparticles and the effects of nanoparticles on CoVs. Finally, we provide our perspective on the potential of using nanoparticles in the future to treat infectious diseases.

2. Mechanism of Entry of Viruses into Host Cells

Virus entry into host cells is required for viral multiplication. The infection process involves several steps including attachment, penetration, uncoating, replication, assembly and release (Figure 1). Viruses enter host cells through specific receptors on the host cell membrane using attachment proteins in the viral capsid or glycoproteins embedded in the viral envelope. The specificity of the interaction determines the kind of virus that infects the host cells. For example, bacteriophages enter the host cell through their nucleic acids and the capsid remains outside of the cell. Some animal and plant viruses enter host cells through endocytosis. Once inside the host cell, RNA viruses such as CoVs, use their genomic for the synthesis of viral genomic RNA as well as Mrna and eventually the release of the new virions produced in the host cells [57–59]. The human respiratory mucosa is the primary site of virus entry for various viruses including the influenza virus, respiratory syncytial virus and parainfluenza virus. These virulent pathogens primarily infect the upper respiratory tract but they may subsequently reach the lower respiratory regions causing more severe illness and ultimately morbidity and mortality. The clinical symptoms of the above pathogen attacks are frequently fever, dyspnea, cough, bronchiolitis and pneumonia [60]. Most respiratory tract infections are caused by SARS-CoV, MERS-CoV and SARS-CoV-2 [61]. A unique feature of CoVs such as SARS-CoV-2 is their mechanism of entry into host cells, initially through binding to the host-receptor cell at the angiotensin converting enzyme 2 (ACE2) protein site. This is followed by fusion with the host cellular membrane and subsequent release of the viral genetic material into the host cytoplasm or nucleus. The released viral RNA from CoVs is transcribed and the viral mRNA directs protein synthesis. Viruses replicate and assemble into new virions and these are released into neighboring cells via exocytosis. COVID-19 is caused by the novel CoV, SARS-CoV-2 [4,5]. SARS-CoV-2 is 80% and 50% homologous with SARS-CoV and MERS-CoV, respectively [4,62].

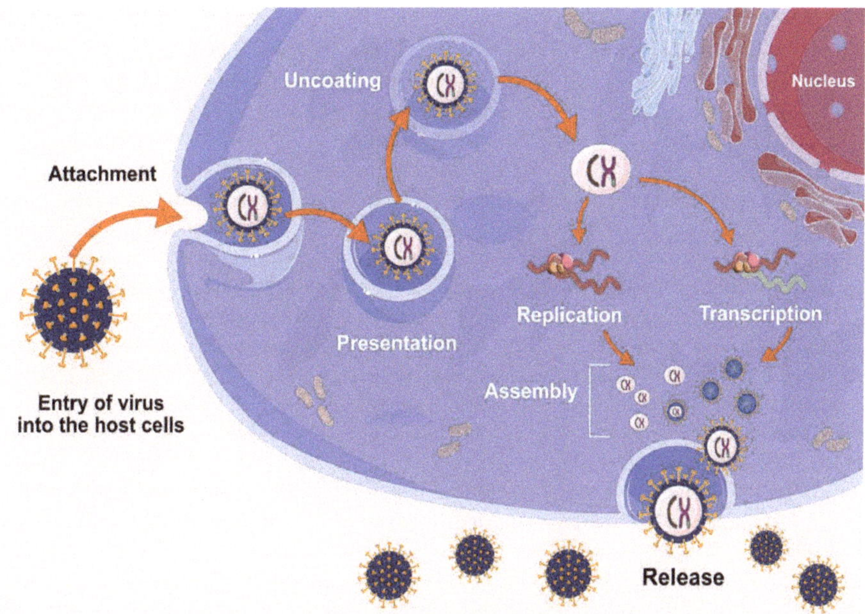

Figure 1. Mechanism of virus entry and replication in host cells.

Viral diseases are major threat to human health and economy; therefore, it is necessary to find suitable, alternative, safe and biocompatible antiviral agents to prevent the spread of infections and reduce economic losses. Generally, nanoparticles including silver nanoparticles (AgNPs), gold NPs (AuNPs), quantum dots (QDs), carbon dots (CDots), graphene oxide (GO), silicon materials, polymeric NPs, dendrimers and polymers possess remarkable antimicrobial and antiviral activities [1,63–67]. Therefore, it is essential to highlight the variety and importance of selective nanoparticles that could be used as antiviral agents and delivery agents (Figure 2).

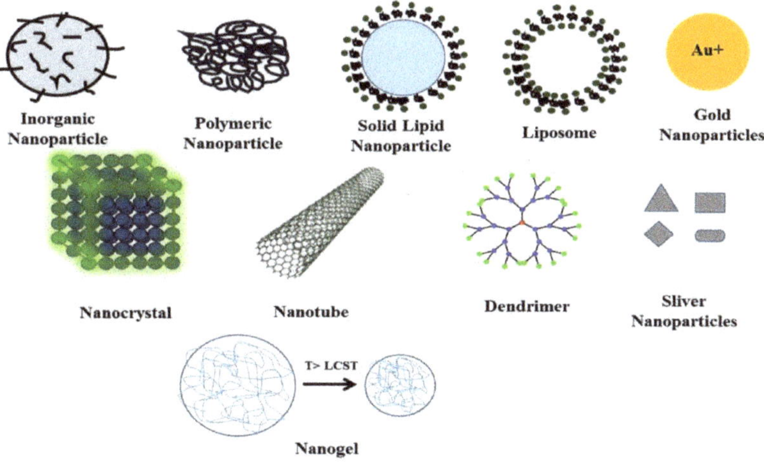

Figure 2. Various types of nanoparticles used for antiviral therapy as antiviral agents and delivery agents.

3. Silver Nanoparticles

Silver nanoparticles (AgNPs) are used as antiviral, antibacterial, anti-inflammatory, anti-angiogenesis, antiplatelet, antifungal and anticancer agents due to their unique physiochemical properties and superior biological functions [54,68,69]. Synthesis of AgNPs is carried out by various physical, chemical and biological methods. Biological methods appear to be environmentally friendly, safe, biocompatible and non-toxic. AgNPs have been used as biomedical therapeutic agents in wound dressings, long-term burn care products and anti-bacterial lotions [70]. Polyvinylpyrrolidone (PVP)-coated AgNPs homogenized in Replens gel (0.15 mg/mL) inhibited HIV-1 transmission of cell-associated and cell-free HIV-1 isolates after 1 min and offered long-lasting protection of cervical tissue from infection after 48 h treatment, with no evidence of cytotoxicity observed in the explants. AgNPs bind to glycoprotein gp120 on the HIV envelope in a manner that prevents CD4-dependent virion binding, fusion and infectivity [53]. AgNP-Coated polyurethane condoms (PUCs) efficiently inactivate HIV-1 and HSV-1/2 and their infectiousness; macrophage (M)-tropic and T lymphocyte (T)-tropic strains of HIV-1 are highly sensitive. The AgNP-coated PUCs can directly inactivate the microbe's infectious ability and provide another line of defense against sexually transmitted microbial infections [71].

Fungi-Mediated synthesis of AgNPs reduced the viral infection dose in a size-dependent manner against HSV-1/2 and with human parainfluenza virus type 3 they blocked interaction of the virus with the cell, which might depend on the size and zeta potential of the AgNPs [72]. Antiviral activity of AgNPs and chitosan composites was evaluated against H1N1 influenza A virus. The composites showed significant antiviral activity in a size-dependent manner; surprisingly, chitosan alone did not show any antiviral effect. Conversely, AgNPs alone did exhibit antiviral activity; however, the composites showed remarkable antiviral activity compared to either AgNPs or chitosan alone [73]. AgNPs prevent transmissible gastroenteritis virus-induced apoptosis by regulating the p38/mitochondria-caspase-3 signaling pathway in swine testicle cells [74]. Curcumin-functionalized AgNPs demonstrated significant inhibitory effects against respiratory syncytial virus (RSV) infection by decreasing viral titers about two-orders of magnitude to non-toxic concentrations in host cells. Further, AgNPs prevented RSV from infecting host cells by inactivating the virus directly [75].

AgNPs showed antiviral and preventive effects against H3N2 influenza virus infection. In the presence of AgNPs, Madin-Darby canine kidney cells infected with H3N2 influenza virus showed better viability and no obvious cytopathic effects compared to an influenza virus control group. Infected mice treated with AgNPs showed lower lung viral titers and minor pathologic lesions in lung tissue and longevity [76]. Graphene oxide (GO)-AgNPs, composed of two nanomaterials in a single platform, were more effective than either single agent. GO-AgNPs inhibited feline CoV (FCoV) infection by 25% and infectious bursal disease virus (IBDV) infection by 23%, whereas GO alone only inhibited FCoV 16% and showed no antiviral activity against IBDV [77]. Huy et al reported antiviral activity of AgNPs against influenza A, HBV, human parainfluenza, HSV and HIV [78]. AgNPs synthesized using a green chemistry ultra-sonication approach exhibited antiviral activity against influenza A [79]. Zanamivir-loaded AgNPs synergistically inhibited H1N1 influenza virus multiplication [80]. Tannic acid-modified AgNP-based muco-adhesive hydrogel effectively reduced HSV-2 infectivity at the vaginal mucosal surface [81]. AgNPs exhibited antiviral activity against human oncogenic γ-herpesviruses, Kaposi's sarcoma-associated herpesvirus and Epstein-Barr virus by reactivating viral lytic replication through the generation of reactive oxygen species (ROS) and autophagy [82]. Children are mostly affected by RSV; however, there is no specific treatment option available. The RSV virion contains two surface glycoproteins (F and G) that are vital for the initial phases of infection, making them critical targets for RSV treatment. AgNPs reduced RSV replication and the levels of pro-inflammatory cytokines (i.e., IL-1α, IL-6 and TNF-α) and pro-inflammatory chemokines (i.e., CCL2, CCL3, CCL5). Mice treated intravaginally with tannic acid (TA)-mediated AgNPs showed better clinical scores and lower virus titers in the vaginal tissues. The TA-mediated AgNP-treated group also showed significantly increased percentages of IFN-gamma+ CD8+ T-cells, activated B cells and plasma cells, while the spleens contained significantly higher percentages of IFN-gamma+ NK cells and effector-memory CD8+ T cells compared to NaCl-treated group. Further, the AgNP-treated animals showed significantly better sera titers of anti-HSV-2 neutralization antibodies than the NaCl-treated animals [83]. AgNPs interaction with HIV-1 in size-dependent manner and that the bound particles exhibit regular spatial relationships. AgNPs undergo preferential binding with the gp120 subunit of the viral envelope glycoprotein. These interaction between AgNPs and glycoproteins inhibit the virus from binding to host cells [84] and AgNPs act as potential antiviral agents against various type of viruses including influenza virus [85,86]. Biologically synthesized AgNPs using plant extracts of *Lampranthus coccineus* and *Malephora lutea* exhibited significant antiviral activity against different types of viruses such as HSV-1, HAV-10 and CoxB4 virus [87]. Altogether, all these studies demonstrated that the antiviral potential of AgNPs.

4. Gold Nanoparticles

Gold nanoparticles (AuNPs) have also drawn great interest in industry and nanomedicine due to their excellent electrical, optical, mechanical and biological properties [88,89]. AuNPs are used to detect DNA sequences, proteins, bacteria and viruses and they are frequently used in cancer studies and as antiviral and antibacterial agents. Gold nanorod, a GNR-5′PPP-ssRNA nanoplex-mediated immune activator, was reported to inhibit H1N1 influenza viral replication by upregulating the expression of IFN-β and other IFN-stimulated genes (ISGs), resulting in decreased viral replication [90]. Hyaluronic acid AuNPs and IFN complex have been used for targeted treatment of hepatitis C (HCV) infection [91] and highly mono-dispersed quasi spherical AuNPs inhibit HSV. Compared with the clinical drug acyclovir, AuNPs are very safe; they do not induce any drug-resistant viral strains and exhibit excellent viricidal properties [92]. Functionalized AuNPs suppress influenza virus, HSV and HIV. AuNPs potentially increase antiviral effects through multivalent interactions; dendronized AuNPs inhibit HIV more effectively than dendrons alone [93,94]. Sialic acid-functionalized AuNPs inhibit influenza virus infection, by multivalent interactions, relatively better than some synthetic clinical drugs such as zanamivir and oseltamivir, which are prone to resistance development by the influenza virus [95,96]. A study was performed to investigate the effect of AuNPs on Schistosoma mansoni-infected mouse liver. Comparing the treated and untreated infected groups, AuNPs significantly decreased the activities of malondialdehyde and nitric oxide and increased the level of glutathione (GSH). Concomitantly, AuNPs ameliorated the inflammatory response by decreasing the mRNA expression of interleukin (IL)-1β, IL-6, tumor necrosis factor (TNF)-α, IFN-γ and inducible nitric oxide synthase [97]. AuNPs also inhibited attachment and penetration of foot-and-mouth disease virus (FMDV) at the early stages of infection and impaired viral replication at later stages at non-toxic concentrations and at early stage [47]. AuNPs conjugated with peptide triazoles (AuNP-PT) exhibited significant antiviral effects against HIV-1 compared to the corresponding peptide triazoles alone. The enhanced virolytic activity and corresponding irreversible HIV-1 inactivation by AuNP-PT is due to multivalent contact between the nanoconjugates and metastable envelope spike (S) proteins on the HIV-1 virus [98]. Conjugation of M2e peptide with AuNPs and CpG as a soluble adjuvant (AuNP-M2e + sCpG) induced lung B cell activation and a robust serum anti-M2e immunoglobulin G (IgG) response in mice. Antibodies generated in response to the highly pathogenic avian influenza virus H5N1 (A/Vietnam/1203/2004) are capable of binding to the homotetrameric form of M2 expressed on infected cells [99]. Nanoparticles functionalized with the FluPep ligand showed enhanced antiviral activity compared to the free peptides. Conjugation of FluPep to AuNPs and AgNPs enhanced antiviral potency [100]. Gold nanoclusters (AuNCs), composed of tens and hundreds of gold atoms with an average size of 1–2 nm, play critical roles in biomedicine. Bai et al. (2018) reported that AuNCs prevented entry of porcine reproductive and respiratory syndrome virus (PRRSV) into host cells [101]. AuNCs selectively inhibited the proliferation and protein expression of PRRSV. Surface modification plays an important role in the antiviral activity of AuNCs; therefore, researchers modified the surface of AuNCs with two different materials—histidine and mercaptoethane sulfonate. Among these, histidine-stabilized AuNCs showed strong inhibition of the proliferation of pseudorabies virus (PRV) [101,102]. Therefore, AuNPs functionalized with different molecules play critical roles in antiviral activity.

5. Quantum Dots

Quantum dots (QDs) are nanocrystals of a semiconducting materials with an average size between 2–10 nm and these are widely used in cell labeling, detection and image tracking with particular size-dependent optical and electronic properties includes carbon, silver, gold, CdSeS/ZnS and so on [103]. However, the uses of QD as an antiviral agent are very limited. Du et al developed GSH-capped cadmium telluride (Cd)Te QDs and demonstrated that they altered the structure of PRV surface proteins inhibiting the virus from entering the host cells [104]. Furthermore, the binding of CdTe QDs to the cell membrane itself also decreased viral numbers. Antiviral effects of stable carbon dots (CDots) with low toxicity were studied with PRV and PRRSV as test models of DNA and RNA viruses, respectively. CDots significantly inhibited the multiplication of both PRV and PRRSV. Furthermore, CDots remarkably induced the production of endogenous IFN and the expression of ISGs, which are responsible for virus replication [105,106]. The effect of various surface functionalizations of CDots was studied using 2,2′-(ethylenedioxy)bis(ethylamine) (EDA) and 3-ethoxypropylamine (EPA) against human norovirus virus-like-particles (VLPs), GI.1 and GII.4. Both EDA- and EPA-CDots (5 μg/mL each) were highly effective in inhibiting the binding of both strains of VLPs to histo-blood group antigen (HBGA) receptors on human cells. EDA-CDots achieved 100% inhibition and EPA CDots achieved 85–99% inhibition [107].

Benzoxazine monomer-derived CDots inhibited the infectivity of flaviviruses and non-enveloped viruses such as porcine parvovirus and adenovirus-associated virus by directly binding to the surface of virions and inhibiting the first step of virus and host cell interaction [108]. Cdots, surface-functionalized with boronic acid or an amine, inhibited the entry of HSV-1 into host cells by way of the alteration of viral coat proteins [109]. The use of biomolecules to prepare CDots provides stable and biocompatible materials. The antiviral effect of curcumin-stabilized cationic carbon dots (CCM-CDots) was evaluated against porcine epidemic diarrhea virus (PEDV) as a CoV model; CCM-CDots were found to inhibit the proliferation of PEDV with much higher efficiency than non-CCM-CDots [105]. CCM-CDots work by changing the viral surface protein structure, suppressing the synthesis of virus negative-strand RNA and virus budding and then inhibiting viral entry by generation of ROS, thus stimulating the production of ISGs and proinflammatory cytokines [105]. Carbon quantum dots (CQDs), derived from hydrothermal carbonization of ethylenediamine/citric acid, inhibited the entry and replication of HCoV-229E by interaction of the functional groups of the CQDs with HCoV-229E entry receptors on the host cell membrane [110]. Curcumin-mediated CQDs (Cur-CQDs) were prepared in a one-step heating process and injected with enterovirus 71 (EV71) in new-born mice. These Cur-CQDs exhibited superior antiviral effects [111]. Similarly, biocompatible CQDs prepared using glycyrrhizic acid (Gly-CDs) inhibited the proliferation of PRRSV by up to five-orders of viral titer [112]. Gly-CDs inhibited PRRSV invasion and replication, stimulated antiviral innate immune responses and suppressed the accumulation of intracellular ROS caused by PRRSV infection. Surface charge of the functionalized materials play critical role in the interactions between positively charged CDots and the negatively charged VLPs Positively charged EDA-CDots exhibited a higher inhibitory effect (~82%) than non-charged EPA-CDots (~60%) [107]. Both types of CDots also exhibited inhibitory effects on the binding of VLPs to their respective antibodies but they were much less effective in blocking the binding to HBGA receptors. After CDot treatment, VLPs remained intact and no degradation of the VLP capsid proteins was observed [107]. To summarize, the observed antiviral effects of CDots on noroviruses were mainly through effective inhibition of VLP binding to HBGA receptors and moderate inhibition of antibody binding, without affecting the integrity of the viral capsid protein and the viral particle. Collectively, QDs play critical roles in inhibiting various types of viruses.

6. Graphene Oxide

Graphene oxide (GO) is a unique single-atom-thick and two-dimensional carbon material arranged in a hexagonal lattice. GO and its derivatives are of immense interest to nanomedicine researchers due to their remarkable electronic, mechanical and thermal properties [113]. GO is widely used as an antibacterial and anticancer agent [114]. GO act as antiviral agent through inactivation of the pathogenic agent of hand-foot-and-mouth disease, EV71 and endemic gastrointestinal avian influenza A virus H9N2 [115]. Nanocomposites consisting of GO and partially reduced sulfonated GO (rGO-SO_3) composite showed antiviral activity against HSV-1 by inhibiting HSV-1 from attaching to host cells [116]. The antiviral activity of GO and reduced GO was evaluated against PRV (a DNA virus) and PEDV (an RNA virus) revealing that GO significantly suppressed the infection of PRV and PEDV by a 2-log reduction in virus titers at non-cytotoxic concentrations. GO inhibited viral entry into the host cells by structural destruction [117]. Nanocomposites consisting of GO and AgNPs showed potential antiviral activity against enveloped and non-enveloped viruses, for example, FCoV and IBDV, respectively. Viral inhibition assays demonstrated that GO-AgNPs inhibited FCoV infection by 25% and IBDV by 23%, whereas GO alone only inhibited FCoV infection by 16% but showed no antiviral activity against IBDV infection. Therefore, the combination of GO and AgNPs exhibited better antiviral activity compared to either GO or silver alone [77]. Curcumin-Loaded GO exhibited a significant inhibitory effect on RSV infection with significant biocompatibility. GO inactivates directly by inhibiting the virus from attaching to host cells and considered to be coupled with prophylactic and therapeutic effects on the virus. The combination of GO and curcumin was more effective than either single agent against RSV infection [118]. Iannazzo et al. (2018) reported that the conjugation of GO and QDs (GQDs) potentially inhibited the replication of HIV [119]. Hypericin (HY)-loaded GO protected against novel duck reovirus (NDRV) disease, which is a serious infectious disease of poultry. The antiviral activity of the complex (GO/HY) was studied in DF-1 cells and in ducklings infected with NDRV TH11 strain. GO/HY showed a dose-dependent inhibition of NDRV replication, which may be attributed to direct virus inactivation or inhibition of virus attachment [120].

7. Zinc Oxide

Zinc oxide nanoparticles is a type of metal nanoparticles exhibited significant microbial activity against various type of microorganisms including viruses. The antiviral activity of zinc oxide (ZnO) micro-nano structures (MNSs) ZnO-MNSs was evaluated in virus infected corneal tissues. Partially negatively charged zinc oxide ZnO-MNSs efficiently trap the virions via a novel virostatic mechanism rendering them unable to enter into human corneal fibroblasts—a natural target cell for HSV-1 infection. Zinc oxide tetrapods (ZnOTs) significantly block the entry of Herpes simplex virus type-2 (HSV-2) into target cells and also stop the spread of the virus. The ZnOTs exhibit the antiviral activity by the ability to neutralize HSV-2 virions that natural target cells such as human vaginal epithelial and HeLa cells showed highly reduced infectivity when infected with HSV-2 virions [121]. Zinc oxide tetrapod nanoparticles (ZOTEN) induce immune system against HSV-2 virus and provide the therapeutic effects [122]. Zinc oxide tetrapods inhibit herpes simplex virus infection of cultured corneas [123]. The antiviral effect of zinc oxide nanoparticles (ZnO-NPs) and polyethylene glycol (PEG)-coated ZnO-NPs (ZnO-PEG-NPs) on herpes simplex virus type 1 (HSV-1). Zinc oxide nanoparticles (ZnO-NPs) and polyethylene glycol (PEG)-coated ZnO–NPs with concentration of 200 μg/ml inhibits at the rate of approximately 92% in copy number of HSV-1 genomic DNA and also it reduces virus titer [124]. Surface modified zinc oxide nanoparticles could modify the infection potential of HSV-1 via neutralizing the virus rather than through interfering with cellular targets by electrostatic interference of H-ZNPs with virus rather than the hydrophobic interaction ZNPs with virus [125]. Ghaffari et al. (2019) determined the antiviral activity of zinc oxide nanoparticles (ZnO-NPs) and PEGylated zinc oxide nanoparticles against H1N1 influenza virus [126]. The findings suggest that post-exposure of influenza virus with PEGylated ZnO-NPs and bare ZnO-NPs at the highest non-toxic concentrations could be led to 2.8 and 1.2 log10 TCID50 reduction in virus titer when compared to the virus control. The inhibition rate was much better in PEGylated ZNPs compared to unPEGylated ZnO-NPs. Zn^{2+} ions potentially inhibited Nidovirus replication and increasing concentration of the intracellular Zn^{2+} concentration can efficiently impair the replication of a variety of RNA viruses [127]. Prior incubation of HSV-2 with zinc oxide tetrapod nanoparticles (ZOTEN) inhibits the ability of the virus to infect vaginal tissue in female Balb/c mice and blocks virus shedding through neutralizing of virions particles in the host cells and also it stimulate a local immune response [128].

8. Organic Nanoparticles

Organic nanoparticles are widely used for drug delivery and therapeutic purposes in humans due to their biocompatibility, biodegradability and easy surface modification. The most common organic nanoparticles are polymeric nanoparticles, which are colloidal solids ranging from 10 to 1000 nm in diameter. The small size can facilitate capillary penetration and uptake by cells resulting in increased concentrations at target sites [129]. Polyhexylcyanoacrylate nanoparticles are loaded with either the HIV protease inhibitor saquinavir (Ro 31-8959) or the nucleoside analog zalcitabine (2′,3′-dideoxycytidine). Both nanoparticulate formulations led to a dose-dependent reduction of HIV-1 antigen production. Nanoparticles as a drug carrier system improved the delivery of antiviral agents to the mononuclear phagocyte system in vivo, overcoming pharmacokinetic problems and enhancing the activities of drugs for the treatment of HIV infection and AIDS [130]. Acyclovir loaded into beta-cyclodextrin-poly(4-acryloylmorpholine) (β-CD-PACM) nanoparticles exhibited significant antiviral activity against two clinical isolates of HSV-1 compared to both the free drug and a soluble β-CD-PACM complex [131]. 3D8 scFv-loaded PLGA (3D8-PLGA) NPs, potentially entered into the cytosolic regions of viral infected cells, were released continuously and hydrolyzed RNAs in the cytoplasm, thereby exhibiting maximal antiviral activity [132]. Diphyllin, a vacuolar ATPase blocker delivered by polymeric nanoparticles consisting of poly(ethylene glycol)-block-poly(lactide-co-glycolide) (PEG-PLGA), was more effective in inhibiting feline infectious peritonitis (FIP), caused by a mutated feline CoV, compared to diphyllin alone. Additionally, mice were more tolerant toward diphyllin-loaded nanoparticles. Therefore, nanoformulation with polymeric nanoparticles yielded potential antiviral activity against FIP [133].

The entry of viruses into host cells is complex and the process of interaction between the virus and host cell typically involves specific protein receptors. For example, multivalent flexible nanogels exhibited broad-spectrum antiviral activity by blocking virus entry and infection [134]. Previously, several studies used nanospheres for the treatment of HSV, HBV and influenza [135–138]. Altogether, organic nanoparticles can serve as drug delivery agents against various types of viral diseases.

Dendrimers are highly branched, symmetrical, macromolecular and hyper-branched structures radiating from a central core via connectors and branching units. Terminal groups are essential for targeting and interactions. Dendrimers are globular and contain three different regions—central core, branches and terminal functional groups. The potential functionality of dendrimers is due to encapsulation of several chemical moieties, interior layers and multiple surface groups [139]. In a study performed to determine the effect of various dendrimers on HSV-1/2, dendrimers BRI-2999 and BRI-6741 showed significant reduction of infection rates [140]. Two different type of peptide-derivatized dendrimers such as SB105 and SB105 A10 completely inhibited human cytomegalovirus (HCMV) replication and also inhibited murine CMV replication, whereas they were not able to inhibit adenovirus or vesicular stomatitis virus. The mechanism of inhibition of peptide-derivatized dendrimers namely SB105_A10 bound to human cells through an interaction with cell surface heparan sulfate and thereby blocked virion attachment to target cells [141]. Mice infected with Japanese encephalitis (JEV, GP78 strain) were administered with Morpholinos (5 mg/kg body weight) via intraperitoneal injection. Administration of Vivo-Morpholinos efficiently increased survival of animals and neuroprotection in a murine model of JEV [142]. SPL7013 is a dendrimer with broad-spectrum activity against HIV-1/2, HSV-1/2 and human papillomavirus. SPL7013 increased viricidal activity against HIV-1 strains that utilize the CXCR4 co-receptor [143]. Jyothi et al. (2015) developed novel liver-targeted cyclosporine

A-encapsulated poly (lactic-co-glycolic) acid (PLGA) nanoparticles that inhibited the replication of HCV both in vitro and in vivo [144]. A liver-specific sustained drug delivery system, generated by conjugating a liver-targeting peptide to PEGylated CsA-encapsulated PLGA nanoparticles, reduced the immunosuppressive effects and toxicity profile of host factor cyclophilin A, which is essential for HCV replication [144].

To reduce the development of resistance by viral mutations, a study was performed with newly designed and efficient entry inhibitors. Antiviral activity dendrimers, as well as fullerene C_{60}-with a unique symmetrical and 3D globular structure-were evaluated against an Ebola pseudotyped infection model. The central alkyne scaffold of fullerene connected to 12 sugar-containing [59] fullerene units (total 120 mannoses)-exhibited outstanding antiviral activity with an IC_{50} in the sub-nanomolar range [145]. The low concentration of camptothecin-loaded dendrimers inhibited HCV replication with very low toxicity. The triple combination of carbosilane dendrimers, tenofovir and maraviroc showed potential for inhibiting HIV sexual transmission [130,131,146]. Polyanionic dendrimers comprising the terminal groups sodium carboxylate, hydroxyl and succinamic acid and polycationic dendrimers containing primary amines were used to assess their antiviral activity in MERS-CoV plaque inhibition assays. The hydroxyl polyanionic set showed a 17.36–29.75% decrease in MERS-CoV plaque formation. All of these dendrimers showed excellent antiviral activity against MERS-CoV [147]. The unique properties of dendrimers are due to the presence of numerous surface functional groups, which facilitate conjugation with multiple drugs or targeting ligands. They also have the ability to encapsulate hydrophobic drugs due to their limited cavity size. VivaGel is a G4-poly-L-lysine dendrimer formed from the divalent benzhydylamine amide of L-lysine and it contains 32 terminal anionic functional groups used as an effective antiviral agent [148].

Polymers have high antiviral capacity due to their long chains and branches and their flexible molecular design. Polymers can be designed as arbitrary standards based on viricidal effects. They can be used not only as effective antiviral drugs but also as co-factors for treatment of viral infectious diseases. Polymers carrying antiviral drugs efficiently increase the solubility of antiviral drugs, thus prolonging the retention time and enhancing the uptake efficiency of drugs into cells. For example, organotin compounds were prepared according to the needs of universal viral agents [149]. Organotin and cisplatin-like polymers effectively kill viruses by inhibiting viral replication [150]. The polymers consist of poly(phenylene ethynylene) (PPE)-based cationic conjugated polyelectrolytes (CPE) and oligo-phenylene ethynylenes (OPE). They act as antiviral agents by the mechanism of oxidative stress, producing singlet oxygen species due to the π bonding system in the backbone of the compound upon exposure to UV-visible light. The oxidative stress induces damage to macromolecules including DNA, RNA and proteins [151]. Nucleic acid polymers containing hepatitis B surface antigen have been used to treat hepatitis D infection by binding to the amphipathic alpha helix in the class I surface glycoprotein [152]. Polymeric nanogels, which are cross-linked hydrogel particles comprising water-soluble and expandable polymers, are easily degradable into smaller sized fragments removed by renal clearance. These polymers can prevent the entry of virus particles into host cells by attaching to the heparan sulfate proteoglycans on the host cell surface. These flexible nanogels serve as robust inhibitors of HSV-2 virus infections [134]. Chun et al. (2018) designed amphiphilic copolymers comprising methoxy-poly(ethylene glycol)-block poly(phenylalanine), which consist of encapsulated mir-323a in the core and favipiravir in the exterior layer as both hydrophilic and hydrophobic antiviral agents [153]. These polymer-carried drugs serve to treat influenza A virus infectious diseases significantly better than naked drug delivery systems without polymers. The specific advantages of polymers are the sustained release of antiviral agents and the improved metabolic stability of the integrated drug. These properties demonstrate the great potential of polymeric particles for the successful delivery of antiviral agents [45]. Nanoviricide is a nanomachine that is armed to destroy a particular kind of virus using various types of nanoparticles. For example, polymeric nanoparticles were used to inhibit Varicella Zoster Virus infection in human skin and also polymeric nanoparticles were used for targeted drug delivery to prevent virus spreading and infections.

9. Liposomes

Liposomes are spherical in shape with an average size of 20 to 30 nm and are composed of a phospholipid bilayer containing an aqueous core [154]. Hydrophilic and lipophilic drugs can be incorporated into the inner aqueous cavity or the phospholipid bilayer, respectively. Liposomes are non-toxic and biocompatible. When the HIV virus envelop region T cell receptor targeted antisense RNA was encapsulated in liposomes, HIV-1 production was completely inhibited [155]. Hematopoietic toxicity and antiviral activity of liposomal-encapsulated 3′-azido-3′-deoxythymidine (AZT) were examined in mice for 5 days following intravenous administration (0.4 to 10 mg/kg body weight); a significantly depressed bone marrow cellularity as well as a corresponding decrease in red blood cells, blood neutrophils and monocyte numbers were observed. Treatment with liposomal AZT significantly reduced virus proliferation using AZT alone [156]. Nanocarbon fullerene lipidosome (NCFL) inhibited influenza virus H1N1 by a direct killing effect, which was comparable to that of rimantadine hydrochloride [157]. To develop broad-spectrum antiviral agents against a variety of viral infections, polyunsaturated liposomes (PLs) were fabricated for use in in vivo drug delivery and to specifically target the endoplasmic reticulum. These PLs contained polyunsaturated fatty acids that exhibit independent antiviral activity by reducing cellular cholesterol. Targeting cholesterol biosynthesis within infected cells is one of the promising targets of many viral systems, including HCV, HBV and HIV [158]. PLs significantly decreased viral infectivity and secretion in HCV, HBV and HIV infections. Pretreatment of cells with PLs reduced the infectivity of both HCV and HIV by suppressing plasma membrane cholesterol levels. Cationic liposomes, containing both a fluorescence marker (calcein) and antiviral drugs HPMPC (Cidofovir®), were internalized in MDBK cells infected with bovine HSV-1 and significantly inhibited viral replication [159]. Norovirus RdRp is used as an antiviral agent and it is a promising target enzyme for the development of new antiviral drugs. The polysulfonated naphthylurea suramin has the potential to inhibit murine and human norovirus polymerases. Suramin-loaded liposomes exhibited significant antiviral activity against murine norovirus cultivated in RAW 264.7 macrophages [160]. Furthermore, surmamin inhibited the replication of various chikungunya viral (CHIKV) isolates including Sindbis virus and Semliki Forest virus. Various studies revealed that the antiviral activity and mechanism of suramin is through interference with (re)initiation of RNA synthesis. The antiviral effect of suramin-containing liposomes potentially prevents and helps treat CHIKV infections [161]. Acyclovir-Loaded flexible membrane vesicles showed significant antiviral activity in a murine model of cutaneous HSV-1 infection. The lipid based system exhibited safe treatment [162]. Cationic liposomes loaded with stearylamine (SA) inhibited viral infectivity without preloaded active pharmaceutical ingredients. SA liposomes suppressed baculoviral infectivity in several mammalian cell lines, including A549 cells. The SA liposomes are non-toxic and could increase antiviral effects by reducing cholesterol content, which intensify concurrently with increased binding of SA liposomes to the cell membrane. SA liposomes potentiate the entry of virus particles into host cells and are compatible with the antiviral drug acyclovir [163]. Hence, liposomes are the best carrier molecules to deliver antiviral drugs.

10. Antiviral Mechanism of Nanoparticles

Antiviral mechanisms of nanoparticles should target attachment, penetration, replication and budding of viruses. Possible mechanisms involve inactivation of the virus directly or indirectly, prevention of attachment of viruses to host cells and blocking viral replication but they also depend on the form and type of nanoparticles used [57]. Most often, nanoparticles block the above steps by altering the structure of the capsid protein and eventually reducing virulence, which can be attributed to both physical and chemical means of decreasing the active viral load. For example, Lara et al. (2010) demonstrated that AgNPs bind to glycoprotein gp120 of the HIV envelope preventing CD4-dependent virion binding, fusion and infectivity in cell-free and cell-associated viral assays [53,164]. Recently, Cagno et al. demonstrated the antiviral mechanisms of broad-spectrum, non-toxic nanoparticles against HSV, human papilloma virus, RSV, dengue and lenti virus [165]. The authors showed that a series of antiviral nanoparticles with long and flexible linkers mimicking heparin sulfate proteoglycans, the highly conserved target of viral attachment ligands (VALs), could achieve efficient viral prevention. This was through effective viral association with binding simulated to be strong and multivalent to the VAL repeating units [165]. Water-Soluble fullerene-polyglycerol sulfates prevented interaction of vesicular stomatitis virus coat glycoprotein with baby hamster kidney cells [166]. Lysenko et al. (2018) proposed that one of the main and direct mechanisms of nanoparticle-mediated antiviral activity is linked to local-field action against the receptors at the virus surface [50]. In this process, the nanoparticles adsorbed on the surface of the cell can greatly alter the membrane potential. On the other hand, the indirect antiviral mechanism of nanoparticles includes blocking the penetration of the virus into the cell due to a change in membrane potential. Collectively, these reports suggest that the main mechanisms of nanoparticle antiviral activity involve interaction with gp120, competitive binding between the virus and nanoparticles for the host cell, interference with viral attachment, inhibition of virus-host cell binding and penetration, binding to the plasma membrane, inactivation of viral particles prior to entry and interaction with double-stranded DNA and/or binding with viral particles [96,135,164,167–170]. While metal nanoparticles interact with viral particles, replication of the virus is blocked (Figure 3).

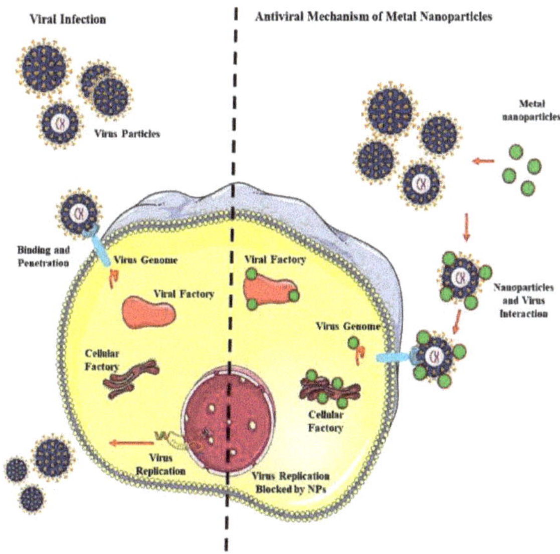

Figure 3. Antiviral mechanism of nanoparticles (NPs).

Recently, Cagno et al. [165] demonstrated the state of the art mechanism of non-toxic nanoparticles against various type of viruses. The authors developed antiviral products, which usually mimic heparan sulfate proteoglycans (HSPG), as well-preserved target of "viral attachment ligands" (VALs). The antiviral effect relies on the binding mechanism of the nanoparticles to the virus surface, thus preventing virus-cell attachment. The most outstanding viricidal effect was found in the AuNPs coated with a 2:1 mixture of decanesulfonic acid (MUS) and 1-octanethiol (OT). MUS allows a multivalent binding as a consequence of its structure comprising a long hydrophobic chain, sulfonic acid terminated. The enhanced activity of MUS:OT-NPs was assigned to the new construct using MUS linker that caused local distortions and then a global virus deformation, leading to irreversible loss of infectivity. The MUS:OT-NPs exhibited efficient virucidal effect against HSV-1 and HSV-2, human papilloma virus (HPV-16), respiratory syncytial virus (RSV), dengue and lenti virus. The synthesized MUS:OT-NPs exhibited toxic to the virus and non-toxic to the host cells [165].

The possible mechanism of antiviral activity of non-toxic nanoparticles such as gold inhibits viral replication and prevent release of viral particles into the host cells. The inhibition can take place by nanoparticles act as blockers of neuraminidase enzyme which cleaves the attachment between hemagglutinin on the progeny virus and sialic acid receptor on the host cell. In this case, nanoparticles prevent this cleavage step and interfere with the release of progeny virus from infected host cells and subsequently, prevent the progression of infection. Therefore, the possible antiviral mechanism could be inhibition of hemagglutinin and neuraminidase activities. In addition, like gold nanoparticles blocking attachment of virus into the host cells.

11. Effects of Nanoparticles on Coronaviruses

Over the last decade, nanoscience and nanotechnology have played a critical role in nanomedicine due to the size, shape and surface charge of nanoparticles as carriers, detection agents and direct inhibitory agents against microbes and cancer cells. Nanoparticles have been shown to be efficient and selective agents for the delivery of therapeutic moieties such as drugs, vaccines, siRNAs and peptides. In addition, nanoparticles have been used to detect and monitor diseases and treat responses using noninvasive imaging modalities. The size of nanomaterials facilitates their involvement in the efficient delivery of antigens due to surface functionalization; they have the capability of co-transporting antigens accompanied by numerous adjuvants. Nanomaterials are able to deliver the drug at suitable concentrations in a precise manner, to the proper place and at the proper time [171].

An important aspect of nanomaterials in relation to CoVs, is that they are able to inhibit or compete with viral binding to the host cell-surface receptor. For example, the ACE2 receptor plays a critical role in the entry of CoVs into host cells, especially in the case of SARS-CoV and SARS-CoV-2 [172]. Hence, blocking and/or decreasing the levels of ACE2 could help in the fight against infection, as well as in the development of antibodies against ACE2. On the other hand, ACE2 has a protective effect against virus-induced lung injury after infection as a result of increasing the production of the vasodilator angiotensin 1–7 [173]. Therefore, preventing COVID-19 in the host can be more effective than fighting against the virus after infection. Instead of developing vaccines, which typically takes a long time and several rounds of protocols and trials, it would be better to try to prevent entry of these viruses into humans using nanotechnology for preparing masks, clothes, gloves and gums by exploiting ACE2-coated nanoflowers and QDs [174]. In addition, researchers have demonstrated that nanoflowers and ACE2-coated QDs can be used as bio-detection probes. Furthermore, they have been used to enhance catalytic activity and stability of ACE2 and also to produce chewing gums, nose filters, masks, clothes and gloves to prevent entry and spread of CoVs [174,175]. Therefore, if materials used in hospitals were prepared with coated nanoparticles, the spread of CoV infection may be inhibited. Herein, we propose a hypothetical antiviral mechanism involving ACE2-coated nanoflowers and ACE2-coated QDs that could block SARS-CoV-2 entry into cells by inhibiting attachment of the virus to the ACE2 protein, while also inhibiting activation of the accessory serine protease TMPRSS2 (Figure 4).

A nanoparticle-based intranasal delivery system is an effective and safe tool to deliver several therapeutic moieties such as vaccines, drugs, siRNAs, peptides and antibodies. Intranasal delivery is noninvasive, practical, simple and cost effective. The intranasal delivery system has been evaluated for vaccination against respiratory viruses such as influenza and CoVs [176,177]. Nanoparticle-Mediated delivery systems have the following benefits—free from enzyme degradation; long-term existence, release and retention within the system; amicable co-delivery with adjuvants; specific targeting of cells through receptor-ligand interactions; and potentiation of the immune system [178,179].

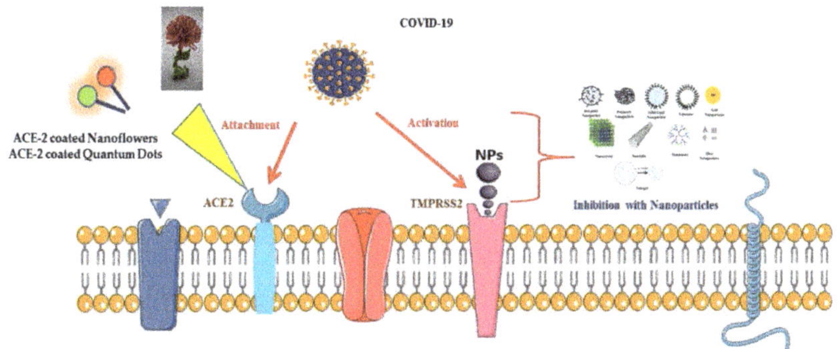

Figure 4. Hypothetical model illustrating the entry of SARS-CoV-2 into a host cell and its pathogenicity. SARS-CoV-2 spike (S) proteins bind to ACE2 receptors and then the virus enters and infects host cells. ACE2 activation may be blocked by ACE2-coated nanoflowers and ACE2-coated quantum dots. Nanoparticles (NPs) may prevent or reduce the activation/expression of the cellular protease TMPRSS2, which primes the S proteins, potentially suppressing SARS-CoV-2 infection.

There are major challenges in combatting infectious diseases such as SARS, MERS and COVID-19, including the fact that there are no effective drugs or vaccines available. Bachmann and Jennings (2010) reported that nanoparticles have the potential to enhance transport in the lymphatic system compared to smaller subunit antigens. Virus-Like nanoparticles (VLNPs) play a significant role in vaccine development as vaccine carriers and they can stimulate host-immune responses [180]. Nanoparticle-based vaccines have shown much promise in improving vaccine efficacy, immunization strategies and targeted delivery. VLNPs improve vaccine efficacy, protect the antigens from premature proteolytic degradation, facilitate antigen uptake, control release and they are non-toxic [181]. VLNPs are composed of a self-assembled viral membrane that forms a monomeric complex displaying a high density of epitopes [182]. VLNPs can accommodate the expression of additional proteins or the endogenous expression of multiple antigens [183,184]. Due to these specific features and unique qualities, VLNPs can provide protection not only against viruses but also against heterologous antigens [185]. A host-immune response was generated after the delivery of an antigen using virus capsid protein SV40 in mammalian cells [186]. VLNPs can increase the immunogenicity of weak antigens including *Salmonella typhi* membrane antigen and influenza A M2 protein. VLNPs containing H1V1 Nef gonadotropin releasing hormone (GnRH) assembled and produced a strong antigen-specific humoral response as well as a cellular immune response [187,188]. VLNPs seem to be better carrier molecules, owing to their multiple surface antigens, compared to antigen presenting cells, which can only present one type of antigen on their surface [189]. Another interesting feature of VLNPs is their high surface energy, which leads to strong adhesion of biomolecules. These qualities contribute to a virus mimicking effect that stimulates the immune system to produce antibodies and immune cells to fight viral infections [190,191]. The combination of AuNPs (with an average size of 100 nm) and spike (S) proteins of infectious bronchitis virus exhibited increased stability when used to developed VLNPs and showed a significant retention of the S proteins compared to viral antigens [192]. VLNPs mediated the inhibitory effects of MERS-CoV S protein nanoparticle vaccine and matrix (M1) protein adjuvant combination on MERS-CoV replication in the lungs of mice. In addition, the MERS-CoV S nanoparticle vaccine produced a high titer of anti-S neutralizing antibodies and protected against MERS-CoV infection in mice in vivo. Altogether, these studies suggest that VLNPs conjugated with S protein seem to be a potential design for a successful vaccine, not only to stimulate the immune system but also to protect humans from MERS-CoV. This approach can also be applied to SARS-CoV-2 as both CoVs use the same mechanism of entry into host cells and similar pathogenicity [193]. Taken together, nanoparticles can be used as antiviral agents against various types of viruses including SARS-CoV-2.

12. Therapeutic Approaches for Coronaviruses

Emerging and reemerging viruses are responsible for a number of recent epidemic outbreaks. A vital part of controlling of viruses are predicting and controlling of spreading and infections. Although we are developing various kinds of antiviral drugs to stop spreading and infections of deadly viruses, the detection of pathogen is very critical in the first place. Therefore, researchers are interested to develop sensitive, rapid, simple technique for accurate characterization of emerging virus strains. Manual magnetic particle-based extraction methods were developed to detect HIV and HCV viral nucleic acids combination with detection by reverse transcriptase-polymerase chain reaction (RT-PCR) one-step. These methods can be used to routinely screen blood donation for viremic donors [194]. Liu et al. (2013) developed a rapid diagnostic platform for pathogen detection based on the acetylcholinesterase-catalyzed hydrolysis reaction which is comparable to that of PCR and easily detectable through changing of color [195]. Silica-Coated magnetic nanoparticles were prepared by co-precipitation method. These Fe_3O_4/SiO_2 nanoparticles were used to isolate genomic DNA of hepatitis virus type B (HBV) and of Epstein-Barr virus (EBV) for detection of the viruses based on polymerase chain reaction (PCR). The results depicted that the purification efficiency of DNA of both HBV and EBV using obtained Fe_3O_4/SiO_2 nanoparticles was significantly better that commercially available reagents [196]. Yeh et al. (2020) developed a portable microfluidic platform containing carbon nanotube arrays with differential filtration porosity for the rapid enrichment and optical identification of viruses [197]. This technique used to characterize various type of viruses including rhinovirus, influenza virus and parainfluenza viruses. This enrichment method could be used to rapidly track and monitor viral outbreaks in real time. Zhao et al. (2020) developed efficient magnetic nanoparticles-based viral RNA extraction method to detect SARS-CoV-2, which is comparable with PCR techniques [198].

Recently, SARS-CoV-2 emerged as a global threat for both healthcare and the economy. Currently, a variety of antiviral agents, including re-purposed drugs, are under testing in clinical trials to assess their efficacy against this new virus but the mission of finding an effective treatment for COVID-19 is ongoing [199–202]. Previously, several modes of treatment were practiced against MERS and SARS infections including the use of inhibitors of viral and host proteases, IFNs and host-directed therapies. Ribavirin, a nucleoside analog that acts as an RNA polymerase inhibitor was used on patients with SARS and MERS [203,204]. Development of new therapies should focus on the CoV S protein because it guides the entry of CoVs into host cells by participating in the binding and fusion of the virus to the ACE2 receptor on the host cell membrane. The S protein is composed of two subunits—S1 recognizes and binds to host receptors and S2 facilitates fusion between the viral envelope and the host cell membrane [205]. Although several agents have been developed using peptide fusion inhibitors, anti-CoV neutralizing monoclonal antibodies and entry receptor antagonists, none of these potentially curative agents is approved for commercial use in humans [206]. Remdesivir, an adenosine analogue, is a broad-spectrum antiviral with potent in vitro efficacy against multiple genetically-unrelated RNA viruses including Ebola, SARS-CoV and MERS-CoV [38]. Remdesivir seems to be an effective antiviral agent against COVID-19 [207]. Another drug, chloroquine, has an immunomodulatory effect and functions at both the entry- and post-entry stages of SARS-CoV-2 infection [208]. However, all of these agents have undesired side effects. Therefore, other agents such as nanoparticles need to be explored as alternative antiviral agents.

Recently, nano and nanomediated combination therapy (nanoparticles plus antiviral drugs) have shown immense promise in nanomedicine. Metal nanoparticles and metal-loaded nanocomposites are known to be extremely effective against microbes and viruses due to their unique property, the controlled release of ions. For example, the controlled release of ionic copper is the fundamental aspect for the antimicrobial and antiviral properties of surfaces [209]. In addition, the controlled release of ions favors the production of ROS. Metal-grafted GO decorated with metals such as Ag, Fe, Cu, Zn, TiO2, CdS and MnS_2 exhibited potential antiviral activity [77,210]. For example, silver and copper decorated GOs are potential antiviral agents for both enveloped and non-enveloped viruses [77]. The development of nanomaterial-mediated therapy is an alternative to conventional therapies in fighting against resistant viruses [211]. AgNPs are able to interact with host receptors and inhibit viral entry. Functionalized AuNPs showed antiviral activity against HIV-1 and also against H1N1, H3N2 and H5N1 [211].

Adaptor protein complex 2 (AP2)-associated protein kinase 1 (AAK1) is a key regulator of endocytosis. Hence, a drug such as baricitinib that inhibits AAK1 may suppress viral entry into target cells. As such, baricitinib could be a potential treatment for COVID-19 [212]. Baricitinib has been shown to bind to another endocytosis regulator, cyclin G-associated kinase and inhibit AAK1, thus preventing viral entry into the cell [213]. HIV protease inhibitors such as lopinavir and ritonavirin suppress 3-chymotrypsin-like protease activity of SARS-CoV and MERS-CoV [214]. Remdesivir, a nucleoside analogue that targets RdRp, suppresses viral RNA synthesis in a broad spectrum of RNA viruses including HCoVs. Remdesivir inhibited RdRp of CoVs in cell cultures and animal models [201,215]. A combination of hydroxychloroquine and azithromycin can potentially increase the recovery of COVID-19 patients. Chloroquine-Based drugs inhibit the fusion of SARS-CoV-2 with host cells by acidifying the lysosomes and thus inhibiting cathepsins that require a low pH for optimal cleavage of SARS-CoV-2 S protein. These drugs can also alter cross-talk between SARS-CoV-2 and host cells and reduce production of pro-inflammatory cytokines, while activating anti-SARS-CoV-2 CD8+ T-cells [216,217].

13. Conclusions and Future Perspectives

Infectious diseases cause immense global mortality and viruses are responsible for about one-third of these deaths. Respiratory infections are among the most common causes of death worldwide, especially due to CoVs. Presently, the outbreak of viral respiratory infections, particularly COVID-19, is widespread and continuing to spread worldwide. As of 27 July 2020, SARS-CoV-2, the infectious agent of COVID-19, had infected 16,430,566 individuals and led to the death of more than 652,434 individuals in 215 countries, while also triggering an exceptional economic crisis. Although there are no specific drugs or vaccines to treat or prevent COVID-19, the available antiviral drugs are re-purposed drugs and active against a limited panel of human pathogens. Therefore, researchers are urgently striving to identify and develop suitable nano-based drugs and nano-vaccines, in addition to conventional approaches. Nanotechnology plays a critical role in both viral disease diagnosis and therapeutics. Nanoparticles show great potential for biomedical applications, especially in patients who relapse after completing conventional antiviral therapy. Antiviral resistance, which is a slowly developing problem of conventional therapeutic approaches, may be addressed using nanoparticles due to their large surface-to-volume ratio, surface charge, size and shape as well as their optical, electronic, biological and functional properties. Furthermore, nano-based approaches are feasible, cost effective, non-toxic, biocompatible and a convenient strategy to deal with various types of viral infections, particularly SARS-CoV-2/COVID-19. In this review, we have provided a brief account of the main mechanism of entry of viruses into host cells. We also discussed in detail the effects of several important types of nanomaterials including AgNPs, AuNPs, QDs, organic nanoparticles, liposomes, dendrimers and polymers against various types of viral infections. These functional nanoparticles can provide a novel platform for fabrication of bio-safe and effective drugs for nanoscale treatment of viral infectious diseases. Further, we discussed antiviral mechanisms, therapeutic approaches of nanoparticles and the effects of nanoparticles on CoVs. Finally, we provide our future perspective of nanoparticles below.

Nanotechnology has become a focal point of research and various types of nanomaterials have been explored and evaluated for their prophylactic and therapeutic activity against different viruses. Therefore, nanoparticle-based therapy seems to be promising but there are still many challenges and barriers to achieve its full potential. The future focus of antiviral nano-based therapy should concentrate on development of new antiviral therapeutics and approaches to challenge the emergence of drug resistance and different secondary effects due to long-term conventional treatments. Antiviral drugs currently available are effective against a few viral diseases such as influenza, hepatitis, HSV and

HIV. Generally, development of an antiviral drug takes a long time (years) and the process involves tedious protocols, particularly for CoVs due to their variability. Most of the antiviral drugs developed through nanoparticle research show immense potential in in vitro and in vivo conditions; however, several variables need to be optimized for a successful translation of nanomaterials from the laboratory to the clinical setting. In addition, one important aspect is the non-toxic nature of nanoparticles, which require long and intense studies to demonstrate their non-toxic nature, let alone their potential activity against specific viruses. The biotech/nanotech industry needs more exposure to demonstrate the effects of nanoparticles on health and their applications in various fields. The crucial points and future focus of nano-based antiviral drugs need to be on the following important issues. First, the system needs to have receptor-based nanoparticles, which could be safely managed as antiviral agents to rapidly target specific viruses. Second, functionalization by specific molecules is needed to facilitate effective targeting of specific sites of viral infections. Third, non- or reduced-toxicity must remain a priority; biocompatibility with no undesired side effects is essential. Fourth, a therapeutic approach that combines nanoparticles and low concentrations of antiviral drugs with excellent efficiency, is needed. Finally, a multidisciplinary consortium is needed to address potential questions related to various types of viruses, variability, frequent mutations and antiviral agents and their use in humans, particularly during pandemic situations as we are now experiencing. We believe that the approach presented here has a chance to produce medically-relevant antiviral drugs against CoVs and other viral diseases.

Author Contributions: S.G. Conceptualization and writing—original manuscript. M.Q. Drawing of figures and reference arrangement, Y.C., J.T.D., C.P., K.H. and J.-H.K. performed all literature surveys, the interpretation of literature; and H.S. Validation and editing. All authors have read and agreed to the published version of the manuscript.

Funding: This paper was supported by Konkuk University in 2018.

Acknowledgments: Although we are the authors of this review, we would never have been able to complete it without the great many people who have contributed to the field of nanomaterials and coronaviruses. We owe our gratitude to all those researchers who have made this review possible. We have cited as many references as permitted and apologize to the authors of those publications that we have not cited due to limitation of references. We apologize to other authors who have worked on these aspects but whom we have unintentionally overlooked. This paper was supported by Konkuk University in 2018.

Conflicts of Interest: The authors declare no conflict of interest.

References

1. Lozano, R.; Naghavi, M.; Foreman, K.; Lim, S.; Shibuya, K.; Aboyans, V.; Abraham, J.; Adair, T.; Aggarwal, R.; Ahn, S.Y.; et al. Global and regional mortality from 235 causes of death for 20 age groups in 1990 and 2010: A systematic analysis for the Global Burden of Disease Study 2010. *Lancet* **2012**, *380*, 2095–2128. [CrossRef]
2. Dube, A. Nanomedicines for infectious diseases. *Pharm. Res.* **2019**, *36*, 1–2. [CrossRef] [PubMed]
3. Qasim, M.; Lim, D.-J.; Park, H.; Na, D. Nanotechnology for diagnosis and treatment of infectious diseases. *J. Nanosci. Nanotechnol.* **2014**, *14*, 7374–7387. [CrossRef] [PubMed]
4. Zhou, P.; Yang, X.-L.; Wang, X.G.; Hu, B.; Zhang, L.; Zhang, W.; Si, H.R.; Zhu, Y.; Li, B.; Huang, C.L.; et al. A pneumonia outbreak associated with a new coronavirus of probable bat origin. *Nature* **2020**, *579*, 270–273. [CrossRef]
5. Zhu, N.; Zhang, D.; Wang, W.; Li, X.; Yang, B.; Song, J.; Zhao, X.; Huang, B.; Shi, W.; Lu, R.; et al. A novel coronavirus from patients with pneumonia in China, 2019. *N. Engl. J. Med.* **2020**, *382*, 727–733. [CrossRef]
6. Brasil, P.; Calvet, G.A.; Siqueira, A.M.; Wakimoto, M.; de Sequeira, P.C.; Nobre, A.; De Quintana, M.S.B.; De Mendonça, M.C.L.; Lupi, O.; de Souza, R.V.; et al. Zika virus outbreak in Rio de Janeiro, Brazil: Clinical characterization, epidemiological and virological aspects. *PLoS Negl. Trop. Dis.* **2016**, *10*, e0004636. [CrossRef]
7. Xie, Y.; Luo, X.; He, Z.; Zheng, Y.; Zuo, Z.; Zhao, Q.; Miao, Y.; Ren, J. VirusMap: A visualization database for the influenza A virus. *J. Genet. Genom.* **2017**, *44*, 281–284. [CrossRef]
8. Spengler, J.R.; Ervin, E.D.; Towner, J.S.; Rollin, P.E.; Nichol, S.T. Perspectives on West Africa ebola virus disease outbreak, 2013–2016. *Emerg. Infect. Dis.* **2016**, *22*, 956–963. [CrossRef]

9. Perlman, S.; Netland, J. Coronaviruses post-SARS: Update on replication and pathogenesis. *Nat. Rev. Microbiol.* **2009**, *7*, 439–450. [CrossRef]
10. Zhang, B.; Zhou, X.; Zhu, C.; Feng, F.; Qiu, Y.; Feng, J.; Jia, Q.; Song, Q.; Zhu, B.; Wang, J. Immune phenotyping based on neutrophil-to-lymphocyte ratio and IgG predicts disease severity and outcome for patients with COVID-19. *medRxiv* **2020**. [CrossRef]
11. Robba, C.; Battaglini, D.; Pelosi, P.; Rocco, P.R.M. Multiple organ dysfunction in SARS-CoV-2: MODS-CoV-2. *Expert Rev. Respir. Med.* **2020**, 1–4. [CrossRef] [PubMed]
12. Zhang, Y.; Xiao, M.; Zhang, S.; Xia, P.; Cao, W.; Jiang, W.; Chen, H.; Ding, X.; Zhao, H.; Zhang, H.; et al. Coagulopathy and Antiphospholipid antibodies in patients with Covid-19. *N. Engl. J. Med.* **2020**, *382*, e38. [CrossRef] [PubMed]
13. Klok, F.A.; Kruip, M.J.H.A.; van der Meer, N.J.M.; Arbous, M.S.; Gommers, D.A.M.P.J.; Kant, K.M.; Kaptein, F.H.J.; van Paassen, J.; Stals, M.A.M.; Huisman, M.V.; et al. Incidence of thrombotic complications in critically ill ICU patients with COVID-19. *Thromb. Res.* **2020**, *191*, 145–147. [CrossRef]
14. Danzi, G.B.; Loffi, M.; Galeazzi, G.; Gherbesi, E. Acute pulmonary embolism and COVID-19 pneumonia: A random association? *Eur. Heart J.* **2020**, *41*, 1858. [CrossRef] [PubMed]
15. Little, S.J.; Holte, S.; Routy, J.-P.; Daar, E.S.; Markowitz, M.; Collier, A.C.; Koup, R.A.; Mellors, J.W.; Connick, E.; Conway, B.; et al. Antiretroviral-drug resistance among patients recently infected with HIV. *N. Engl. J. Med.* **2002**, *347*, 385–394. [CrossRef]
16. Shafer, R.W.; Rhee, S.-Y.; Pillay, D.; Miller, V.; Sandstrom, P.; Schapiro, J.M.; Kuritzkes, D.R.; Bennett, D. HIV-1 protease and reverse transcriptase mutations for drug resistance surveillance. *AIDS* **2007**, *21*, 215–223. [CrossRef]
17. Cosgrove, S.E. The Relationship between Antimicrobial resistance and patient outcomes: Mortality, length of hospital stay, and health care costs. *Clin. Infect. Dis.* **2006**, *42*, S82–S89. [CrossRef]
18. Lu, H. Drug treatment options for the 2019-new coronavirus (2019-nCoV). *Biosci. Trends* **2020**, *14*, 69–71. [CrossRef]
19. Cinatl, J.; Morgenstern, B.; Bauer, G.; Chandra, P.; Rabenau, H.; Doerr, H.W. Treatment of SARS with human interferons. *Lancet* **2003**, *362*, 293–294. [CrossRef]
20. Stockman, L.J.; Bellamy, R.; Garner, P. SARS: Systematic review of treatment effects. *PloS Med.* **2006**, *3*, e343. [CrossRef]
21. Toots, M.; Yoon, J.J.; Cox, R.M.; Hart, M.; Sticher, Z.M.; Makhsous, N.; Plesker, R.; Barrena, A.H.; Reddy, P.G.; Mitchell, D.G.; et al. Characterization of orally efficacious influenza drug with high resistance barrier in ferrets and human airway epithelia. *Sci. Transl. Med.* **2019**, *11*, eaax5866. [CrossRef] [PubMed]
22. Aguiar, A.C.C.; Murce, E.; Cortopassi, W.A.; Pimentel, A.S.; Almeida, M.M.F.S.; Barros, D.C.S.; Guedes, J.S.; Meneghetti, M.R.; Krettli, A.U. Chloroquine analogs as antimalarial candidates with potent in vitro and in vivo activity. *Int. J. Parasitol. Drugs Drug Resist.* **2018**, *8*, 459–464. [CrossRef]
23. Liu, J.; Cao, R.; Xu, M.; Wang, X.; Zhang, H.; Hu, H.; Li, Y.; Hu, Z.; Zhong, W.; Wang, M. Hydroxychloroquine, a less toxic derivative of chloroquine, is effective in inhibiting SARS-CoV-2 infection in vitro. *Cell Discov.* **2020**, *6*, 1–4. [CrossRef]
24. Matsuyama, S.; Kawase, M.; Nao, N.; Shirato, K.; Ujike, M.; Kamitani, W.; Shimojima, M.; Fukushi, S. The inhaled corticosteroid ciclesonide blocks coronavirus RNA replication by targeting viral NSP15. *BioRxiv* **2020**. [CrossRef]
25. Kimura, H.; Kurusu, H.; Sada, M.; Kurai, D.; Murakami, K.; Kamitani, W.; Tomita, H.; Katayama, K.; Ryo, A. Molecular pharmacology of ciclesonide against SARS-CoV-2. *J. Allergy Clin. Immunol.* **2020**, *146*, 330–331. [CrossRef] [PubMed]
26. Caly, L.; Druce, J.D.; Catton, M.G.; Jans, D.A.; Wagstaff, K.M. The FDA-approved drug ivermectin inhibits the replication of SARS-CoV-2 in vitro. *Antivir. Res.* **2020**, *178*, 104787. [CrossRef] [PubMed]
27. Heidary, F.; Gharebaghi, R. Ivermectin: A systematic review from antiviral effects to COVID-19 complementary regimen. *J. Antibiot.* **2020**, *73*, 593–602. [CrossRef] [PubMed]
28. Ianevski, A.; Yao, R.; Fenstad, M.H.; Biza, S.; Zusinaite, E.; Reisberg, T.; Lysvand, H.; Løseth, K.; Landsem, V.M.; Malmring, J.F.; et al. Potential antiviral options against SARS-CoV-2 infection. *Viruses* **2020**, *12*, 642. [CrossRef] [PubMed]
29. Wang, X.; Cao, R.; Zhang, H.; Liu, J.; Xu, M.; Hu, H.; Li, Y.; Zhao, L.; Li, W.; Sun, X.; et al. The anti-influenza virus drug, arbidol is an efficient inhibitor of SARS-CoV-2 in vitro. *Cell Discov.* **2020**, *6*, 28. [CrossRef]

30. Agostini, M.L.; Andres, E.L.; Sims, A.C.; Graham, R.L.; Sheahan, T.P.; Lu, X.; Smith, E.C.; Case, J.B.; Feng, J.Y.; Jordan, R.; et al. Coronavirus susceptibility to the antiviral remdesivir (GS-5734) is mediated by the viral polymerase and the proofreading exoribonuclease. *MBio* **2018**, *9*. [CrossRef]
31. Gordon, C.J.; Tchesnokov, E.P.; Feng, J.Y.; Porter, D.P.; Götte, M. The antiviral compound remdesivir potently inhibits RNAdependent RNA polymerase from Middle East respiratory syndrome coronavirus. *J. Biol. Chem.* **2020**, *295*, 4773–4779. [CrossRef] [PubMed]
32. De Wit, E.; Rasmussen, A.L.; Falzarano, D.; Bushmaker, T.; Feldmann, F.; Brining, D.L.; Fischer, E.R.; Martellaro, C.; Okumura, A.; Chang, J.; et al. Middle East respiratory syndrome coronavirus (MERSCoV) causes transient lower respiratory tract infection in rhesus macaques. *Proc. Natl. Acad. Sci. USA* **2013**, *110*, 16598–16603. [CrossRef] [PubMed]
33. Holshue, M.L.; DeBolt, C.; Lindquist, S.; Lofy, K.H.; Wiesman, J.; Bruce, H.; Spitters, C.; Ericson, K.; Wilkerson, S.; Tural, A.; et al. First case of 2019 novel coronavirus in the United States. *N. Engl. J. Med.* **2020**, *382*, 929–936. [CrossRef] [PubMed]
34. Chu, C.M.; Cheng, V.C.C.; Hung, I.F.N.; Wong, M.M.L.; Chan, K.H.; Chan, K.S.; Kao, R.Y.T.; Poon, L.L.M.; Wong, C.L.P.; Guan, Y.; et al. Role of lopinavir/ritonavir in the treatment of SARS: Initial virological and clinical findings. *Thorax* **2004**, *59*, 252–256. [CrossRef] [PubMed]
35. Cvetkovic, R.S.; Goa, K.L. Lopinavir/ritonavir: A review of its use in the management of HIV infection. *Drugs* **2003**, *63*, 769–802. [CrossRef] [PubMed]
36. Arabi, Y.M.; Fowler, R.; Hayden, F.G. Critical care management of adults with community-acquired severe respiratory viral infection. *Intensive Care Med.* **2020**, *46*, 315–328. [CrossRef]
37. Lim, J.; Jeon, S.; Shin, H.Y.; Kim, M.J.; Seong, Y.M.; Lee, W.J.; Choe, K.W.; Kang, Y.M.; Lee, B.; Park, S.J. Case of the index patient who caused tertiary transmission of coronavirus disease 2019 in Korea: The application of lopinavir/ritonavir for the treatment of COVID-19 pneumonia monitored by quantitative RT-PCR. *J. Korean Med. Sci.* **2020**, *35*. [CrossRef]
38. Wang, Z.; Chen, X.; Lu, Y.; Chen, F.; Zhang, W. Clinical characteristics and therapeutic procedure for four cases with 2019 novel coronavirus pneumonia receiving combined Chinese and Western medicine treatment. *Biosci. Trends* **2020**, *14*. [CrossRef]
39. Xia, J.; Rong, L.; Sawakami, T.; Inagaki, Y.; Song, P.; Hasegawa, K.; Sakamoto, Y.; Tang, W. Capsule and its active ingredients induce apoptosis, inhibit migration and invasion, and enhances doxorubicin therapeutic efficacy in hepatocellular carcinoma. *Biomed. Pharmacother.* **2018**, *99*, 921–930. [CrossRef]
40. Wu, C.J.; Huang, H.W.; Liu, C.Y.; Hong, C.F.; Chan, Y.L. Inhibition of SARS-CoV replication by siRNA. *Antivir. Res.* **2005**, *65*, 45–48. [CrossRef]
41. Abbott, T.R.; Dhamdhere, G.; Liu, Y.; Lin, X.; Goudy, L.; Zeng, L.; Chemparathy, A.; Chmura, S.; Heaton, N.S.; Debs, R.; et al. Development of CRISPR as an antiviral strategy to combat SARS-CoV-2 and Influenza. *Cell* **2020**, *181*, 865–876.e12. [CrossRef] [PubMed]
42. Singh, R.K.; Chang, H.W.; Yan, D.; Lee, K.M.; Ucmak, D.; Wong, K.; Abrouk, M.; Farahnik, B.; Nakamura, M.; Zhu, T.H.; et al. Influence of diet on the gut microbiome and implications for human health. *J. Transl. Med.* **2017**, *15*, 1–17. [CrossRef]
43. Jiao, Y.; Tibbits, A.; Gillman, A.; Hsiao, M.S.; Buskohl, P.; Drummy, L.F.; Vaia, R.A. Deformation behavior of Polystyrene-grafted nanoparticle assemblies with low grafting density. *Macromolecules* **2018**, *51*, 7257–7265. [CrossRef]
44. Gurunathan, S.; Kang, M.H.; Qasim, M.; Kim, J.H. Nanoparticle-mediated combination therapy: Two-in-one approach for cancer. *Int. J. Mol. Sci.* **2018**, *19*, 3264. [CrossRef] [PubMed]
45. Szunerits, S.; Barras, A.; Khanal, M.; Pagneux, Q.; Boukherroub, R. Nanostructures for the inhibition of viral infections. *Molecules* **2015**, *20*, 14051–14081. [CrossRef] [PubMed]
46. Lei, L.; Sun, R.W.Y.; Chen, R.; Hui, C.K.; Ho, C.M.; Luk, J.M.; Lau, G.K.; Che, C.M. Silver nanoparticles inhibit hepatitis B virus replication. *Antivir. Ther.* **2008**, *13*, 252–262.
47. Rafiei, S.; Rezatofighi, S.E.; Ardakani, M.R.; Rastegarzadeh, S. Gold nanoparticles impair foot-and-mouth disease virus replication. *IEEE Trans. Nanobioscience* **2016**, *15*, 34–40. [CrossRef]
48. Levina, A.S.; Repkova, M.N.; Mazurkova, N.A.; Zarytova, V.F. Nanoparticle-Mediated Nonviral DNA Delivery for Effective Inhibition of Influenza a Viruses in Cells. *IEEE Trans. Nanotechnol.* **2016**, *15*, 248–254. [CrossRef]

49. Botequim, D.; Maia, J.; Lino, M.M.F.; Lopes, L.M.F.; Simões, P.N.; Ilharco, L.M.; Ferreira, L. Nanoparticles and surfaces presenting antifungal, antibacterial and antiviral properties. *Langmuir* **2012**, *28*, 7646–7656. [CrossRef]
50. Lysenko, V.; Lozovski, V.; Lokshyn, M.; Gomeniuk, Y.V.; Dorovskih, A.; Rusinchuk, N.; Pankivska, Y.; Povnitsa, O.; Zagorodnya, S.; Tertykh, V.; et al. Nanoparticles as antiviral agents against adenoviruses. *Adv. Nat. Sci. Nanosci. Nanotechnol.* **2018**, *9*, 025021. [CrossRef]
51. Sucipto, T.H.; Churrotin, S.; Setyawati, H.; Kotaki, T.; Martak, F.; Soegijanto, S. Antiviral activity of copper(ii)chloride dihydrate against dengue virus type-2 in vero cell. *Indones. J. Trop. Infect. Dis.* **2017**, *6*, 84. [CrossRef]
52. Mazurkova, N.A.; Spitsyna, Y.E.; Shikina, N.V.; Ismagilov, Z.R.; Zagrebel'nyi, S.N.; Ryabchikova, E.I. Interaction of titanium dioxide nanoparticles with influenza virus. *Nanotechnol. Russ.* **2010**, *5*, 417–420. [CrossRef]
53. Lara, H.H.; Ayala-Nuñez, N.V.; Ixtepan-Turrent, L.; Rodriguez-Padilla, C. Mode of antiviral action of silver nanoparticles against HIV-1. *J. Nanobiotechnol.* **2010**, *8*, 1–10. [CrossRef]
54. Hu, R.; Li, S.; Kong, F.; Hou, R.; Guan, X.; Guo, F. Inhibition effect of silver nanoparticles on herpes simplex virus 2. *Genet. Mol. Res.* **2014**, *13*, 7022–7028. [CrossRef] [PubMed]
55. Lokshyn, M.; Lozovski, V.; Lysenko, V.; Ushenin, Y.; Rusinchuk, N.; Shydlovska, O.; Spivak, M.; Zholobak, N. Purification of bioliquids from viruses by surface plasmon-polaritons. *J. Bionanosci.* **2015**, *9*, 431–438. [CrossRef]
56. Sportelli, M.C.; Izzi, M.; Kukushkina, E.A.; Hossain, S.I.; Picca, R.A.; Ditaranto, N.; Cioff, N. Can nanotechnology and materials science help the fight against sars-cov-2? *Nanomaterials* **2020**, *10*, 802. [CrossRef] [PubMed]
57. Chen, L.; Liang, J. An overview of functional nanoparticles as novel emerging antiviral therapeutic agents. *Mater. Sci. Eng. C* **2020**, *112*, 110924. [CrossRef]
58. Yang, M.; Sunderland, K.; Mao, C. Virus-Derived Peptides for Clinical Applications. *Chem. Rev.* **2017**, *117*, 10377–10402. [CrossRef]
59. Oswald, M.; Geissler, S.; Goepferich, A. Targeting the Central Nervous System (CNS): A review of rabies virus-targeting strategies. *Mol. Pharm.* **2017**, *14*, 2177–2196. [CrossRef]
60. Kutter, J.S.; Spronken, M.I.; Fraaij, P.L.; Fouchier, R.A.; Herfst, S. Transmission routes of respiratory viruses among humans. *Curr. Opin. Virol.* **2018**, *28*, 142–151. [CrossRef]
61. Cui, J.; Li, F.; Shi, Z.L. Origin and evolution of pathogenic coronaviruses. *Nat. Rev. Microbiol.* **2019**, *17*, 181–192. [CrossRef] [PubMed]
62. Kim, D.; Quinn, J.; Pinsky, B.; Shah, N.H.; Brown, I. Rates of Co-infection between SARS-CoV-2 and other respiratory pathogens. *JAMA J. Am. Med. Assoc.* **2020**, *323*, 2085–2086. [CrossRef] [PubMed]
63. Qasim, M.; Baipaywad, P.; Udomluck, N.; Na, D.; Park, H. Enhanced therapeutic efficacy of lipophilic amphotericin B against Candida albicans with amphiphilic poly(N-isopropylacrylamide) nanogels. *Macromol. Res.* **2014**, *22*, 1125–1131. [CrossRef]
64. Gurunathan, S.; Qasim, M.; Park, C.; Yoo, H.; Kim, J.-H.; Hong, K. Cytotoxic potential and molecular pathway analysis of silver nanoparticles in human colon cancer cells HCT116. *Int. J. Mol. Sci.* **2018**, *19*, 2269. [CrossRef]
65. Jeyaraj, M.; Gurunathan, S.; Qasim, M.; Kang, M.H.; Kim, J.H. A comprehensive review on the synthesis, characterization, and biomedical application of platinum nanoparticles. *Nanomaterials* **2019**, *9*, 1719. [CrossRef]
66. Ivan, F.D.; Botezat, D.; Gardikiotis, I.; Uritu, C.M.; Dodi, G.; Trandafir, L.; Rezuș, C.; Rezuș, E.; Tamba, B.-I.; Mihai, C. Nanomaterials Designed for Antiviral Drug Delivery Transport across Biological Barriers. *Pharmaceutics* **2020**, *12*, 171. [CrossRef]
67. Skov, C.; Gurunathan, S.; Qasim, M.; Park, C.H.; Iqbal, M.A.; Yoo, H.; Hwang, J.H.; Uhm, S.J. Cytotoxicity and transcriptomic analyses of biogenic palladium nanoparticles in human ovarian cancer. *Nanomaterials* **2019**, *9*, 787.
68. Gurunathan, S.; Qasim, M.; Park, C.; Yoo, H.; Choi, D.Y.; Song, H.; Park, C.; Kim, J.H.; Hong, K. Cytotoxicity and transcriptomic analysis of silver nanoparticles in mouse embryonic fibroblast cells. *Int. J. Mol. Sci.* **2018**, *19*, 3618. [CrossRef]

69. Zhang, X.-F.; Liu, Z.-G.; Shen, W.; Gurunathan, S. Silver nanoparticles: Synthesis, characterization, properties, applications, and therapeutic approaches. *Int. J. Mol. Sci.* **2016**, *17*, 1534. [CrossRef]
70. Rosa, R.M.; Silva, J.C.; Sanches, I.S.; Henriques, C. Simultaneous photo-induced cross-linking and silver nanoparticle formation in a PVP electrospun wound dressing. *Mater. Lett.* **2017**, *207*, 145–148. [CrossRef]
71. Mohammed Fayaz, A.; Ao, Z.; Girilal, M.; Chen, L.; Xiao, X.; Kalaichelvan, P.T.; Yao, X. Inactivation of microbial infectiousness by silver nanoparticles-coated condom: A new approach to inhibit HIV- and HSV-transmitted infection. *Int. J. Nanomed.* **2012**, *7*, 5007–5018. [CrossRef]
72. Gaikwad, S.; Ingle, A.; Gade, A.; Rai, M.; Falanga, A.; Incoronato, N.; Russo, L.; Galdiero, S.; Galdiero, M. Antiviral activity of mycosynthesized silver nanoparticles against herpes simplex virus and human parainfluenza virus type 3. *Int. J. Nanomed.* **2013**, *8*, 4303–4314. [CrossRef]
73. Mori, Y.; Ono, T.; Miyahira, Y.; Nguyen, V.Q.; Matsui, T.; Ishihara, M. Antiviral activity of silver nanoparticle/chitosan composites against H1N1 influenza A virus. *Nanoscale Res. Lett.* **2013**, *8*, 93. [CrossRef] [PubMed]
74. Lv, X.; Wang, P.; Bai, R.; Cong, Y.; Suo, S.; Ren, X.; Chen, C. Inhibitory effect of silver nanomaterials on transmissible virus-induced host cell infections. *Biomaterials* **2014**, *35*, 4195–4203. [CrossRef]
75. Yang, X.X.; Li, C.M.; Huang, C.Z. Curcumin modified silver nanoparticles for highly efficient inhibition of respiratory syncytial virus infection. *Nanoscale* **2016**, *8*, 3040–3048. [CrossRef]
76. Xiang, D.; Zheng, C.; Zheng, Y.; Li, X.; Yin, J.; O' Conner, M.; Marappan, M.; Miao, Y.; Xiang, B.; Duan, W.; et al. Inhibition of A/Human/Hubei/3/2005 (H3N2) influenza virus infection by silver nanoparticles in vitro and in vivo. *Int. J. Nanomed.* **2013**, *8*, 4103. [CrossRef]
77. Chen, Y.N.; Hsueh, Y.H.; Hsieh, C.-T.; Tzou, D.Y.; Chang, P.L. Antiviral activity of graphene–silver nanocomposites against non-enveloped and enveloped viruses. *Int. J. Environ. Res. Public Health* **2016**, *13*, 430. [CrossRef]
78. Huy, T.Q.; Hien Thanh, N.T.; Thuy, N.T.; Van Chung, P.; Hung, P.N.; Le, A.T.; Hong Hanh, N.T. Cytotoxicity and antiviral activity of electrochemical—Synthesized silver nanoparticles against poliovirus. *J. Virol. Methods* **2017**, *241*, 52–57. [CrossRef]
79. Sreekanth, T.V.M.; Nagajyothi, P.C.; Muthuraman, P.; Enkhtaivan, G.; Vattikuti, S.V.P.; Tettey, C.O.; Kim, D.H.; Shim, J.; Yoo, K. Ultra-sonication-assisted silver nanoparticles using Panax ginseng root extract and their anti-cancer and antiviral activities. *J. Photochem. Photobiol. B Biol.* **2018**, *188*, 6–11. [CrossRef]
80. Lin, Z.; Li, Y.; Guo, M.; Xu, T.; Wang, C.; Zhao, M.; Wang, H.; Chen, T.; Zhu, B. The inhibition of H1N1 influenza virus-induced apoptosis by silver nanoparticles functionalized with zanamivir. *RSC Adv.* **2017**, *7*, 742–750. [CrossRef]
81. Szymańska, E.; Orłowski, P.; Winnicka, K.; Tomaszewska, E.; Bąska, P.; Celichowski, G.; Grobelny, J.; Basa, A.; Krzyżowska, M. Multifunctional Tannic Acid/Silver nanoparticle-based mucoadhesive hydrogel for improved local treatment of HSV infection: In vitro and in vivo studies. *Int. J. Mol. Sci.* **2018**, *19*, 387. [CrossRef] [PubMed]
82. Wan, C.; Tai, J.; Zhang, J.; Guo, Y.; Zhu, Q.; Ling, D.; Gu, F.; Gan, J.; Zhu, C.; Wang, Y.; et al. Silver nanoparticles selectively induce human oncogenic γ-herpesvirus-related cancer cell death through reactivating viral lytic replication. *Cell Death Dis.* **2019**, *10*, 1–16. [CrossRef] [PubMed]
83. Orłowski, P.; Kowalczyk, A.; Tomaszewska, E.; Ranoszek-Soliwoda, K.; Węgrzyn, A.; Grzesiak, J.; Celichowski, G.; Grobelny, J.; Eriksson, K.; Krzyzowska, M. Antiviral activity of tannic acid modified silver nanoparticles: Potential to activate immune response in herpes genitalis. *Viruses* **2018**, *10*, 524. [CrossRef] [PubMed]
84. Elechiguerra, J.L.; Burt, J.L.; Morones, J.R.; Camacho-Bragado, A.; Gao, X.; Lara, H.H.; Yacaman, M.J. Interaction of silver nanoparticles with HIV-1. *J. Nanobiotechnol.* **2005**, *3*, 6. [CrossRef] [PubMed]
85. Mehrbod, P.; Motamed, N.; Tabatabaian, M.; Estyar, R.S.; Amini, E.; Shahidi, M.; Kheiri, M. In vitro antiviral effect of "nanosilver" on influenza virus. *Daru J. Pharm. Sci.* **2009**, *17*, 88–93.
86. Galdiero, S.; Falanga, A.; Vitiello, M.; Cantisani, M.; Marra, V.; Galdiero, M. Silver nanoparticles as potential antiviral agents. *Molecules* **2011**, *16*, 8894–8918. [CrossRef]
87. Haggag, E.G.; Elshamy, A.M.; Rabeh, M.A.; Gabr, N.M.; Salem, M.; Youssif, K.A.; Samir, A.; Bin Muhsinah, A.; Alsayari, A.; Abdelmohsen, U.R. Antiviral potential of green synthesized silver nanoparticles of lampranthus coccineus and malephora lutea. *Int. J. Nanomed.* **2019**, *14*, 6217–6229. [CrossRef]

88. Gupta, A.; Moyano, D.F.; Parnsubsakul, A.; Papadopoulos, A.; Wang, L.S.; Landis, R.F.; Das, R.; Rotello, V.M. Ultrastable and biofunctionalizable gold nanoparticles. *ACS Appl. Mater. Interfaces* **2016**, *8*, 14096–14101. [CrossRef]
89. Bartczak, D.; Muskens, O.L.; Sanchez-Elsner, T.; Kanaras, A.G.; Millar, T.M. Manipulation of in vitro angiogenesis using peptide-coated gold nanoparticles. *ACS Nano* **2013**, *7*, 5628–5636. [CrossRef]
90. Chakravarthy, K.V.; Bonoiu, A.C.; Davis, W.G.; Ranjan, P.; Ding, H.; Hu, R.; Bowzard, J.B.; Bergey, E.J.; Katz, J.M.; Knight, P.R.; et al. Gold nanorod delivery of an ssRNA immune activator inhibits pandemic H1N1 influenza viral replication. *Proc. Natl. Acad. Sci. USA* **2010**, *107*, 10172–10177. [CrossRef]
91. Lee, M.Y.; Yang, J.A.; Jung, H.S.; Beack, S.; Choi, J.E.; Hur, W.; Koo, H.; Kim, K.; Yoon, S.K.; Hahn, S.K. Hyaluronic acid-gold nanoparticle/interferon α complex for targeted treatment of hepatitis C virus infection. *ACS Nano* **2012**, *6*, 9522–9531. [CrossRef] [PubMed]
92. Halder, A.; Das, S.; Ojha, D.; Chattopadhyay, D.; Mukherjee, A. Highly monodispersed gold nanoparticles synthesis and inhibition of herpes simplex virus infections. *Mater. Sci. Eng. C* **2018**, *89*, 413–421. [CrossRef] [PubMed]
93. Andresen, H.; Mager, M.; Grießner, M.; Charchar, P.; Todorova, N.; Bell, N.; Theocharidis, G.; Bertazzo, S.; Yarovsky, I.; Stevens, M.M. Single-step homogeneous immunoassays utilizing epitope-tagged gold nanoparticles: On the mechanism, feasibility, and limitations. *Chem. Mater.* **2014**, *26*, 4696–4704. [CrossRef]
94. Bowman, M.C.; Ballard, T.E.; Ackerson, C.J.; Feldheim, D.L.; Margolis, D.M.; Melander, C. Inhibition of HIV fusion with multivalent gold nanoparticles. *J. Am. Chem. Soc.* **2008**, *130*, 6896–6897. [CrossRef]
95. Wen, W.H.; Lin, M.; Su, C.Y.; Wang, S.Y.; Cheng, Y.S.E.; Fang, J.M.; Wong, C.H. Synergistic effect of zanamivir-porphyrin conjugates on inhibition of neuraminidase and inactivation of influenza virus. *J. Med. Chem.* **2009**, *52*, 4903–4910. [CrossRef]
96. Papp, I.; Sieben, C.; Ludwig, K.; Roskamp, M.; Böttcher, C.; Schlecht, S.; Herrmann, A.; Haag, R. Inhibition of influenza virus infection by multivalent sialic-acid- functionalized gold nanoparticles. *Small* **2010**, *6*, 2900–2906. [CrossRef] [PubMed]
97. Dkhil, M.A.; Bauomy, A.A.; Diab, M.S.; Al-Quraishy, S. Antioxidant and hepatoprotective role of gold nanoparticles against murine hepatic schistosomiasis. *Int. J. Nanomed.* **2015**, *10*, 7467. [CrossRef]
98. Bastian, A.R.; Nangarlia, A.; Bailey, L.D.; Holmes, A.; Sundaram, R.V.K.; Ang, C.; Moreira, D.R.M.; Freedman, K.; Duffy, C.; Contarino, M.; et al. Mechanism of multivalent nanoparticle encounter with HIV-1 for potency enhancement of peptide triazole virus inactivation. *J. Biol. Chem.* **2015**, *290*, 529–543. [CrossRef]
99. Tao, W.; Hurst, B.L.; Shakya, A.K.; Uddin, M.J.; Ingrole, R.S.J.; Hernandez-Sanabria, M.; Arya, R.P.; Bimler, L.; Paust, S.; Tarbet, E.B.; et al. Consensus M2e peptide conjugated to gold nanoparticles confers protection against H1N1, H3N2 and H5N1 influenza A viruses. *Antivir. Res.* **2017**, *141*, 62–72. [CrossRef]
100. Alghrair, Z.K.; Fernig, D.G.; Ebrahimi, B. Enhanced inhibition of influenza virus infection by peptide-noble-metal nanoparticle conjugates. *Beilstein J. Nanotechnol.* **2019**, *10*, 1038–1047. [CrossRef] [PubMed]
101. Bai, Y.; Zhou, Y.; Liu, H.; Fang, L.; Liang, J.; Xiao, S. Glutathione-stabilized fluorescent gold nanoclusters vary in their influences on the proliferation of pseudorabies virus and porcine reproductive and respiratory syndrome virus. *ACS Appl. Nano Mater.* **2018**, *1*, 969–976. [CrossRef]
102. Feng, C.; Fang, P.; Zhou, Y.; Liu, L.; Fang, L.; Xiao, S.; Liang, J. Different Effects of His-Au NCs and MES-Au NCs on the propagation of pseudorabies virus. *Glob. Chall.* **2018**, *2*, 1800030. [CrossRef] [PubMed]
103. Michalet, X.; Pinaud, F.F.; Bentolila, L.A.; Tsay, J.M.; Doose, S.; Li, J.J.; Sundaresan, G.; Wu, A.M.; Gambhir, S.S.; Weiss, S. Quantum dots for live cells, in vivo imaging, and diagnostics. *Science* **2005**, *307*, 538–544. [CrossRef] [PubMed]
104. Du, T.; Cai, K.; Han, H.; Fang, L.; Liang, J.; Xiao, S. Probing the interactions of CdTe quantum dots with pseudorabies virus. *Sci. Rep.* **2015**, *5*, 1–10. [CrossRef]
105. Du, T.; Liang, J.; Dong, N.; Lu, J.; Fu, Y.; Fang, L.; Xiao, S.; Han, H. Glutathione-Capped Ag_2S Nanoclusters Inhibit Coronavirus Proliferation through Blockage of Viral RNA Synthesis and Budding. *ACS Appl. Mater. Interfaces* **2018**, *10*, 4369–4378. [CrossRef]
106. Du, T.; Liang, J.; Dong, N.; Liu, L.; Fang, L.; Xiao, S.; Han, H. Carbon dots as inhibitors of virus by activation of type I interferon response. *Carbon* **2016**, *110*, 278–285. [CrossRef]
107. Dong, X.; Moyer, M.M.; Yang, F.; Sun, Y.P.; Yang, L. Carbon dots' antiviral functions against noroviruses. *Sci. Rep.* **2017**, *7*, 1–10. [CrossRef]

108. Huang, S.; Gu, J.; Ye, J.; Fang, B.; Wan, S.; Wang, C.; Ashraf, U.; Li, Q.; Wang, X.; Shao, L.; et al. Benzoxazine monomer derived carbon dots as a broad-spectrum agent to block viral infectivity. *J. Colloid Interface Sci.* **2019**, *542*, 198–206. [CrossRef]
109. Barras, A.; Pagneux, Q.; Sane, F.; Wang, Q.; Boukherroub, R.; Hober, D.; Szunerits, S. High efficiency of functional carbon nanodots as entry inhibitors of herpes simplex virus type 1. *ACS Appl. Mater. Interfaces* **2016**, *8*, 9004–9013. [CrossRef]
110. Łoczechin, A.; Séron, K.; Barras, A.; Giovanelli, E.; Belouzard, S.; Chen, Y.T.; Metzler-Nolte, N.; Boukherroub, R.; Dubuisson, J.; Szunerits, S. Functional carbon quantum dots as medical countermeasures to human coronavirus. *ACS Appl. Mater. Interfaces* **2019**, *11*, 42964–42974. [CrossRef]
111. Lin, C.; Chang, L.; Chu, H.; Lin, H.; Chang, P.; Wang, R.Y.L.; Unnikrishnan, B.; Mao, J.; Chen, S.; Huang, C. High amplification of the antiviral activity of curcumin through transformation into carbon quantum dots. *Small* **2019**, *15*, 1902641. [CrossRef]
112. Tong, T.; Hu, H.; Zhou, J.; Deng, S.; Zhang, X.; Tang, W.; Fang, L.; Xiao, S.; Liang, J. Glycyrrhizic-acid-based carbon dots with high antiviral activity by multisite inhibition mechanisms. *Small* **2020**, *16*, 1906206. [CrossRef]
113. Ghosal, K.; Sarkar, K. Biomedical Applications of Graphene Nanomaterials and beyond. *ACS Biomater. Sci. Eng.* **2018**, *4*, 2653–2703. [CrossRef]
114. Gurunathan, S.; Iqbal, M.A.; Qasim, M.; Park, C.H.; Yoo, H.; Hwang, J.H.; Uhm, S.J.; Song, H.; Park, C.; Do, J.T.; et al. Evaluation of graphene oxide induced cellular toxicity and transcriptome analysis in human embryonic kidney cells. *Nanomaterials* **2019**, *9*, 969. [CrossRef]
115. Song, Z.; Wang, X.; Zhu, G.; Nian, Q.; Zhou, H.; Yang, D.; Qin, C.; Tang, R. Virus capture and destruction by label-free graphene oxide for detection and disinfection applications. *Small* **2015**, *11*, 1171–1176. [CrossRef]
116. Sametband, M.; Kalt, I.; Gedanken, A.; Sarid, R. Herpes simplex virus type-1 attachment inhibition by functionalized graphene oxide. *ACS Appl. Mater. Interfaces* **2014**, *6*, 1228–1235. [CrossRef]
117. Ye, S.; Shao, K.; Li, Z.; Guo, N.; Zuo, Y.; Li, Q.; Lu, Z.; Chen, L.; He, Q.; Han, H. Antiviral activity of graphene oxide: How sharp edged structure and charge matter. *ACS Appl. Mater. Interfaces* **2015**, *7*, 21578–21579. [CrossRef]
118. Yang, X.X.; Li, C.M.; Li, Y.F.; Wang, J.; Huang, C.Z. Synergistic antiviral effect of curcumin functionalized graphene oxide against respiratory syncytial virus infection. *Nanoscale* **2017**, *9*, 16086–16092. [CrossRef]
119. Iannazzo, D.; Pistone, A.; Salamò, M.; Galvagno, S.; Romeo, R.; Giofré, S.V.; Branca, C.; Visalli, G.; Di Pietro, A. Graphene quantum dots for cancer targeted drug delivery. *Int. J. Pharm.* **2017**, *518*, 185–192. [CrossRef]
120. Du, X.; Xiao, R.; Fu, H.; Yuan, Z.; Zhang, W.; Yin, L.; He, C.; Li, C.; Zhou, J.; Liu, G.; et al. Hypericin-loaded graphene oxide protects ducks against a novel duck reovirus. *Mater. Sci. Eng. C* **2019**, *105*, 110052. [CrossRef]
121. Antoine, T.E.; Mishra, Y.K.; Trigilio, J.; Tiwari, V.; Adelung, R.; Shukla, D. Prophylactic, therapeutic and neutralizing effects of zinc oxide tetrapod structures against herpes simplex virus type-2 infection. *Antivir. Res.* **2012**, *96*, 363–375. [CrossRef] [PubMed]
122. Antoine, T.E.; Hadigal, S.R.; Yakoub, A.M.; Mishra, Y.K.; Bhattacharya, P.; Haddad, C.; Valyi-Nagy, T.; Adelung, R.; Prabhakar, B.S.; Shukla, D. Intravaginal zinc oxide tetrapod nanoparticles as novel immunoprotective agents against genital herpes. *J. Immunol.* **2016**, *196*, 4566–4575. [CrossRef]
123. Duggal, N.; Jaishankar, D.; Yadavalli, T.; Hadigal, S.; Mishra, Y.K.; Adelung, R.; Shukla, D. Zinc oxide tetrapods inhibit herpes simplex virus infection of cultured corneas. *Mol. Vis.* **2017**, *23*, 26–38. [PubMed]
124. Tavakoli, A.; Ataei-Pirkooh, A.; Mm Sadeghi, G.; Bokharaei-Salim, F.; Sahrapour, P.; Kiani, S.J.; Moghoofei, M.; Farahmand, M.; Javanmard, D.; Monavari, S.H. Polyethylene glycol-coated zinc oxide nanoparticle: An efficient nanoweapon to fight against herpes simplex virus type 1. *Nanomedicine* **2018**, *13*, 2675–2690. [CrossRef] [PubMed]
125. Farouk, F.; Sgebl, R.I. Comparing surface chemical modifications of zinc oxide nanoparticles for modulating their antiviral activity against herpes simplex virus type-1. *Int. J. Nanopart. Nanotechnol.* **2018**, *4*, 21. [CrossRef]
126. Ghaffari, H.; Tavakoli, A.; Moradi, A.; Tabarraei, A.; Bokharaei-Salim, F.; Zahmatkeshan, M.; Farahmand, M.; Javanmard, D.; Kiani, S.J.; Esghaei, M.; et al. Inhibition of H1N1 influenza virus infection by zinc oxide nanoparticles: Another emerging application of nanomedicine. *J. Biomed. Sci.* **2019**, *26*, 70. [CrossRef]
127. Ishida, T. Review on the role of Zn2+ Ions in viral pathogenesis and the effect of Zn2+ Ions for host cell-virus growth inhibition. *Am. J. Biomed. Sci. Res.* **2019**, *2*, 28–37. [CrossRef]

128. Agelidis, A.; Koujah, L.; Suryawanshi, R.; Yadavalli, T.; Mishra, Y.K.; Adelung, R.; Shukla, D. An intra-vaginal zinc oxide tetrapod nanoparticles (ZOTEN) and genital herpesvirus cocktail can provide a novel platform for live virus vaccine. *Front. Immunol.* **2019**, *10*, 500. [CrossRef]
129. Ochekpe, N.A.; Olorunfemi, P.O.; Ngwuluka, N.C. Nanotechnology and drug delivery part 2: Nanostructures for drug delivery. *Trop. J. Pharm. Res.* **2009**, *8*, 275–287. [CrossRef]
130. Bender, A.R.; Von Briesen, H.; Kreuter, J.; Duncan, I.B.; Rubsamen-Waigmann, H. Efficiency of nanoparticles as a carrier system for antiviral agents in human immunodeficiency virus-infected human monocytes/macrophages in vitro. *Antimicrob. Agents Chemother.* **1996**, *40*, 1467–1471. [CrossRef]
131. Cavalli, R.; Donalisio, M.; Civra, A.; Ferruti, P.; Ranucci, E.; Trotta, F.; Lembo, D. Enhanced antiviral activity of Acyclovir loaded into β-cyclodextrin-poly(4-acryloylmorpholine) conjugate nanoparticles. *J. Control. Release* **2009**, *137*, 116–122. [CrossRef] [PubMed]
132. Lee, J.; Park, H.; Kim, M.; Seo, Y.; Lee, Y.; Byun, S.J.; Lee, S.; Kwon, M.H. Functional stability of 3D8 scFv, a nucleic acid-hydrolyzing single chain antibody, under different biochemical and physical conditions. *Int. J. Pharm.* **2015**, *496*, 561–570. [CrossRef] [PubMed]
133. Hu, C.M.J.; Chang, W.S.; Fang, Z.S.; Chen, Y.T.; Wang, W.L.; Tsai, H.H.; Chueh, L.L.; Takano, T.; Hohdatsu, T.; Chen, H.W. Nanoparticulate vacuolar ATPase blocker exhibits potent host-targeted antiviral activity against feline coronavirus. *Sci. Rep.* **2017**, *7*, 1–11. [CrossRef]
134. Dey, P.; Bergmann, T.; Cuellar-Camacho, J.L.; Ehrmann, S.; Chowdhury, M.S.; Zhang, M.; Dahmani, I.; Haag, R.; Azab, W. Multivalent flexible nanogels exhibit broad-spectrum antiviral activity by blocking virus entry. *ACS Nano* **2018**, *12*, 6429–6442. [CrossRef] [PubMed]
135. Baram-Pinto, D.; Shukla, S.; Perkas, N.; Gedanken, A.; Sarid, R. Inhibition of herpes simplex virus type 1 infection by silver nanoparticles capped with mercaptoethane sulfonate. *Bioconjug. Chem.* **2009**, *20*, 1497–1502. [CrossRef] [PubMed]
136. Zeng, P.; Xu, Y.; Zeng, C.; Ren, H.; Peng, M. Chitosan-modified poly(d,l-lactide-co-glycolide) nanospheres for plasmid DNA delivery and HBV gene-silencing. *Int. J. Pharm.* **2011**, *415*, 259–266. [CrossRef]
137. Dehghan, S.; Kheiri, M.T.; Tabatabaiean, M.; Darzi, S.; Tafaghodi, M. Dry-powder form of chitosan nanospheres containing influenza virus and adjuvants for nasal immunization. *Arch. Pharm. Res.* **2013**, *36*, 981–992. [CrossRef]
138. Mohajer, M.; Khameneh, B.; Tafaghodi, M. Preparation and characterization of PLGA nanospheres loaded with inactivated influenza virus, CpG–ODN and quillaja saponin. *Iran. J. Basic Med. Sci.* **2014**, *17*, 553–559. [CrossRef]
139. Caminade, A.M.; Laurent, R.; Majoral, J.P. Characterization of dendrimers. *Adv. Drug Deliv. Rev.* **2005**, *57*, 2130–2146. [CrossRef]
140. Bourne, N.; Stanberry, L.R.; Kern, E.R.; Holan, G.; Matthews, B.; Bernstein, D.I. Dendrimers, a new class of candidate topical microbicides with activity against herpes simplex virus infection. *Antimicrob. Agents Chemother.* **2000**, *44*, 2471–2474. [CrossRef]
141. Luganini, A.; Giuliani, A.; Pirri, G.; Pizzuto, L.; Landolfo, S.; Gribaudo, G. Peptide-derivatized dendrimers inhibit human cytomegalovirus infection by blocking virus binding to cell surface heparan sulfate. *Antivir. Res.* **2010**, *85*, 532–540. [CrossRef] [PubMed]
142. Nazmi, A.; Dutta, K.; Basu, A. Antiviral and neuroprotective role of octaguanidinium dendrimer-conjugated Morpholino oligomers in Japanese encephalitis. *PLoS Negl. Trop. Dis.* **2010**, *4*, e892. [CrossRef] [PubMed]
143. Telwatte, S.; Moore, K.; Johnson, A.; Tyssen, D.; Sterjovski, J.; Aldunate, M.; Gorry, P.R.; Ramsland, P.A.; Lewis, G.R.; Paull, J.R.A.; et al. Virucidal activity of the dendrimer microbicide SPL7013 against HIV-1. *Antivir. Res.* **2011**, *90*, 195–199. [CrossRef] [PubMed]
144. Jyothi, K.R.; Beloor, J.; Jo, A.; Nguyen, M.N.; Choi, T.G.; Kim, J.H.; Akter, S.; Lee, S.K.; Maeng, C.H.; Baik, H.H.; et al. Liver-targeted cyclosporine A-encapsulated poly (lactic-co-glycolic) acid nanoparticles inhibit hepatitis C virus replication. *Int. J. Nanomed.* **2015**, *10*, 903–921. [CrossRef]
145. Illescas, B.M.; Rojo, J.; Delgado, R.; Martín, N. Multivalent glycosylated nanostructures to inhibit ebola virus infection. *J. Am. Chem. Soc.* **2017**, *139*, 6018–6025. [CrossRef]
146. Sepúlveda-Crespo, D.; Sánchez-Rodríguez, J.; Serramía, M.J.; Gómez, R.; De La Mata, F.J.; Jiménez, J.L.; Muñoz-Fernández, M.Á. Triple combination of carbosilane dendrimers, tenofovir and maraviroc as potential microbicide to prevent HIV-1 sexual transmission. *Nanomedicine* **2015**, *10*, 899–914. [CrossRef]

147. Kandeel, M.; Al-Taher, A.; Park, B.K.; Kwon, H.; Al-Nazawi, M. A pilot study of the antiviral activity of anionic and cationic polyamidoamine dendrimers against the Middle East respiratory syndrome coronavirus. *J. Med. Virol.* **2020**, jmv.25928. [CrossRef]
148. Rupp, R.; Rosenthal, S.L.; Stanberry, L.R. *VivaGel™ (SPL7013 Gel): A Candidate Dendrimer-Microbicide for the Prevention of HIV and HSV Infection*; Dove Press: Macclesfield, UK, 2007; Volume 2.
149. Lee, K.J.; Angulo, A.; Ghazal, P.; Janda, K.D. Soluble-polymer supported synthesis of a prostanoid library: Identification of antiviral activity. *Org. Lett.* **1999**, *1*, 1859–1862. [CrossRef]
150. Roner, M.R.; Carraher, C.E., Jr.; Shahi, K.; Barot, G. Antiviral activity of metal-containing polymers—Organotin and cisplatin-like polymers. *Materials* **2011**, *4*, 991–1012. [CrossRef]
151. Wang, Y.; Canady, T.D.; Zhou, Z.; Tang, Y.; Price, D.N.; Bear, D.G.; Chi, E.Y.; Schanze, K.S.; Whitten, D.G. Cationic phenylene ethynylene polymers and oligomers exhibit efficient antiviral activity. *ACS Appl. Mater. Interfaces* **2011**, *3*, 2209–2214. [CrossRef] [PubMed]
152. Wranke, A.; Wedemeyer, H. Antiviral therapy of hepatitis delta virus infection—Progress and challenges towards cure. *Curr. Opin. Virol.* **2016**, *20*, 112–118. [CrossRef] [PubMed]
153. Chun, H.; Yeom, M.; Kim, H.O.; Lim, J.W.; Na, W.; Park, G.; Park, C.; Kang, A.; Yun, D.; Kim, J.; et al. Efficient antiviral co-delivery using polymersomes by controlling the surface density of cell-targeting groups for influenza A virus treatment. *Polym. Chem.* **2018**, *9*, 2116–2123. [CrossRef]
154. Singh, L.; Kruger, H.G.; Maguire, G.E.M.; Govender, T.; Parboosing, R. The role of nanotechnology in the treatment of viral infections. *Adv. Infect. Dis.* **2017**, *4*, 105–131. [CrossRef] [PubMed]
155. Renneisen, K.; Leserman, L.; Matthes, E.; Schröder, H.C.; Müller, W.E. Inhibition of expression of human immunodeficiency virus-1 in vitro by antibody-targeted liposomes containing antisense RNA to the env region. *J. Biol. Chem.* **1990**, *265*, 16337–16342.
156. Phillips, N.; Tsoukas, C. Liposomal encapsulation of azidothymidine results in decreased hematopoietic toxicity and enhanced activity against murine acquired immunodeficiency syndrome. *Blood* **1992**, *79*, 1137–1143. [CrossRef]
157. Ji, H.; Yang, Z.; Jiang, W.; Geng, C.; Gong, M.; Xiao, H.; Wang, Z.; Cheng, L. Antiviral activity of nano carbon fullerene lipidosome against influenza virus in vitro. *J. Huazhong Univ. Sci. Technol. Med. Sci.* **2008**, *28*, 243–246. [CrossRef]
158. Pollock, S.; Branza Nichita, N.; Böhmer, A.; Radulescu, C.; Dwek, R.A.; Zitzmann, N. Polyunsaturated liposomes are antiviral against hepatitis B and C viruses and HIV by decreasing cholesterol levels in infected cells. *Proc. Natl. Acad. Sci. USA* **2010**, *107*, 17176–17181. [CrossRef]
159. Korvasová, Z.; Drašar, L.; Mašek, J.; Knotigová, P.T.; Kulich, P.; Matiašovic, J.; Kovařčík, K.; Bartheldyová, E.; Koudelka, Š.; Škrabalová, M.; et al. Antiviral effect of HPMPC (Cidofovir®), entrapped in cationic liposomes: In vitro study on MDBK cell and BHV-1 virus. *J. Control. Release* **2012**, *160*, 330–338. [CrossRef]
160. Mastrangelo, E.; Mazzitelli, S.; Fabbri, J.; Rohayem, J.; Ruokolainen, J.; Nykänen, A.; Milani, M.; Pezzullo, M.; Nastruzzi, C.; Bolognesi, M. Delivery of suramin as an antiviral agent through liposomal systems. *ChemMedChem* **2014**, *9*, 933–939. [CrossRef]
161. Albulescu, I.C.; Van Hoolwerff, M.; Wolters, L.A.; Bottaro, E.; Nastruzzi, C.; Yang, S.C.; Tsay, S.C.; Hwu, J.R.; Snijder, E.J.; Van Hemert, M.J. Suramin inhibits chikungunya virus replication through multiple mechanisms. *Antivir. Res.* **2015**, *121*, 39–46. [CrossRef]
162. Sharma, G.; Thakur, K.; Setia, A.; Amarji, B.; Singh, M.P.; Raza, K.; Katare, O.P. Fabrication of acyclovir-loaded flexible membrane vesicles (FMVs): Evidence of preclinical efficacy of antiviral activity in murine model of cutaneous HSV-1 infection. *Drug Deliv. Transl. Res.* **2017**, *7*, 683–694. [CrossRef]
163. Tahara, K.; Kobayashi, M.; Yoshida, S.; Onodera, R.; Inoue, N.; Takeuchi, H. Effects of cationic liposomes with stearylamine against virus infection. *Int. J. Pharm.* **2018**, *543*, 311–317. [CrossRef]
164. Sim, W.; Barnard, R.; Blaskovich, M.; Ziora, Z. Antimicrobial Silver in Medicinal and Consumer Applications: A Patent Review of the Past Decade (2007–2017). *Antibiotics* **2018**, *7*, 93. [CrossRef] [PubMed]
165. Cagno, V.; Andreozzi, P.; D'Alicarnasso, M.; Silva, P.J.; Mueller, M.; Galloux, M.; Le Goffic, R.; Jones, S.T.; Vallino, M.; Hodek, J.; et al. Broad-spectrum non-toxic antiviral nanoparticles with a virucidal inhibition mechanism. *Nat. Mater.* **2018**, *17*, 195–203. [CrossRef] [PubMed]
166. Donskyi, I.; Drüke, M.; Silberreis, K.; Lauster, D.; Ludwig, K.; Kühne, C.; Unger, W.; Böttcher, C.; Herrmann, A.; Dernedde, J.; et al. Interactions of Fullerene-Polyglycerol sulfates at viral and cellular interfaces. *Small* **2018**, *14*, 1800189. [CrossRef] [PubMed]

167. Yildirimer, L.; Thanh, N.T.K.; Loizidou, M.; Seifalian, A.M. Toxicological considerations of clinically applicable nanoparticles. *Nano Today* **2011**, *6*, 585–607. [CrossRef]
168. Speshock, J.L.; Murdock, R.C.; Braydich-Stolle, L.K.; Schrand, A.M.; Hussain, S.M. Interaction of silver nanoparticles with Tacaribe virus. *J. Nanobiotechnol.* **2010**, *8*, 19. [CrossRef]
169. Rogers, J.V.; Parkinson, C.V.; Choi, Y.W.; Speshock, J.L.; Hussain, S.M. A preliminary assessment of silver nanoparticle inhibition of monkeypox virus plaque formation. *Nanoscale Res. Lett.* **2008**, *3*, 129–133. [CrossRef]
170. Sun, R.W.Y.; Chen, R.; Chung, N.P.Y.; Ho, C.M.; Lin, C.L.S.; Che, C.M. Silver nanoparticles fabricated in Hepes buffer exhibit cytoprotective activities toward HIV-1 infected cells. *Chem. Commun.* **2005**, 5059–5061. [CrossRef]
171. Kingsley, J.D.; Dou, H.; Morehead, J.; Rabinow, B.; Gendelman, H.E.; Destache, C.J. Nanotechnology: A focus on nanoparticles as a drug delivery system. *J. Neuroimmune Pharmacol.* **2006**, *1*, 340–350. [CrossRef]
172. Li, W.; Hulswit, R.J.G.; Kenney, S.P.; Widjaja, I.; Jung, K.; Alhamo, M.A.; van Dieren, B.; van Kuppeveld, F.J.M.; Saif, L.J.; Bosch, B.J. Broad receptor engagement of an emerging global coronavirus may potentiate its diverse cross-species transmissibility. *Proc. Natl. Acad. Sci. USA* **2018**, *115*, E5135–E5143. [CrossRef] [PubMed]
173. Imai, Y.; Kuba, K.; Rao, S.; Huan, Y.; Guo, F.; Guan, B.; Yang, P.; Sarao, R.; Wada, T.; Leong-Poi, H.; et al. Angiotensin-converting enzyme 2 protects from severe acute lung failure. *Nature* **2005**, *436*, 112–116. [CrossRef] [PubMed]
174. Aydemir, D.; Ulusu, N.N. Correspondence: Angiotensin-converting enzyme 2 coated nanoparticles containing respiratory masks, chewing gums and nasal filters may be used for protection against COVID-19 infection. *Travel Med. Infect. Dis.* **2020**, 101697, in press. [CrossRef] [PubMed]
175. Kenney, P.; Hilberg, O.; Laursen, A.C.; Peel, R.G.; Sigsgaard, T. Preventive effect of nasal filters on allergic rhinitis: A randomized, double-blind, placebo-controlled crossover park study. *J. Allergy Clin. Immunol.* **2015**, *136*, 1566–1572.e5. [CrossRef] [PubMed]
176. Costantino, H.R.; Illum, L.; Brandt, G.; Johnson, P.H.; Quay, S.C. Intranasal delivery: Physicochemical and therapeutic aspects. *Int. J. Pharm.* **2007**, *337*, 1–24. [CrossRef]
177. Al-Halifa, S.; Gauthier, L.; Arpin, D.; Bourgault, S.; Archambault, D. Nanoparticle-based vaccines against respiratory viruses. *Front. Immunol.* **2019**, *10*, 22. [CrossRef]
178. Alshweiat, A.; Csóka, I.; Tömösi, F.; Janáky, T.; Kovács, A.; Gáspár, R.; Sztojkov-Ivanov, A.; Ducza, E.; Márki, Á.; Szabó-Révész, P.; et al. Nasal delivery of nanosuspension-based mucoadhesive formulation with improved bioavailability of loratadine: Preparation, characterization, and in vivo evaluation. *Int. J. Pharm.* **2020**, *579*, 119166. [CrossRef]
179. Zhao, Z.X.; Huang, Y.Z.; Shi, S.G.; Tang, S.H.; Li, D.H.; Chen, X.L. Cancer therapy improvement with mesoporous silica nanoparticles combining photodynamic and photothermal therapy. *Nanotechnology* **2014**, *25*, 285701. [CrossRef]
180. Roldão, A.; Mellado, M.C.M.; Castilho, L.R.; Carrondo, M.J.T.; Alves, P.M. Virus-like particles in vaccine development. *Expert Rev. Vaccines* **2010**, *9*, 1149–1176. [CrossRef]
181. Pati, R.; Shevtsov, M.; Sonawane, A. Nanoparticle vaccines against infectious diseases. *Front. Immunol.* **2018**, *9*, 2224. [CrossRef]
182. Zeltins, A. Construction and characterization of virus-like particles: A review. *Mol. Biotechnol.* **2013**, *53*, 92–107. [CrossRef] [PubMed]
183. Strable, E.; Finn, M.G. Chemical modification of viruses and virus-like particles. *Curr. Top. Microbiol. Immunol.* **2009**, *327*, 1–21. [PubMed]
184. Patel, K.G.; Swartz, J.R. Surface functionalization of virus-like particles by direct conjugation using azide-alkyne click chemistry. *Bioconjug. Chem.* **2011**, *22*, 376–387. [CrossRef] [PubMed]
185. Grgacic, E.V.L.; Anderson, D.A. Virus-like particles: Passport to immune recognition. *Methods* **2006**, *40*, 60–65. [CrossRef] [PubMed]
186. Kawano, M.; Matsui, M.; Handa, H. SV40 virus-like particles as an effective delivery system and its application to a vaccine carrier. *Expert Rev. Vaccines* **2013**, *12*, 199–210. [CrossRef] [PubMed]
187. Tissot, A.C.; Renhofa, R.; Schmitz, N.; Cielens, I.; Meijerink, E.; Ose, V.; Jennings, G.T.; Saudan, P.; Pumpens, P.; Bachmann, M.F. Versatile Virus-Like Particle Carrier for Epitope Based Vaccines. *PLoS ONE* **2010**, *5*, e9809. [CrossRef] [PubMed]

188. Gao, Y.; Wijewardhana, C.; Mann, J.F.S. Virus-like particle, liposome, and polymeric particle-based vaccines against HIV-1. *Front. Immunol.* **2018**, *9*, 345. [CrossRef]
189. Moon, J.J.; Suh, H.; Li, A.V.; Ockenhouse, C.F.; Yadava, A.; Irvine, D.J. Enhancing humoral responses to a malaria antigen with nanoparticle vaccines that expand T fh cells and promote germinal center induction. *Proc. Natl. Acad. Sci. USA* **2012**, *109*, 1080–1085. [CrossRef]
190. Tenzer, S.; Docter, D.; Kuharev, J.; Musyanovych, A.; Fetz, V.; Hecht, R.; Schlenk, F.; Fischer, D.; Kiouptsi, K.; Reinhardt, C.; et al. Rapid formation of plasma protein corona critically affects nanoparticle pathophysiology. *Nat. Nanotechnol.* **2013**, *8*, 772–781. [CrossRef]
191. Schöttler, S.; Becker, G.; Winzen, S.; Steinbach, T.; Mohr, K.; Landfester, K.; Mailänder, V.; Wurm, F.R. Protein adsorption is required for stealth effect of poly(ethylene glycol)- and poly(phosphoester)-coated nanocarriers. *Nat. Nanotechnol.* **2016**, *11*, 372–377. [CrossRef]
192. Chen, H.W.; Huang, C.Y.; Lin, S.Y.; Fang, Z.S.; Hsu, C.H.; Lin, J.C.; Chen, Y.I.; Yao, B.Y.; Hu, C.M.J. Synthetic virus-like particles prepared via protein corona formation enable effective vaccination in an avian model of coronavirus infection. *Biomaterials* **2016**, *106*, 111–118. [CrossRef] [PubMed]
193. Coleman, C.M.; Venkataraman, T.; Liu, Y.V.; Glenn, G.M.; Smith, G.E.; Flyer, D.C.; Frieman, M.B. MERS-CoV spike nanoparticles protect mice from MERS-CoV infection. *Vaccine* **2017**, *35*, 1586–1589. [CrossRef] [PubMed]
194. Albertoni, G.A.; Arnoni, C.P.; Barboza Araujo, P.R.; Andrade, S.S.; Carvalho, F.O.; Castello Girão, M.J.B.; Schor, N.; Barreto, J.A. Magnetic bead technology for viral RNA extraction from serum in blood bank screening. *Braz. J. Infect. Dis.* **2011**, *15*, 547–552. [CrossRef]
195. Liu, D.; Wang, Z.; Jin, A.; Huang, X.; Sun, X.; Wang, F.; Yan, Q.; Ge, S.; Xia, N.; Niu, G.; et al. Acetylcholinesterase-catalyzed hydrolysis allows ultrasensitive detection of pathogens with the naked eye. *Angew. Chem.* **2013**, *52*, 14065–14069. [CrossRef] [PubMed]
196. Quy, D.V.; Hieu, N.M.; Tra, O.T.; Nam, N.H.; Hai, N.H.; Son, N.T.; Nghia, P.T.; Anh, N.T.V.; Hong, T.T.; Luong, N.H. Synthesis of silica-coated magnetic nanoparticles and application in the detection of pathogenic viruses. *J. Nanomater.* **2013**, *2013*, 6. [CrossRef]
197. Yeh, Y.T.; Gulino, K.; Zhang, Y.H.; Sabestien, A.; Chou, T.W.; Zhou, B.; Lin, Z.; Albert, I.; Lu, H.; Swaminathan, V.; et al. A rapid and label-free platform for virus capture and identification from clinical samples. *Proc. Natl. Acad. Sci. USA* **2020**, *117*, 895–901. [CrossRef]
198. Zhao, Z.; Cui, H.; Song, W.; Ru, X.; Zhou, W.; Yu, X. A simple magnetic nanoparticles-based viral RNA extraction method for efficient detection of SARS-CoV-2. *bioRxiv* **2020**, *518055*, 2020.02.22.961268. [CrossRef]
199. Gao, J.; Tian, Z.; Yang, X. Breakthrough: Chloroquine phosphate has shown apparent efficacy in treatment of COVID-19 associated pneumonia in clinical studies. *Biosci. Trends* **2020**, *14*, 72–73. [CrossRef]
200. Colson, P.; Rolain, J.-M.; Lagier, J.-C.; Brouqui, P.; Raoult, D. Chloroquine and hydroxychloroqunie as available weapons to fight COVID-19. *Int. J. Antimicrob. Agents* **2020**, *55*, 105932. [CrossRef]
201. Wang, M.; Cao, R.; Zhang, L.; Yang, X.; Liu, J.; Xu, M.; Shi, Z.; Hu, Z.; Zhong, W.; Xiao, G. Remdesivir and chloroquine effectively inhibit the recently emerged novel coronavirus (2019-nCoV) in vitro. *Cell Res.* **2020**, *30*, 269–271. [CrossRef]
202. Chang, Y.; Tung, Y.; Lee, K.; Chen, T.; Hsiao, Y.; Chang, C.; Hsieh, T.; Su, C.; Wang, S.; Yu, J.; et al. Potential therapeutic agents for COVID-19 based on the analysis of protease and RNA polymerase docking. *Preprints* **2020**, 1–7. [CrossRef]
203. de Wit, E.; van Doremalen, N.; Falzarano, D.; Munster, V.J. SARS and MERS: Recent insights into emerging coronaviruses. *Nat. Rev. Microbiol.* **2016**, *14*, 523–534. [CrossRef] [PubMed]
204. Mercorelli, B.; Palù, G.; Loregian, A. Drug Repurposing for Viral Infectious Diseases: How Far Are We? *Trends Microbiol.* **2018**, *26*, 865–876. [CrossRef] [PubMed]
205. Du, L.; He, Y.; Zhou, Y.; Liu, S.; Zheng, B.J.; Jiang, S. The spike protein of SARS-CoV—A target for vaccine and therapeutic development. *Nat. Rev. Microbiol.* **2009**, *7*, 226–236. [CrossRef] [PubMed]
206. Song, Z.; Xu, Y.; Bao, L.; Zhang, L.; Yu, P.; Qu, Y.; Zhu, H.; Zhao, W.; Han, Y.; Qin, C. From SARS to MERS, thrusting coronaviruses into the spotlight. *Viruses* **2019**, *11*, 59. [CrossRef] [PubMed]
207. Savarino, A.; Di Trani, L.; Donatelli, I.; Cauda, R.; Cassone, A. New insights into the antiviral effects of chloroquine. *Lancet Infect. Dis.* **2006**, *6*, 67–69. [CrossRef]
208. Touret, F.; de Lamballerie, X. Of chloroquine and COVID-19. *Antivir. Res.* **2020**, *177*, 104762. [CrossRef]

209. Cioffi, N.; Torsi, L.; Ditaranto, N.; Tantillo, G.; Ghibelli, L.; Sabbatini, L.; Bleve-Zacheo, T.; D'Alessio, M.; Zambonin, P.G.; Traversa, E. Copper nanoparticle/polymer composites with antifungal and bacteriostatic properties. *Chem. Mater.* **2005**, *17*, 5255–5262. [CrossRef]
210. Hang, X.; Peng, H.; Song, H.; Qi, Z.; Miao, X.; Xu, W. Antiviral activity of cuprous oxide nanoparticles against Hepatitis C virus in vitro. *J. Virol. Methods* **2015**, *222*, 150–157. [CrossRef]
211. Kerry, R.G.; Malik, S.; Redda, Y.T.; Sahoo, S.; Patra, J.K.; Majhi, S. Nano-based approach to combat emerging viral (NIPAH virus) infection. *Nanomed. Nanotechnol. Biol. Med.* **2019**, *18*, 196–220. [CrossRef]
212. Richardson, P.; Griffin, I.; Tucker, C.; Smith, D.; Oechsle, O.; Phelan, A.; Stebbing, J. Baricitinib as potential treatment for 2019-nCoV acute respiratory disease. *Lancet* **2020**, *395*, e30–e31. [CrossRef]
213. Chen, N.; Zhou, M.; Dong, X.; Qu, J.; Gong, F.; Han, Y.; Qiu, Y.; Wang, J.; Liu, Y.; Wei, Y.; et al. Epidemiological and clinical characteristics of 99 cases of 2019 novel coronavirus pneumonia in Wuhan, China: A descriptive study. *Lancet* **2020**, *395*, 507–513. [CrossRef]
214. Li, G.; De Clercq, E. Therapeutic options for the 2019 novel coronavirus (2019-nCoV). *Nat. Rev. Drug Discov.* **2020**, *19*, 149–150. [CrossRef] [PubMed]
215. de Wit, E.; Feldmann, F.; Cronin, J.; Jordan, R.; Okumura, A.; Thomas, T.; Scott, D.; Cihlar, T.; Feldmann, H. Prophylactic and therapeutic remdesivir (GS-5734) treatment in the rhesus macaque model of MERS-CoV infection. *Proc. Natl. Acad. Sci. USA* **2020**, *117*, 6771–6776. [CrossRef] [PubMed]
216. Simmons, G.; Bertram, S.; Glowacka, I.; Steffen, I.; Chaipan, C.; Agudelo, J.; Lu, K.; Rennekamp, A.J.; Hofmann, H.; Bates, P.; et al. Different host cell proteases activate the SARS-coronavirus spike-protein for cell-cell and virus-cell fusion. *Virology* **2011**, *413*, 265–274. [CrossRef] [PubMed]
217. Devaux, C.A.; Rolain, J.M.; Colson, P.; Raoult, D. New insights on the antiviral effects of chloroquine against coronavirus: What to expect for COVID-19? *Int. J. Antimicrob. Agents* **2020**, *55*, 105938. [CrossRef]

 nanomaterials

MDPI

Review

Biopolymer-Based Multilayer Capsules and Beads Made via Templating: Advantages, Hurdles and Perspectives

Anna S. Vikulina [1,2,*] and Jack Campbell [3]

1 Department of Theory and Bio-Systems, Max Planck Institute of Colloids and Interfaces, Am Mühlenberg, 1, 14476 Potsdam, Germany
2 Bavarian Polymer Institute, Friedrich-Alexander University Erlangen-Nürnberg (FAU), Dr.-Mack-Straße, 77, 90762 Fürth, Germany
3 School of Science and Technology, Nottingham Trent University, Clifton Lane, Nottingham NG11 8NS, UK; jack.campbell@ntu.ac.uk
* Correspondence: anna.vikulina@mpikg.mpg.de or anna.vikulina@fau.de

Abstract: One of the undeniable trends in modern bioengineering and nanotechnology is the use of various biomolecules, primarily of a polymeric nature, for the design and formulation of novel functional materials for controlled and targeted drug delivery, bioimaging and theranostics, tissue engineering, and other bioapplications. Biocompatibility, biodegradability, the possibility of replicating natural cellular microenvironments, and the minimal toxicity typical of biogenic polymers are features that have secured a growing interest in them as the building blocks for biomaterials of the fourth generation. Many recent studies showed the promise of the hard-templating approach for the fabrication of nano- and microparticles utilizing biopolymers. This review covers these studies, bringing together up-to-date knowledge on biopolymer-based multilayer capsules and beads, critically assessing the progress made in this field of research, and outlining the current challenges and perspectives of these architectures. According to the classification of the templates, the review sequentially considers biopolymer structures templated on non-porous particles, porous particles, and crystal drugs. Opportunities for the functionalization of biopolymer-based capsules to tailor them toward specific bioapplications is highlighted in a separate section.

Keywords: polyelectrolyte multilayers; encapsulation; calcium carbonate; drug delivery; shrinkage

Citation: Vikulina, A.S.; Campbell, J. Biopolymer-Based Multilayer Capsules and Beads Made via Templating: Advantages, Hurdles and Perspectives. *Nanomaterials* **2021**, *11*, 2502. https://doi.org/10.3390/nano11102502

Academic Editors: Ivan Stoikov and Pavel Padnya

Received: 24 August 2021
Accepted: 20 September 2021
Published: 26 September 2021

1. Biopolymer-Based Multilayers

1.1. The LbL Technique

With the introduction of the versatile approach of the layer-by-layer (LbL) assembly of oppositely charged polyelectrolytes some decades ago, the research field has grown exponentially, owing to the ease of control over the layered film properties (i.e., film thickness, homogeneity, stability, porosity, etc.) [1–10] and the ability to coat surfaces of varying geometries [11–14]. The LbL approach is based upon the build-up of oppositely charged polyelectrolytes, resulting in the build-up of polyelectrolyte multilayer (PEM) films, as illustrated in Figure 1(4) [15–18]. The capability to fine-tune their properties at both the macro- and nano-scale [19–22] has since spurred the rapid development of biocompatible materials for use in a plethora of bioapplications for both 2D (e.g., implant coatings [23,24], bio-scaffolds [25,26], biosensors/electronics [27], cell patterning [28] etc.), and 3D PEMs (i.e., polyelectrolyte multilayer capsules (PEMCs) for drug delivery [29–31] and tissue engineering [32] applications), including those used to mimic the cellular environment [33]. LbL assembly can be achieved by a variety of methods, including traditional dip-coating [34], spray-coating [35,36], and spin-coating [37] approaches (Figure 1(1)–(3)). More recent emerging methodologies include that of microfluidic assisted approaches via the 2D coating of channels and the packing of 3D capsules [32,38,39], as well as "brushing" approaches [19].

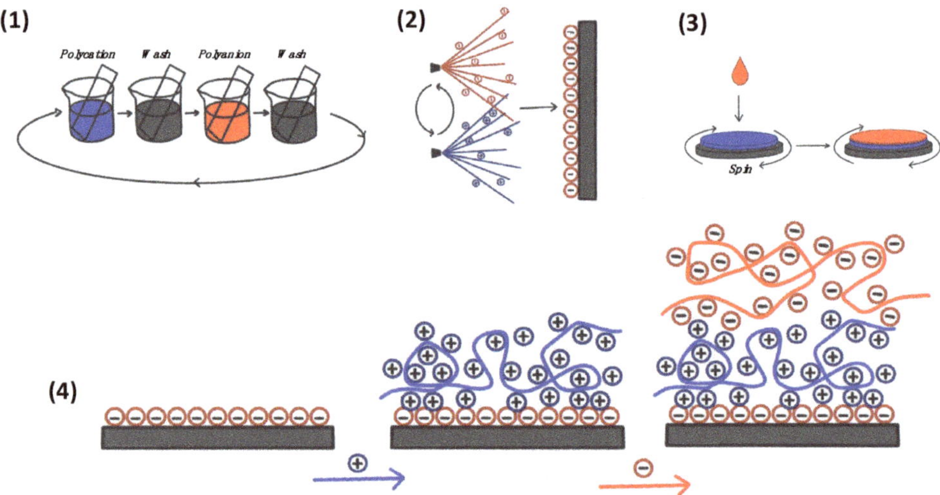

Figure 1. Schematic of classical approaches to form LbL films: (**1**) dip-coating, (**2**) spray-coating, and (**3**) spin-coating; (**4**) schematic of the build-up of LbL films upon a negatively charged substrate. Blue: polycation, red: polyanion. Adapted from reference [21].

Interactions responsible for the build-up of these LbL films are numerous; the most common cause for the build-up are electrostatic interactions as, typically, oppositely charged polyelectrolytes are employed. Hydrogen bonding plays a significant role in multilayer formation, with some LbL films formed fully via hydrogen bonding interactions [40–42]. Hydrophobic interactions also play a role in the formation of films due to the hydrophobic nature of the hydrocarbon backbone of many of the polymers used [43], and they, therefore, cannot be neglected. Studies of the planar biopolymer-based LbL films demonstrated approaches to control the build-up of PEMs via temperature [44], ionic strength, and pH [45]. Covalent bond formation was manipulated to form LbL films via click-chemistry and they are much more robust in terms of withstanding drastic environmental change [46]. Free-standing 3D PEM structures were seen in recent work; biopolymer-based micro- and nano-capsules/gels are popular tools for the encapsulation and targeted delivery of fragile bioactives, including proteins [47], enzymes [48], and small molecules (such as ibuprofen [49], cisplatin [50], and polyphenol [51]). Polyelectrolyte multilayer capsules offer targeted delivery as well as a variety of release methods, including enzymatic degradation [52,53], magnetic field interactions [54,55], and ultrasound [56]. They are particularly attractive due to their tunable properties, whether via ionic strength, pH, or temperature, to control release dynamics and permeability [57–61].

1.2. The Classification of Sacrificial Templates and Issues of Biocompatability

A plethora of sacrificial templates are utilised to form such 3D structures and all host a variety of properties related to the final capsule structure, stability, and application. The variation in capsule internal structure is dependent on both the porosity of the template, as well as the size of the biopolyelectrolytes utilised for multilayer coating [62]. Templates may be categorized as porous, such as carbonates (i.e., calcium [63,64] and manganese [65,66] carbonates), mesoporous silica [67–69], and, potentially, calcium phosphate [70], or non-porous templates, such as polystyrene latex [71,72] and melamine formaldehyde [73]. Biological entities (e.g., erythrocytes [74,75] or bacteria (*Escherichia coli* (further *E. Coli*) for instance [76,77])) are also utilised as templates.

When utilizing a porous template, varying capsule structures are yielded. If the pore diameters are larger than that of the biopolymers used, during the initial deposition stages of LbL assembly the biopolymers adsorb to the template surface, however, in this case, the polymers are also able to permeate the template internal structure through surface pores and form an internal polymeric matrix [78]. Once the desired number of deposition stages is achieved, the template undergoes dissolution, leaving a polyelectrolyte multilayer capsule (PEMC) with an internal gel-like matrix; this capsule is coined as a matrix-type capsule, or a microgel [78–80] (Figure 2). During LbL deposition upon a non-porous template, the polyelectrolytes adsorb to the template surface and, following this, the template is eliminated, leaving a polymeric shell and a hollow lumen; these are known as hollow-type capsules. These shells may also form if no polymer is able to permeate the pores of a porous template, as illustrated in Figure 2, using vaterite $CaCO_3$ as an example template. The terminology of such "matrix-type" capsules is largely in question; for instance, if less polymer is present in the capsule lumen compared to one completely filled, it is often overlooked due to its insignificance, and a suitable threshold between hollow and matrix should be established [79]. Moreover, the control over the capsule internal structure allows for control of the capsule release properties; a burst or sustained release can both be achieved with hollow- and matrix-type capsules, respectively [78].

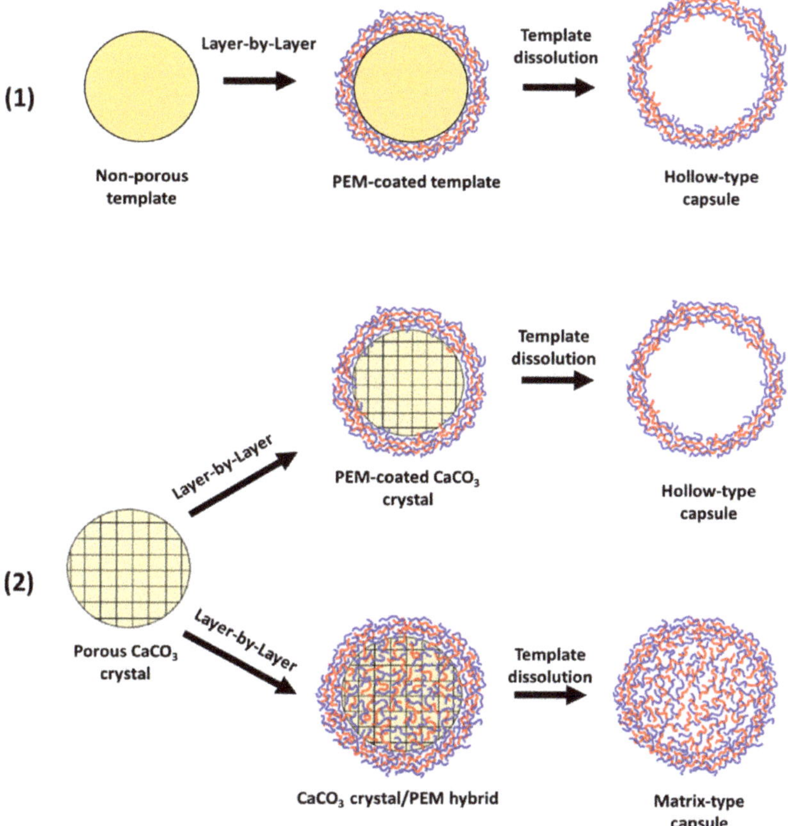

Figure 2. (**1**) The formation of hollow-type PEMCs upon a porous template. (**2**) The formation of hollow- and matrix-type PEMCs via either the formation of a polymer shell or $CaCO_3$/PEM hybrid, respectively. (**2**) adapted with a permission from reference [79]. Copyright © 2021 American Chemical Society.

For stable biopolymer-based capsules to form, a sacrificial core that undergoes dissolution at mild conditions, and gives rise to the minimal amount of osmotic pressure upon dissolution, is required to prevent the rupture of multilayers during dissolution. However, many sacrificial cores possess specific drawbacks typically related to dissolution conditions and their toxicity in vivo. For example, melamine formaldehyde (MF) cores undergo dissolution at low pH levels [81] and are, hence, not physiologically relevant, but upon dissolution there is an increase in osmotic pressure within the capsule, leading to high degrees of swelling and, sometimes, capsule rupture; if the capsule is able to form, residues of MF often remain within [82]. The carbonates $MnCO_3$ and $CaCO_3$ are particularly attractive; the vaterite polymorph of $CaCO_3$ holds a well-developed porous structure [83], with pore sizes in the typical range of 5-35 nm, which can be further manipulated to control capsule internal structure [62]. It is biocompatible, low cost to produce [84], and undergoes dissolution at mild conditions [85], e.g., with ethylenediaminetetraacetic acid (EDTA) or citric acid, resulting in little osmotic pressure [83]. Importantly, biomacromolecules preserve their biological activities when encapsulated in the vaterite crystals [86–88].

A range of polymers were utilized to produce PEMCs and largely synthetic polymers [89,90]. Synthetic polyelectrolytes, despite their numerous advantages, including intrinsic robustness, large working windows of ionic strength and pH, as well as high levels of control over synthesis conditions (i.e., molecular weight), are typically toxic and have issues with biodegradability in the body [91,92]. As a result, there was a boom in the development of biopolymer-based materials for drug delivery and bioapplications in general. Biopolymers, however, are intrinsically liable and largely sensitive to their microenvironment, thereby making them onerous, but easily manipulated components to work with in the field of functional biomaterials [91,93]. Despite this slight drawback, they present numerous advantages due to their little-to-no toxicity, ease of biodegradability, and intrinsic biocompatibility [94]. Polysaccharides (typically glycosaminoglycans [95,96], functionalised chitosans [96,97], alginates [29,98], etc.) and polyamino acids (poly-L-glutamic acid (PGLU), poly-L-lysine (PLL), etc.) [30,99,100] are typically used for the formulation of 3D bio-based multilayer structures (i.e., PEMCs). Moreover, components of the extracellular matrix are also utilised, including, for instance, hyaluronic acid [101] and chondroitin sulphate [26,102]. With such biopolymers, we are able to exert control over the functionality of capsules, as well as achieve the mimicking of extracellular matrices and intracellular environment with ease in both 2D and 3D materials. This provides the opportunity to develop films and capsules to act as protein and bioactive reservoirs or anchors for encapsulation and cell adhesion/drug delivery [45,103–106]. Given the scope of biopolymer manipulation one can now achieve, the next sections will discuss the variety of sacrificial templates utilized for biopolymer-PEMC formulation and their specific drawbacks and advantages, along with the variety of PEMC functionalization performed within the literature.

2. Biopolymer-Based Capsules Templated on Non-Porous Templates

The advantages of the non-porous templates typically used to form PEMCs are the resultant properties following particle synthesis, including the range of controllable sizes available, starting from as small as tens of nanometres up to a few millimetres, and their typical monodispersity and stability as colloidal particles, which is ideal for PEMC formation and delivery carriers. Despite this, some of these carriers are hindered by their lack of biocompatibility both before and after template dissolution. For instance, non-porous silica templates require hydrofluoric acid to undergo dissolution, which is not ideal for biopolymeric materials.

DS/PR-based PEMCs were templated upon MF cores and, following treatment of MF with pH 1.7 HCl, MF residues were retained within the PEMC interior, as shown via Raman spectroscopy, likely due to DS-MF complex formation [82]. Although beneficial for the binding of encapsulated peroxidase, more biocompatible templates are preferred due to the harsh acidic removal conditions. The same was observed for ALG/CHI PEMs in the form of ALG-MF complexes and the binding of positively charged insulin at low pH levels [47], as well as within DNA/SP and ALG/PLL microcapsules [107]. The stable PEMC formation is dependent upon the osmotic pressure build-up during core dissolution and the ability of the MF oligomers to diffuse outwards [81], which may also be responsible for PEMC swelling during the core dissolution. Moreover, MF resin may become irreversibly adsorbed into the capsule shell and can contribute up to 20% of the capsule mass.

Further to this, polystyrene latex (PS) templates were also employed for the formation of biopolymer PEMCs (fucoidan (FC)/CHI, for instance); the templates were treated with tetrahydrofuran (THF) [108] for two hours for removal and the PEMCs shrank by almost 50% from their template size. Of note, calcination of the coated PS core is also used for template removal [71], which is not ideal for many sensitive biogenic capsules. As PEMs formed on smooth non-porous templates are typically thinner and well defined [109], the loading and release of bioactives post-template dissolution can be well controlled via alterations in pH, ionic strength, and cross-linking of the shell. For instance, silica-templated DS/CHI PEMCs demonstrate a reversible permeability phenomenon; PEMCs were impermeable to dextrans from 4000–250,000 Da at pH < 6.8, but were permeable above pH 8. These permeable capsules were then reduced to pH 5.6 and the dextran was entrapped within the capsule interior (Figure 3(1)) [110].

This is attributed to the electrostatics of the biopolymers within the PEMs; at higher pH levels, there is an increased repulsion between the sulfonate groups upon DS and a reduced cationic charge on CHI, thus causing the polymers to change their conformation and increasing PEM permeability [110], as illustrated in Figure 3(2). Similar behaviour is observed with HS/PR [111] and pGLU/CHI [112] PEMCs, as well as Pectin/CHI PEMCs with the release of DOX; DOX is not released at pH 7.4, but is rapidly released at pH levels 5 and 6. This was attributed to the swelling of the PEMCs upon decrease in pH due to the increased protonation state of CHI below pH 6.5 (pKa value), with a size increase of 418 to 527 nm from pH 7.4 to 5, respectively [113]. A schematic of this is shown in Figure 3(3).

This pH-responsive property is useful as a way to control the release of the bioactive encapsulated, especially in-terms of low intracellular or tumorous pHs in which the PEMCs may be uptaken. Further to this, metal nanoparticles may also be used as PEMC templates. Gold nanoparticles in particular were utilised as templates for glucose-sensitive ALG/phenylboronic-modified PLL PEMCs; the gold core was removed via the addition of potassium cyanide, followed by dialysis for the removal of the gold complex formed (the authors confirmed the lack of gold via inductively coupled plasma analysis) [114], as demonstrated in Figure 3(4),(5). The use of such well-established monodisperse metal nanoparticles as templates allows for the formulation of PEMCs on the nanoscale (~40 nm in this study). Further examples of non-porous templates utilised for the formation of biopolymer PEMCs can be found in Table 1.

However, a disadvantage of these non-porous templates is related to the bioactive loading capacity of the final PEMC formed. Only post-loading approaches may be applied to such PEMCs, which may hinder the relative amount of bioactive material encapsulated. Porous templates, for instance, can be pre-loaded prior to PEMC formation and their loading potential is greatly increased compared to non-porous templates. These templates will now be discussed.

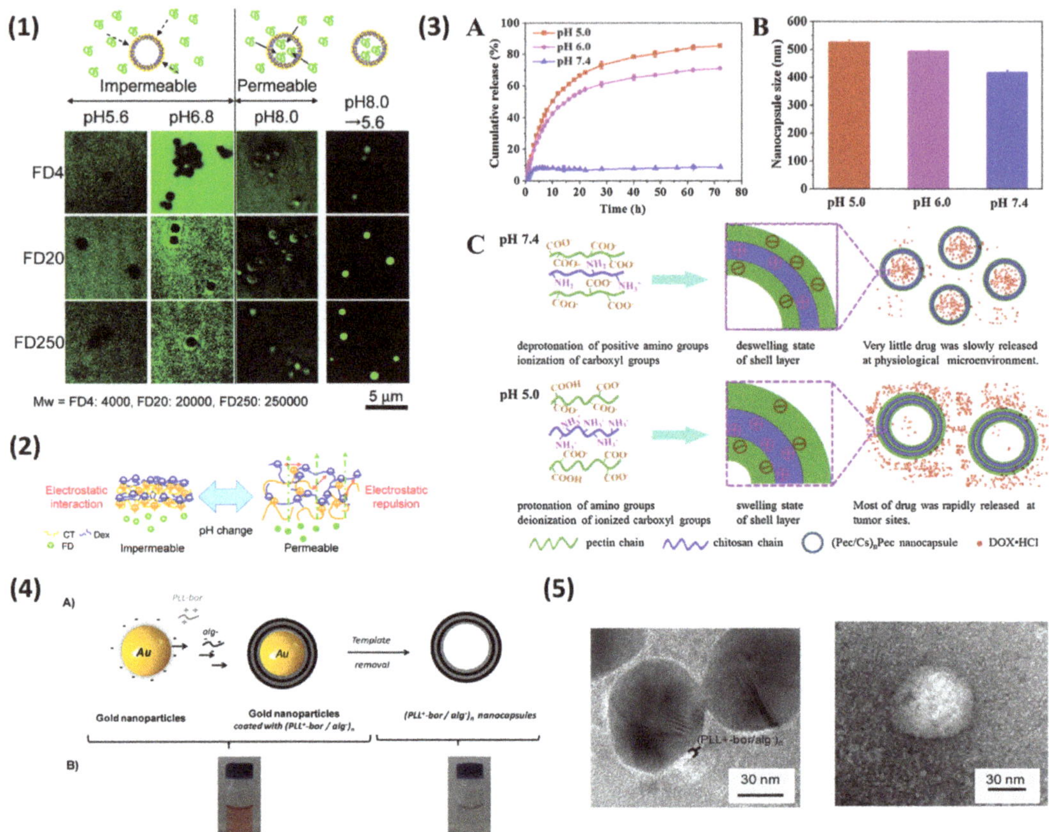

Figure 3. (**1**) Schematic and confocal microscopy images of the relative permeability of dextrans of varying molecular weight; FD4–4000, FD20–20,000, and FD250–250,000 in buffer solutions (pH 5.6, 0.05 M acetic acid buffer; and pH 6.8 and 8.0, 0.05 M TRIS buffer) containing dissolved fluorescein isothiocyanate-labelled dextran. A (**2**) schematic illustrating the change in permeability within the PEM when changing pH. Reprinted with permission from reference [110] copyright © 2008 American Chemical Society. The (**3**) cumulative release of DOX from (pectin/CHI)$_3$/pectin nanocapsules (**A**), the hydrodynamic size of nanocapsules at varying pH (**B**), and the schematic illustration of the pH responsive nanocapsules (**C**). Reprinted with permission from reference [113] copyright © 2017 Elsevier. The (**4**) schematic illustration of the formation of glucose-responsive nanocapsules (**A**), and the suspensions of coated-gold nanoparticles (**left**) and nanocapsules following removal of the core (**right**). (**5**) TEM micrographs of (**left**) (PLL-bor)/alg−)4-coated gold nanoparticles and (**right**) nanocapsules. Reprinted with permission from reference [114] copyright © 2019 Elsevier.

Table 1. Summary of successfully fabricated nano- and microcapsules templated on porous templates as reported in literature. MF: melamine formaldehyde; NPs: nanoparticles; PS: polystyrene; ALG: alginate; CAR: carrageenan; CHI: chitosan; FC: fucoidan; HS: heparin sulphate; PR: protamine; pGLU: poly-L-glutamic acid; PLL: poly-L-lysine; PLL-pb: phenylboronic modified poly-l-lysine; and SP: spermidine.

Polyanion	Polycation	Template, Size	#Layers	References
ALG	CHI	MF, 2.1 μm	10	[47]
	PR	MF, 6.5 μm	8 and 16	[115]
	PLL	MF, 5.7 μm	5	[107]
	PLL-pb	Gold NPs, ~40 nm	4 and 8	[114]
CAR	CHI	SiO_2-NH_2, 100 nm	11	[116]
DS	CHI	Silica, 3 μm	14	[110]
		Silica, 330 nm	10	[117]
		Silica, 220 nm	8	[118]
	PR	MF, ~5 μm	8	[73,82]
HS	PR	Silica, 180 nm	6	[111]
	CHI	Silica, 220 nm	6	[96]
pGLU	CHI	MF, ~1 μm	10	[99]
		Silica, 330 nm	8	[112]
FC	CHI	PS, 90 nm	10	[108]
Pectin	CHI	SiO_2-NH_2, ~100 nm	7	[113]
DNA	SP	MF, 1.8 & 5.7 μm	5	[107]

3. Biopolymer-Based Capsules and Beads Templated on Porous Templates

3.1. Capsules Formed from Porous Templates

The use of porous templates (i.e., $MnCO_3$, $CaCO_3$, and MS) for the formation of PEMCs holds numerous advantages over that of non-porous templates. One of the most prevalent being the ability to load bioactive material into the template's intrinsic porous structure. Porous vaterite $CaCO_3$ has demonstrated enormous loading capacities [11,119–122]; lysozyme, for instance, can be loaded up to 500 ± 128 mg/g $CaCO_3$ within sub-micron crystals [123], and superoxide dismutase can reach up to 240 ± 8 mg/g $CaCO_3$ in crystals of ~4 μm [124]. Moreover, the loading of such bioactives may be enhanced via the pre-encapsulation of a polyelectrolyte of affinity to the material of interest. Shi et al. (2018) [125] enhanced the loading of lysozyme by the co-synthesis of heparin sulphate (HS) prior. Of great interest, this can also be applied to the encapsulation of low molecular weight bioactives, which are difficult to encapsulate due to their small size in comparison to the large pore sizes of such templates [126]. This was applied with lentinan [127] with DOX, and CMC [128] with daunorubicin. Balabushevich et al. (2019) [120] successfully encapsulated DOX via the electrostatic binding to a gel-like mucin matrix pre-encapsulated within the vaterite crystal, reaching doxorubicin content of up to 1.3 mg/g $CaCO_3$. This provides scope for a variety of clinically-relevant drug delivery applications and the potential use of vaterite as vehicles with mucoadhesive properties [11,129]. Moreover, due to the variety of loading mechanisms, one is able to tailor the loading technique to the biomaterial of interest; for instance, certain macromolecules or nanoparticles may be sensitive to salt solutions forming the crystal template and, hence, adsorption may be preferred. In addition, freezing-induced loading was recently demonstrated with the loading of TiO_2 nanoparticles [130].

The encapsulation of large sensitive macromolecules is possible due to the ability to form and load vaterite at close-to-physiological conditions [124], thereby maintaining the bioactivity of such materials [88]. Vaterite $CaCO_3$ also proved itself a diverse material as both a stand-alone drug delivery vehicle [131–134] and as a material for surface coatings [123] for functional use as an antimicrobial carrier, for instance [135,136]. However, there can be an issue with the aggregation of vaterite $CaCO_3$ crystals, making it difficult to form monodisperse templates without the need of additives, such as polypeptides [137].

Moreover, upon introduction of vaterite $CaCO_3$ to aqueous solutions, the transformation from the porous vaterite polymorph to the non-porous calcite polymorph can take place over the course of hours or days. This means that bare-vaterite should be handled quickly when in the presence of water [138], however, polyelectrolyte coatings and stabilizers can prevent this [139,140]. This, however, can be beneficial for the release of encapsulated cargo via a re-crystallisation mechanism [120,124]. Despite this, due to its low cost production ($0.2–0.4/g dried weight), potential ease of scalability [85], and soft dissolution conditions, vaterite $CaCO_3$ presents itself as an attractive template for PEMCs, the morphology of which is presented in Figure 4(2) with example biopolymer PEMCs below. Alternate templates, such as MS, for example, although advantageous with regard to pre-loading capability, are typically removed via hydrofluoric acid, much like non-porous silica, which means they are not suitable for bioapplications nor ideal for pre-encapsulated bioactives. Recently, however, PEMCs were formed via the removal of MS at physiological conditions via dissolution in a buffered salt solution [67], as demonstrated in Figure 4(1,3). Due to these advantageous properties, many more examples of biopolymer-based PEMCs are emerging, examples of such can be found in Table 2.

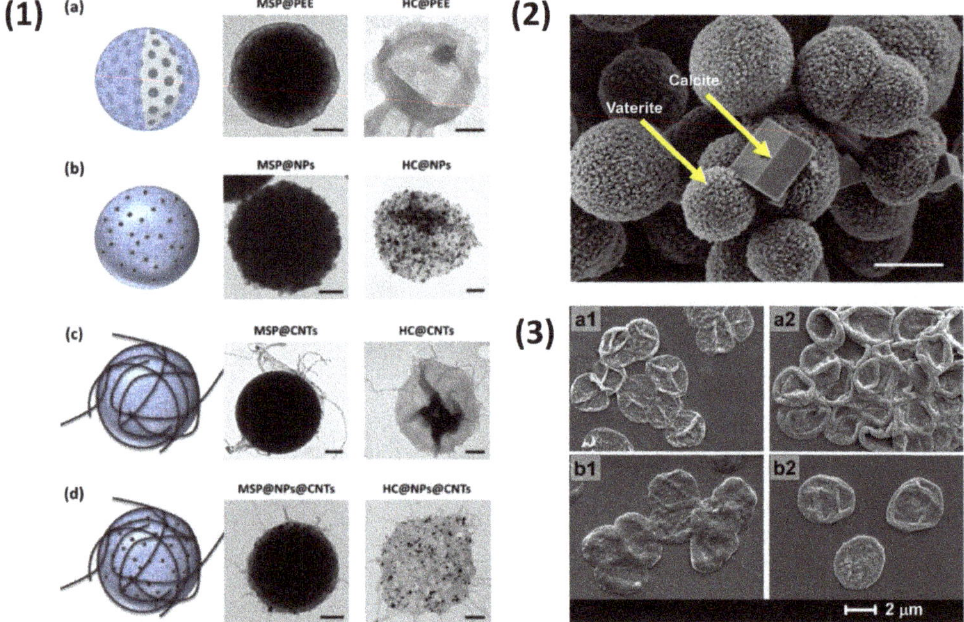

Figure 4. (**1**) Illustrations and transmission electron microscopy images of polyelectrolyte capsules before and after silica core dissolution. Hollow-capsules (**a**), $CoFe_2O_4$ NP-functionalised capsules (**b**), carbon nanotube (CNTs)-functionalised capsules (**c**), and capsules functionalised with both NPs and CNTs (**d**). Reprinted with permission from reference [67]. (**2**) Typical morphology of $CaCO_3$ vaterite and calcite crystals via SEM, indicated by yellow arrow. Adapted with permission from reference [124] copyright © 2019 Elsevier. (**3**) SEM images of empty PEMCs, $(DS/pARG)_3$–a1 and $(ALG/pARG)_3$–b1, and post-loaded capsules with TRITC-BSA $(DS/pARG)_3$–a2 and $(ALG/pARG)_3$–b2. Reprinted with permission from reference [141] copyright © 2010 American Chemical Society.

Table 2. Summary of successfully fabricated nano- and microcapsules reported in literature. V-$CaCO_3$: vaterite $CaCO_3$; ALG: alginate; BSA: bovine serum albumin; CHI: chitosan; CS: chondroitin sulphate; CMC: carboxymethylcellulose; COL: collagen; DOX: doxorubicin; ELR: elastin-like recombinamer; FG2: basic fibroblast growth factor; GA: glutaraldehyde; HA: hyaluronic acid; HS: heparin sulphate; Hgb: hemoglobin; IgY: egg yolk immunoglobulin; LF: lactoferrin; MNP: magnetic nanoparticle; pARG: poly-L-arginine; pASP: poly(L-aspartic acid); pGLU: poly-L-glutamicacid; PLL: poly-L-lysine; pONT: poly-L-ornithine; PAH: poly(allylhydrochloride); PSS: poly(styrene sulfonate); and TA: tannic acid.

Polyanion	Polycation	Template, Size	#Layers	Reference
		Polysaccharide-Based PEMCs		
CS	pARG	$MnCO_3$, 4 μm	8	[142]
	PLL	V-$CaCO_3$ pre-loaded with CS, 3–6 μm	10	[143]
		V-$CaCO_3$, 7-9 μm	5	[79]
	PR	V-$CaCO_3$ pre-loaded with PSS, 5 μm	4	[102]
		V-$CaCO_3$, 7-9 μm	5	[79]
DS	pARG	V-$CaCO_3$ pre-loaded with DEX, 3 μm	8	[144]
		V-$CaCO_3$ pre-loaded with FG2	6–14	[145]
		V-$CaCO_3$, 550 nm	6	[146]
		V-$CaCO_3$, 10 μm	7/8	[147]
	PR	V-$CaCO_3$, 7–9 μm	5	[79]
		V-$CaCO_3$ pre-loaded with PSS,~4 μm	12	[148]
HA	CHI	V-$CaCO_3$ pre-loaded with penicillin, ampicillin or ciprofloxacin, 5 μm	1-6	[136]
	PLL	V-$CaCO_3$, 7–9 μm	5	[79]
		V-$CaCO_3$, 5 μm	9	[149]
	COL	V-$CaCO_3$ pre-loaded with BSA, 3–6 μm	12	[101]
HS	PLL	V-$CaCO_3$, 7–9 μm	5	[79]
	CHI	V-$CaCO_3$ pre-loaded with DOX, 4 μm	10	[150]
		V-$CaCO_3$ coated with $(PSS/PAH)_4$, 3–4 μm	9	[52]
	PR	V-$CaCO_3$, 7–9 μm	5	[79]
	pARG	V-$CaCO_3$ pre-loaded with HS, ~4 μm	4	[30]
		V-$CaCO_3$, 3–5 μm	4	[151]
ALG	CHI	V-$CaCO_3$ pre-loaded with CMC, 3–5 μm	10	[29]
	pARG	V-$CaCO_3$ pre-loaded with PSS, 850 nm	4	[152]
ELR	CHI	V-$CaCO_3$ pre-loaded with ovalbumin, 4 μm	4	[153]
		Protein-based PEMCs		
TA	Pepsin and BSA	PLL-coated V-$CaCO_3$, 3 μm	8	[154]
	BSA	PLL-coated V-$CaCO_3$ pre-loaded with LF, 3 μm	8/16	[155]
	BSA	V-$CaCO_3$ pre-loaded with BSA and MNPs, 3 μm	<6	[156]
GA-crosslinked BSA		$MnCO_3$, 7.4 μm	10	[157]
GA-crosslinked Hgb		$MnCO_3$, 5 μm	10	[158]
		Polyamino acid-based PEMCs		
pASP	pARG	V-$CaCO_3$ pre-loaded with pronase, 3–6 μm	7	[159]
pGLU	pONT	MS pre-loaded with DOX, 2 μm	8	[160]
	PLL	MS bare or pre-loaded with lysozyme or catalase, 2–4 μm	6	[161]
		V-$CaCO_3$ pre-loaded with PLL or pGLU, 6 μm	7	[162]
		V-$CaCO_3$ pre-loaded with IgY, 2–10 μm	10	[163]

Apart from typical polysaccharide/amino acid PEMCs, vaterite $CaCO_3$ and $MnCO_3$ were also used as a template for DNA-based capsules. Tetramethylrhodamine-modified dextran (TMR-D)-co-synthesised vaterite crystals were coated with a primary layer of positively charged poly(allylamine hydrochloride) (PAH), followed by the LbL-build-up of two hybrid nucleic acids (for full sequences see [164]), of which one sequence contained the anti-adenosine triphosphate (ATP) aptamer. Following addition of EDTA and subsequent dissolution of $CaCO_3$, hollow aptamer-cross-linked capsules were formed (Figure 5(1,2)) and underwent a shrinkage phenomenon (~3.2 um to ~2.5 um) [164], which was noted in other bio-capsule systems [79,150,162]. TMR-D was then subsequently released via the exposure of the capsules to ATP, which complexes with its aptamer, disrupting the bridging of DNA within the layers (Figure 5(3)). $CaCO_3$-templated DNA-based capsules with size-selective macromolecule permeation were also produced using the LbL approach; 56 kDa dextran permeated the capsules, while 155 kDa dextran was inaccessible to the capsule [165,166]. Such DNA-based systems may prove useful for the size-selective encapsulation of macromolecules, as well as potentially low molecular weight drugs and genetic material [167,168].

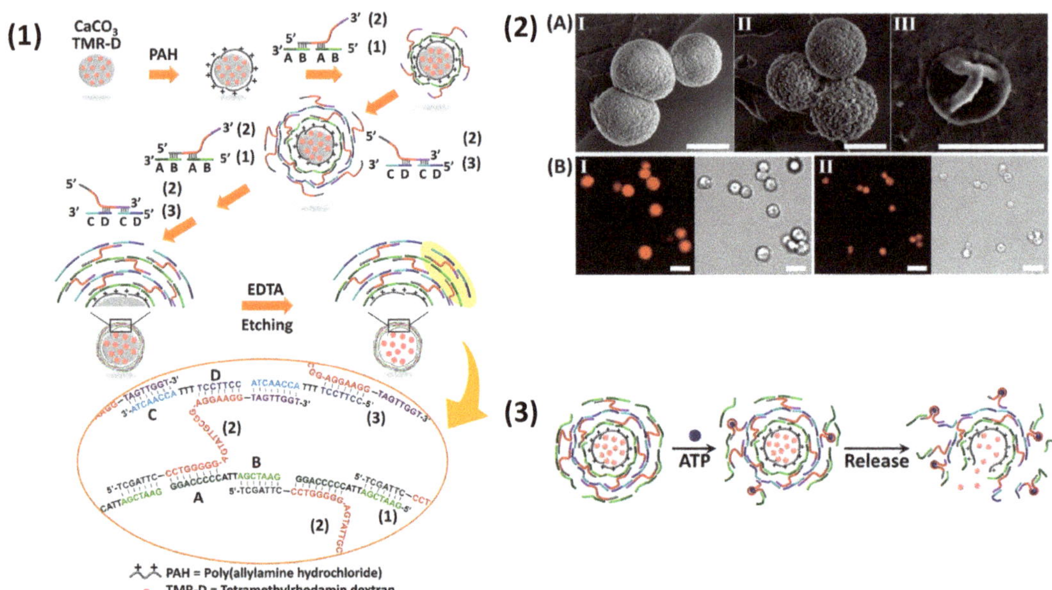

Figure 5. (**1**) Schematic illustration of the preparation of DNA microcapsules via $CaCO_3$ templating. ATP-binding aptamer sequences, labelled in red colour, are embedded into DNA films as stimuli-sensitive switches. (**2**) SEM images of uncoated (**A**, I), DNA-coated (**A**, II) $CaCO_3$ vaterite crystals and DNA PEMCs following EDTA addition (**A**, III). Below are corresponding confocal and brightfield confocal images of DNA-coated crystals (**B**, I) and DNA PEMCs (**B**, II). (**3**) Representation of ATP-induced PEMC rupture and release of TMR-D. Reprinted with permission from reference [164] copyright © 2015 American Chemical Society.

The formation of nano-sized PEMCs, formed upon porous templates with varying biopolymers, can still be challenging as the typical size of $CaCO_3$ PEMC templates fall within the range of 3–10 μm (Table 2). However, recent progress was made in the formulation of vaterite $CaCO_3$ templates in the sub-micron and nano-regions [84,123,169,170]. Both additive [170] and additive-free [123] approaches emerged as facile methods to synthesise sub-micron vectors. For instance, when using a glycerol/gelatin formulation with ultrasonic treatment of the pre-cursor salt solutions, sizes of 54 ± 9 nm were achieved, along with a range of other sizes up to ~800 nm. Using additive-free methods, it is possible to reach sizes of close to 720 nm. Of note, we are able to reach sizes of up to 55 um $CaCO_3$ crystals [123] for potential use as porogens [171] for tissue engineering scaffolds (Figure 6(3,4)). Smaller sub-micron and nano-sized functional delivery vehicles are necessary for the effective treatment of ailments, including cancer, in which the enhanced permeation and retention effect is prevalent. Approaches concerning the shrinkage of PEMCs, with regard to reaching the necessary sizes for advanced drug delivery, are now emerging. Recent studies and their applications will be discussed in the next section.

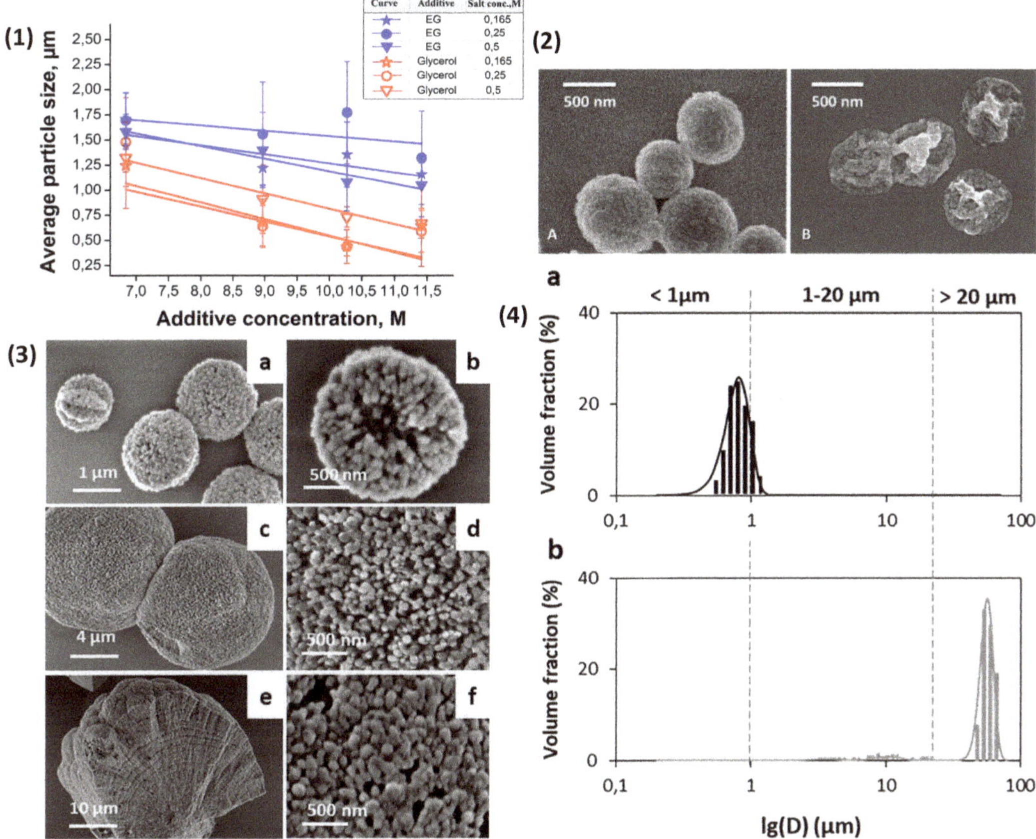

Figure 6. (**1**) Graphical representation of the effect of additive concentration on the size of vaterite $CaCO_3$ crystals at different salt concentrations. (**2**) SEM images of sub-micron $CaCO_3$ coated with (pARG/DS) 4.5 layers (**A**) and hollow PEMCs (**B**). Reprinted with permission from reference [84] copyright © 2016 American Chemical Society. (**3**) SEM images of vaterite $CaCO_3$ sub-micron crystals (**a**,**b**), middle-sized $CaCO_3$ crystals (**c**), and their typical surface (**d**). The spall of sub-millimetre vaterite $CaCO_3$ crystal (**e**), and typical surface (**f**). (**4**) Typical size distribution of sub-micron-$CaCO_3$ (**a**) and giant $CaCO_3$ (**b**) crystals grown by the mixing of $CaCl_2$ and Na_2CO_3 salts in water. Bars represent experimental data; lines show the fitting with Gaussian function. Reprinted with permission from reference [123] copyright © 2021 Elsevier.

Shrunken Biopolymer-Based Capsules

The shrinkage of PEMCs formed upon vaterite $CaCO_3$ templates was reported numerous times [58,79,146,149], typically via the thermal treatment of capsules [61,172,173]; The observed shrinkage of PEM films, especially within capsule systems, holds important applications in drug delivery. Controlled shrinkage allows us to readily tune the size of delivery vehicles depending upon the targeted area; for example, the delivery of nano-capsules to tumorous cells saw an increase in research as of late due to their biodegradability and their ability to host large amounts of biological cargo. For instance, DS/pARG capsules were seen in recent work regarding the shrinkage of PEM capsules. Trushina et al. [172,174] subjected DS/PARG capsules to heat treatment at different temperatures (up to 90° C); the heightened temperature results in the shrinkage of nano-capsules by factors of up to 42%. This was due to the temperature-induced annealing of biopolymers forming a more compact capsule shell upon shrinkage. This compaction is presented in Figure 7(1), wherein the shell clearly becomes denser upon compaction (shown via TEM imaging).

Furthermore, it was demonstrated that increasing the ionic strength to that of physiological systems (0.15 M NaCl) and subjecting the capsules to heat treatment causes the capsules to collapse after 30 min of incubation. The authors attributed this to the drastic effect of ionic cross-linking upon the increase in ionic strength, thereby inhibiting the shrinking capability of DS/PARG capsules.

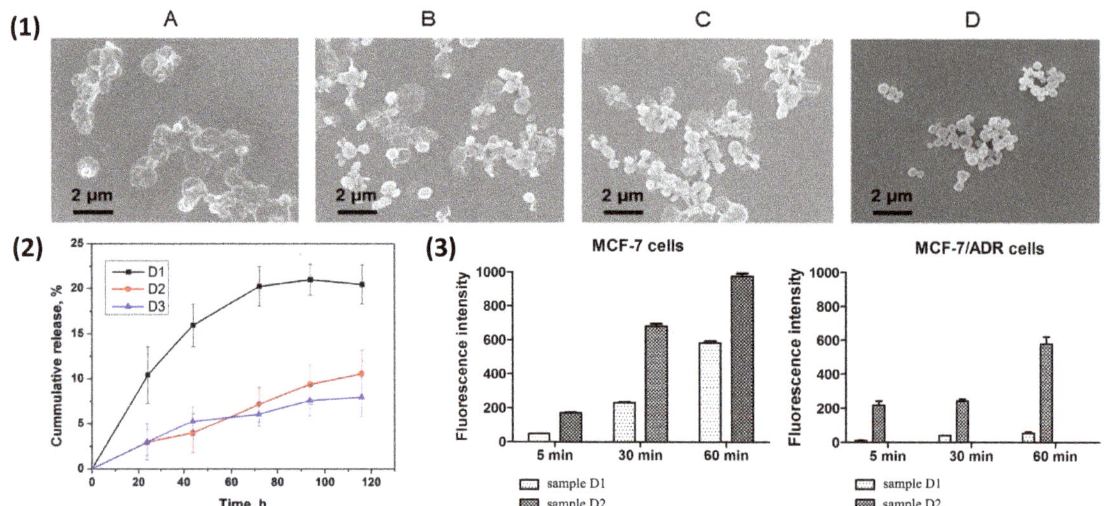

Figure 7. (**1**) SEM images of (Parg/DS)$_{4.5}$ PEMCs: freshly prepared (**A**) and after the heat treatment at 50 °C for 15 min (**B**), at 50 °C for 120 min (**C**), and at 90 °C for 60 min (**D**). Reprinted with permission from reference [172] copyright © 2018 Elsevier. (**2**) DOX release from intact (initial size: 550 nm, final size: 550 nm, D1) and shrunken (initial size: 550 nm, final size: 290 nm, D2; and initial size: 290 nm, final size: 290 nm, D3) PEMCs. (**3**) Uptake of DOX-loaded PEMCs by human breast adenocarcinoma MCF-7 cells (left) and DOX-resistant MCF-7/ADR cells (right) after 5, 30, and 60 min incubation. Reprinted taken with permission from reference [146] copyright © 2019 Elsevier.

These capsules were also utilised in the encapsulation of chemotherapeutic drugs, including gemcitabine, clodronate [175], and doxorubicin [146].The shrunken capsules held a significantly higher rate of cellular uptake in vitro, and both gemcitabine and clodronate reduced the viability of lung cancer cells and the tumour-promoting function of bone marrow-derived macrophages, respectively. Doxorubicin-loaded capsules showed a sustained release profile, as opposed to a burst-release system (Figure 7(2)), which is typically observed. This was attributed to the thicker capsule shell, slowing the diffusion of the drug through the polymeric network. Indeed, a thicker capsule shell may be compared to the internal structure of that of matrix-type capsules, which are known to alter the release profiles due to the dense polyelectrolyte network within the capsule lumen [78]. Moreover, shrunken doxorubicin-loaded (DS/PARG)$_3$ capsules were shown to accumulate within human breast adenocarcinoma MCF-7 and MCF-7/ADR (drug resistant) cells and managed to overcome the drug resistance of MCF-7/ADR cells (Figure 7(3)). Such sub-micron capsules may prove useful for future chemotherapeutic applications by taking advantage of advanced PEM shrinkage via highly dynamic biopolymers [21].

Biopolymer PEMCs with the ability to shrink at room temperature were also reported. Szarpak et al. (2010) [149] demonstrated the room temperature shrinkage of HA/PLL capsules upon dissolution of the $CaCO_3$ core by ~50%, and demonstrated the inhibition of shrinkage upon cross-linking with EDC/NHS, as is demonstrated in Figure 8(1). This ambient shrinkage may potentially be used to entrap molecules of interest within the PEMC interior. Campbell et al. (2021) [79] reported the room temperature shrinkage (up to a factors of ~7) of a variety of PEMCs, as shown in Figure 8(2) (PLL and PR paired with

HA/CS/DS/HS). The shrinkage was explained via the rearrangement of the biopolymers within the multilayers upon $CaCO_3$ dissolution. Increasing the molecular weight of PLL resulted in a reduction in the shrinkage of PEMCs (Figure 8(3)), and a similar trend was observed upon the increase in polyanion charge density, giving scope for the facile control of final PEMC size depending upon the properties of the polymers used. Of interest, the majority of these PEMCs also adhere to the surface on which they are formed, allowing for the potential patterning of surfaces such as implants.

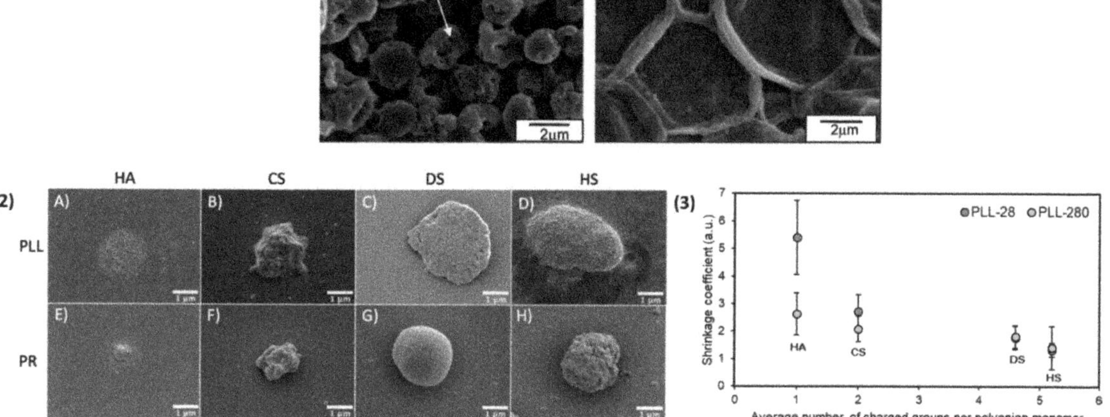

Figure 8. (**1**) SEM images of vaterite $CaCO_3$-templated dried (HA/PLL) 4.5 PEMCs, the arrow indicates holes in the PEMC shell (**A**), and cross-linked via means of 200 mM EDC (HA/PLL) 4.5 PEMCs (**B**). Reprinted with permission from reference [149] copyright © 2010 American Chemical Society. (**2**) Typical SEM images of vaterite $CaCO_3$-templated PEMCs consisting of 2.5 bilayers of PLL-based (top row) capsules consisting of HA, CS, DS, and HS are shown in images (**A–D**), respectively. PR-based (bottom row) capsules, consisting of HA, CS, DS, and HS, are shown in images (**E–H**), respectively. (**3**) Effect of number of charged groups upon the polyanion monomer unit on the shrinkage coefficient of PLL-28 (average molecular weight ~28 kDa) and PLL-280 (average molecular weight 280 kDa) PEMCs. Reprinted with permission from reference [79] copyright © 2021 American Chemical Society.

3.2. *Beads Formed from Porous Templates*

Simultaneously, the porous sacrificial templates were also utilized for the fabrication of polymer-based beads. Methodologically simplistic and robust, these particulate structures are formed via the loading of the respective molecule into a porous template via adsorption, followed the by cross-linking and elimination of this template, resulting in the formation of a pure polymeric particle within which is an inverted replica of the template utilized. Moreover, dependent upon the physiochemical properties of the biopolymer used and the extent of cross-linking, varying resultant properties in the final particles are seen, for instance, if strong inter-polymer interactions or a high degree of cross-linking occurs and free-standing porous particles are produced. Where there are weaker interactions, the particles may collapse to non-porous beads to minimize their contact with surrounding water [176]. Multicompartment heterogenous particles may also be produced [64]. Free-standing protein particles may be formed if the protein is insoluble at conditions where the template is soluble [176]. Through manipulation of the large loading capacity of vaterite $CaCO_3$ crystals, insulin particles were formed. Insulin was incubated with $CaCO_3$ crystals at pH 9.5, where $CaCO_3$ is insoluble and insulin is soluble, and the pH was then slowly reduced down to pH 5.2. During this decrease, insulin solubility is reduced and precipitates within the crystal pores. At these lower pHs, $CaCO_3$ also undergoes dissolution and pure insulin particles are formed (Figure 9(1)) [122,177]. Polyelectrolytes may also be deposited upon protein particles via LbL to control release, as with α-chymotrypsin for instance [178,179]. Haemoglobin microparticles were also formed, but via co-synthesis into $CaCO_3$ crystals. Loading of the haemoglobin, followed by cross-linking with glutaraldehyde and dissolution of the carbonate template with 0.2 M EDTA (Figure 9(2)), resulted in smooth, spherical particles with an average diameter of 3.2 μm. These particles were successfully used as oxygen carriers, with each particle's haemoglobin content consisting of at least one third of that found in an erythrocyte cell [180]. Successful fabrication of submicron haemoglobin particles was also achieved, via the use of peanut-shaped $MnCO_3$ particles (Figure 9(4)) with very high haemoglobin uptake efficiency, utilising the same approach [181]. Alternatively to glutaraldehyde, particles of pure soy protein were formed via adsorption into $CaCO_3$, utilising cross-linking with transglutaminase and dissolution of the template [182]. Polyelectrolyte-bridged protein particles were also templated upon mesoporous silica [183]. Moreover, dithioretinol was used to promote the opening of disulfide bonds of BSA entrapped within $CaCO_3$ by co-synthesis and the subsequent removal of dithioretinol allowed for the formation of new disulfide bonds, both inter- and intra-molecularly, contributing to stable BSA particle formation [184]. UV-induced release of macromolecules from $MnCO_3$-templated.

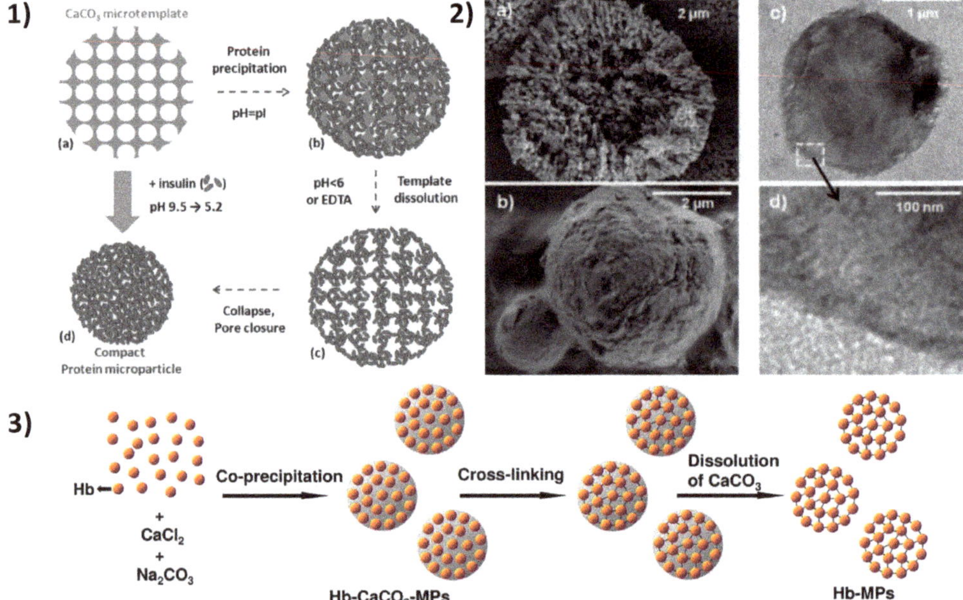

Figure 9. *Cont.*

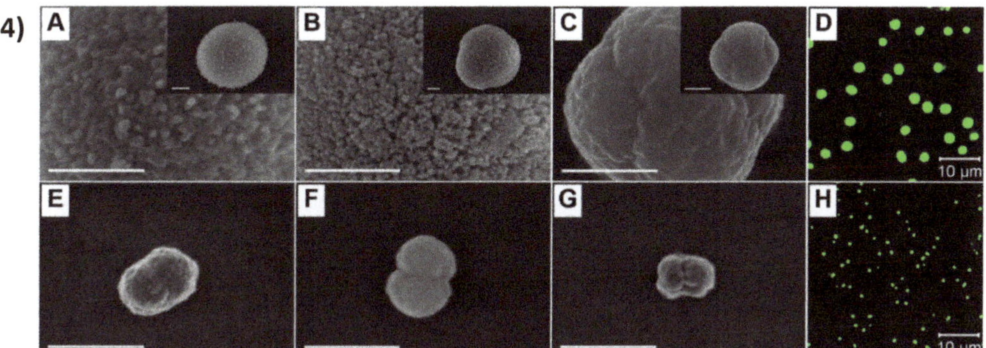

Figure 9. (**1**) Schematic of the formation of protein microparticles; a–b: the loading of porous $CaCO_3$ templates with protein by isoelectric precipitation; b–c: dissolution of the $CaCO_3$ template; and c-d: the shrinkage of protein matrix to a compact bead. Reprinted with permission from reference [122] copyright © 2012 John Wiley & Sons. (**2**) SEM images of vaterite $CaCO_3$ crystals (**a**), and insulin microparticles (**b**). TEM images of insulin microparticle (**c**), and a magnified section in (**d**). Reprinted with permission from reference [185] copyright © 2010 John Wiley & Sons. (**3**) Schematic illustration of the production of haemoglobin microparticles. Reprinted with permission from reference [180] copyright © 2012 American Chemical Society. (**4**) SEM images of bare-$CaCO_3$ crystal (**A**), haemoglobin-encapsulated $CaCO_3$ crystal (**B**), and haemoglobin particle following $CaCO_3$ dissolution (**C**). Corresponding confocal image of protein particles (**D**). SEM image of bare-$MnCO_3$ particle (**E**), haemoglobin-encapsulated $MnCO_3$ particle (**F**), and haemoglobin particle following $MnCO_3$ dissolution (**G**). Corresponding confocal image of protein particles (**H**). Reprinted with permission from reference [181] copyright © 2018 Elsevier.

BSA particles can be achieved. The cross-linking of BSA with *ortho*-nitrobenzyl derivative 4-bromomethyl-3-nitrobenzoic acid under activation with 4-(4,6-dimethoxy-1,3,5-triazin-2-yl)-4-methylmorpholinium chloride, followed by $MnCO_3$ dissolution, results in particulate BSA, which contains photo-cleavable C-N bonds, allowing for subsequent UV-controlled release of the cargo [186].

Recently, antibody-containing protein microparticles for the ELISA-based detection of human immunoglobulin G were produced by Neumann and Volodkin (2020) [187]. BSA and goat anti-human immunoglobulin G were co-synthesised into $CaCO_3$ crystals, reaching loading efficiencies of up to 70%, followed by cross-linking with glutaraldehyde and dissolution with EDTA. Due to hydrophobic effects, the final porous protein particles shrank by 31% of the original crystal size, similarly to that of other protein-based particles [122,182]. With increasing control over the original porous template size (e.g., $CaCO_3$, both with [170] and without [123] additives, $MnCO_3$ [188], and mesoporous silica [189]), one may exert fine control over the final polymer particle size. Furthermore, through the use of thermo-responsive polymers, temperature-controlled shrinkage/swelling of pure polymeric particles can be achieved [190]. Recently reported, through the use of vaterite-templated poly(N-isopropylacrylamide) microgels, the temperature-controlled release of macromolecular drugs was demonstrated [191]. Ionic strength- and pH-induced swelling/shrinkage was also reported in gelatin-based microgels templated upon $CaCO_3$ crystals [192]. Furthermore, cross-linked enzyme aggregates (CLEAs) may also be fabricated from this hard-templating approach via the immobilisation of the enzyme [193] within porous particles. CLEAs are popular materials in the field of biocatalysis as reusable catalysts [194], and the control of final aggregate size and porosity are important for the relative catalytic activity and practical use in industrial catalysis. Hard-templating may provide an appropriate means of controlling these properties [195–197].

4. Soft-Templated Biopolymer-Based Capsules

Liposomes are spherical micro-/nano-structures that are formed from phospholipid bilayers with an aqueous compartment within. A plethora of lipids can be used to form liposomes; for instance, variations of phosphatidylglycerol (anionic at pH 7) and phosphatidylethanolamine (zwitterionic at pH 7), and phosphatidylcholine (zwitterionic at pH 7) (see review [198] and structures therein). Due to this, they possess amphiphilic character and can encapsulate hydrophilic molecules within their core and hydrophobic molecules within their lipid membranes. The lipid character of their structure gives them high biocompatibility and are, hence, very attractive as drug delivery vehicles due to their versatility. However, liposomes can possess poor stability unless in a buffered environment and can sometimes have poor drug-trapping potential, with some cargo elution [199,200]. This has led to extensive studies regarding the functionalisation of liposomes with different molecular species (i.e., antibodies, proteins, carbohydrates, and PEG) [201–205]. Moreover, multilayer coatings of biopolyelectrolytes to further protect the encapsulated cargo, during the delivery phase, and functionalise the outer-shell, have become attractive [206,207]; for instance, with coatings of PLL/pGLU [208], CHI with DS [209], and ALG [206], as well as protein-based BSA/lactoferrin [210]. A small number of studies report the soft-templating of biopolymer PEMCs upon liposomal structures. Cuomo et al. [211,212] demonstrated the formation of ALG/CHI PEMCs upon 300 nm phosphatidylcholine/ didodecyldimethylammonium bromide (DDAB) liposomes, a schematic of which is shown in Figure 10(1). The removal of the core was performed with a non-ionic surfactant (Triton X-100 in this study) via inducing a liposome-to-micelle transition, which was monitored with Nile red dye-sensitive to its microenvironment (whether a micelle or lipid bilayer; for example, fluorescence maxima in Figure 10(2)). Loaded liposomes may also be coated and immobilised onto functionalised surfaces (i.e., multilayer coated surfaces) to act as controlled delivery vectors [213,214].

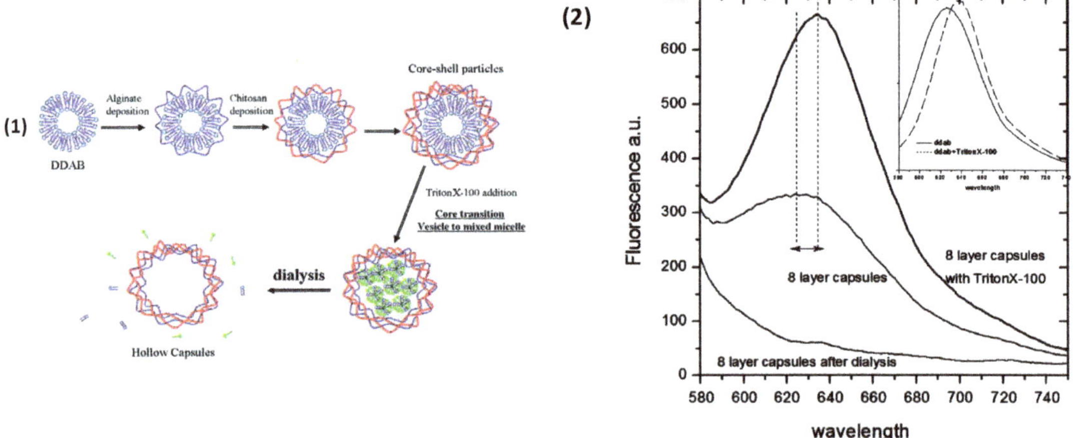

Figure 10. (**1**) Schematic illustration of the formation of liposome templated PEMCs. (**2**) Nile red fluorescence emission spectra (excitation at 530 nm) in 4 bilayer-coated liposomes (maximum at 625 nm) and following the addition of Triton X (maximum at 635 nm). Following dialysis, the fluorescence disappears. In the inset, a red shift for the bare DDAB vesicles is reported. The solid line refers to the fluorescence of Nile red in DDAB vesicles; the dashed line is the fluorescence spectra in mixed micelles. Reprinted with permission from reference [212] copyright © 2010 American Chemical Society.

Further to liposomes, microgels may be used as soft templating materials, for instance dextran hydroxyethylmethacrylate (DEX-HEMA) microgels, which can be synthesised in a broad size range, with so-called giant microgels (150 μm) used for the formation of synthetic capsules [215]. Biopolymer-based PEMCs templated on DEX-HEMA were reported with pARG paired with CS, pASP, pGLU, and DS. The dissolution of the core microgel was performed with 0.1 M NaOH via hydrolysis of the cross-linking carbonate esters between dextran chains. Upon degradation, only pARG/DS produced stable hollow PEMCs, while the rest self-ruptured, which was attributed to the build-up of osmotic pressure within the polyelectrolyte shell; demonstrated in Figure 11(3). However, upon increasing the molecular weight of pGLU, a percentage of PEMCs remained intact (Figure 11), suggesting a larger polymer chain length may increase PEMC mechanical strength or permeability of the shell. Of note, HA, CHI, PLL, pONT, and ALG were tested as LbL components, but the authors reported instantaneous microgel aggregation upon dispersion into the biopolymer solution, whereas pARG did not cause this [216]. Disulfide-crosslinked HA gels (~16 μm) was also coated with HA/PLL multilayers, followed by the addition of dithiothretinol (at neutral pH), to cleave the sulfide linkages in order to remove the microgel core [217], as shown in Figure 11(1,2). Using such gels allows us to pre-encapsulate bioactive material pre-LbL deposition, much like that of porous inorganic templates ($CaCO_3$/$MnCO_3$/MS), and produce PEMCs under mild conditions suitable for biopolymers. Other potential soft materials may include that of PNIPAM and alginate gels [218,219].

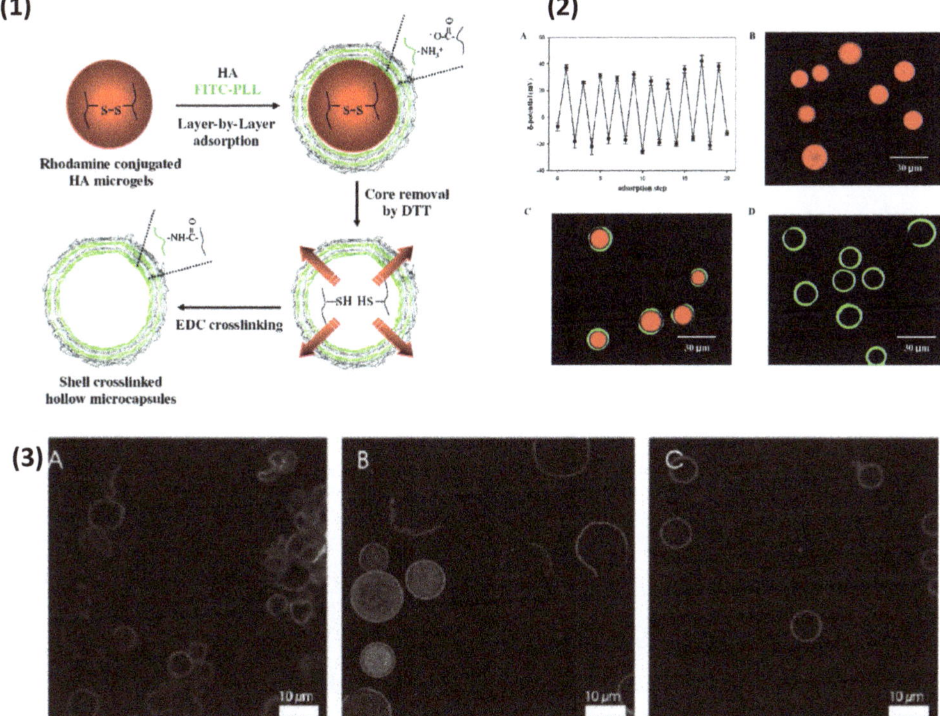

Figure 11. **(1)** Schematic illustration of the formation of cross-linked hollow HA/PLL PEMCs and **(2)** zeta potential as a function of layer number during LbL deposition upon HA microgels (**A**), confocal images of rhodamine-conjugated HA microgels (**B**), FITC-labelled HA/PLL shell containing rhodamine (**C**), and following core removal (**D**). Figures taken with permission from reference [217] copyright © 2007 American Chemical Society. **(3)** Confocal images of dex-HEMA microgels coated with C /pARG)$_4$ (**A**), (pGLU(high molecular weight)/pARG)$_4$ (**B**), and (DS/pARG)$_4$ (**C**) after degradation of the microgel core. In (**A**), all microcapsules were broken and released their contents. In (**B**), both broken as well as intact (still filled with 150 kDa FITC-dextran) microcapsules were observed. The capsules in (**C**) remained intact, but had released their contents by diffusion through the bio-polyelectrolyte coating. Figure taken with permission from reference [216] copyright © 2007 John Wiley & Sons.

Biological templates were also used as PEMC templates, with erythrocytes used in particular as templates for synthetic PEMCs [74,76]. It was reported that the removal of an erythrocyte core via NaOCl could lead to changes in the chemical nature of the polyelectrolytes used, with amino groups in PSS/PAH oxidised to nitro-, nitroso- and nitrile- groups, causing the subsequent cross-linking of PAH. Although advantageous in terms of mechanical stabilisation, this may not be suitable for such sensitive biopolyelectrolytes [220]. However, as of recently, live *E. coli* was used as PEMC template via the coating of ALG/CHI biopolymers, following dissolution via cell lysis (incubation in lysis buffer-(0.1% Triton X-100, 2 mM EDTA in 10 mM Tris-pH 8) with 100 μg/mL lysozyme overnight) [221]. Stable hollow PEMCs were formed with a slight increase in shell thickness, from 10–20 nm to 20–50 nm, which was attributed to alterations in polymer conformation on *E.coli* degradation. This demonstrated the novel formulation of biopolymer PEMCs on bacterial cells formed at soft conditions. This also leaves no harmful polymer residue in the shell, meaning it is suitable for release of cargo for biotherapeutic applications, perhaps opening up for a wider variety of biological PEMC templates.

5. Drug Crystal-Templated Biopolymer-Based Capsules

Key properties for the formulation of drug delivery vehicles include bioavailability, biodegradability, and high control over the drug content. A facile method to control drug loading is to use pure drug nano-crystals themselves. A number of approaches were successfully applied to reduce the size of such pure drug particles from the micro- to the nano-region, including both top-down [222,223] and bottom-up [224] approaches. The top-down approach involves the sonication or milling of coarse drug crystals to aid in the production of nano-sized crystals, whereas bottom-up involves the precipitation of nanocrystals from the dissolved drug via a solvent-induced supersaturation state, which is then followed by drug nucleation and subsequent growth. Precipitation may also be induced via pH change, if the drug of choice holds pH-dependent solubility, as well as via emulsification into organic solvent nanodroplets, in which nanocrystals may grow (for further details see chapter [225]). Many drugs, however, are typically poorly soluble in aqueous environments and, upon reduction of size, their solubility dramatically increases. This is due to the increase in the surface area to volume ratio of the crystal, giving increased solvent-crystal contact [226]. Despite this increasing drug bioavailability, such drug nanocrystals may solubilise and release drug molecules instantaneously at the target site or en route. Hence, LbL coatings present themselves as popular methods to control the drug release rate, as well as potentially the biodistribution if functionalised. Furthermore, one must be sure to deposit the polyelectrolytes at appropriate conditions, wherein the solubility of the drug is not increased nor decreased for the polyelectrolytes used, and retain their stability. Table 3 includes examples of biopolymer-coated drug nanocrystals.

Table 3. Summary of successfully fabricated nano- and microcapsules templated via drug crystal cores reported in literature. ALG: alginate; CHI: chitosan; DMPA: dimyristoylphosphatidic acid; DOX: doxorubicin; DS: dextran sulphate; HA: hyaluronic acid; HAS: human serum albumin; HS: heparin sulphate; GEL: gelatin; LS: lignosulfonate; PLL: poly-L-lysine; SF: silk fibronectin.

Polyanion	Polycation	Drug Crystal Core	Reference
ALG	CHI	Resveratol, ~200 nm	[227]
		Curcumin, 420 ± 17	[228]
	GEL	Naproxen, 11–20 um	[229]
Pectin	CHI	Indomethacin, ca. 200 nm	[230]
DS	CHI	Naproxen, 11–20 um	[229]
HA	-	DOX-coated cellulose nano-crystal, 134 ± 17 nm	[231]
HAS	DMPA	Ibuprofen, 15/36 um	[232]
HS	CHI	Insulin, ca 1 um	[233]
	PLL	Paclitaxel, 170–180 nm	[234]
		Camtothecin, <150 nm	[234]
SF	PLL	Dexamethasone, Side length 7.68 um, thickness of 750 nm	[231]
LS	CHI	Picloram, 1–4 um	[235]

Nano-crystals of indomethacin, an anti-inflammatory drug, were prepared via the top-down approach using mortar and pestle grinding, followed by sonication. LbL coating of chitosan and pectin was performed at pH 4.5, where the drug is almost insoluble, and the CHI was fully dissociated, while the pectin was dissociated at 80%. Four layers were alternately deposited, and the subsequent release of indomethacin was studied at pH 7. The release was slowed for those drug crystals coated with CHI/pectin [230] (Figure 12(1)) to a saturation point within 5 h, compared to 1 h for uncoated crystals. This same effect was observed for picloram, a herbicide, coated with LS/CHI PEMs [235], (Figure 12(1)) as well as in SA/Gelitin, DS/Gelitin, and DS/CHI multilayers with naproxen crystals [229]. Furthermore, following the dissolution of the drug nano-crystal, a PEMC may remain if the shell has not ruptured. Following the dissolution of picloram, for instance, a hollow PEMC remains [235], as observed in Figure 12(3). Moreover, the coating of such nanocrystals with biopolymer PEMs is of particular interest due to their typically intrinsic biocompatibility and controllable permeability. Further, from drug crystals, those particles formed fully from bioactive material, such as the biopolymer beads, enzyme crystals [48,236], or protein aggregates discussed previously, are of interest to coat in order to control the release of protein or enzyme, as well as to protect such fragile cargo [178]. To ensure the safe travel of these bioactives to their target site, functionalisation of the PEMC shell can be of great importance and interest. This will be discussed in the next section.

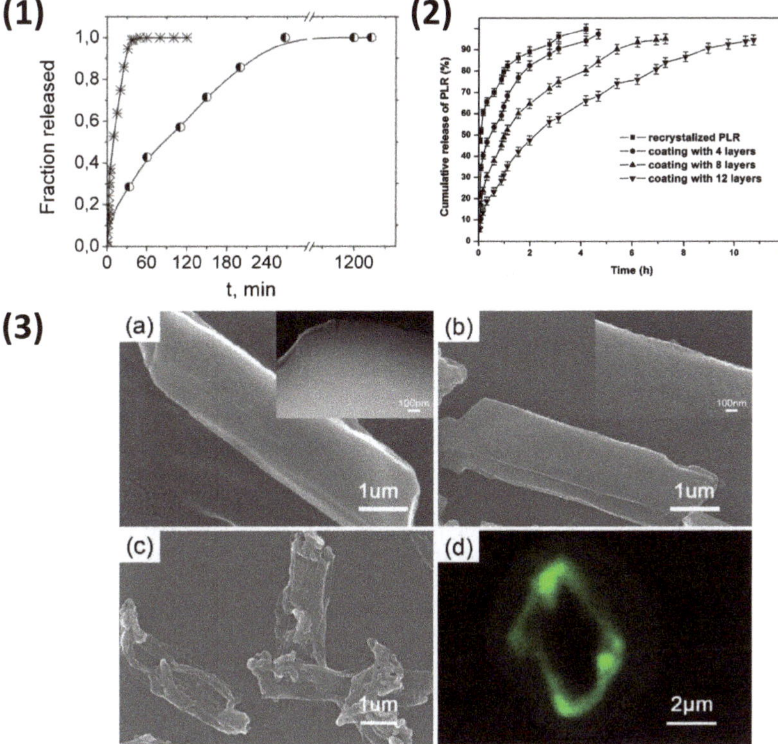

Figure 12. (**1**) Release profile of bare (*) and CHI/pectin coated (◐) indomethacin particles in phosphate buffer pH 7. Figure taken with permission from reference [230] copyright © 2017 Elsevier. (**2**) Release profiles of picloram from LS/CHI PEMCs with 0, 4, 8, and 12 polyelectrolyte layers, and (**3**) SEM images of recrystallised picloram (**a**), picloram coated with 5 bilayers (**b**), and hollow PEMCs after picloram release (**c**), with corresponding confocal image (**d**). Figures taken with permission from reference [235] copyright © 2013 American Chemical Society.

6. Functionalisation of Biopolymer-Based Capsules

6.1. Functionalisation with Nanoparticles

The ability to functionalise the PEMC shell gives us the opportunity to form hybrid-PEMCs and tailor them toward specific bioapplications [237]. Functionalisation of PEM biocoatings with plasmonic nanoparticles was demonstrated as a powerful tool for controlled release of the payload [238] and for control over cell adhesion [239]. Functionalisation of PEMCs with nanoparticles serves similar purposes. Magnetic nanoparticles were embedded into the multilayer shell for increased accumulation at the target site, within DS/pARG capsules, for example [174,240]. For instance, Fe_3O_4 and doxorubicin-loaded PR-carboxymethylcellulose PEMCs showed greater efficacy against doxorubicin-resistant HeLa cells in mice, resulting in greater apoptotic effects in tumour cells. The authors attributed this to the controlled in vivo distribution of doxorubicin, which was controlled via an external magnetic field [241]. This gives scope for magnetic bio-capsules to be used in theranostics, both as drug delivery vehicles and contrast agents [240]. Superparamagnetic iron oxide nanoparticles were also adsorbed onto $CaCO_3$-templated DS/pARG multilayer capsules through replacing a negatively charged layer. It was demonstrated the viability of HeLa and 293T cells was not altered after 24 h of exposure to the PEMCs. Following the application of a magnetic field, the phagocytosed capsules can be retained at the target site under physiologically relevant shear stress conditions, following HeLa-EGFP engulfment [242]. Nanoparticle functionalised PEMCs may also be used in photothermal therapy. Gold nanorods, encapsulated within GA-crosslinked CHI and ALG PEMCs, exhibit a photothermal effect, inducing a temperature increase and the collapse of the PEMC following near-infrared irradiation [243]. The authors attributed this to the thermal degradation of the aldehyde groups of glutaraldehyde, leading to multilayer instability. Of note, this system also acted as a dual-therapeutic agent with phototherapy combined with chemotherapy. The PEMC was also able to host doxorubicin (10.56% *w*/*w*), which was released upon the irradiation and subsequent collapse of the capsule.

X-ray radioprotective nanoceria was also previously incorporated into the polyelectrolyte shell of DS/pARG PEMCs. The PEMC acts to preserve the antioxidant effects of cerium and release it intracellularly in a controlled manner [244–246]. Cerium oxide acts as a reactive oxygen species scavenger and may also act as a "filter" within the PEMC shell, filtering out ROS and protecting its cargo, as shown through the protection of the encapsulated enzyme luciferase [247]. Furthermore, (haemoglobin/poly (ethylene glycol))$_4$-coated $CaCO_3$ crystals loaded with magnetic iron oxide nanoparticles showed an affinity for the adsorption of UO_2^{2+}, with the presence of the protein bilayers increasing percent uranyl sorption by ~90%. The authors proposed that this system may be used as an approach to separate damaging UO_2^{2+} from contaminated bodies as a radioprotectant, due to the simple removal of the $CaCO_3$ template and biocompatibility of the components used [248]).

SiO_2-coated DS/pARG PEMCs were used as non-viral vectors for the delivery of CRISPR-Cas9 components, as the coated PEMCs act to protect the bioactive via greatly reduced shell permeability. Superior transfection was observed compared to commercial reagents and the authors demonstrated the ability of the $CaCO_3$-templated PEMCs to overcome extra- and intracellular barriers to deliver genetic material [170,249]. Moreover, photoluminescent near-infrared emitting (750–1200 nm) $MnCO_3$-templated PEMCs of ALG/PR with a capping layer DS and CHI were formed via the adsorption of $Cd_xHg_{1-x}Te$ nanocrystals into the polymer matrix. The authors proposed these PEMCs may be useful for the monitoring luminescence of the material in tissue as part of the drug delivery process [250]. Further to this, the incorporation of nanoparticles into such PEMs can result in the alterations of the mechanical properties. For instance, gold nanoparticles were shown to increase the Young's modulus 16-fold in 2D HA/PLL PEMs [251], and increases in this value were also seen with silver nanoparticles and graphene flakes [252] as well as SiO_2-coated PEMCs [249]. This holds positive implications in terms of tissue engineering applications with regard to the packing of biocompatible functionalised PEMCs together with varying surface properties and, hence, tunable cell adhesion and differentiation.

6.2. Functionalisation with Ligands and Antibodies

Alongside nanoparticles, LbL particles may be functionalised with certain molecules, or radiolabelled to either enhance the efficacy of the encapsulated agent, provide a means of in vivo imaging, and increase therapeutic effects, for example. Recently, biopolymer-coated $CaCO_3$ particles were applied as local radionuclide therapy agents for melanoma treatment (Figure 13). Human serum albumin/TA-coated sub-micron $CaCO_3$ particles were radiolabelled with actinium-225 to increase the efficacy of α-radionuclide therapy in mice bearing B16-F10 tumours [253], as well as zirconium-89 within the shell of PEMCs and coated-$CaCO_3$ templates for the positron emission tomographic imaging of vehicles in vivo [253,254]. Figure 13(1) demonstrates a schematic for the radiotherapy treatment and subsequent tumour growth inhibition, while Figure 13(2) gives representative PET images of mice treated with ^{89}Zr-labelled LbL particles, demonstrating the retained high resolution after 14 days [253]. The incorporation of biorelevant ligands to promote the targeting or cellular uptake of the PEMC is also of great interest. Scheffler et al. (2020) [255] demonstrated the assembly of the fusion protein (glycoprotein G) of the vesicular stomatitis virus (VSV-G) onto lipid bilayer coate SiO_2 particles, pre-coated with DS/PR multilayers. The virus particles were isolated from the supernatant of infected BHK cells and, following UV inactivation, were incorporated onto the particle via the conformational reversibility and activity of the fusion protein in varying pH (Figure 14(1)) [256]. Fusion proteins mediate the entry of viruses into cells and were shown to enhance the uptake of these LbL particles in Vero cells compared to the PEM and lipid bilayer alone. Its potential pathing can be observed in Figure 14(2). This may also be applied to PEMCs with encapsulated cargo.

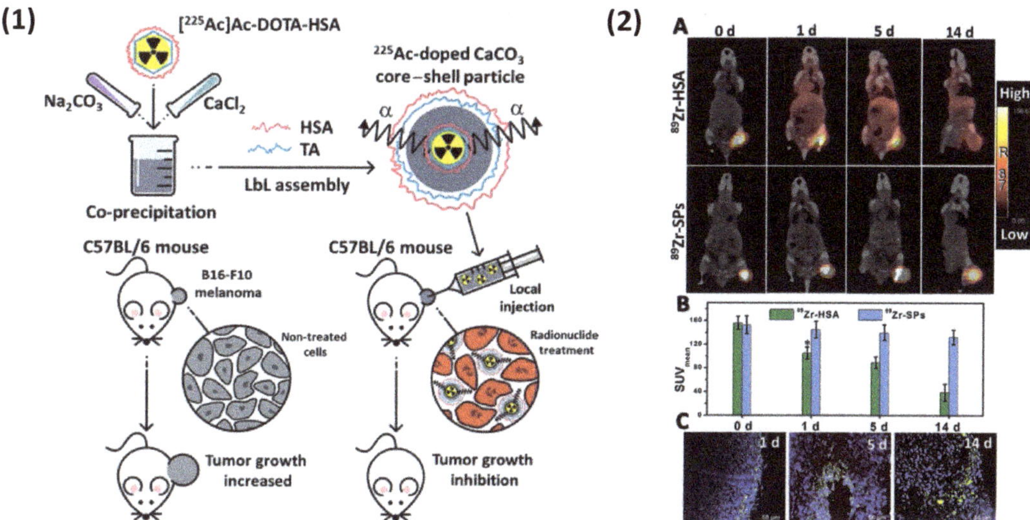

Figure 13. (**1**) Schematic illustration of the formation of sub-micron ^{225}Ac-doped TA/HSA-coated $CaCO_3$ particles and their application as local radionuclide therapy agents for melanoma treatment. (**2**) In vivo biodistribution studies after intratumoral injection of ^{89}Zr-HSA and ^{89}Zr-particles in B16-F10 melanoma-bearing mice: PET/CT images of mice treated with ^{89}Zr-HSA (0.5 μCi, V = 50 μL) and ^{89}Zr-particles (0.5 μCi, V = 50 μL, cSPs = 50 μg/mL, NSPs ~18 × 10^7) at different time points (**A**). The biodistribution of radiolabelled test samples measured (standardized uptake value, SUV) (**B**). Confocal images of tumour tissue after intratumoral injection of FITC-labelled particles (blue colour corresponds to DAPI stained cell nuclei, and green colour to particles) (**C**). Reprinted with permission from reference [253] copyright © 2021 American Chemical Society.

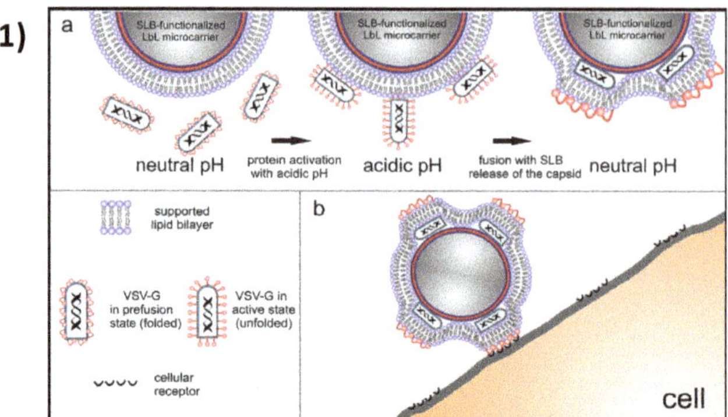

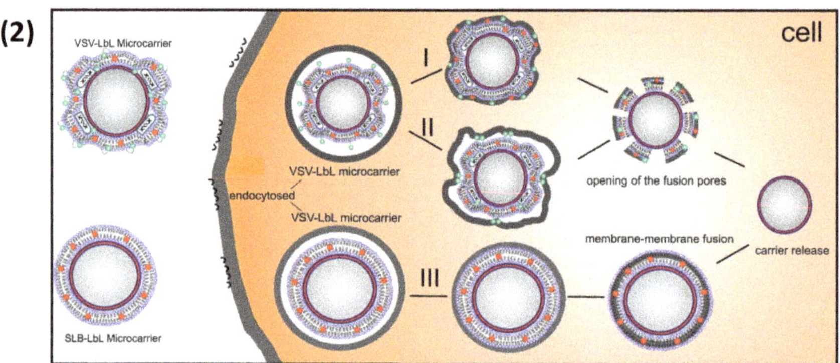

Figure 14. (**1**) Schematic illustration of the optimal virus assembly on lipid bilayer-coated -LbL microparticles using the unique conformational reversibility of the VSV-G protein. Assembly strategy is shown, making use of the unfolded state of the protein at pH 4, to initiate fusion with the lipid bilayer (**a**). After neutralization, VSV-G regains its prefusion state, allowing the protein to be available for specific cell interaction via receptor binding (**b**). Reprinted with permission from reference [256] copyright © 2018 American Chemical Society. (**2**) Schematic illustration of the possible fusion mechanisms between the lipid-bilayer-coated LbL coated particles and the endolysosomal membrane. In the upper panel, I and II, virus protein-mediated membrane fusion with endolysosome is shown, while the lower panel (III) illustrates the membrane-membrane fusion of nonfunctionalized particles and endolysosome. Red dots represent the fluorescent label of the lipid layer, and green dots demonstrate the formation of the fusion protein. Reprinted with permission from reference [255] copyright © 2020 American Chemical Society.

The antibody-functionalisation of the PEMC surface is also known. Recently, Ferrari et al. (2021) [257] encapsulated BSA into $CaCO_3$ nanoparticles, with subsequent LbL of three bilayers of DS/pARG, followed by surface adsorption of anti-intracellular adhesion molecule-1 monoclonal antibody or donkey anti-mouse IgG secondary antibody. These LbL particles were uptaken by EA.hy926 endothelial cells and did not reduce cell viability, demonstrating the potential targeting ability of these antibody-functionalised particles. As aforementioned, biopolymer PEMCs are attractive in terms of their potential to mimic the extracellular matrix (ECM) as well as host bioactive molecules. Single- and multi-domain peptides (SDPs and MDPs, respectively) are emerging as materials for use in tissue engineering/regeneration applications within the development of nanofibrous scaffolds [258,259]. The incorporation of these bioactive-mimicking SDP/MDPs into PEMCs are of interest. For instance, MDPs based upon ECM proteins, which have neurite outgrowth promoting activity, may hold important applications in neuromuscular tissue regeneration [260]. Such programmed intrinsic bioactive responses, such as anti-inflammatory [261,262], osteogenic [263], and neural development responses [260], provide an efficient means of developing functional PEMC-based [38]/hydrogel scaffolds [219,259] to both guide cell growth and proliferation, as well as control cellular differentiation. Conjugating a functional molecule to the polyelectrolyte chain or varying the final layer of the multilayer are also options for the functionalisation of PEMCs. For instance, use of folic acid-conjugated chitosan upon DS/PRO-curcumin PEMCs [264] to improve mucoadhesive properties, and the conjugation incorporation of cyclodextrins [265,266] within multilayers and the PEMC interior. For instance, HA-conjugated β-cyclodextrin entities played host to the hydrophobic drug paclitaxel [265], as demonstrated in Figure 15. Polycationic cyclodextrins were also paired with ALG to act as antimicrobial agents with the capability to host guest molecules within the cyclodextrin and PEMC interior [267], demonstrating the PEMCs potential for use as a multifunctional, or even multicompartmental, drug delivery vehicle.

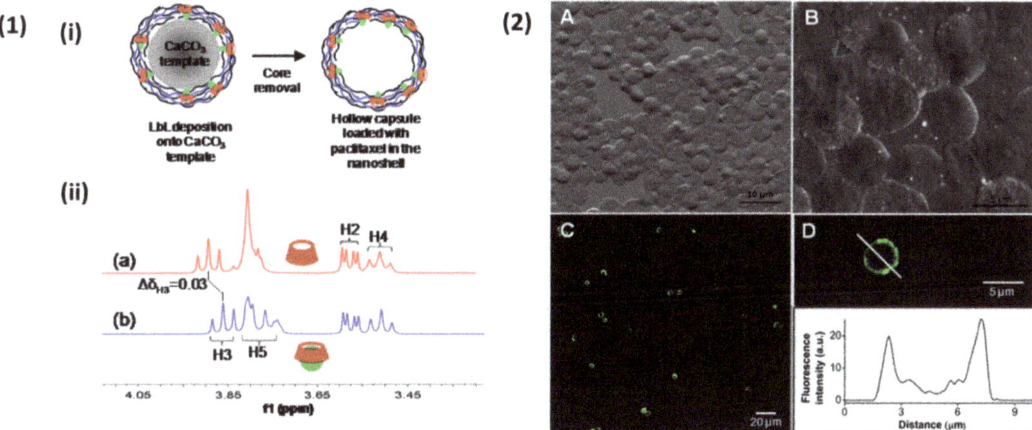

Figure 15. (**1**) Schematic illustration of the formation of cyclodextrin-modified HA/PLL PEMCs via dissolution of the $CaCO_3$ template (**i**), the partial H NMR spectra of cyclodextrin (**a**), and mixture of cyclodextrin with paclitaxel (**b**) (**ii**). (**2**) SEM images of hollow PEMCs (**A**,**B**) and confocal images of PEMCs (**C**) with corresponding fluorescence profile. (**D**) The fluorescence arises from the complexation of Oregon-green labelled paclitaxel with the cyclodextrin the PEMC shell. Reprinted with permission from reference [265] copyright © 2013 American Chemical Society.

7. Conclusions and Future Outlook

Recent years have witnessed the versatile templating approach expand its scope of applications and it was successfully used for the architecting of biogenic-made materials. Providing one of the most accessible strategies for the synthesis of various porous materials with internal structure designed at the nanoscale, soft- and hard-templating became powerful tools for the production of capsules, beads, hybrid nano- and microparticles and their assemblies. Templating of such inherently complex and labile molecules as biopolymers is rather challengeable due to their nature, but it is worth the cost: the use of the biopolymers versus traditional synthetic analogues for the design of biomaterials manifests in excellent biocompatibility, biodegradability, no-to-low toxicity, the possibility of replicating the composition and structure of natural cellular microenvironments, and their intrinsic biofunctionality. Templating of biopolymers largely became possible due to the application of "bio-friendly" soft and hard templates, which not only do not require harsh dissolution agents for their elimination, but are also biocompatible and/or biodegradable themselves. In this context, it is hard to underestimate the importance of mesoporous carbonate templates, which are prone to dissolution at truly mild conditions and are fabricated in one step that can optionally be combined with the functionalisation of the particles via co-synthesis. Together with soft templating and coating of the crystal drugs, these approaches demonstrate the ability to generate biomaterials capable of controlled, optionally targeted, synergistic, and sustained multi-drug release, which can be driven by external or internal stimuli. From this point of view, the most promising clinical and biological applications are in (i) controlled drug delivery, (ii) sensing and bioimaging, (iii) theranostics, and (iv) tissue engineering and regenerative medicine. Among them, current drug delivery applications seem to be closest to the stage of lab-to-clinic transition. Other fascinating applications include cell patterning, functionalisation of biomaterial surfaces, functionalisation of the coating, etc.

Up-to-date reports in this field, including the first in vitro and in vivo studies, showed the high promise of biopolymer-based multilayer capsules and beads, made via templating, as biomaterials of the fourth generation. However, fundamental research on the mechanisms of interaction between these templated materials and biological cells, tissues and translational research are still needed. Remaining challenges in the field include the control over biomechanical properties, optimisation of synthetic procedures, approaches to enable up-scaled production and long-term storage of biopolymer-based particles, exploration of new cargo release mechanisms, and ways of particle functionalisation to decipher the interaction with other biomaterials and tissues. We believe that the next years of research will open opportunities for true biomedical translation of biopolymer-based multilayer capsules and beads made via the templating approach.

Author Contributions: Conceptualization, A.S.V.; investigation, writing—original draft preparation, A.S.V. and J.C.; writing—review and editing, A.S.V. and J.C.; supervision, A.S.V. All authors have read and agreed to the published version of the manuscript.

Funding: A.V. acknowledges the support from the Alexander von Humboldt Foundation (in the frame of the Humboldt Postdoctoral Fellowship) and from the Staedtler Foundation. J.C. acknowledges NTU for funding his PhD.

Conflicts of Interest: The authors declare no conflict of interest.

References

1. Mauquoy, S.; Dupont-Gillain, C. Combination of Collagen and Fibronectin to Design Biomimetic Interfaces: Do These Proteins Form Layer-by-Layer Assemblies? *Colloids Surf. B Biointerfaces* **2016**, *147*, 54–64. [CrossRef] [PubMed]
2. Löhmann, O.; Zerball, M.; Von Klitzing, R. Water Uptake of Polyelectrolyte Multilayers Including Water Condensation in Voids. *Langmuir* **2018**, *34*, 11518–11525. [CrossRef] [PubMed]
3. Lee, H. Effect of Polyelectrolyte Size on Multilayer Conformation and Dynamics at Different Temperatures and Salt Concentrations. *J. Mol. Graph. Model.* **2016**, *70*, 246–252. [CrossRef]
4. Boulmedais, F.; Ball, V.; Schwinte, P.; Frisch, B.; Schaaf, P.; Voegel, J.C. Buildup of Exponentially Growing Multilayer Polypeptide Films with Internal Secondary Structure. *Langmuir* **2003**, *19*, 440–445. [CrossRef]
5. Picart, C.; Mutterer, J.; Richert, L.; Luo, Y.; Prestwich, G.D.; Schaaf, P.; Voegel, J.C.; Lavalle, P. Molecular Basis for the Explanation of the Exponential Growth of Polyelectrolyte Multilayers. *Proc. Natl. Acad. Sci. USA* **2002**, *99*, 12531–12535. [CrossRef]
6. Kilan, K.; Warszyński, P. Thickness and Permeability of Multilayers Containing Alginate Cross-Linked by Calcium Ions. *Electrochim. Acta* **2014**, *144*, 254–262. [CrossRef]
7. Antipov, A.A.; Sukhorukov, G.B. Polyelectrolyte Multilayer Capsules as Vehicles with Tunable Permeability. *Adv. Colloid Interface Sci.* **2004**, *111*, 49–61. [CrossRef] [PubMed]
8. Shiratori, S.S.; Rubner, M.F. PH-Dependent Thickness Behavior of Sequentially Adsorbed Layers of Weak Polyelectrolytes. *Macromolecules* **2000**, *33*, 4213–4219. [CrossRef]
9. Park, S.; Choi, D.; Jeong, H.; Heo, J.; Hong, J. Drug Loading and Release Behavior Depending on the Induced Porosity of Chitosan/Cellulose Multilayer Nanofilms. *Mol. Pharm.* **2017**, *14*, 3322–3330. [CrossRef]
10. Andreeva, T.D.; Hartmann, H.; Taneva, S.G.; Krastev, R. Regulation of the Growth, Morphology, Mechanical Properties and Biocompatibility of Natural Polysaccharide-Based Multilayers by Hofmeister Anions. *J. Mater. Chem. B* **2016**, *4*, 7092–7100. [CrossRef]
11. Balabushevich, N.G.; Sholina, E.A.; Mikhalchik, E.V.; Filatova, L.Y.; Vikulina, A.S.; Volodkin, D. Self-Assembled Mucin-Containing Microcarriers via Hard Templating on $CaCO_3$ Crystals. *Micromachines* **2018**, *9*, 307. [CrossRef]
12. Sustr, D.; Duschl, C.; Volodkin, D. A FRAP-Based Evaluation of Protein Diffusion in Polyelectrolyte Multilayers. *Eur. Polym. J.* **2015**, *68*, 665–670. [CrossRef]

13. Uhlig, K.; Madaboosi, N.; Schmidt, S.; Jäger, M.S.; Rose, J.; Duschl, C.; Volodkin, D.V. 3D Localization and Diffusion of Proteins in Polyelectrolyte Multilayers. *Soft Matter* **2012**, *8*, 11786–11789. [CrossRef]
14. Madaboosi, N.; Uhlig, K.; Schmidt, S.; Jäger, M.S.; Möhwald, H.; Duschl, C.; Volodkin, D.V. Microfluidics Meets Soft Layer-by-Layer Films: Selective Cell Growth in 3D Polymer Architectures. *Lab Chip* **2012**, *12*, 1434–1436. [CrossRef] [PubMed]
15. Boddohi, S.; Killingsworth, C.E.; Kipper, M.J. Polyelectrolyte Multilayer Assembly as a Function of PH and Ionic Strength Using the Polysaccharides Chitosan and Heparin. *Biomacromolecules* **2008**, *9*, 2021–2028. [CrossRef] [PubMed]
16. Crouzier, T.; Picart, C. Ion Pairing and Hydration in Polyelectrolyte Multilayer Films Containing Polysaccharides. *Biomacromolecules* **2009**, *10*, 433–442. [CrossRef] [PubMed]
17. Picart, C.; Lavalle, P.; Hubert, P.; Cuisinier, F.J.G.; Decher, G.; Schaaf, P.; Voegel, J.C. Buildup Mechanism for Poly(L-Lysine)/Hyaluronic Acid Films onto a Solid Surface. *Langmuir* **2001**, *17*, 7414–7424. [CrossRef]
18. Tang, K.; Besseling, N.A.M. Formation of Polyelectrolyte Multilayers: Ionic Strengths and Growth Regimes. *Soft Matter* **2016**, *12*, 1032–1040. [CrossRef]
19. Park, K.; Choi, D.; Hong, J. Nanostructured Polymer Thin Films Fabricated with Brush-Based Layer-by-Layer Self-Assembly for Site- Selective Construction and Drug Release. *Sci. Rep.* **2018**, *8*. [CrossRef]
20. Rivero, P.J.; Goicoechea, J.; Arregui, F.J. Layer-by-Layer Nano-Assembly: A Powerful Tool for Optical Fiber Sensing Applications. *Sensors* **2019**, *19*, 683. [CrossRef]
21. Campbell, J.; Vikulina, A.S. Layer-by-Layer Assemblies of Biopolymers: Build-up, Mechanical Stability and Molecular Dynamics. *Polymers* **2020**, *12*, 1949. [CrossRef]
22. Parakhonskiy, B.V.; Parak, W.J.; Volodkin, D.; Skirtach, A.G. Hybrids of Polymeric Capsules, Lipids, and Nanoparticles: Thermodynamics and Temperature Rise at the Nanoscale and Emerging Applications. *Langmuir* **2019**, *35*, 8574–8583. [CrossRef]
23. Shi, Q.; Qian, Z.; Liu, D.; Liu, H. Surface Modification of Dental Titanium Implant by Layer-by-Layer Electrostatic Self-Assembly. *Front. Physiol.* **2017**, *8*. [CrossRef]
24. Macdonald, M.L.; Samuel, R.E.; Shah, N.J.; Padera, R.F.; Beben, Y.M.; Hammond, P.T. Tissue Integration of Growth Factor-Eluting Layer-by-Layer Polyelectrolyte Multilayer Coated Implants. *Biomaterials* **2011**, *32*, 1446–1453. [CrossRef]
25. Gribova, V.; Auzely-Velty, R.; Picart, C. Polyelectrolyte Multilayer Assemblies on Materials Surfaces: From Cell Adhesion to Tissue Engineering. *Chem. Mater.* **2012**, *24*, 854–869. [CrossRef] [PubMed]
26. Silva, J.M.; Georgi, N.; Costa, R.; Sher, P.; Reis, R.L.; van Blitterswijk, C.A.; Karperien, M.; Mano, J.F. Nanostructured 3D Constructs Based on Chitosan and Chondroitin Sulphate Multilayers for Cartilage Tissue Engineering. *PLoS ONE* **2013**, *8*. [CrossRef]
27. Iost, R.M.; Crespilho, F.N. Layer-by-Layer Self-Assembly and Electrochemistry: Applications in Biosensing and Bioelectronics. *Biosens. Bioelectron.* **2012**, *31*. [CrossRef]
28. Volodkin, D.; Klitzing, R.V.; Moehwald, H. Polyelectrolyte Multilayers: Towards Single Cell Studies. *Polymers* **2014**, *6*, 1502–1527. [CrossRef]
29. Zhao, Q.; Han, B.; Wang, Z.; Gao, C.; Peng, C.; Shen, J. Hollow Chitosan-Alginate Multilayer Microcapsules as Drug Delivery Vehicle: Doxorubicin Loading and in Vitro and in Vivo Studies. *Nanomed. Nanotechnol. Biol. Med.* **2007**, *3*, 63–74. [CrossRef]
30. De Cock, L.J.; De Wever, O.; Van Vlierberghe, S.; Vanderleyden, E.; Dubruel, P.; De Vos, F.; Vervaet, C.; Remon, J.P.; De Geest, B.G. Engineered (Hep/PARG) 2 Polyelectrolyte Capsules for Sustained Release of Bioactive TGF-B1. *Soft Matter* **2012**, *8*, 1146–1154. [CrossRef]
31. Volodkin, D.V.; Balabushevitch, N.G.; Sukhorukov, G.B.; Larionova, N.I. Model System for Controlled Protein Release: PH-Sensitive Polyelectrolyte Microparticles. *S.T.P. Pharma Sci.* **2003**, *13*, 163–170.
32. Paulraj, T.; Feoktistova, N.; Velk, N.; Uhlig, K.; Duschl, C.; Volodkin, D. Microporous Polymeric 3D Scaffolds Templated by the Layer-by-Layer Self-Assembly. *Macromol. Rapid Commun.* **2014**, *35*, 1408–1413. [CrossRef] [PubMed]
33. Vikulina, A.S.; Skirtach, A.G.; Volodkin, D. Hybrids of Polymer Multilayers, Lipids, and Nanoparticles: Mimicking the Cellular Microenvironment. *Langmuir* **2019**, *35*, 8565–8573. [CrossRef] [PubMed]
34. Srisang, S.; Nasongkla, N. Layer-by-Layer Dip Coating of Foley Urinary Catheters by Chlorhexidine-Loaded Micelles. *J. Drug Deliv. Sci. Technol.* **2019**, *49*, 235–242. [CrossRef]
35. Blell, R.; Lin, X.; Lindstro, T.; Ankerfors, M.; Pauly, M.; Felix, O.; Decher, G. Generating In-Plane Orientational Order in Multilayer Films Prepared by Spray-Assisted Layer-by-Layer Assembly. *ACS Nano* **2017**, *11*, 84–94. [CrossRef] [PubMed]
36. Schaaf, P.; Voegel, J.C.; Jierry, L.; Boulmedais, F. Spray-Assisted Polyelectrolyte Multilayer Buildup: From Step-by-Step to Single-Step Polyelectrolyte Film Constructions. *Adv. Mater.* **2012**, *24*, 1001–1016. [CrossRef]
37. Zhuk, A.; Selin, V.; Zhuk, I.; Belov, B.; Ankner, J.F.; Sukhishvili, S.A. Chain Conformation and Dynamics in Spin-Assisted Weak Polyelectrolyte Multilayers. *Langmuir* **2015**, *31*, 3889–3896. [CrossRef]
38. Yola, A.M.; Campbell, J.; Volodkin, D. Microfluidics Meets Layer-by-Layer Assembly for the Build-up of Polymeric Scaffolds. *Appl. Surf. Sci. Adv.* **2021**, *5*, 100091. [CrossRef]
39. Madaboosi, N.; Uhlig, K.; Jäger, M.S.; Möhwald, H.; Duschl, C.; Volodkin, D.V. Microfluidics as a Tool to Understand the Build-up Mechanism of Exponential-like Growing Films. *Macromol. Rapid Commun.* **2012**, *33*, 1775–1779. [CrossRef]
40. Manna, U.; Bharani, S.; Patil, S. Layer-by-Layer Self-Assembly of Modified Hyaluronic Acid/Chitosan Based on Hydrogen Bonding. *Biomacromolecules* **2009**, *10*, 2632–2639. [CrossRef]
41. Belbekhouche, S.; Charaabi, S.; Carbonnier, B. Glucose-Sensitive Capsules Based on Hydrogen-Bonded (Polyvinylpyrrolidone/Phenylboronic—Modified Alginate) System. *Colloids Surf. B Biointerfaces* **2019**, *177*, 416–424. [CrossRef] [PubMed]

42. Lee, H. Effects of Temperature, Salt Concentration, and the Protonation State on the Dynamics and Hydrogen-Bond Interactions of Polyelectrolyte Multilayers on Lipid Membranes. *Phys. Chem. Chem. Phys.* **2016**, *18*, 6691–6700. [CrossRef] [PubMed]
43. Schönhoff, M. Layered Polyelectrolyte Complexes: Physics of Formation and Molecular Properties. *J. Phys. Condens. Matter* **2003**, *15*. [CrossRef]
44. Vikulina, A.S.; Anissimov, Y.G.; Singh, P.; Prokopović, V.Z.; Uhlig, K.; Jaeger, M.S.; Von Klitzing, R.; Duschl, C.; Volodkin, D. Temperature Effect on the Build-up of Exponentially Growing Polyelectrolyte Multilayers. An Exponential-to-Linear Transition Point. *Phys. Chem. Chem. Phys.* **2016**, *18*, 7866–7874. [CrossRef]
45. Velk, N.; Uhlig, K.; Vikulina, A.; Duschl, C.; Volodkin, D. Mobility of Lysozyme in Poly(L-Lysine)/Hyaluronic Acid Multilayer Films. *Colloids Surf. B Biointerfaces* **2016**, *147*, 343–350. [CrossRef] [PubMed]
46. Multilayers, E.D.; Ochs, C.J.; Such, G.K.; Yan, Y.; Koeverden, M.P.V.; Caruso, F. Biodegradable Click Capsules with Engineered Drug-Loaded Multilayers. *ACS Nano* **2010**, *4*, 1653–1663.
47. Ye, S.; Wang, C.; Liu, X.; Tong, Z.; Ren, B.; Zeng, F. New Loading Process and Release Properties of Insulin from Polysaccharide Microcapsules Fabricated through Layer-by-Layer Assembly. *J. Control. Release* **2006**, *112*, 79–87. [CrossRef]
48. Caruso, F.; Trau, D.; Mo, H.; Renneberg, R.; Möhwald, H.; Renneberg, R. Enzyme Encapsulation in Layer-by-Layer Engineered Polymer Multilayer Capsules. *Langmuir* **2000**, *16*, 1485–1488. [CrossRef]
49. Wang, C.; He, C.; Tong, Z.; Liu, X.; Ren, B.; Zeng, F. Combination of Adsorption by Porous $CaCO_3$ Microparticles and Encapsulation by Polyelectrolyte Multilayer Films for Sustained Drug Delivery. *Int. J. Pharm.* **2006**, *308*, 160–167. [CrossRef] [PubMed]
50. Vergaro, V.; Papadia, P.; Leporatti, S.; De Pascali, S.A.; Fanizzi, F.P.; Ciccarella, G. Synthesis of Biocompatible Polymeric Nano-Capsules Based on Calcium Carbonate: A Potential Cisplatin Delivery System. *J. Inorg. Biochem.* **2015**, *153*, 284–292. [CrossRef] [PubMed]
51. Paini, M.; Aliakbarian, B.; Casazza, A.A.; Perego, P.; Ruggiero, C.; Pastorino, L. Chitosan/Dextran Multilayer Microcapsules for Polyphenol Co-Delivery. *Mater. Sci. Eng. C* **2015**, *46*, 374–380. [CrossRef]
52. Sun, L.; Xiong, X.; Zou, Q.; Ouyang, P.; Krastev, R. Controlled Heparinase-Catalyzed Degradation of Polyelectrolyte Multilayer Capsules with Heparin as Responsive Layer. *J. Appl. Polym. Sci.* **2017**, *134*. [CrossRef]
53. Lomova, M.V.; Brichkina, A.I.; Kiryukhin, M.V.; Vasina, E.N.; Pavlov, A.M.; Gorin, D.A.; Sukhorukov, G.B.; Antipina, M.N. Multilayer Capsules of Bovine Serum Albumin and Tannic Acid for Controlled Release by Enzymatic Degradation. *ACS Appl. Mater. Interfaces* **2015**, *7*, 11732–11740. [CrossRef]
54. Hu, S.H.; Tsai, C.H.; Liao, C.F.; Liu, D.M.; Chen, S.Y. Controlled Rupture of Magnetic Polyelectrolyte Microcapsules for Drug Delivery. *Langmuir* **2008**, *24*, 11811–11818. [CrossRef]
55. Zheng, C.; Ding, Y.; Liu, X.; Wu, Y.; Ge, L. Highly Magneto-Responsive Multilayer Microcapsules for Controlled Release of Insulin. *Int. J. Pharm.* **2014**, *475*, 17–24. [CrossRef]
56. Pavlov, A.M.; Saez, V.; Cobley, A.; Graves, J.; Sukhorukov, G.B.; Mason, T.J. Controlled Protein Release from Microcapsules with Composite Shells Using High Frequency Ultrasound—Potential for in Vivo Medical Use. *Soft Matter* **2011**, *7*, 4341–4347. [CrossRef]
57. Sato, K.; Yoshida, K.; Takahashi, S.; Anzai, J. ichi. PH- and Sugar-Sensitive Layer-by-Layer Films and Microcapsules for Drug Delivery. *Adv. Drug Deliv. Rev.* **2011**, *63*, 809–821. [CrossRef] [PubMed]
58. She, S.; Shan, B.; Li, Q.; Tong, W.; Gao, C. Phenomenon and Mechanism of Capsule Shrinking in Alkaline Solution Containing Calcium Ions. *J. Phys. Chem. B* **2012**, *116*, 13561–13567. [CrossRef] [PubMed]
59. Köhler, K.; Biesheuvel, P.M.; Weinkamer, R.; Möhwald, H.; Sukhorukov, G.B. Salt-Induced Swelling-to-Shrinking Transition in Polyelectrolyte Multilayer Capsules. *Phys. Rev. Lett.* **2006**, *97*, 188301. [CrossRef] [PubMed]
60. Sukhorukov, G.B.; Antipov, A.A.; Voigt, A.; Donath, E.; Möhwald, H. PH-Controlled Macromolecule Encapsulation in and Release from Polyelectrolyte Multilayer Nanocapsules. *Macromol. Rapid Commun.* **2001**, *22*, 44–46. [CrossRef]
61. Köhler, K.; Shchukin, D.G.; Möhwald, H.; Sukhorukov, G.B. Thermal Behavior of Polyelectrolyte Multilayer Microcapsules. 1. The Effect of Odd and Even Layer Number. *J. Phys. Chem. B* **2005**, *109*, 18250–18259. [CrossRef]
62. Feoktistova, N.; Rose, J.; Prokopović, V.Z.; Vikulina, A.S.; Skirtach, A.; Volodkin, D. Controlling the Vaterite $CaCO_3$ Crystal Pores. Design of Tailor-Made Polymer Based Microcapsules by Hard Templating. *Langmuir* **2016**, *32*, 4229–4238. [CrossRef] [PubMed]
63. Volodkin, D.V.; Larionova, N.I.; Sukhorukov, G.B. Protein Encapsulation via Porous $CaCO_3$ Microparticles Templating. *Biomacromolecules* **2004**, *5*, 1962–1972. [CrossRef]
64. Schmidt, S.; Volodkin, D. Microparticulate Biomolecules by Mild $CaCO_3$ Templating. *J. Mater. Chem. B* **2013**, *1*, 1210–1218. [CrossRef]
65. Abreu, S.; Carvalho, J.A.; Tedesco, A.C.; Junior, M.B.; Simioni, A.R. *Fabrication of Polyelectrolyte Microspheres Using Porous Manganese Carbonate as Sacri Fi Cial Template for Drug Delivery Application*; Cambridge University Press: Cambridge, UK, 2019. [CrossRef]
66. Marchenko, I.; Borodina, T.; Trushina, D.; Rassokhina, I.; Shirinian, V.; Zavarzin, I.; Gogin, A.; Bukreeva, T.; Marchenko, I.; Borodina, T.; et al. Mesoporous Particle-Based Microcontainers for Intranasal Delivery of Imidazopyridine Drugs. *J. Microencapsul.* **2019**, *35*, 657–666. [CrossRef] [PubMed]
67. Rodríguez-Ramos, A.; Marín-Caba, L.; Iturrioz-Rodríguez, N.; Padín-González, E.; García-Hevia, L.; Oliveira, T.M.; Corea-Duarte, M.A.; Fanarraga, M.L. Design of Polymeric and Biocompatible Delivery Systems by Dissolving Mesoporous Silica Templates. *Int. J. Mol. Sci.* **2020**, *21*, 9573. [CrossRef]

68. Castillo, R.R.; Lozano, D.; Vallet-Regí, M. Mesoporous Silica Nanoparticles as Carriers for Therapeutic Biomolecules. *Pharmaceutics* **2020**, *12*, 432. [CrossRef]
69. Nasab, N.A.; Kumleh, H.H.; Beygzadeh, M.; Teimourian, S.; Kazemzad, M. Delivery of Curcumin by a PH-Responsive Chitosan Mesoporous Silica Nanoparticles for Cancer Treatment. *Artif. Cells Nanomed. Biotechnol.* **2018**, *46*, 75–81. [CrossRef]
70. Trofimov, A.D.; Ivanova, A.A.; Zyuzin, M.V.; Timin, A.S. Porous Inorganic Carriers Based on Silica, Calcium Carbonate and Calcium Phosphate for Controlled/Modulated Drug Delivery: Fresh Outlook and Future Perspectives. *Pharmaceutics* **2018**, *10*, 167. [CrossRef] [PubMed]
71. Caruso, F. Hollow Capsule Processing through Colloidal Templating and Self-Assembly. *Chem.-A Eur. J.* **2000**, *6*, 413–419. [CrossRef]
72. Ye, S.; Wang, C.; Liu, X.; Tong, Z. Multilayer Nanocapsules of Polysaccharide Chitosan and Alginate through Layer-by-Layer Assembly Directly on PS Nanoparticles for Release. *J. Biomater. Sci. Polym. Ed.* **2005**, *16*. [CrossRef] [PubMed]
73. Balabushevich, N.G.; Larionova, N.I. Protein-Loaded Microspheres Prepared by Sequential Adsorption of Dextran Sulphate and Protamine on Melamine Formaldehyde Core. *J. Microencapsul.* **2009**, *26*, 571–579. [CrossRef] [PubMed]
74. Georgieva, R.; Moya, S.; Donath, E.; Baumer, H. Permeability and Conductivity of Red Blood Cell Templated Polyelectrolyte Capsules Coated with Supplementary Layers. *Langmuir* **2004**, *20*, 1895–1900. [CrossRef]
75. Estrela-lopis, I.; Leporatti, S.; Typlt, E.; Clemens, D.; Donath, E. Small Angle Neutron Scattering Investigations (SANS) of Polyelectrolyte Multilayer Capsules Templated on Human Red Blood Cells. *Langmuir* **2007**, *23*, 7209–7215. [CrossRef]
76. Neu, B.; Voigt, A.; Mitlöhner, R.; Leporatti, S.; Gao, C.Y.; Donath, E.; Kiesewetter, H.; Möhwald, H.; Meiselman, H.J.; Bäumler, H. Biological Cells as Templates for Hollow Microcapsules. *J. Microencapsul.* **2001**, *18*, 385–395. [CrossRef] [PubMed]
77. Lederer, F.L.; Günther, T.J.; Weinert, U.; Raff, J.; Pollmann, K. Development of Functionalised Polyelectrolyte Capsules Using Filamentous Escherichia Coli Cells. *Microb. Cell Fact.* **2012**, *11*, 163. [CrossRef] [PubMed]
78. Jeannot, L.; Bell, M.; Ashwell, R.; Volodkin, D.; Vikulina, A.S. Internal Structure of Matrix-Type Multilayer Capsules Templated on Porous Vaterite $CaCO_3$ Crystals as Probed by Staining with a Fluorescence Dye. *Micromachines* **2018**, *9*, 547. [CrossRef] [PubMed]
79. Campbell, J.; Abnett, J.; Kastania, G.; Volodkin, D.; Vikulina, A.S. Which Biopolymers Are Better for the Fabrication of Multilayer Capsules? A Comparative Study Using Vaterite $CaCO_3$ as Templates. *ACS Appl. Mater. Interfaces* **2021**, *13*, 3259–3269. [CrossRef] [PubMed]
80. Volodkin, D.V.; Petrov, A.I.; Prevot, M.; Sukhorukov, G.B. Matrix Polyelectrolyte Microcapsules: New System for Macromolecule Encapsulation. *Langmuir* **2004**, *20*, 3398–3406. [CrossRef] [PubMed]
81. Gao, C.; Moya, S.; Lichtenfeld, H.; Casoli, A.; Fiedler, H.; Donath, E.; Möhwald, H. The Decomposition Process of Melamine Formaldehyde Cores: The Key Step in the Fabrication of Ultrathin Polyelectrolyte Multilayer Capsules. *Macromol. Mater. Eng.* **2001**, *286*, 355–361. [CrossRef]
82. Balabushevich, N.G.; Tiourina, O.P.; Volodkin, D.V.; Larionova, N.I.; Sukhorukov, G.B. Loading the Multilayer Dextran Sulfate/Protamine Microsized Capsules with Peroxidase. *Biomacromolecules* **2003**, *4*, 1191–1197. [CrossRef]
83. Volodkin, D. $CaCO_3$ Templated Micro-Beads and -Capsules for Bioapplications. *Adv. Colloid Interface Sci.* **2014**, *207*, 306–324. [CrossRef] [PubMed]
84. Trushina, D.B.; Bukreeva, T.V.; Antipina, M.N. Size-Controlled Synthesis of Vaterite Calcium Carbonate by the Mixing Method: Aiming for Nanosized Particles. *Cryst. Growth Des.* **2016**, *16*, 1311–1319. [CrossRef]
85. Vikulina, A.; Voronin, D.; Fakhrullin, R.; Vinokurov, V.; Volodkin, D. Naturally Derived Nano- And Micro-Drug Delivery Vehicles: Halloysite, Vaterite and Nanocellulose. *New J. Chem.* **2020**, *44*, 5638–5655. [CrossRef]
86. Balabushevich, N.G.; Guerenu, A.V.L.D.; Feoktistova, N.A.; Skirtach, A.G.; Volodkin, D. Protein-Containing Multilayer Capsules by Templating on Mesoporous $CaCO_3$ Particles: POST- and PRE-Loading Approaches. *Macromol. Biosci.* **2016**, *16*, 95–105. [CrossRef] [PubMed]
87. Balabushevich, N.G.; Guerenu, A.V.L.D.; Feoktistova, N.A.; Volodkin, D. Protein Loading into Porous $CaCO_3$ Microspheres: Adsorption Equilibrium and Bioactivity Retention. *Phys. Chem. Chem. Phys.* **2015**, *17*, 2523–2530. [CrossRef] [PubMed]
88. Feoktistova, N.A.; Vikulina, A.S.; Balabushevich, N.G.; Skirtach, A.G.; Volodkin, D. Bioactivity of Catalase Loaded into Vaterite $CaCO_3$ Crystals via Adsorption and Co-Synthesis. *Mater. Des.* **2020**, *185*, 108223. [CrossRef]
89. Pechenkin, M.A.; Möhwald, H.; Volodkin, D.V. PH- and Salt-Mediated Response of Layer-by-Layer Assembled PSS/PAH Microcapsules: Fusion and Polymer Exchange. *Soft Matter* **2012**, *8*, 8659–8665. [CrossRef]
90. Tong, W.; Dong, W.; Gao, C.; Möhwald, H. Charge-Controlled Permeability of Polyelectrolyte Microcapsules. *J. Phys. Chem. B* **2005**, *109*, 13159–13165. [CrossRef] [PubMed]
91. Hamid, M.; Rehman, K.; Chen, S. Natural and Synthetic Polymers as Drug Carriers for Delivery of Therapeutic Proteins. *Polym. Rev.* **2015**, *55*, 371–406. [CrossRef]
92. Silva, J.M.; Reis, R.L.; Mano, J.F. Biomimetic Extracellular Environment Based on Natural Origin Polyelectrolyte Multilayers. *Small* **2016**, *12*, 4308–4342. [CrossRef]
93. Mtibe, A.; Motloung, M.P.; Bandyopadhyay, J.; Ray, S.S. Synthetic Biopolymers and Their Composites: Advantages and Limitations—An Overview. *Macromol. Rapid Commun.* **2021**, *42*, 2100130. [CrossRef]
94. Gopi, S.; Amalraj, A.; Sukumaran, N.P.; Haponiuk, J.; Thomas, S. Biopolymers and Their Composites for Drug Delivery: A Brief Review. *Macromol. Symp.* **2018**, *380*, 180014. [CrossRef]

95. Sun, L.; Xiong, X.; Zou, Q.; Ouyang, P.; Burkhardt, C.; Krastev, R. Design of Intelligent Chitosan/Heparin Hollow Microcapsules for Drug Delivery. *J. Appl. Polym. Sci.* **2017**, *134*. [CrossRef]
96. Thomas, M.B.; Radhakrishnan, K.; Gnanadhas, D.P.; Chakravortty, D.; Raichur, A.M. Intracellular Delivery of Doxorubicin Encapsulated in Novel PH-Responsive Chitosan/Heparin Nanocapsules. *Int. J. Nanomed.* **2013**, *8*, 267–273. [CrossRef]
97. Caridade, S.G.; Monge, C.; Gilde, F.; Boudou, T.; Mano, J.F.; Picart, C. Free-Standing Polyelectrolyte Membranes Made of Chitosan and Alginate. *Biomacromolecules* **2013**, *14*, 1653–1660. [CrossRef]
98. Silva, J.M.; Duarte, A.R.C.; Custódio, C.A.; Sher, P.; Neto, A.I.; Pinho, A.C.M.; Fonseca, J.; Reis, R.L.; Mano, J.F. Nanostructured Hollow Tubes Based on Chitosan and Alginate Multilayers. *Adv. Healthc. Mater.* **2014**, *3*, 433–440. [CrossRef] [PubMed]
99. Yan, S.; Zhu, J.; Wang, Z.; Yin, J.; Zheng, Y.; Chen, X. Layer-by-Layer Assembly of Poly(L-Glutamic Acid)/Chitosan Microcapsules for High Loading and Sustained Release of 5-Fluorouracil. *Eur. J. Pharm. Biopharm.* **2011**, *78*, 336–345. [CrossRef]
100. Bünger, C.M.; Gerlach, C.; Freier, T.; Schmitz, K.P.; Pilz, M.; Werner, C.; Jonas, L.; Schareck, W.; Hopt, U.T.; De Vos, P. Biocompatibility and Surface Structure of Chemically Modified Immunoisolating Alginate-PLL Capsules. *J. Biomed. Mater. Res.-Part A* **2003**, *67*, 1219–1227. [CrossRef]
101. Sousa, F.; Kreft, O.; Sukhorukov, G.B.; Möhwald, H.; Kokol, V. Biocatalytic Response of Multi-Layer Assembled Collagen/Hyaluronic Acid Nanoengineered Capsules. *J. Microencapsul.* **2014**, *31*, 270–276. [CrossRef]
102. Radhakrishnan, K.; Tripathy, J.; Raichur, A.M. Dual Enzyme Responsive Microcapsules Simulating an "OR" Logic Gate for Biologically Triggered Drug Delivery Applications. *Chem. Commun.* **2013**, *49*, 5390–5392. [CrossRef]
103. Vladimirov, G.K.; Vikulina, A.S.; Volodkin, D.; Vladimirov, Y.A. Structure of the Complex of Cytochrome c with Cardiolipin in Non-Polar Environment. *Chem. Phys. Lipids* **2018**, *214*, 35–45. [CrossRef]
104. Niepel, M.S.; Ekambaram, B.K.; Schmelzer, C.E.H.; Groth, T. Polyelectrolyte Multilayers of Poly (l-Lysine) and Hyaluronic Acid on Nanostructured Surfaces Affect Stem Cell Response. *Nanoscale* **2019**, *11*, 2878–2891. [CrossRef]
105. Vikulna, A.S.; Alekseev, A.V.; Proskurnina, E.V.; Vladimirova, G.A.; Vladimirov, Y.A. The Complexs of Cytochrome c with Cardiolipin in Non-Polar Environment. *Biochem. Mosc.* **2015**, *80*, 1298–1302. [CrossRef]
106. Niepel, M.S.; Almouhanna, F.; Ekambaram, B.K.; Menzel, M.; Heilmann, A.; Groth, T. Cross-Linking Multilayers of Poly-l-Lysine and Hyaluronic Acid: Effect on Mesenchymal Stem Cell Behavior. *Int. J. Artif. Organs* **2018**, *41*, 223–235. [CrossRef] [PubMed]
107. Schüler, C.; Caruso, F. Decomposable Hollow Biopolymer-Based Capsules. *Biomacromolecules* **2001**, *2*, 921–926. [CrossRef] [PubMed]
108. Pinheiro, A.C.; Bourbon, A.I.; Cerqueira, M.A.; Maricato, É.; Nunes, C.; Coimbra, M.A.; Vicente, A.A. Chitosan/Fucoidan Multilayer Nanocapsules as a Vehicle for Controlled Release of Bioactive Compounds. *Carbohydr. Polym.* **2015**, *115*, 1–9. [CrossRef] [PubMed]
109. Parakhonskiy, B.V.; Yashchenok, A.M.; Konrad, M.; Skirtach, A.G. Colloidal Micro- and Nano-Particles as Templates for Polyelectrolyte Multilayer Capsules. *Adv. Colloid Interface Sci.* **2014**, *207*, 253–264. [CrossRef]
110. Itoh, Y.; Matsusaki, M.; Kida, T.; Akashi, M. Locally Controlled Release of Basic Fibroblast Growth Factor from Multilayered Capsules. *Biomacromolecules* **2008**, *9*, 2202–2206. [CrossRef]
111. Radhakrishnan, K.; Thomas, M.B.; Pulakkat, S.; Gnanadhas, D.P.; Chakravortty, D.; Raichur, A.M. Stimuli-Responsive Protamine-Based Biodegradable Nanocapsules for Enhanced Bioavailability and Intracellular Delivery of Anticancer Agents. *J. Nanopart. Res.* **2015**, *17*, 341. [CrossRef]
112. Imoto, T.; Kida, T.; Matsusaki, M.; Akashi, M. Preparation and Unique PH-Responsive Properties of Novel Biodegradable Nanocapsules Composed of Poly (g-Glutamic Acid) and Chitosan as Weak Polyelectrolytes. *Macromol. Biosci.* **2010**, *10*, 271–277. [CrossRef] [PubMed]
113. Ji, F.; Li, J.; Qin, Z.; Yang, B.; Zhang, E.; Dong, D.; Wang, J.; Wen, Y.; Yao, F. Engineering Pectin-Based Hollow Nanocapsules for Delivery of Anticancer Drug. *Carbohydr. Polym.* **2017**, *177*, 86–96. [CrossRef] [PubMed]
114. Mansour, O.; Peker, T.; Hamadi, S.; Belbekhouche, S. Glucose-Responsive Capsules Based on (Phenylboronic-Modified Poly(Lysine)/ Alginate) System. *Eur. Polym. J.* **2019**, *120*, 109248. [CrossRef]
115. Tiourina, O.P.; Sukhorukov, G.B. Multilayer Alginate/Protamine Microsized Capsules: Encapsulation of α-Chymotrypsin and Controlled Release Study. *Int. J. Pharm.* **2002**, *242*, 155–161. [CrossRef]
116. Yuxi, L.; Jing, Y.; Ziqi, Z.; Junjie, L.; Rui, Z.; Fanglian, Y. Formation and Characterization of Natural Polysaccharide Hollow Nanocapsules via Template Layer-by-Layer Self-Assembly. *J. Colloid Interface Sci.* **2012**, *379*, 130–140. [CrossRef]
117. Itoh, Y.; Matsusaki, M.; Kida, T.; Akashi, M. Preparation of Biodegradable Hollow Nanocapsules by Silica Template Method. *Chem. Lett.* **2004**, *33*, 1552–1553. [CrossRef]
118. Gnanadhas, D.P.; Ben Thomas, M.; Elango, M.; Raichur, A.M.; Chakravortty, D. Chitosan-Dextran Sulphate Nanocapsule Drug Delivery System as an Effective Therapeutic against Intraphagosomal Pathogen Salmonella. *J. Antimicrob. Chemother.* **2013**, *68*, 2576–2586. [CrossRef]
119. Petrov, A.I.; Volodkin, D.V.; Sukhorukov, G.B. Protein-Calcium Carbonate Coprecipitation: A Tool for Protein Encapsulation. *Biotechnol. Prog.* **2005**, *21*, 918–925. [CrossRef]
120. Balabushevich, N.G.; Kovalenko, E.A.; Le-Deygen, I.M.; Filatova, L.Y.; Volodkin, D.; Vikulina, A.S. Hybrid $CaCO_3$-Mucin Crystals: Effective Approach for Loading and Controlled Release of Cationic Drugs. *Mater. Des.* **2019**, *182*, 108020. [CrossRef]
121. Vikulina, A.S.; Feoktistova, N.A.; Balabushevich, N.G.; Skirtach, A.G.; Volodkin, D. The Mechanism of Catalase Loading into Porous Vaterite $CaCO_3$ Crystals by Co-Synthesis. *Phys. Chem. Chem. Phys.* **2018**, *20*, 8822–8831. [CrossRef] [PubMed]

122. Volodkin, D.V.; Schmidt, S.; Fernandes, P.; Larionova, N.I.; Sukhorukov, G.B.; Duschl, C.; Möhwald, H.; Von Klitzing, R. One-Step Formulation of Protein Microparticles with Tailored Properties: Hard Templating at Soft Conditions. *Adv. Funct. Mater.* **2012**, *22*, 1914–1922. [CrossRef]
123. Vikulina, A.; Webster, J.; Voronin, D.; Ivanov, E.; Fakhrullin, R.; Vinokurov, V.; Volodkin, D. Mesoporous Additive-Free Vaterite $CaCO_3$ Crystals of Untypical Sizes: From Submicron to Giant. *Mater. Des.* **2021**, *197*, 109220. [CrossRef]
124. Binevski, P.V.; Balabushevich, N.G.; Uvarova, V.I.; Vikulina, A.S.; Volodkin, D. Bio-Friendly Encapsulation of Superoxide Dismutase into Vaterite $CaCO_3$ Crystals. Enzyme Activity, Release Mechanism, and Perspectives for Ophthalmology. *Colloids Surf. B Biointerfaces* **2019**, *181*, 437–449. [CrossRef] [PubMed]
125. Shi, P.; Luo, S.; Voit, B.; Appelhans, D.; Zan, X. A Facile and Efficient Strategy to Encapsulate the Model Basic Protein Lysozyme into Porous $CaCO_3$. *J. Mater. Chem. B* **2018**, *6*, 4205–4215. [CrossRef] [PubMed]
126. Campbell, J.; Kastania, G.; Volodkin, D. Encapsulation of Low-Molecular-Weight Drugs into Polymer Multilayer Capsules Templated on Vaterite $CaCO_3$ Crystals. *Micromachines* **2020**, *11*, 717. [CrossRef] [PubMed]
127. Ma, X.; Yuan, S.; Yang, L.; Li, L.; Zhang, X.; Su, C.; Wang, K. Fabrication and Potential Applications of $CaCO_3$—Lentinan Hybrid Materials with Hierarchical Composite Pore Structure Obtained by Self-Assembly of Nanoparticles. *CrystEngComm* **2013**, 8288–8299. [CrossRef]
128. Han, B.; Baiyong, S.; Wang, Z.; Shi, M.; Li, H.; Peng, C.; Zhao, Q.; Gao, C. Layered Microcapsules for Danorubicin Loading and Release as Well as in Vitro and in Vivo Studies. *Polym. Adv. Technol.* **2008**, *19*, 36–46. [CrossRef]
129. Balabushevich, N.G.; Kovalenko, E.A.; Mikhalchik, E.V.; Filatova, L.Y.; Volodkin, D.; Vikulina, A.S. Mucin Adsorption on Vaterite $CaCO_3$ Microcrystals for the Prediction of Mucoadhesive Properties. *J. Colloid Interface Sci.* **2019**, *545*, 330–339. [CrossRef]
130. Demina, P.A.; Voronin, D.V.; Lengert, E.V.; Abramova, A.M.; Atkin, V.S.; Nabatov, B.V.; Semenov, A.P.; Shchukin, D.G.; Bukreeva, T.V. Freezing-Induced Loading of TiO2 into Porous Vaterite Microparticles: Preparation of $CaCO_3/TiO_2$ Composites as Templates to Assemble UV-Responsive Microcapsules for Wastewater Treatment. *ACS Omega* **2020**, *5*, 4115–4124. [CrossRef]
131. Gusliakova, O.; Atochina-vasserman, E.N.; Sindeeva, O.; Sindeev, S.; Pinyaev, S.; Pyataev, N.; Revin, V.; Sukhorukov, G.B.; Gorin, D.; Gow, A.J. Use of Submicron Vaterite Particles Serves as an Effective Delivery Vehicle to the Respiratory Portion of the Lung. *Front. Pharmacol.* **2018**, *9*. [CrossRef]
132. Bukreeva, T.V.; Marchenko, I.V.; Borodina, T.N.; Degtev, I.V.; Sitnikov, S.L.; Moiseeva, Y.V.; Gulyaeva, N.V.; Kovalchuk, M.V. Calcium Carbonate and Titanium Dioxide Particles as a Basis for Container Fabrication for Brain Delivery of Compounds. *Phys. Chem.* **2011**, *440*, 165–167. [CrossRef]
133. Borodina, T.N.; Trushina, D.B.; Marchenko, I.V.; Bukreeva, T.V. Calcium Carbonate-Based Mucoadhesive Microcontainers for Intranasal Delivery of Drugs Bypassing the Blood–Brain Barrier. *Bionanoscience* **2016**, *6*, 261–268. [CrossRef]
134. Gusliakova, O.; Verkhovskii, R.; Abalymov, A.; Lengert, E.; Kozlova, A.; Atkin, V.; Nechaeva, O.; Morrison, A.; Tuchin, V.; Svenskaya, Y. Transdermal Platform for the Delivery of the Antifungal Drug Naftifine Hydrochloride Based on Porous Vaterite Particles. *Mater. Sci. Eng. C* **2021**, *119*, 111428. [CrossRef]
135. Ferreira, A.M.; Vikulina, A.S.; Volodkin, D. $CaCO_3$ Crystals as Versatile Carriers for Controlled Delivery of Antimicrobials. *J. Control. Release* **2020**, *328*, 470–489. [CrossRef]
136. Ali, F.; Bousserrhine, N.; Alphonse, V.; Michely, L.; Belbekhouche, S. Antibiotic Loading and Development of Antibacterial Capsules by Using Porous $CaCO_3$ Microparticles as Starting Material. *Int. J. Pharm.* **2020**, *579*, 119175. [CrossRef]
137. Konopacka-Łyskawa, D. Synthesis Methods and Favorable Conditions for Spherical Vaterite Precipitation: A Review. *Crystals* **2019**, *9*, 223. [CrossRef]
138. Rodriguez-Blanco, J.D.; Shaw, S.; Benning, L.G. The Kinetics and Mechanisms of Amorphous Calcium Carbonate (ACC) Crystallization to Calcite, via Vaterite. *Nanoscale* **2011**, *3*, 265–271. [CrossRef]
139. Ismaiel Saraya, M.E.-S.; Rokbaa, H.H.A.E.-L. Formation and Stabilization of Vaterite Calcium Carbonate by Using Natural Polysaccharide. *Adv. Nanopart.* **2017**, *6*, 158–182. [CrossRef]
140. Sergeeva, A.; Sergeev, R.; Lengert, E.; Zakharevich, A.; Parakhonskiy, B.; Gorin, D.; Sergeev, S.; Volodkin, D. Composite Magnetite and Protein Containing $CaCO_3$ Crystals. External Manipulation and Vaterite $\rightarrow$ Calcite Recrystallization-Mediated Release Performance. *ACS Appl. Mater. Interfaces* **2015**, *7*, 21315–21325. [CrossRef] [PubMed]
141. She, Z.; Antipina, M.N.; Li, J.; Sukhorukov, G.B. Mechanism of Protein Release from Polyelectrolyte Multilayer Microcapsules. *Biomacromolecules* **2010**, *11*, 1241–1247. [CrossRef]
142. Shchukin, D.G.; Patel, A.A.; Sukhorukov, G.B.; Lvov, Y.M. Nanoassembly of Biodegradable Microcapsules for DNA Encasing. *J. Am. Chem. Soc.* **2004**, *126*, 3374–3375. [CrossRef] [PubMed]
143. Zhao, Q.; Li, B. PH-Controlled Drug Loading and Release from Biodegradable Microcapsules. *Nanomed. Nanotechnol. Biol. Med.* **2008**, *4*, 302–310. [CrossRef] [PubMed]
144. De Geest, B.G.; Vandenbroucke, R.E.; Guenther, A.M.; Sukhorukov, G.B.; Hennink, W.E.; Sanders, N.N.; Demeester, J.; De Smedt, S.C. Intracellularly Degradable Polyelectrolyte Microcapsules. *Adv. Mater.* **2006**, *18*, 1005–1009. [CrossRef]
145. She, Z.; Wang, C.; Li, J.; Sukhorukov, G.B.; Antipina, M.N. Encapsulation of Basic Fibroblast Growth Factor by Polyelectrolyte Multilayer Microcapsules and Its Controlled Release for Enhancing Cell Proliferation. *Biomacromolecules* **2012**, *13*, 2174–2180. [CrossRef]

146. Trushina, D.B.; Akasov, R.A.; Khovankina, A.V.; Borodina, T.N.; Bukreeva, T.V.; Markvicheva, E.A. Doxorubicin-Loaded Biodegradable Capsules: Temperature Induced Shrinking and Study of Cytotoxicity in Vitro. *J. Mol. Liq.* **2019**, *284*, 215–224. [CrossRef]
147. Strehlow, V.; Lessig, J.; Göse, M.; Reibetanz, U. Development of LbL Biopolymer Capsules as a Delivery System for the Multilayer-Assembled Anti-Inflammatory Substance A1-Antitrypsin. *J. Mater. Chem. B* **2013**, *1*, 3633–3643. [CrossRef] [PubMed]
148. Jin, Y.; Liu, W.; Wang, J.; Fang, J.; Gao, H. (Protamine/Dextran Sulfate)6 Microcapules Templated on Biocompatible Calcium Carbonate Microspheres. *Colloids Surf. A Physicochem. Eng. Asp.* **2009**, *342*, 40–45. [CrossRef]
149. Szarpak, A.; Cui, D.; Dubreuil, F.; De Geest, B.G.; De Cock, L.J.; Picart, C.; Auzély-Velty, R. Designing Hyaluronic Acid-Based Layer-by-Layer Capsules as a Carrier for Intracellular Drug Delivery. *Biomacromolecules* **2010**, *11*, 713–720. [CrossRef]
150. Chen, J.; Liang, Y.; Liu, W.; Huang, J.; Chen, J. Fabrication of Doxorubicin and Heparin Co-Loaded Microcapsules for Synergistic Cancer Therapy. *Int. J. Biol. Macromol.* **2014**, *69*, 554–560. [CrossRef]
151. De Cock, L.J.; Lenoir, J.; De Koker, S.; Vermeersch, V.; Skirtach, A.G.; Dubruel, P.; Adriaens, E.; Vervaet, C.; Remon, J.P.; De Geest, B.G. Mucosal Irritation Potential of Polyelectrolyte Multilayer Capsules. *Biomaterials* **2011**, *32*, 1967–1977. [CrossRef]
152. Roy, S.; Elbaz, N.M.; Parak, W.J.; Feliu, N. Biodegradable Alginate Polyelectrolyte Capsules As Plausible Biocompatible Delivery Carriers. *ACS Appl. Bio Mater.* **2019**, *2*, 3245–3256. [CrossRef]
153. Costa, R.R.; Girotti, A.; Santos, M.; Arias, F.J.; Mano, J.F.; Rodríguez-cabello, J.C. Cellular Uptake of Multilayered Capsules Produced with Natural and Genetically Engineered Biomimetic Macromolecules. *Acta Biomater.* **2014**, *10*, 2653–2662. [CrossRef] [PubMed]
154. Hong, H.; Murney, R.; Yakovlev, N.L.; Novoselova, M.V.; Hui, S.; Roy, N.; Singh, H.; Sukhorukov, G.B.; Haigh, B.; Kiryukhin, M.V. Protein-Tannic Acid Multilayer Films: A Multifunctional Material for Microencapsulation of Food-Derived Bioactives. *J. Colloid Interface Sci.* **2017**, *505*, 332–340. [CrossRef]
155. Kilic, E.; Novoselova, M.V.; Lim, S.H.; Pyataev, N.A.; Pinyaev, S.I.; Kulikov, O.A.; Sindeeva, O.A.; Mayorova, O.A.; Murney, R.; Antipina, M.N.; et al. Formulation for Oral Delivery of Lactoferrin Based on Bovine Serum Albumin and Tannic Acid Multilayer Microcapsules. *Sci. Rep.* **2017**, *7*. [CrossRef] [PubMed]
156. Novoselova, M.V.; Voronin, D.V.; Abakumova, T.O.; Demina, P.A.; Petrov, A.V.; Petrov, V.V.; Zatsepin, T.S.; Sukhorukov, G.B. Focused Ultrasound-Mediated Fluorescence of Composite Microcapsules Loaded with Magnetite Nanoparticles: In Vitro and in Vivo Study. *Colloids Surf. B Biointerfaces* **2019**, *181*, 680–687. [CrossRef]
157. Tong, W.; Gao, C.; Möhwald, H. PH-Responsive Protein Microcapsules Fabricated via Glutaraldehyde Mediated Covalent Layer-by-Layer Assembly. *Colloid Polym. Sci.* **2008**, *286*, 1103–1109. [CrossRef]
158. Duan, L.; He, Q.; Yan, X.; Cui, Y.; Wang, K.; Li, J. Hemoglobin Protein Hollow Shells Fabricated through Covalent Layer-by-Layer Technique. *Biochem. Biophys. Res. Commun.* **2007**, *354*, 357–362. [CrossRef] [PubMed]
159. Microcapsules, S.; Borodina, T.; Markvicheva, E.; Kunizhev, S.; Sukhorukov, G.B.; Kreft, O. Controlled Release of DNA from Self-Degrading Microcapsules. *Macromol. Mater. Eng.* **2007**, *28*, 1894–1899. [CrossRef]
160. Huang, W.; Zhang, T.; Shi, P.; Yang, D.; Luo, S. The Construction and Effect of Physical Properties on Intracellular Drug Delivery of Poly(Amino Acid) Capsules. *Colloids Surf. B Biointerfaces* **2019**, *177*, 178–187. [CrossRef]
161. Yu, A.; Wang, Y.; Barlow, E.; Caruso, F. Mesoporous Silica Particles as Templates for Preparing Enzyme-Loaded Biocompatible Microcapsules. *Adv. Mater.* **2005**, *17*, 1737–1741. [CrossRef]
162. Zhang, S.; Xing, M.; Li, B. Capsule-Integrated Polypeptide Multilayer Films for Effective PH-Responsive Multiple Drug Co-Delivery. *ACS Appl. Mater. Interfaces* **2018**, *10*, 44267–44278. [CrossRef]
163. Wu, X.; Zhao, S.; Zhang, J.; Wu, P.; Peng, C. Encapsulation of EV71-Specific IgY Antibodies by Multilayer Polypeptide Microcapsules and Its Sustained Release for Inhibiting Enterovirus 71 Replication. *RSC Adv.* **2014**, *4*, 14603–14612. [CrossRef]
164. Liao, W.-C.; Lu, C.-H.; Hartmann, R.; Wang, F.; Sohn, Y.S.; Parak, W.J.; Willner, I. Adenosine Triphosphate-Triggered Release of Macromolecular and Nanoparticle Loads from Aptamer/DNA-Cross-Linked Microcapsules. *ACS Nano* **2015**, *9*, 9078–9086. [CrossRef]
165. Fujii, A.; Maruyama, T.; Ohmukai, Y.; Kamio, E.; Sotani, T.; Matsuyama, H. Cross-Linked DNA Capsules Templated on Porous Calcium Carbonate Microparticles. *Colloids Surf. A Physicochem. Eng. Asp.* **2010**, *356*, 126–133. [CrossRef]
166. Fujii, A.; Ohmukai, Y.; Maruyama, T.; Sotani, T.; Matsuyama, H. Preparation of DNA Capsules Cross-Linked through NeutrAvidin—Biotin Interaction. *Colloids Surf. A Physicochem. Eng. Asp.* **2011**, *384*, 529–535. [CrossRef]
167. Zhao, D.; Wang, C.; Zhuo, R.; Cheng, S. Biointerfaces Modification of Nanostructured Calcium Carbonate for Efficient Gene Delivery. *Colloids Surf. B Biointerfaces* **2014**, *118*, 111–116. [CrossRef] [PubMed]
168. Zhao, P.; Wu, S.; Cheng, Y.; You, J.; Chen, Y.; Li, M.; He, C.; Zhang, X.; Yang, T.; Lu, Y.; et al. MiR-375 Delivered by Lipid-Coated Doxorubicin-Calcium Carbonate Nanoparticles Overcomes Chemoresistance in Hepatocellular Carcinoma. *Nanomed. Nanotechnol. Biol. Med.* **2017**, *13*, 2507–2516. [CrossRef] [PubMed]
169. Svenskaya, Y.I.; Fattah, H.; Inozemtseva, O.A.; Ivanova, A.G.; Shtykov, S.N.; Gorin, D.A.; Parakhonskiy, B.V. Key Parameters for Size- and Shape-Controlled Synthesis of Vaterite Particles. *Cryst. Growth Des.* **2018**, *18*, 331–337. [CrossRef]
170. Tarakanchikova, Y.; Muslimov, A.; Sergeev, I.; Lepik, K.; Yolshin, N.; Goncharenko, A.; Vasilyev, K.; Eliseev, I.; Bukatin, A.; Sergeev, V.; et al. A Highly Efficient and Safe Gene Delivery Platform Based on Polyelectrolyte Core-Shell Nanoparticles for Hard-to-Transfect Clinically Relevant Cell Types. *J. Mater. Chem. B* **2020**, *8*, 9576–9588. [CrossRef] [PubMed]

171. Kastania, G.; Campbell, J.; Mitford, J.; Volodkin, D. Polyelectrolyte Multilayer Capsule (PEMC)-Based Scaffolds for Tissue Engineering. *Micromachines* **2020**, *11*, 797. [CrossRef]
172. Trushina, D.B.; Bukreeva, T.V.; Borodina, T.N.; Belova, D.D.; Belyakov, S.; Antipina, M.N. Heat-Driven Size Reduction of Biodegradable Polyelectrolyte Multilayer Hollow Capsules Assembled on $CaCO_3$ Template. *Colloids Surf. B Biointerfaces* **2018**, *170*, 312–321. [CrossRef]
173. Van der Meeren, L.; Li, J.; Konrad, M.; Skirtach, A.G.; Volodkin, D.; Parakhonskiy, B.V. Temperature Window for Encapsulation of an Enzyme into Thermally Shrunk, $CaCO_3$ Templated Polyelectrolyte Multilayer Capsules. *Macromol. Biosci.* **2020**, *20*, 2000081. [CrossRef] [PubMed]
174. Trushina, D.B.; Burova, A.S.; Borodina, T.N.; Soldatov, M.A.; Klochko, T.Y.; Bukreeva, T.V. Thermo-Induced Shrinking of "Dextran Sulfate/Polyarginine" Capsules with Magnetic Nanoparticles in the Shell. *Colloid J.* **2018**, *80*, 710–715. [CrossRef]
175. Novoselova, M.V.; Loh, H.M.; Trushina, D.B.; Ketkar, A.; Abakumova, T.O.; Zatsepin, T.S.; Kakran, M.; Brzozowska, A.M.; Lau, H.H.; Gorin, D.A.; et al. Biodegradable Polymeric Multilayer Capsules for Therapy of Lung Cancer. *ACS Appl. Mater. Interfaces* **2020**, *12*, 5610–5623. [CrossRef] [PubMed]
176. Volodkin, D. Colloids of Pure Proteins by Hard Templating. *Colloid Polym. Sci.* **2014**, *292*, 1249–1259. [CrossRef]
177. Schmidt, S.; Uhlig, K.; Duschl, C.; Volodkin, D. Stability and Cell Uptake of Calcium Carbonate Templated Insulin Microparticles. *Acta Biomater.* **2014**, *10*, 1423–1430. [CrossRef] [PubMed]
178. Balabushevitch, N.G.; Sukhorukov, G.B.; Moroz, N.A.; Volodkin, D.V.; Larionova, N.I.; Donath, E.; Mohwald, H. Encapsulation of Proteins by Layer-by-Layer Adsorption of Polyelectrolytes onto Protein Aggregates: Factors Regulating the Protein Release. *Biotechnol. Bioeng.* **2001**, *76*, 207–213. [CrossRef] [PubMed]
179. Volodkin, D.V.; Balabushevitch, N.G.; Sukhorukov, G.B.; Larionova, N.I. Inclusion of Proteins into Polyelectrolyte Microparticles by Alternative Adsorption of Polyelectrolytes on Protein Aggregates. *Biochem.* **2003**, *68*, 236–241.
180. Xiong, Y.; Steffen, A.; Andreas, K.; Müller, S.; Sternberg, N.; Georgieva, R.; Bäumler, H. Hemoglobin-Based Oxygen Carrier Microparticles: Synthesis, Properties, and in Vitro and in Vivo Investigations. *Biomacromolecules* **2012**, *13*, 3292–3300. [CrossRef] [PubMed]
181. Xiong, Y.; Georgieva, R.; Steffen, A.; Smuda, K.; Bäumler, H. Structure and Properties of Hybrid Biopolymer Particles Fabricated by Co-Precipitation Cross-Linking Dissolution Procedure. *J. Colloid Interface Sci.* **2018**, *514*, 156–164. [CrossRef] [PubMed]
182. Dong, Y.; Lan, T.; Wang, X.; Zhang, Y.; Jiang, L.; Sui, X. Preparation and Characterization of Soy Protein Microspheres Using Amorphous Calcium Carbonate Cores. *Food Hydrocoll.* **2020**, *107*, 105953. [CrossRef]
183. Wang, Y.; Caruso, F. Nanoporous Protein Particles through Templating Mesoporous Silica Spheres. *Adv. Mater.* **2006**, *18*, 795–800. [CrossRef]
184. Mak, W.C.; Georgieva, R.; Renneberg, R.; Bäumler, H. Protein Particles Formed by Protein Activation and Spontaneous Self-Assembly. *Adv. Funct. Mater.* **2010**, *20*, 4139–4144. [CrossRef]
185. Volodkin, D.V.; Von Klitzing, R.; Möhwald, H. Pure Protein Microspheres by Calcium Carbonate Templating. *Angew. Chemie-Int. Ed.* **2010**, *49*, 9258–9261. [CrossRef] [PubMed]
186. Li, H.; Xiong, Y.; Zhang, Y.; Tong, W.; Georgieva, R.; Bäumler, H.; Gao, C. Photo-Decomposable Sub-Micrometer Albumin Particles Cross-Linked by Ortho-Nitrobenzyl Derivatives. *Macromol. Chem. Phys.* **2017**, *218*, 1–6. [CrossRef]
187. Neumann, M.M.; Volodkin, D. Porous Antibody-Containing Protein Microparticles as Novel Carriers for ELISA. *Analyst* **2020**, *145*, 1202–1206. [CrossRef]
188. Zhu, H.; Stein, E.W.; Lu, Z.; Lvov, Y.M.; McShane, M.J. Synthesis of Size-Controlled Monodisperse Manganese Carbonate Microparticles as Templates for Uniform Polyelectrolyte Microcapsule Formation. *Chem. Mater.* **2005**, *17*, 2323–2328. [CrossRef]
189. Narayan, R.; Nayak, U.Y.; Raichur, A.M.; Garg, S. Mesoporous Silica Nanoparticles: A Comprehensive Review on Synthesis and Recent Advances. *Pharmaceutics* **2018**, *10*, 118. [CrossRef]
190. Natalia, F.; Stoychev, G.; Puretskiy, N.; Leonid, I.; Dmitry, V. Porous Thermo-Responsive PNIPAM Microgels. *Eur. Polym. J.* **2015**, *68*, 650–656. [CrossRef]
191. Vikulina, A.S.; Feoktistova, N.A.; Balabushevich, N.G.; Von Klitzing, R.; Volodkin, D. Cooling-Triggered Release from Mesoporous Poly(N-Isopropylacrylamide) Microgels at Physiological Conditions. *ACS Appl. Mater. Interfaces* **2020**, *12*, 57401–57409. [CrossRef]
192. Wang, A.; Cui, Y.; Li, J.; Van Hest, J.C.M. Fabrication of Gelatin Microgels by a "Cast" Strategy for Controlled Drug Release. *Adv. Funct. Mater.* **2012**, *22*, 2673–2681. [CrossRef]
193. Velasco-Lozano, S.; López-Gallego, F.; Mateos-Díaz, J.C.; Favela-Torres, E. Cross-Linked Enzyme Aggregates (CLEA) in Enzyme Improvement—A Review. *Biocatalysis* **2016**, *1*, 166–177. [CrossRef]
194. Sheldon, R.A. CLEAs, Combi-CLEAs and 'Smart' Magnetic CLEAs: Biocatalysis in a Bio-Based Economy. *Catalysts* **2019**, *9*, 261. [CrossRef]
195. Miao, C.; Li, H.; Zhuang, X.; Wang, Z.; Yang, L.; Lv, P.; Luo, W. Synthesis and Properties of Porous CLEAs Lipase by the Calcium Carbonate Template Method and Its Application in Biodiesel Production. *RSC Adv.* **2019**, *9*, 29665–29675. [CrossRef]
196. Cui, J.; Zhao, Y.; Tan, Z.; Zhong, C.; Han, P.; Jia, S. Mesoporous Phenylalanine Ammonia Lyase Microspheres with Improved Stability through Calcium Carbonate Templating. *Int. J. Biol. Macromol.* **2017**, *98*, 887–896. [CrossRef]
197. Cui, J.; Zhao, Y.; Feng, Y.; Lin, T.; Zhong, C.; Tan, Z.; Jia, S. Encapsulation of Spherical Cross-Linked Phenylalanine Ammonia Lyase Aggregates in Mesoporous Biosilica. *J. Agric. Food Chem.* **2017**, *65*, 618–625. [CrossRef] [PubMed]

198. Rideau, E.; Dimova, R.; Schwille, P.; Wurm, F.R.; Landfester, K. Liposomes and Polymersomes: A Comparative Review towards Cell Mimicking. *Chem. Soc. Rev.* **2018**, *47*, 8527–8610. [CrossRef] [PubMed]
199. Has, C.; Sunthar, P. A Comprehensive Review on Recent Preparation Techniques of Liposomes. *J. Liposome Res.* **2020**, *30*, 336–365. [CrossRef] [PubMed]
200. Mateos-maroto, A.; Abelenda-Núñez, I.; Ortega, F.; Rubio, R.; Guzmán, E. Polyelectrolyte Multilayers on Soft Colloidal Nanosurfaces: A New Life for the Layer-By-Layer Method. *Polymers* **2021**, *13*, 1221. [CrossRef] [PubMed]
201. Riaz, M.K.; Riaz, M.A.; Zhang, X.; Lin, C.; Wong, K.H.; Chen, X.; Id, G.Z.; Lu, A.; Yang, Z. Surface Functionalization and Targeting Strategies of Liposomes in Solid Tumor Therapy: A Review. *Int. J. Mol. Sci.* **2018**, *19*, 195. [CrossRef] [PubMed]
202. Weber, C.; Voigt, M.; Simon, J.; Danner, A.; Frey, H.; Mailänder, V.; Helm, M.; Morsbach, S.; Landfester, K. Functionalization of Liposomes with Hydrophilic Polymers Results in Macrophage Uptake Independent of the Protein Corona. *Biomacromolecules* **2019**, *20*, 2989–2999. [CrossRef]
203. Vabbilisetty, P.; Sun, X. Liposome Surface Functionalization Based on Different Anchoring Lipids via Staudinger Ligation. *Org. Biomol. Chem.* **2014**, *12*, 1237–1244. [CrossRef]
204. Volodkin, D.; Mohwald, H.; Voegel, J.C.; Ball, V. Coating of Negatively Charged Liposomes by Polylysine: Drug Release Study. *J. Control. Release* **2007**, *117*, 111–120. [CrossRef]
205. Volodkin, D.; Ball, V.; Schaaf, P.; Voegel, J.C.; Mohwald, H. Complexation of Phosphocholine Liposomes with Polylysine. Stabilization by Surface Coverage versus Aggregation. *Biochim. Biophys. Acta* **2007**, *1768*, 280–290. [CrossRef] [PubMed]
206. Liu, W.; Liu, J.; Liu, W.; Li, T.; Liu, C. Improved Physical and in Vitro Digestion Stability of a Polyelectrolyte Delivery System Based on Layer-by-Layer Self- Assembly Alginate—Chitosan-Coated Nanoliposomes. *J. Agric. Food Chem.* **2013**, *61*, 4133–4144. [CrossRef] [PubMed]
207. Eivazi, A.; Medronho, B.; Lindman, B.; Norgren, M. On the Development of All-Cellulose Capsules by Vesicle-Templated Layer-by-Layer Assembly. *Polymers* **2021**, *13*, 589. [CrossRef] [PubMed]
208. Hermal, F.; Frisch, B.; Specht, A.; Bourel-bonnet, L.; Heurtault, B. Development and Characterization of Layer-by-Layer Coated Liposomes with Poly(L-Lysine) and Poly(L-Glutamic Acid) to Increase Their Resistance in Biological Media. *Int. J. Pharm.* **2020**, *586*, 119568. [CrossRef]
209. Madrigal-carballo, S.; Lim, S.; Rodriguez, G.; Vila, A.O.; Krueger, C.G.; Gunasekaran, S.; Reed, J.D. Biopolymer Coating of Soybean Lecithin Liposomes via Layer-by-Layer Self-Assembly as Novel Delivery System for Ellagic Acid. *J. Funct. Foods* **2010**, *2*, 99–106. [CrossRef]
210. Liu, W.; Kong, Y.; Tu, P.; Lu, J.; Liu, C.; Liu, W.; Han, J.; Liu, J. Physical–Chemical Stability and in Vitro Digestibility of Hybrid Nanoparticles Based on the Layer-by-Layer Assembly of Lactoferrin and BSA on Liposomes. *Food Funct.* **2017**, *8*, 1688–1697. [CrossRef]
211. Cuomo, F.; Lopez, F.; Ceglie, A.; Maiuro, L.; Miguel, G. PH-Responsive Liposome-Templated Polyelectrolyte Nanocapsules. *Soft Matter* **2012**, 4415–4420. [CrossRef]
212. Cuomo, F.; Lopez, F.; Miguel, M.G. Vesicle-Templated Layer-by-Layer Assembly for the Production of Nanocapsules. *Langmuir* **2010**, *26*, 10555–10560. [CrossRef]
213. Volodkin, D.; Arntz, Y.; Schaaf, P.; Moehwald, H.; Voegel, J.; Ball, V. Composite Multilayered Biocompatible Polyelectrolyte Films with Intact Liposomes: Stability and Temperature Triggered Dye Release. *Soft Matter* **2008**, 122–130. [CrossRef]
214. Volodkin, D.V.; Schaaf, P.; Mohwald, H.; Voegel, J.; Ball, V. Effective Embedding of Liposomes into Polyelectrolyte Multilayered Films: The Relative Importance of Lipid-Polyelectrolyte and Interpolyelectrolyte Interactions. *Soft Matter* **2009**, 1394–1405. [CrossRef]
215. Bédard, M.; De Geest, B.G.; Helmuth, M.; Sukhorukov, G.B.; Skirtach, A.G. Direction Specific Release from Giant Microgel-Templated Polyelectrolyte Microcontainers. *Soft Matter* **2009**, *5*, 3927–3931. [CrossRef]
216. De Geest, B.G.; Déjugnat, C.; Prevot, M.; Sukhorukov, G.B.; Demeester, J.; De Smedt, S.C. Self-Rupturing and Hollow Microcapsules Prepared from Bio-Polyelectrolyte- Coated Microgels. *Adv. Funct. Mater.* **2007**, *17*, 531–537. [CrossRef]
217. Lee, H.; Jeong, Y.; Park, T.G. Shell Cross-Linked Hyaluronic Acid/Polylysine Layer-by-Layer Polyelectrolyte Microcapsules Prepared by Removal of Reducible Hyaluronic Acid Microgel Cores. *Biomacromolecules* **2007**, *8*, 3705–3711. [CrossRef] [PubMed]
218. Díez-Pascual, A.M.; Shuttleworth, P.S. Layer-by-Layer Assembly of Biopolyelectrolytes onto Thermo/PH-Responsive Micro/Nano-Gels. *Materials* **2014**, *7*, 7472–7512. [CrossRef]
219. Sergeeva, A.; Feoktistova, N.; Prokopovic, V.; Gorin, D.; Volodkin, D. Design of Porous Alginate Hydrogels by Sacrificial $CaCO_3$ Templates: Pore Formation Mechanism. *Adv. Mater. Interfaces* **2015**, *2*, 1500386. [CrossRef]
220. Moya, S.; Dähne, L.; Voigt, A.; Leporatti, S.; Donath, E.; Möhwald, H. Polyelectrolyte Multilayer Capsules Templated on Biological Cells: Core Oxidation Influences Layer Chemistry. *Colloids Surf. A Physicochem. Eng. Asp.* **2001**, *183–185*, 27–40. [CrossRef]
221. Yitayew, M.Y.; Tabrizian, M. Hollow Microcapsules through Layer-by- Layer Self-Assembly of Chitosan/Alginate on «E. Coli». *MRS Adv.* **2020**, *5*, 2401–2407. [CrossRef]
222. Santos, A.C.; Pattekari, P.; Jesus, S.; Veiga, F.; Lvov, Y. Sonication-Assisted Layer-by-Layer Assembly for Low Solubility Drug Nanoformulation. *ACS Appl. Mater. Interfaces* **2015**, *10*, 11972–11983. [CrossRef]
223. Agarwal, A.; Lvov, Y.; Sawant, R.; Torchilin, V. Stable Nanocolloids of Poorly Soluble Drugs with High Drug Content Prepared Using the Combination of Sonication and Layer-by-Layer Technology. *J. Control. Release* **2008**, *128*, 255–260. [CrossRef]

224. Du, B.; Shen, G.; Wang, D.; Pang, L.; Chen, Z.; Liu, Z. Development and Characterization of Glimepiride Nanocrystal Formulation and Evaluation of Its Pharmacokinetic in Rats. *Drug Deliv.* **2013**, 7544. [CrossRef]
225. Santos, A.; Caldas, M.; Pattekari, P.; Fontes, C.; Lvov, Y.; Veiga, F. *Layer-by-Layer Coated Drug-Core Nanoparticles as Versatile Delivery Platforms*; Alexandru Mihai Grumezescu, Ed.; Elsevier: Amsterdam, The Netherlands, 2018. [CrossRef]
226. Khadka, P.; Ro, J.; Kim, H.; Kim, I.; Tae, J.; Kim, H.; Min, J.; Yun, G.; Lee, J. Pharmaceutical Particle Technologies: An Approach to Improve Drug Solubility, Dissolution and Bioavailability. *Asian J. Pharm. Sci.* **2014**, *9*, 304–316. [CrossRef]
227. Lvov, Y.M.; Pattekari, P.; Zhang, X.; Torchilin, V. Converting Poorly Soluble Materials into Stable Aqueous Nanocolloids. *Langmuir* **2011**, *27*, 1212–1217. [CrossRef] [PubMed]
228. Oshi, M.A.; Lee, J.; Naeem, M.; Hasan, N.; Kim, J.; Kim, H.J.; Lee, E.H.; Jung, Y.; Yoo, J. Curcumin Nanocrystal/PH-Responsive Polyelectrolyte Multilayer Core – Shell Nanoparticles for Inflammation-Targeted Alleviation of Ulcerative Colitis. *Biomacromolecules* **2020**, *21*, 3571–3581. [CrossRef] [PubMed]
229. Shenoy, D.B.; Sukhorukov, G.B. Engineered Microcrystals for Direct Surface Modification with Layer-by-Layer Technique for Optimized Dissolution. *Eur. J. Pharm. Biopharm.* **2004**, *58*, 521–527. [CrossRef]
230. Kamburova, K.; Mitarova, K.; Radeva, T. Polysaccharide-Based Nanocapsules for Controlled Release of Indomethacin. *Colloids Surf. A Physicochem. Eng. Asp.* **2017**, *519*, 199–204. [CrossRef]
231. Seo, J.; Yi, S.; Hwang, C.; Yang, M.; Lee, J.; Lee, S.; Cho, H. Multi-Layered Cellulose Nanocrystal System for CD44 Receptor-Positive Tumor-Targeted Anticancer Drug Delivery. *Int. J. Biol. Macromol.* **2020**, *162*, 798–809. [CrossRef]
232. An, Z.; Lu, G.; Möhwald, H.; Li, J. Self-Assembly of Human Serum Albumin (HSA) and L-a-Dimyristoylphosphatidic Acid (DMPA) Microcapsules for Controlled Drug Release. *Chem.-A Eur. J.* **2004**, *5*, 5848–5852. [CrossRef]
233. Song, L.; Zhi, Z.; Pickup, J. Nanolayer Encapsulation of Insulin- Chitosan Complexes Improves Efficiency of Oral Insulin Delivery. *Int. J. Nano* **2014**, *9*, 2127–2136.
234. Shutava, T.G.; Pattekari, P.; Arapov, K.A.; Torchilin, V.P.; Lvov, Y.M. Architectural Layer-by-Layer Assembly of Drug Nanocapsules with PEGylated Polyelectrolytes. *Soft Matter* **2012**, *8*, 9418–9427. [CrossRef]
235. Wang, X.; Zhao, J. Encapsulation of the Herbicide Picloram by Using Polyelectrolyte Biopolymers as Layer-by-Layer Materials. *J. Agric. Food Chem.* **2013**, *61*, 3789–3796. [CrossRef] [PubMed]
236. Städler, B.; Price, A.D.; Chandrawati, R.; Hosta-Rigau, L.; Zelikin, A.N.; Caruso, F. Polymer hydrogel capsules: En route toward synthetic cellular systems. *Nanoscale* **2009**, *1*, 68–73. [CrossRef] [PubMed]
237. Lengert, E.V.; Koltsov, S.I.; Li, J.; Ermakov, A.V.; Parakhonskiy, B.V.; Skorb, E.V.; Skirtach, A.G. Nanoparticles in Polyelectrolyte Multilayer Layer-by-Layer (LbL) Films and Capsules—Key Enabling Components of Hybrid Coatings. *Coatings* **2020**, *10*, 1131. [CrossRef]
238. Volodkin, D.; Skirtach, A.; Madaboosi, N.; Blacklock, J.; von Klitzing, R.; Lankenau, A.; Duschl, C.; Möhwald, H. IR-Light Triggered Drug Delivery from Micron-Sized Polymer Biocoatings. *J. Control. Release* **2010**, *148*, 70–71. [CrossRef] [PubMed]
239. Prokopović, V.Z.; Vikulina, A.S.; Sustr, D.; Duschl, C.; Volodkin, D. Biodegradation-Resistant Multilayers Coated with Gold Nanoparticles. Toward a Tailor-Made Artificial Extracellular Matrix. *ACS Appl. Mater. Interfaces* **2016**, *8*, 24345–24349. [CrossRef]
240. Svenskaya, Y.; Garello, F.; Lengert, E.; Kozlova, A.; Verkhovskii, R.; Bitonto, V.; Ruggiero, M.; German, S.; Gorin, D.; Terreno, E. Nanotheranostics Biodegradable Polyelectrolyte/Magnetite Capsules for MR Imaging and Magnetic Targeting of Tumors. *Nanotheranostics* **2021**, *5*, 362–377. [CrossRef]
241. Elumalai, R.; Patil, S.; Maliyakkal, N.; Rangarajan, A.; Kondaiah, P.; Raichur, A.M. Protamine-Carboxymethyl Cellulose Magnetic Nanocapsules for Enhanced Delivery of Anticancer Drugs against Drug Resistant Cancers. *Nanomed. Nanotechnol. Biol. Med.* **2015**, *11*, 969–981. [CrossRef]
242. Read, J.E.; Luo, D.; Chowdhury, T.; Flower, R.; Poston, R.N.; Sukhorukov, G.B.; Gould, D.J. Magnetically Responsive Layer-by-Layer Microcapsules Can Be Retained in Cells and under Flow Conditions to Promote Local Drug Release without Triggering ROS Production. *Nanoscale* **2020**, *12*, 7735–7748. [CrossRef]
243. Chen, H.; Di, Y.; Chen, D.; Madrid, K.; Zhang, M.; Tian, C.; Tang, L.; Gu, Y. Combined Chemo- and Photo-Thermal Therapy Delivered by Multifunctional Theranostic Gold Nanorod-Loaded Microcapsules. *Nanoscale* **2015**, *7*, 8884–8897. [CrossRef]
244. Popova, N.R.; Popov, A.; Ermakov, A.; Reukov, V.; Ivanov, V. Ceria-Containing Hybrid Multilayered Microcapsules for Enhanced Cellular Internalisation with High Radioprotection Efficiency. *Molecules* **2020**, *25*, 2957. [CrossRef]
245. Hanafy, B.I.; Cave, G.W.V.; Barnett, Y.; Pierscionek, B. Treatment of Human Lens Epithelium with High Levels of Nanoceria Leads to Reactive Oxygen Species Mediated Apoptosis. *Molecules* **2020**, *24*, 441. [CrossRef] [PubMed]
246. Popov, A.L.; Popova, N.R.; Tarakina, N.V.; Ivanova, O.S.; Ermakov, A.M.; Ivanov, V.K.; Sukhorukov, G.B. Intracellular Delivery of Antioxidant CeO2 Nanoparticles via Polyelectrolyte Microcapsules. *ACS Biomater. Sci. Eng.* **2018**, *4*, 2453–2462. [CrossRef] [PubMed]
247. Popov, A.L.; Popova, N.; Gould, D.J.; Shcherbakov, A.B.; Sukhorukov, G.B.; Ivanov, V.K. Ceria Nanoparticles-Decorated Microcapsules as a Smart Drug Delivery/Protective System: Protection of Encapsulated P. Pyralis Luciferase. *Appl. Mater. Interfaces* **2018**, *10*, 14367–14377. [CrossRef]
248. Duan, L.; Du, L.; Jia, Y.; Liu, W.; Liu, Z.; Li, J. High Impact of Uranyl Ions on Carrying—Releasing Oxygen Capability of Hemoglobin-Based Blood Substitutes. *Chem.-A Eur. J.* **2015**, *21*, 520–525. [CrossRef]

249. Timin, A.S.; Muslimov, A.R.; Lepik, K.V.; Epifanovskaya, O.S.; Shakirova, A.I.; Mock, U.; Riecken, K.; Okilova, M.V.; Sergeev, V.S.; Afanasyev, B.V.; et al. Efficient Gene Editing via Non-Viral Delivery of CRISPR–Cas9 System Using Polymeric and Hybrid Microcarriers. *Nanomed, Nanotechnol, Biol. Med.* **2018**, *14*, 97–108. [CrossRef] [PubMed]
250. Gaponik, N.; Radtchenko, I.L.; Gerstenberger, M.R.; Fedutik, Y.A.; Sukhorukov, G.B.; Rogach, A.L. Labeling of Biocompatible Polymer Microcapsules with Near-Infrared Emitting Nanocrystals. *Nano Lett.* **2003**, *4*, 4–7. [CrossRef]
251. Schmidt, S.; Madaboosi, N.; Uhlig, K.; Köhler, D.; Skirtach, A.; Duschl, C.; Möhwald, H.; Volodkin, D.V. Control of Cell Adhesion by Mechanical Reinforcement of Soft Polyelectrolyte Films with Nanoparticles. *Langmuir* **2012**, *28*, 7249–7257. [CrossRef] [PubMed]
252. Mzyk, A.; Lackner, J.M.; Wilczek, P.; Lipińska, L.; Niemiec-Cyganek, A.; Samotus, A.; Morenc, M. Polyelectrolyte Multilayer Film Modification for Chemo-Mechano-Regulation of Endothelial Cell Response. *RSC Adv.* **2016**, *6*, 8811–8828. [CrossRef]
253. Muslimov, A.R.; Antuganov, D.O.; Tarakanchikova, Y.V.; Zhukov, M.V.; Nadporojskii, M.A.; Zyuzin, M.V.; Timin, A.S. Calcium Carbonate Core−Shell Particles for Incorporation of and Their Application in Local α-Radionuclide Therapy. *ACS Appl. Mater. Interfaces* **2021**, *13*, 25599–25610. [CrossRef]
254. Kozlovskaya, V.; Alford, A.; Dolmat, M.; Ducharme, M.; Caviedes, R.; Radford, L.; Lapi, S.E.; Kharlampieva, E. Multilayer Microcapsules with Shell-Chelated and Controlled Delivery 89Zr for PET Imaging. *ACS Appl. Mater. Interfaces* **2020**, *12*, 56792–56804. [CrossRef]
255. Scheffler, K.; Bilz, N.C.; Brueckner, M.; Stanifer, M.L.; Boulant, S.; Claus, C.; Reibetanz, U. Enhanced Uptake and Endosomal Release of LbL Microcarriers Functionalized with Reversible Fusion Proteins. *ACS Appl. Bio Mater.* **2020**, *3*, 1553–1567. [CrossRef]
256. Sche, K.; Claus, C.; Stanifer, M.L.; Boulant, S.; Reibetanz, U. Reversible Fusion Proteins as a Tool to Enhance Uptake of Virus-Functionalized LbL Microcarriers. *Biomacromolecules* **2018**, *19*, 3212–3232. [CrossRef]
257. Francesco, P.; Zattera, E.; Pastorino, L.; Perego, P.; Palombo, D. Dextran/Poly-L-Arginine Multi-Layered $CaCO_3$-Based Nanosystem for Vascular Drug Delivery. *Int. J. Biol. Macromol.* **2021**, *177*, 548–558. [CrossRef]
258. Yang, S.; Dong, H. Modular Design and Self-Assembly of Multidomain Peptides towards Cytocompatible Supramolecular Cell Penetrating Nanofibers. *RSC Adv.* **2020**, *10*, 29469–29474. [CrossRef]
259. Leach, D.G.; Newton, J.M.; Florez, M.A.; Lopez-silva, T.L.; Jones, A.A.; Young, S.; Sikora, A.G.; Hartgerink, J.D. Drug-Mimicking Nanofibrous Peptide Hydrogel for Inhibition of Inducible Nitric Oxide Synthase. *ACS Biomater. Sci. Eng.* **2019**, *5*, 6755–6765. [CrossRef] [PubMed]
260. Lopez-silva, T.L.; Cristobal, C.D.; Lai, C.; Leyva-aranda, V.; Lee, H.; Hartgerink, J.D. Biomaterials Self-Assembling Multidomain Peptide Hydrogels Accelerate Peripheral Nerve Regeneration after Crush Injury. *Biomaterials* **2021**, *265*, 120401. [CrossRef]
261. Lopez-silva, T.L.; Leach, D.G.; Azares, A.; Li, I.; Woodside, D.G.; Hartgerink, J.D. Chemical Functionality of Multidomain Peptide Hydrogels Governs Early Host Immune Response. *Biomaterials* **2020**, *231*, 119667. [CrossRef]
262. Al-Khoury, H.; Espinosa-Cano, E.; Aguilar, M.R.; Román, J.S.; Syrowatka, F.; Schmidt, G.; Groth, T. Anti-Inflammatory Surface Coatings Based on Polyelectrolyte Multilayers of Heparin and Polycationic Nanoparticles of Naproxen-Bearing Polymeric Drugs. *Biomacromolecules* **2019**, *20*, 4015–4025. [CrossRef]
263. Guduru, D.; Niepel, M.S.; Gonzalez-Garcia, C.; Salmeron-Sanchez, M.; Groth, T. Comparative Study of Osteogenic Activity of Multilayers Made of Synthetic and Biogenic Polyelectrolytes. *Macromol. Biosci.* **2017**, *17*. [CrossRef]
264. Hanafy, N.A.N. Optimally Designed Theranostic System Based Folic Acids and Chitosan as a Promising Mucoadhesive Delivery System for Encapsulating Curcumin LbL Nano-Template against Invasiveness of Breast Cancer. *Int. J. Biol. Macromol.* **2021**, *182*, 1981–1993. [CrossRef] [PubMed]
265. Jing, J.; Szarpak-Jankowska, A.; Guillot, R.; Pignot-Paintrand, I.; Picart, C.; Auzély-Velty, R. Cyclodextrin/Paclitaxel Complex in Biodegradable Capsules for Breast Cancer Treatment. *Chem. Mater.* **2013**, *25*, 3867–3873. [CrossRef]
266. Kurapati, R.; Raichur, A.M. Composite Cyclodextrin-Calcium Carbonate Porous Microparticles and Modified Multilayer Capsules: Novel Carriers for Encapsulation of Hydrophobic Drugs. *J. Mater. Chem. B* **2013**, *1*, 3175–3184. [CrossRef]
267. Belbekhouche, S.; Bousserrhine, N.; Alphonse, V.; Carbonnier, B. From Beta-Cyclodextrin Polyelectrolyte to Layer-by-Layer Self-Assembly Microcapsules: From Inhibition of Bacterial Growth to Bactericidal Effect. *Food Hydrocoll.* **2019**, *95*, 219–227. [CrossRef]

 nanomaterials

Article

Towards Polymeric Nanoparticles with Multiple Magnetic Patches

Elham Yammine [1,2], Laurent Adumeau [1], Maher Abboud [3], Stéphane Mornet [1], Michel Nakhl [2] and Etienne Duguet [1,*]

1 Univ. Bordeaux, CNRS, Bordeaux INP, ICMCB, UMR 5026, 33600 Pessac, France; elhamyammine93@gmail.com (E.Y.); laurent.adumeau@cbni.ucd.ie (L.A.); stephane.mornet@icmcb.cnrs.fr (S.M.)
2 LCPM/PR2N (EDST), Lebanese University, Faculty of Sciences II, Jdeidet El Met 90656 , Lebanon; mnakhl@ul.edu.lb
3 Unité Environnement Génomique et Protéomique, U-EGP, Faculté des Sciences, Université Saint-Joseph, Campus des Sciences et Technologies Mar Roukos–B.P. 1514, Riad El Solh, Beirut 1107 2050, Lebanon; maher.abboud@usj.edu.lb
* Correspondence: etienne.duguet@icmcb.cnrs.fr; Tel.: +33-540-002-651

Abstract: Fabricating future materials by self-assembly of nano-building blocks programmed to generate specific lattices is among the most challenging goals of nanotechnology and has led to the recent concept of patchy particles. We report here a simple strategy to fabricate polystyrene nanoparticles with several silica patches based on the solvent-induced self-assembly of silica/polystyrene monopods. The latter are obtained with morphological yields as high as 99% by seed-growth emulsion polymerization of styrene in the presence of 100 nm silica seeds previously modified with an optimal surface density of methacryloxymethyl groups. In addition, we fabricate "magnetic" silica seeds by silica encapsulation of preformed maghemite supraparticles. The polystyrene pod, i.e., surface nodule, serves as a sticky point when the monopods are incubated in a bad/good solvent mixture for polystyrene, e.g., ethanol/tetrahydrofuran mixtures. After self-assembly, mixtures of particles with two, three, four silica or magnetic silica patches are mainly obtained. The influence of experimental parameters such as the ethanol/tetrahydrofuran volume ratio, monopod concentration and incubation time is studied. Further developments would consist of obtaining pure batches by centrifugal sorting and optimizing the relative position of the patches in conventional repulsion figures.

Keywords: silica; polystyrene; maghemite supraparticles; patchy particles; seeded-growth emulsion polymerization; solvent-induced self-assembly

Citation: Yammine, E.; Adumeau, L.; Abboud, M.; Mornet, S.; Nakhl, M.; Duguet, E. Towards Polymeric Nanoparticles with Multiple Magnetic Patches. *Nanomaterials* **2021**, *11*, 147. https://doi.org/10.3390/nano11010147

Received: 21 December 2020
Accepted: 6 January 2021
Published: 9 January 2021

1. Introduction

Building hierarchical structures with unique properties for both fundamental studies and technological applications is among the most active research areas in nanotechnology. For bottom-up approaches, the main current goal is the preparation of nanoparticles serving as building blocks for controlled assembly [1]. However, isotropic spherical nanoparticles lead to only a low number of close-packed colloidal lattices, while anisotropic nanoparticles are able to order into many novel crystalline or non-crystalline structures [2]. In this context, there is great interest in patchy particles envisioned as valence-endowed colloidal atoms for targeting specifically more open lattices and low-coordination architectures [3–8]. These surface-patterned particles are indeed designed to experience specific directional interactions with neighboring particles. Among the investigated attractive forces, magnetic dipolar interactions have received specific attention over the past fifteen years [9–15]. Particles with a single magnetic patch were essentially investigated, because the dipole moment is shifted out of the particle center, and original assemblies were obtained according to the predictions of simulation studies. These monopatch particles are rather easily fabricated, on a wide range of scales from a few hundred nanometers to a few millimeters, through conventional dissymmetrisation routes used to produce Janus particles [15]. Previous studies have neglected the option of particles carrying multiple magnetic patches. The difficulty of manufacturing such objects is not the only problem, since to our knowledge, the simulators have not published such studies either, while studies of particle assembly having several nonmagnetic patches abound [8]. Nevertheless, the possibility of making particles with two magnetic patches was reported twice. A microfluidic route using double emulsion templates encapsulated two droplets of magnetic particles in hydrogel microparticles by increasing the inner flow [16]. Two-patch particles were also reported as by-products when manufacturing organosilica microparticles carrying one hematite cube [17]. In both cases, the relative position of the two patches was not finely tuned, and there is, therefore, a need to find synthesis routes of particles having 2, 3, 4, etc. magnetic patches, if possible, to position these according to predefined geometric figures.

The paradigm shift that we propose here is to consider the synthesis of particles with several magnetic patches through the self-assembly of monopatch particles by fusing their nonmagnetic components. It concerns polymer-based particles, and we drew inspiration from some reports that relate to nonmagnetic colloids. Micron-sized asymmetric dumbbells made of a polystyrene (PS) sphere and a poly(n-butylacrylate) lobe self-assemble into well-defined clusters when dispersed in aqueous media, when the desorption of the poly(vinylpyrrolidone) stabilizer from the particles triggers the merging of the soft poly(n-butylacrylate) lobes upon contact through collision [18]. Later, other researchers fabricated by a microfluidic technology 200 μm hydrophilic poly(ethylene glycol) diacrylate particles with a 160 μm ethoxylated trimethylolpropane tri-acrylate lobe [19]. They obtained quite regular submillimeter-sized clusters by drying Pickering-like water droplets stabilized from these particles. More recently, we reported that dumbbell-shaped silica/PS nanoparticles can self-assemble in regular clusters by incubation in mixtures of bad/good solvents for PS, i.e., ethanol/dimethylformamide (DMF) mixtures [20]. In such conditions, the partial swelling of the PS patch makes it sticky enough to attract and merge to the PS patches of other particles.

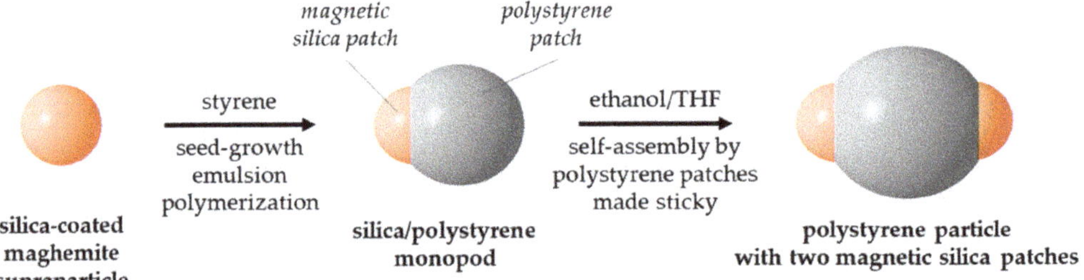

Figure 1. Scheme showing the two-step synthesis pathway to obtain polystyrene particles with several magnetic silica patches, exemplified here for two patches.

In the present study, we extent this concept to magnetic silica/PS asymmetric nanoparticles in order to obtain PS particles bearing several patches made of "magnetic" silica, i.e., silica nanoparticle encapsulation of a supraparticle obtained by superparamagnetic maghemite nanocrystal close-packing. As shown in Figure 1, the first step consists in the dissymmetrisation of the magnetic silica nanoparticle by a PS nodule obtained after seed-growth emulsion polymerization of styrene. This silica/PS monopod can in fact be described twice as a monopatch particle depending on whether we consider the PS nodule, i.e., PS patch, which will be used for solvent-induced self-assembly (2nd step), or the magnetic silica particle, i.e., magnetic patch, whose forthcoming studies will take benefit for subsequent assembly assisted by magnetic fields. Our strategy consisted of (i) optimizing the operating conditions of the two steps with model silica particles, (ii) finding an indisputable way to check that the silica patches protrude out of the surface of the PS particles throughout the process and (iii) showing that the process is robust enough to work for magnetic silica patches. Our results are essentially supported by statistical studies carried out using transmission electron microscopy (TEM) images.

2. Materials and Methods

2.1. Materials

Tetraethoxysilane (TEOS, ≥99%), styrene (≥99%, with ca. 50 ppm 4-tert-butylcatechol as stabilizer), sodium persulfate ($Na_2S_2O_8$, ≥99%), polyethylene glycol nonylphenyl ether (Synperonic® NP30), sodium dodecylsulfate (SDS, 99%), hydroquinone (≥99%), cyclohexane (≥99.7%) and tetrahydrofuran (THF, ≥99% contains 250 ppm butylhydroxytoluene as inhibitor) were purchased from Sigma-Aldrich (Saint-Quentin Fallavier, France). 3-methacryloxypropyl(trimethoxy)silane (MPS, 98%) was purchased from ABCR (Karlsruhe, Germany). Ammonium hydroxide (NH_4OH, 28–30% in water) and ethanol (99%) were provided by Atlantic Labo (Bruges, France). Oleic acid (65–88%) was purchased from Fisher Scientific (Illkirch-Graffenstaden, France). Deionized water with a resistivity of 18.2 MΩ·cm at 25 °C was obtained from a Milli-Q system (Merck Millipore, Darmstadt, Germany). All chemicals were used without further purification.

2.2. Colloid Synthesis

Silica nanoparticles with an average diameter of 100 nm and polydispersity index of 1.003 were obtained by TEOS hydrolysis/polycondensation according to an already published protocol [21]. At the end of the synthesis, the silica surface was functionalized with methacryloxypropyl functions by reacting with MPS at room temperature for 3 h and then one more hour at 90 °C under stirring. The MPS volume (V_{MPS}) was calculated for tuning the nominal grafting surface density (d_{MPS}) between 0.3 and 2 funct./nm^2 according to Equation (1):

$$V_{MPS} = \frac{6 \cdot d_{MPS} \cdot M_{MPS} \cdot V_{silica} \cdot N_{silica}}{N_A \cdot \rho_{MPS} \cdot \rho_{silica} \cdot D_{silica}} \quad (1)$$

where N, M, D, N_A and ρ symbolize particle number, molar mass, particle diameter, Avogadro number and density, respectively. After dialysis against MilliQ water, the mass concentration of silica particles was checked by the dry-extract method.

Magnetic silica nanoparticles were prepared by silica strengthening and coating of maghemite supraparticles whose synthesis by the emulsion evaporation route using SDS as surfactant was previously reported by Simard and coworkers [22]. For this purpose, we used a hydrophobic ferrofluid synthesized from 7.5 nm-sized maghemite nanoparticles prepared in a conventional manner by alkaline coprecipitation [23] and stabilized by oleic acid in cyclohexane according to a protocol previously reported by van Ewijk et al. [24]. We describe here in detail only the silica strengthening and coating stages. First, 250 mg of 79 ± 10 nm-sized maghemite supraparticles dispersed in 20 mL SDS aqueous solution (50 mM) were diluted with 30 mL water, before adding successively 2.5 mL NH_4OH and 250 μL TEOS. After stirring for 2 h at 40 °C, 200 mL water was added, before two centrifugation stages: a first one (8000× *g*; 15 min) to recover the pellet and redispersion in water and a second (500× *g*; 15 min) to recover the supernatant. The final volume of the dispersion was adjusted to 20 mL with water. Secondly, we added 72 mL ethanol, 4.4 mL ammonia and then a known volume of TEOS (0.1 mL/h). The reaction medium was stirred at room temperature for at least 4 h. The ammonia and some of the ethanol were removed using a rotary evaporator before washing the particles in a water/ethanol mixture (1:1) by two centrifugation/redispersion cycles (10,000× *g*; 15 min). The TEOS volume was calculated as a function of the targeted diameter D of the magnetic silica nanoparticles from the initial average diameter d of the maghemite supraparticles according to Equation (2):

$$V_{TEOS} = \frac{\pi \cdot \rho_{SiO_2} \cdot M_{TEOS}}{6 \cdot M_{SiO_2} \cdot \rho_{TEOS}} \cdot N_{Fe_2O_3\ supraparticle} \cdot \left(D^3 - d^3\right) \quad (2)$$

Nevertheless, as shown in Section 3.3, Equation (2) underestimates the amount of TEOS serving to fill the space between the 7.5 nm maghemite nanoparticles within the supraparticles. Therefore, future researchers are invited to use instead Equation (3), which includes further a corrective term considering a random close packing density of 0.64 for the maghemite nanoparticles inside each supraparticle [25]:

$$V_{TEOS} = \frac{\pi \cdot \rho_{SiO_2} \cdot M_{TEOS}}{6 \cdot M_{SiO_2} \cdot \rho_{TEOS}} \cdot N_{Fe_2O_3\ supraparticle} \cdot \left[\left(D^3 - d^3\right) + 0.36\, d^3\right] \quad (3)$$

For this study, magnetic silica seeds with a diameter of 105 ± 10 nm were obtained corresponding to a silica thickness of 13 nm. The maghemite weight fraction was estimated to be about 48% using Equation (4):

$$wt.\ \%(Fe_2O_3) = \frac{0.64\, d^3 \cdot \rho_{Fe_2O_3}}{\left[\left(D^3 - d^3\right) + 0.36\, d^3\right] \cdot \rho_{SiO_2} + 0.64\, d^3 \cdot \rho_{Fe_2O_3}} \cdot 100 \quad (4)$$

The silica/PS monopods were obtained through seed-growth emulsion polymerization of styrene, according to a first protocol already fully described in our previous publications [26]. Briefly, 50 mL emulsion was prepared by mixing Synperonic® NP30 (2.85 g/L), SDS (0.15 g/L), the MPS-modified silica nanoparticles (7.3×10^{15} part./L), styrene (89 g/L) and $Na_2S_2O_8$ (0.46 g/L) in MilliQ water. Then, the polymerization was triggered by increasing the temperature to 70 °C for 6 h. Monomer-to-polymer conversion was determined by the dry-extract method and comprised between 60% and 95%. When necessary, free PS particles were removed by discarding supernatant after two centrifugation cycles (2000 g; 15 min) after dilution 10 times in water or ethanol. The monopods were redispersed in water or ethanol, and the dispersion concentration was determined by the dry-extract method.

An alternative emulsion polymerization protocol was also used, inspired by the work of Ge et al. [27], and especially well adapted for low seed concentrations. Briefly, a 15 mL emulsion was prepared by mixing Synperonic® NP30 (0.19 g/L), SDS (0.01 g/L), the MPS-modified silica nanoparticles (4.8×10^{14} part./L), styrene (56.8 g/L) and $Na_2S_2O_8$ (0.78 g/L) in MilliQ water. The polymerization temperature and time employed were identical to those of the previous protocol.

2.3. Solvent-Induced Colloidal Self-Assembly

We were inspired by our own work concerning the assembly of silica/PS monopods but displaying another morphology [20]. The self-assembly of each type of monopod was carried out in a 2.5 mL glass bottle closed with a polypropylene cap, gently stirred on a roller-mixer at 60 rpm. The quantities of monopod (dispersed in water or ethanol), of good (DMF or THF) and bad solvent (water or ethanol), for PS were carefully measured to reach a final volume of 1 mL.

2.4. Characterization by Electron Microscopy

The morphology and size of colloids were studied for each batch by transmission electron microscopy (TEM) using a JEM 1400+ LaB_6 apparatus sourced from JEOL Europe SAS (Croissy-sur-Seine, France) operating at 120 kV. The samples were prepared by directly depositing a drop of the colloid dispersion on the TEM grids. Statistical image analyses were performed over at least 150 nanoparticles. Size polydispersity indexes were calculated from the weight average $\overline{D_w}$ and number average $\overline{D_n}$ diameter values according to Equation (5):

$$\text{size polydispersity index} = \overline{D_w}/\overline{D_n} = \frac{\sum n_i D_i^4}{\sum n_i D_i} / \frac{\sum n_i D_i^3}{\sum n_i} \quad (5)$$

To check that the surface of the silica was not covered by even a very thin PS layer, we carried out a step of silica regrowth before TEM observation, according to the following operating mode [28]. To 1 mL of the particle dispersion in ethanol or water, we added 45.5 mL ethanol and 3.5 mL NH_4OH. Then, we added dropwise a solution of 3.8 mL TEOS (10 vol.% in absolute ethanol) using a single-syringe pump (1 mL/h). The particles were washed using two centrifugation/redispersion cycles in ethanol (12,000× *g*; 10 min).

Elemental mapping by scanning transmission electron microscopy coupled to energy dispersive X-ray spectrometry (STEM-EDX) was carried out with a JEM 2200 FS apparatus sourced from JEOL Europe SAS (Croissy-sur-Seine, France) equipped with a field emissive gun operating at 200 kV. The samples were prepared as for conventional TEM experiments.

3. Results and Discussion

3.1. Preparation of Silica/Polystyrene Monopods

We prepared batches of size-monodisperse silica seeds with a diameter of 100 nm and surface methacryloxypropyl groups at nominal grafting densities from 0.3 to 2 funct./nm^2. Each one was then used as seeds in an emulsion polymerization process of styrene [21,29] in order to find the optimized conditions to promote nucleation and growth of a single PS nodule on each silica seed. It is worth mentioning that high amounts of free PS particles, i.e., which do not contain a silica core, were also observed in the TEM images (Figure S1). Nevertheless, they may be easily removed by two centrifugation cycles. The features of the as-obtained batches and their statistical morphological analysis from TEM images are summarized in Table 1. The lower the nominal grafting density (entries #1.1 and #1.2), the higher the number of PS nodules (up to 4 and 3, respectively). Therefore, the number of monopods was too low. From 0.6 to 2 funct./nm^2 (entries #1.3 to #1.6), monopods were systematically obtained with morphological yield as high as 99%. This phenomenon may be explained as follows: the higher the nominal grafting density, the more organophilic the silica surface, the smaller the contact angle of the growing PS nodules on the silica seed and, therefore, the lower their surface mobility and the easier it is for them to merge into a single nodule. This is particularly evidenced by the diameter of the PS nodule of the monopods, which is systematically much higher than that of the free PS particles that have grown independently from the silica seeds.

Table 1. Morphological distribution and geometrical parameters of the silica/PS particles as a function of the nominal grafting surface density d_{MPS}. Experimental conditions: [100-nm silica] = 7.3 × 10^{15} part./L; [styrene] = 89 g/L; [Synperonic®NP30] = 2.85 g/L; [SDS] = 0.15 g/L; [$Na_2S_2O_8$] = 0.46 g/L; 70 °C and 6 h. Representative TEM images are shown in Figure S1.

Entry	d_{MPS} (funct./nm^2)	Monomer-to-Polymer Conversion (%)	Final Batch Composition (%) [1]					PS Nodule Diameter (nm)	
			Monopods	Bipods	Tripods	Tetrapods	Multisilica [2]	Monopods	Free Particles
#1.1	0.3	95	17	38	32	11	2	210	128
#1.2	0.5	88	63	29	6	-	2	214	126
#1.3	0.6	63	99	-	-	-	1	165	98
#1.4	0.7	60	98	-	-	-	2	163	94
#1.5	1.0	64	99	-	-	-	1	127	82
#1.6	2.0	70	99	-	-	-	1	145	113
#1.7	0.6	88	97	-	-	-	3	220	151
#1.8	0.6	68	98	1	-	-	1	148	102

[1] Fractions expressed with respect to the silica seeds, i.e., free PS particles were not considered. [2] Multisilica are silica/PS particles made of two or more silica seeds.

As seen in the TEM images (Figure S1), even if the difference in contrast between silica and PS is sufficient to distinguish both phases, the exact particle morphological interpretation is strongly dependent on its orientation on the TEM grid. Indeed, it is not so easy to determine whether the silica core protrudes out of the surface of the PS particle or else it is just below the surface covered at this position by a thin polymer layer. Because this criterion is critical to our end goal of obtaining accessible inorganic patches, we collected 1 mL of each monopod batch (entries #1.3 to #1.6) and added ethanol NH_4OH and TEOS, i.e., experimental conditions expected to promote the silica core regrowth [28]. Figure 2 shows the TEM images of each batch taken after this treatment. We can observe for the three batches prepared from seeds with $0.6 \leq d_{MPS} \leq 1$ funct./nm^2 the presence of a silica protrusion attached to the silica core indicating its accessibility to the reagents (Figure 2a–c). However, this was not true for the last batch (d_{MPS} = 2 funct./nm^2): new silica particles, circled in red in Figure 2d, did appear but separately from the silica/PS nanoparticles. Therefore, the latter is not composed of monopods as expected, because the silica core is fully encapsulated in the PS particle. Such decentered core-shell particles result from a degree of PS on silica wettability that is too high, as reported by Reculusa et al. [30] and Lang and coworkers [31,32] under quite similar conditions.

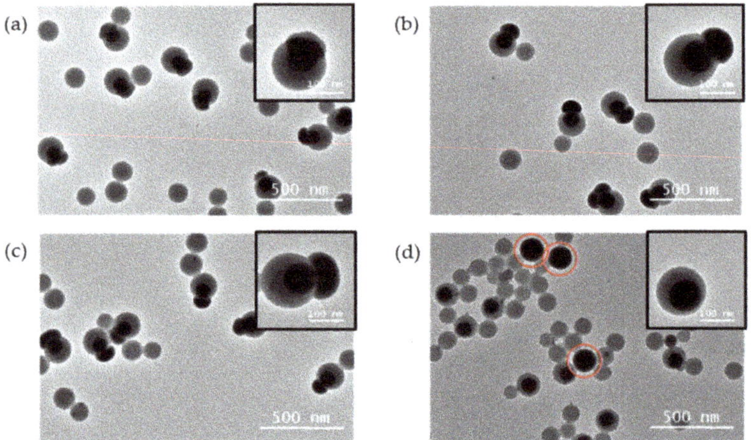

Figure 2. TEM images of silica/PS nanoparticles obtained for different nominal grafting surface densities d_{MPS} and after silica regrowth: (**a**) 0.6 funct./nm^2; (**b**) 0.7 funct./nm^2; (**c**) 1 funct./nm^2; (**d**) 2 funct./nm^2. The red circles show free silica particles obtained after the silica regrowth. The free PS particles observed in the images result from the polymerization step, because they were not removed by centrifugation for this series of experiments.

To further understand the mechanism of monopod formation during the seeded emulsion polymerization, we collected samples over time, taking care to immediately deactivate the polymerization reaction by adding hydroquinone as inhibitor, and storing the samples at 4 °C. Each sample was observed by TEM to perform statistical analyses of the morphological evolution.

We first studied the polymerization experiment carried out on seeds with d_{MPS} = 0.5 funct./nm^2 (Figure 3). We observed that after 30 min, monopods, bipods, tripods and tetrapods coexisted and that bipods were the main population. As the polymerization progressed, the size of the PS nodules increased, and the tetrapods, tripods and bipods gradually disappeared in favor of the monopods whose abundance increased from 33% to 63% after 6 h. This growth mechanism is consistent with Thill's model that we reported a few years ago [29,33]. This means that the nucleation of PS particles leads to the coexistence of several PS nodules on the silica seed. As they grow, i.e., as the seed surface is increasingly crowded, they can move, coalesce and/or leave the seed. Here, coalescence seemed to be the preferred scenario because the proportion of free particles appeared to be constant and the bipods after 6 h were dissymmetric with both nodules of different size. Thill's model also makes it possible to explain the non-negligible quantity of bipods, which remain at the end of the polymerization. Indeed, when there are only two nodules of PS left on the surface, the coalescence probability is almost zero if they are positioned at an angle of 180°, because they can thus grow to an infinite size independently. Thus, this growth mechanism is not favorable to obtain pure batches of monopods.

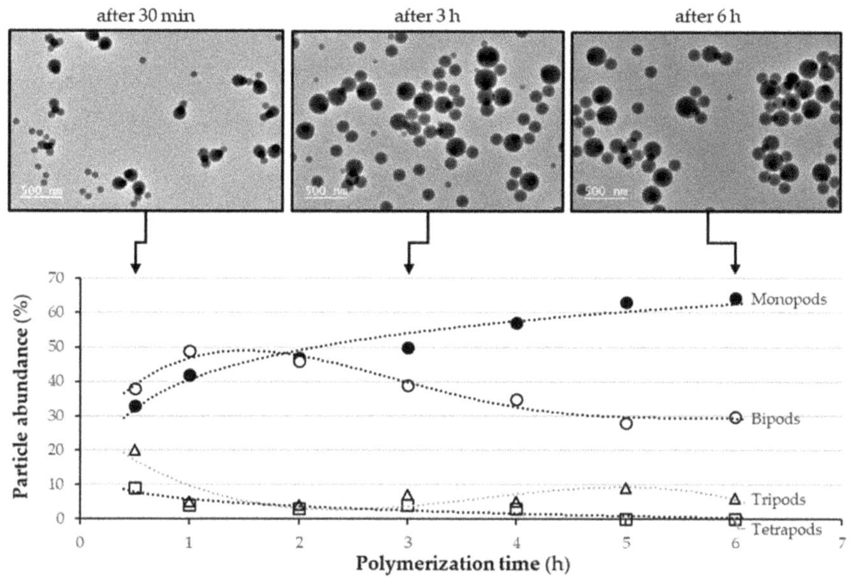

Figure 3. Graph showing the evolution with polymerization time of the abundance of the four main types of silica/PS nanoparticles as observed by TEM from silica seeds with d_{MPS} = 0.5 funct./nm^2, and representative TEM images taken after 30 min, 3 h and 6 h. Experimental conditions: [100-nm silica] = 7.3 × 10^{15} part./L; [styrene] = 89 g/L; [Synperonic® NP30] = 2.85 g/L; [SDS] = 0.15 g/L; [$Na_2S_2O_8$] = 0.46 g/L and 70 °C. The dash-dotted curves are a guide for the eye.

When we studied the growth mechanism from silica seeds with d_{MPS} = 0.6 funct./nm^2, we observed a different process of monopod formation (Figure 4): after 30 min of polymerization, the reactor contained exclusively one type of PS/silica nanoparticle with one PS nodule which almost entirely wrapped around the silica core. As the polymerization time and consequently the polymer nodule size increased, this very low contact angle was maintained and led, after 6 h, to a monopod yield of 99%, as already mentioned in Table 1, entry #1.3.

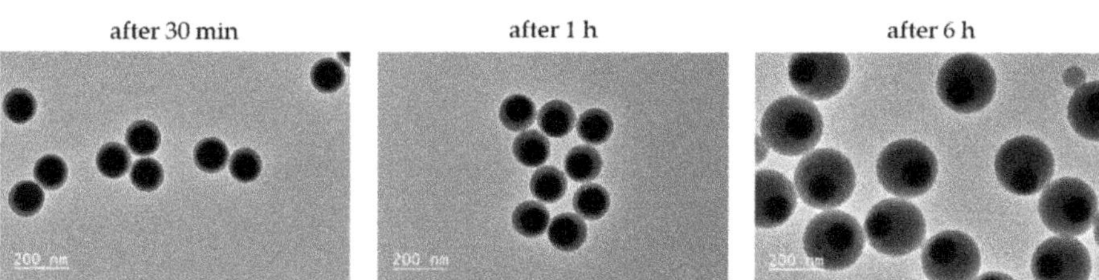

Figure 4. TEM images showing the evolution with time of the morphology of the silica/PS monopods obtained from silica seeds with d_{MPS} = 0.6 funct./nm^2. Experimental conditions: [100-nm silica] = 7.3 × 10^{15} part./L; [styrene] = 89 g/L; [Synperonic® NP30] = 2.85 g/L; [SDS] = 0.15 g/L; [$Na_2S_2O_8$] = 0.46 g/L and 70 °C. Before observation, the free PS particles were removed by one centrifugation/redispersion cycle in ethanol (2000× g; 15 min).

Therefore, we concluded the existence of three main regimes for monopod formation as a function of the surface grafting density of the methacryloxypropyl functions on the seed surface, which dictates the contact angle of the PS nodule on the silica core:

- For $d_{MPS} \leq 0.5$ funct./nm^2, the fairly high value of the contact angle leads to the coexistence of several nodules on the surface of the silica seed and the phenomenon of coalescence is insufficient for all the initially generated structures to transform ultimately into monopods;

- For $0.6 \leq d_{MPS} \leq 1$ funct./nm^2, the contact angle is sufficiently low to form a single PS nodule, while leaving the silica core partially protruding, i.e., part of the silica surface remains accessible;

- For $d_{MPS} \geq 2$ funct./nm^2, the contact angle is too low and the silica core is engulfed in the polymer nodule giving decentered core-shell nanoparticles.

For the purpose of our study, only the second growth regime is relevant. To check the repeatability of the experiments, the synthesis of monopods from silica seeds with d_{MPS} = 0.6 funct/nm^2 (Table 1, entry #1.3) was reproduced twice (Table 1, entries #1.7 and #1.8). Only slight differences were observed in the diameter of the PS nodule consistently with differences measured in the monomer-to-polymer conversion. The self-assembly experiments discussed in the next section were essentially performed with the monopod batch described in Table 1 (entry #1.3). The fraction of free PS particles was easily lowered below 5% thanks to two centrifugation/redispersion cycles (Figure 5).

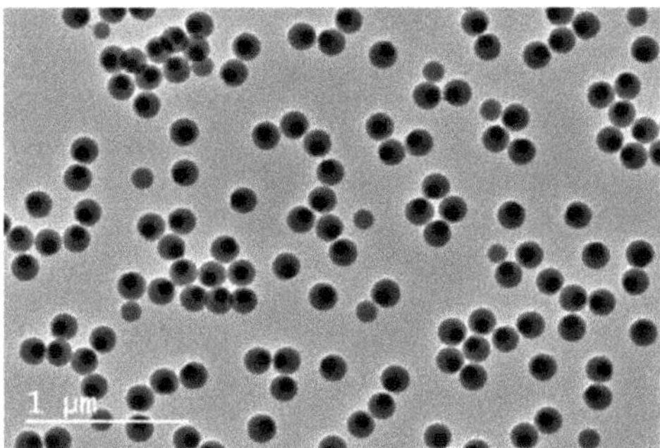

Figure 5. TEM images of silica/PS monopods (Table 1, entry #1.3) after two sorting cycles by centrifugation/redispersion (2000× *g*; 15 min).

3.2. Solvent-Induced Self-Assembly of Silica/Polystyrene Monopods

We carried out a protocol similar to one we recently reported about differently-shaped monopods [20]. It consists of making the PS patches sticky by dispersing the monopods in a good solvent, e.g., THF or DMF, in order to swell the nodules, without dissolving them entirely. To avoid the dissolution, the quality of the solvent was lowered by the presence of a second liquid, which is both miscible with the good solvent and non-solvent for PS, e.g., water or ethanol. For each experiment, we used 7.3×10^{11} monopods in a total liquid volume of 1 mL. The good solvent fraction was introduced last, and the samples were continuously stirred. The self-assembly was investigated by TEM statistical analyses after collection, deposition on the grid and evaporation of a sample microdroplet. By counting the number of silica particles henceforth combined in the same PS particle, we determined the number N of silica patches, meaning that N = 1 for unassembled monopods. N may be also considered here as the aggregation number.

Preliminary trials were performed with the following bad/good solvent systems: ethanol/DMF, water/THF and ethanol/THF mixtures in 20/80 volume ratio. Only the latter system led to significant self-assembly of the monopods. That is why we focused our study on the THF/ethanol system, and we performed a systematic study by varying the volume fraction of THF (Table 2, entries #2.1–#2.5). When the volume fraction of THF was lower than 70%, the amount of good solvent was too low for making the PS nodule sticky enough, and no multipatch particles was observed. From 70%, PS particles with two or more silica patches were observed, but their quantities remained low, resulting from the self-assembly of no more than one fourth of the monopods. Moreover, the higher the THF fraction, the more numerous the tiny and coalescing PS particles resulting from the partial extraction by solubilization and precipitation of PS macromolecules. This is why we decided that a 20/80 ethanol/THF volume ratio is optimal for our experimental conditions. In order to increase the quantities of multipatch PS particles, we increased the concentration of monopods to increase the probability of the particles meeting (Table 2, entries #2.4 and #2.6–#2.7). Indeed, quadruple the concentration made it possible, all other parameters being equal, to increase the assembled fraction of monopods from 20% to 62%. In such conditions, 35% led to two-patch particles, 17% led to three-patch particles and 10% to particles with higher patch numbers.

Table 2. Final distribution of the number of silica patches on the PS particles as a function of the composition of the ethanol/THF mixture and the concentration of monopods. Experimental conditions: [monopods] = C = 7.3×10^{14} part./L at 25 °C for 24 h. Representative TEM images are shown in Figure S2 for entries #2.3 to #2.7.

Entry.	Ethanol/THF Volume Ratio	Monopod Concentration	Final Distribution of the Number of Silica Patches (%) [1]				
			N = 1 [2]	N = 2	N = 3	N = 4	N ≥ 5
#2.1	50/50	C	100	-	-	-	-
#2.2	40/60	C	100	-	-	-	-
#2.3	30/70	C	91	7	2	-	-
#2.4	20/80	C	80	17	3	-	-
#2.5	10/90	C	76	13	4	1	6
#2.6	20/80	2C	77	19	3	1	-
#2.7	20/80	4C	38	35	17	5	5

[1] Expressed in fraction of monopods assembled in PS particles of similar patch number; [2] corresponds to the fraction of unassembled monopods.

With the optimized conditions of this last experiment (Table 2, entry #2.7), we studied the kinetics of self-assembly by collecting and analyzing samples over time (Figure 6). It can be observed that within the first minutes, a large number of particles with two patches was formed as well as some others with 3 or 4 patches. Then, these proportions changed more slowly in favor of a progressive increase in the average number of patches, mainly to the detriment of the monopod population. The two-patch particle fraction remained relatively stable. Beyond 18 h, the variations in the composition of the system were of the same order of magnitude as the margin of error in the statistical analysis, which showed that an equilibrium was probably reached. We also studied the impact of an increase in temperature over the 25–45 °C range, and we showed that it has a relatively small influence on the distribution of the number of patches on the PS particles—we observed a slight increase in the average number—but this significantly complicated the experimental setup. Thus, we carried out the next experiments at room temperature.

Typical TEM images of as-obtained PS particles with several silica patches are displayed in Figure 7a. To check that the surface of the silica patches was still accessible, we performed the test of silica regrowth (Figure 7b). We can observe that silica protrusions appeared on the surface of the silica patches proving as previously that they protrude out of the surface of the PS particles. To verify that the self-assembly protocol was robust, we proceeded to assemble the monopods on the surface of which we had previously created a silica protrusion. Figure 7c shows TEM images very similar to those of Figure 7b, which proves on the one hand that the regrowth of the silica can be carried out as well before or after the assembly, and on the other hand that the assembly of the monopods is primarily controlled by the PS patch and not the silica patch. One would have expected that the steric hindrance of the silica protrusion would lead to particles that are more symmetrical. However, the comparison of TEM images shows that the deviations from the repulsion figures, e.g., straight line and equilateral triangle for particles with two and three patches, respectively, were very similar.

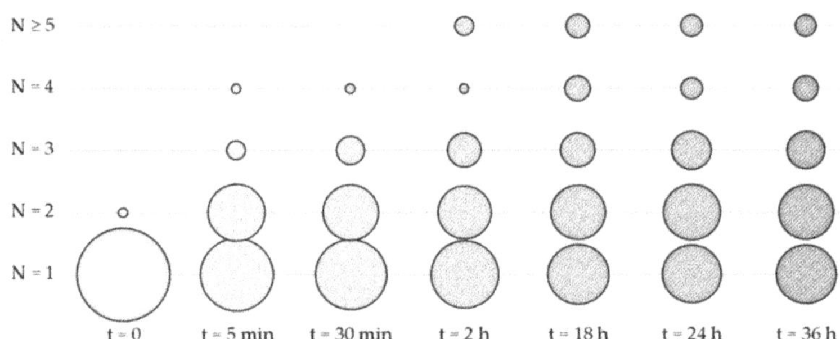

Figure 6. Evolution of the distribution of the number of silica patches with time. The surface of the disks is proportional to the fraction of the monopods assembled in PS particles of similar patch number. Self-assembly experimental conditions: [monopods] = 4C = 2.9×10^{15} part./L at 25 °C in ethanol/THF volume ratio 20/80. The data for t = 0 and t = 24 h correspond to the Entry #1.3 of Table 1 and Entry #2.7 of Table 2, respectively.

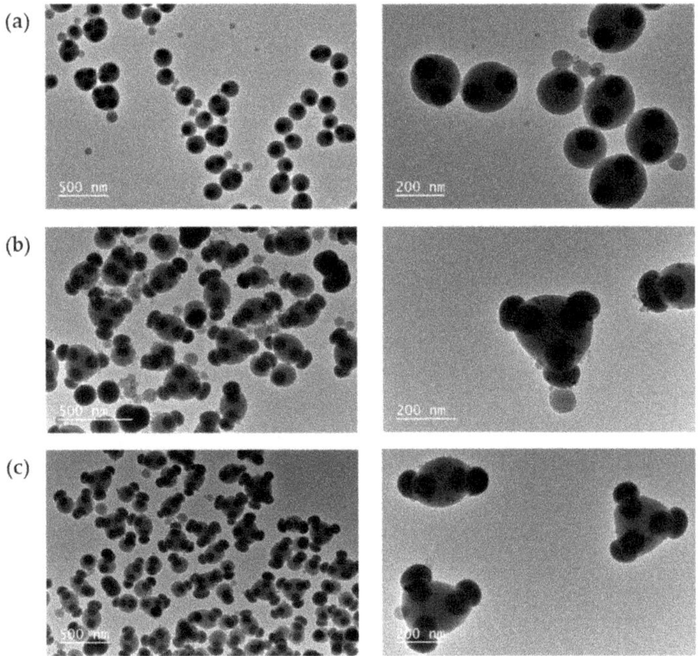

Figure 7. Representative TEM images of (**a**) PS particles with silica patches as obtained according to experimental conditions described in Table 2, entry #2.7; (**b**) the same particles after silica regrowth; and (**c**) the particles obtained after assembly in similar conditions of the silica/PS monopods after silica regrowth.

3.3. Extending the Strategy to Magnetic Silica Patches

To return to our primary objective of making particles with several magnetic patches, we tried to apply the experimental conditions, optimized in the previous sections, using magnetic silica particles thereafter. They are in fact maghemite supraparticles encapsulated in a silica shell so that the final size is around 100 nm, that is, equivalent to those of the model silica patches used earlier. The maghemite supraparticles were prepared through the evaporation-induced emulsion route [22] that we adapted from the original work to replace the polyacrylate-based coating by silica. This multistep procedure includes:

- The preparation of a hydrophobic ferrofluid from 7.5 nm maghemite nanoparticles surface-coated with oleic acid and dispersed in cyclohexane. Compared to the initial protocol [22], cyclohexane was preferred to toluene because its viscosity is slightly higher and makes it possible to minimize the formation of small satellite droplets during droplet rupturing;
- Its emulsification in water in the presence of SDS as surfactant;
- The cyclohexane evaporation to obtain 79 $\pm$ 10 nm sized maghemite supraparticles dispersed in water thanks to the outer layer of SDS;
- Their mechanical strengthening with silica which first requires a delicate transfer of the supra-particles in a hydro alcoholic medium—made possible by the addition of SDS—and then is performed in two steps. We showed by TEM analysis of the as-obtained magnetic silica particles that the thickness of the observed silica layer was always about 8 nm less than the expected thickness (Figure S3a). This demonstrates that part of the produced silica serves above all to fill the space between the 7.5 nm maghemite nanoparticles within the supraparticles, as confirmed by infrared absorption spectroscopy analysis, which showed no traces of SDS or oleic acid in the magnetic silica particles (Figure S3b).

Before describing the preparation of the magnetic silica/PS monopods, two notable differences with regard to the pure silica seeds shall be noted. First, the size polydispersity was greater, despite the fact that the size distribution was narrowed by two centrifugation steps, one to remove the smallest and the other to remove the largest. Thus, the diameter of the magnetic silica particles used was 105 $\pm$ 10 nm (Figure S3c), i.e., a size polydispersity index of 1.750 against 1.003 for the pure silica seeds. Second, the available quantities of magnetic silica particles were quite low, leading us to adapt the seed-growth emulsion polymerization protocol.

We used the seed-growth emulsion polymerization protocol reported by Yin and coworkers [27], which made it possible to reduce by a factor of 15 the seed concentration. We first checked that these new experimental conditions led to the preparation of monopods from pure silica seeds with d_{MPS} = 0.6 funct./nm^2 corresponding to the optimal value determined previously for the original polymerization protocol (Table 3, entry #3.0). Then, we used magnetic silica seeds with the same nominal MPS grafting density (Table 3, entry #3.1; Figure S4a). In that case, the observed fraction of monopods was very low (4%), and two thirds of the multipods were undesired hexapods. Therefore, we increased progressively the nominal MPS grafting density and obtained for d_{MPS} = 1.9 funct./nm^2 essentially monopods with morphological yield as high as 90% for experiments repeated three times under the same conditions (Table 3, entries #3.4 to #3.6). The main side-product was PS particles containing several magnetic silica seeds and/or bipods. Two reasons can explain why we needed a nominal MPS grafting density more than 3 times greater. First, we can not exclude different silica roughness and porosity between the pure seeds and the magnetic seeds; therefore, d_{MPS} was probably underestimated for magnetic seeds. Second, their size polydispersity could also be another source of calculation error.

During the synthesis of the batch of magnetic monopods corresponding to entry #3.6 of Table 3, we collected two samples at t = 30 min and t = 3 h to compare the size of the PS nodule with that obtained at the end of reaction, i.e., for t = 6 h. We then observed an average PS nodule-to-magnetic seed size ratio of 1.6, 2.2 and 2.4, respectively. Moreover, we performed EDX analysis in STEM mode to show the presence of carbon, silicon and iron elements and to confirm the presence of iron oxide nanoparticles within the silica patch while this patch protrudes slightly out of the PS particle (Figure 8a).

Table 3. Morphological distribution and geometrical parameters of the magnetic silica/PS particles as a function of the nominal grafting surface density d_{MPS}. Experimental conditions: [100-nm silica] = 4.8 × 10^{14} part./L; [styrene] = 56.8 g/L; [Synperonic® NP30] = 0.19 g/L; [SDS] = 0.01 g/L; [$Na_2S_2O_8$] = 0.78 g/L; 70 °C and 6 h. Representative TEM images are shown in Figure S4 for entries #3.1, #3.2 and #3.4.

Entry	d_{MPS} (funct./nm^2)	Monomer-to-Polymer Conversion (%)	Final Batch Composition (%) [1]							
			Monopods	Bipods	Tripods	Tetrapods	Pentapods	Hexapods	Other Multipods	Multisilica [2]
#3.0 [3]	0.6	95	96	-	-	-	-	-	-	4
#3.1	0.6	88	4	9	4	4	6	65	7	1
#3.2	1.0	76	3	6	14	30	10	23	11	3
#3.3	1.5	66	44	30	19	6	-	-	-	1
#3.4	1.9	93	90	1	-	-	-	-	-	9
#3.5	1.9	85	96	-	-	-	-	-	-	4
#3.6	1.9	91	94	2	-	-	-	-	-	4

[1] Fractions expressed with respect to the magnetic silica seeds, i.e., free PS particles were not considered; [2] multisilica are magnetic silica/PS particles made of two or more seeds; [3] reference experiment carried out from pure silica seeds.

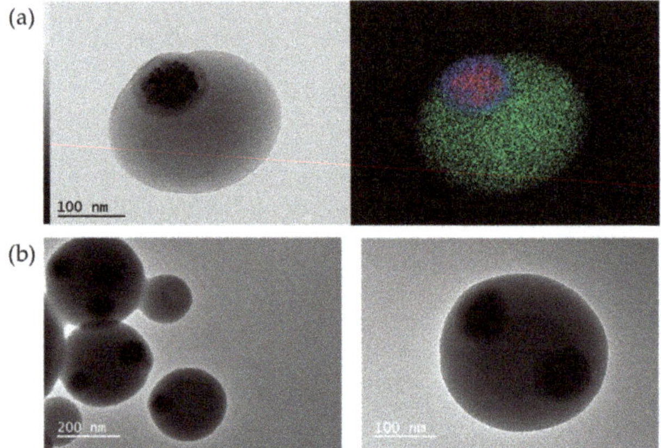

Figure 8. (**a**) STEM image (left) and EDX elemental mapping (right) of a representative magnetic silica/PS monopod (red, blue and green pixels represent iron, silicon and carbon, respectively) corresponding to the entry #3.6 of Table 3; (**b**) TEM images of PS particles with several magnetic patches as obtained after assembly of the magnetic monopods, corresponding to entry #3.6 of Table 3 collected at t = 30 min, using the optimal conditions as defined in entry #2.4 of Table 2.

Lastly, we performed preliminary experiments of solvent-induced self-assembly with the magnetic monopods corresponding to entry #3.6 of Table 3 collected at t = 30 min, i.e., with an average PS nodule-to-magnetic seed size ratio of 1.6. We used the optimal assembly conditions as defined in entry #2.4 (Table 2). After 24 h, we observed that 10% and 11% of the magnetic monopods led to two-patch and three-patch PS particles, respectively (Figure 8b). These results are quite close to those observed with non-magnetic monopods (17% and 3%, respectively). Consequently, experiments—which remain to be performed—under conditions with at least a four-fold increase in concentration should make it possible to significantly improve the yield and obtain particles with higher magnetic patch number.

4. Conclusions

In the present study, we showed that the solvent-induced self-assembly route is efficient and robust for obtaining PS particles of a few hundred nanometers with several 100 nm silica or magnetic silica patches emerging at their surface. The magnetic silica patches are silica particles, which each encapsulate a single maghemite supraparticle. Thus, the proposed paradigm shift is realistic: it is possible to envision the fabrication of polymer particles with several patches by solvent-induced self-assembly of single-patch particles, which are usually easier to synthesize. We can nevertheless identify two avenues for improving both the process and the morphology of the as-obtained multipatch particles. Firstly, the process only makes it possible to obtain mixtures of particles with different numbers of patches. Sorting steps should, therefore, be considered to separate the particles with 2 patches, 3 patches, etc. This can be envisaged by density gradient centrifugation [34], or by sorting using a magnetic field gradient for particles with magnetic patches, as previously developed for narrowing size distribution in ferrofluids [35]. Secondly, it must be admitted that the particles with several patches are not more symmetrical than those reported elsewhere [16,17]. One undoubtedly effective way for the patches to fit into repulsion figures would be both to reduce the volume of the PS nodules on the monopods and/or to promote repulsive forces between the patches, for example, by grafting ionic groups on the surface of the patches.

When homogeneous batches of particles with several magnetic patches are obtained, then assembly studies under a magnetic field will be possible and will have to be correlated with studies by simulation integrating the exact geometry of these multipatch particles. We can nevertheless imagine that these particles with several inorganic patches, magnetic or not, will also be of interest in applications independently of any assembly, such as nanomotors, nanostirrers, nanoswimmers, e-paper pixels, targeted drug delivery and water treatment.

Supplementary Materials: The following are available online at https://www.mdpi.com/2079-4991/11/1/147/s1, Figure S1: Representative TEM images of silica/PS nanoparticles obtained for different nominal MPS grafting surface densities; Figure S2: Representative TEM images of silica/PS monopods after solvent-induced assembly as a function of the composition of the ethanol/THF mixture and the concentration of monopods; Figure S3: Silica strengthening and coating of the maghemite supraparticles with a graph showing the thickness shift, infrared spectrum and representative TEM image of maghemite supraparticles and size distribution; Figure S4: Representative TEM images of silica/PS monopods after centrifugal sorting.

Author Contributions: Conceptualization, E.D.; investigation, E.Y. and L.A.; supervision, M.A., S.M., M.N. and E.D.; writing—original draft preparation, E.Y.; writing—review and editing, E.Y., L.A., M.A., S.M., M.N. and E.D.; funding acquisition, M.N. and E.D. All authors have read and agreed to the published version of the manuscript.

Funding: This research received no external funding.

Institutional Review Board Statement: Not applicable.

Informed Consent Statement: Not applicable.

Data Availability Statement: The data presented in this study are available on request from the corresponding author.

Acknowledgments: Electron microscopy experiments were performed at the Plateforme de Caractérisation des Matériaux (UMS CNRS 5636) at the University of Bordeaux. We acknowledge Jérôme Majimel for STEM-EDX mapping and Weiya Li for helpful discussions about solvent-induced assembly. EY is indebted to the Lebanese University and Azm & Saade association for their support.

Conflicts of Interest: The authors declare no conflict of interest.

References

1. Boles, M.A.; Engel, M.; Talapin, D.V. Self-Assembly of Colloidal Nanocrystals: From Intricate Structures to Functional Materials. *Chem. Rev.* **2016**, *116*, 11220–11289. [CrossRef] [PubMed]
2. Wu, N.; Lee, D.; Striolo, A. (Eds.) *Anisotropic Particle Assemblies*; Elsevier: Amsterdam, The Netherlands, 2018; ISBN 9780128040690.
3. Zhang, Z.; Glotzer, S.C. Self-Assembly of Patchy Particles. *Nano Lett.* **2004**, *4*, 1407–1413. [CrossRef] [PubMed]
4. Rodríguez-Fernández, D.; Liz-Marzán, L.M. Metallic Janus and Patchy Particles. *Part. Part. Syst. Charact.* **2013**, *30*, 46–60. [CrossRef]
5. Yi, G.-R.; Pine, D.J.; Sacanna, S. Recent progress on patchy colloids and their self-assembly. *J. Phys. Condens. Matter* **2013**, *25*, 193101. [CrossRef] [PubMed]

6. Van Anders, G.; Klotsa, D.; Karas, A.S.; Dodd, P.M.; Glotzer, S.C. Digital Alchemy for Materials Design: Colloids and Beyond. *ACS Nano* **2015**, *9*, 9542–9553. [CrossRef] [PubMed]
7. Ravaine, S.; Duguet, E. Synthesis and assembly of patchy particles: Recent progress and future prospects. *Curr. Opin. Colloid Interface Sci.* **2017**, *30*, 45–53. [CrossRef]
8. Li, W.; Palis, H.; Mérindol, R.; Majimel, J.; Ravaine, S.; Duguet, E. Colloidal molecules and patchy particles: Complementary concepts, synthesis and self-assembly. *Chem. Soc. Rev.* **2020**, *49*, 1955–1976. [CrossRef]
9. Tierno, P. Recent advances in anisotropic magnetic colloids: Realization, assembly and applications. *Phys. Chem. Chem. Phys.* **2014**, *16*, 23515–23528. [CrossRef]
10. Teo, B.M.; Young, D.J.; Loh, X.J. Magnetic Anisotropic Particles: Toward Remotely Actuated Applications. *Part. Part. Syst. Charact.* **2016**, *33*, 709–728. [CrossRef]
11. Rossi, L. Magnetic Colloids as Building Blocks for Complex Structures: Preparation and Assembly. *Front. Nanosci.* **2019**, *13*, 1–22. [CrossRef]
12. Walther, A.; Müller, A.H.E. Janus Particles: Synthesis, Self-Assembly, Physical Properties, and Applications. *Chem. Rev.* **2013**, *113*, 5194–5261. [CrossRef] [PubMed]
13. Zhang, J.; Grzybowski, B.A.; Granick, S. Janus Particle Synthesis, Assembly, and Application. *Langmuir* **2017**, *33*, 6964–6977. [CrossRef] [PubMed]
14. Ma, F.; Yang, X.; Wu, N. Directed assembly of anisotropic particles under external fields. In *Anisotropic Particle Assemblies*; Elsevier: Amsterdam, The Netherlands, 2018; pp. 131–165. ISBN 9780128040690.
15. Yammine, E.; Souaid, E.; Youssef, S.; Abboud, M.; Mornet, S.; Nakhl, M.; Duguet, E. Particles with Magnetic Patches: Synthesis, Morphology Control, and Assembly. *Part. Part. Syst. Charact.* **2020**, *37*, 2000111. [CrossRef]
16. Chen, C.-H.; Abate, A.R.; Lee, D.; Terentjev, E.M.; Weitz, D.A. Microfluidic Assembly of Magnetic Hydrogel Particles with Uniformly Anisotropic Structure. *Adv. Mater.* **2009**, *21*, 3201–3204. [CrossRef]
17. Sacanna, S.; Rossi, L.; Pine, D.J. Magnetic Click Colloidal Assembly. *J. Am. Chem. Soc.* **2012**, *134*, 6112–6115. [CrossRef]
18. Skelhon, T.S.; Chen, Y.; Bon, S.A.F. Hierarchical self-assembly of 'hard–soft' Janus particles into colloidal molecules and larger supracolloidal structures. *Soft Matter* **2014**, *10*, 7730–7735. [CrossRef]
19. Ge, X.-H.; Geng, Y.-H.; Chen, J.; Xu, J.-H. Smart Amphiphilic Janus Microparticles: One-Step Synthesis and Self-Assembly. *ChemPhysChem* **2018**, *19*, 2009–2013. [CrossRef]
20. Li, W.; Ravaine, S.; Duguet, E. Clustering of asymmetric dumbbell-shaped silica/polystyrene nanoparticles by solvent-induced self-assembly. *J. Colloid Interface Sci.* **2020**, *560*, 639–648. [CrossRef]
21. Désert, A.; Chaduc, I.; Fouilloux, S.; Taveau, J.-C.; Lambert, O.; Lansalot, M.; Bourgeat-Lami, E.; Thill, A.; Spalla, O.; Ravaine, S.; et al. High-Yield preparation of polystyrene/silica clusters of controlled morphology. *Polym. Chem.* **2012**, *3*, 1130. [CrossRef]
22. Paquet, C.; Pagé, L.; Kell, A.; Simard, B. Nanobeads Highly Loaded with Superparamagnetic Nanoparticles Prepared by Emulsification and Seeded-Emulsion Polymerization. *Langmuir* **2010**, *26*, 5388–5396. [CrossRef]
23. Massart, R. Preparation of aqueous magnetic liquids in alkaline and acidic media. *IEEE Trans. Magn.* **1981**, *17*, 1247–1248. [CrossRef]
24. Van Ewijk, G.A.; Vroege, G.J.; Philipse, A.P. Convenient preparation methods for magnetic colloids. *J. Magn. Magn. Mater.* **1999**, *201*, 31–33. [CrossRef]
25. Torquato, S.; Truskett, T.M.; Debenedetti, P.G. Is Random Close Packing of Spheres Well Defined? *Phys. Rev. Lett.* **2000**, *84*, 2064–2067. [CrossRef] [PubMed]
26. Hubert, C.; Chomette, C.; Désert, A.; Sun, M.; Treguer-Delapierre, M.; Mornet, S.; Perro, A.; Duguet, E.; Ravaine, S. Synthesis of multivalent silica nanoparticles combining both enthalpic and entropic patchiness. *Faraday Discuss.* **2015**, *181*, 139–146. [CrossRef] [PubMed]
27. Ge, J.; Hu, Y.; Zhang, T.; Yin, Y. Superparamagnetic Composite Colloids with Anisotropic Structures. *J. Am. Chem. Soc.* **2007**, *129*, 8974–8975. [CrossRef] [PubMed]
28. Désert, A.; Hubert, C.; Fu, Z.; Moulet, L.; Majimel, J.; Barboteau, P.; Thill, A.; Lansalot, M.; Bourgeat-Lami, E.; Duguet, E.; et al. Synthesis and Site-Specific Functionalization of Tetravalent, Hexavalent, and Dodecavalent Silica Particles. *Angewandte Chemie Int. Ed.* **2013**, *52*, 11068–11072. [CrossRef]
29. Désert, A.; Morele, J.; Taveau, J.-C.; Lambert, O.; Lansalot, M.; Bourgeat-Lami, E.; Thill, A.; Spalla, O.; Belloni, L.; Ravaine, S.; et al. Multipod-Like silica/polystyrene clusters. *Nanoscale* **2016**, *8*, 5454–5469. [CrossRef]
30. Reculusa, S.; Poncet-Legrand, C.; Perro, A.; Duguet, E.; Bourgeat-Lami, E.; Mingotaud, C.; Ravaine, S. Hybrid Dissymmetrical Colloidal Particles. *Chem. Mater.* **2005**, *17*, 3338–3344. [CrossRef]
31. Bourgeat-Lami, E.; Lang, J. Encapsulation of Inorganic Particles by Dispersion Polymerization in Polar Media. *J. Colloid Interface Sci.* **1998**, *197*, 293–308. [CrossRef]
32. Corcos, F.; Bourgeat-Lami, E.; Novat, C.; Lang, J. Poly(styrene-b-ethylene oxide) copolymers as stabilizers for the synthesis of silica-polystyrene core-shell particles. *Colloid Polym. Sci.* **1999**, *277*, 1142–1151. [CrossRef]
33. Thill, A.; Désert, A.; Fouilloux, S.; Taveau, J.-C.; Lambert, O.; Lansalot, M.; Bourgeat-Lami, E.; Spalla, O.; Belloni, L.; Ravaine, S.; et al. Spheres Growing on a Sphere: A Model to Predict the Morphology Yields of Colloidal Molecules Obtained through a Heterogeneous Nucleation Route. *Langmuir* **2012**, *28*, 11575–11583. [CrossRef] [PubMed]

34. Wang, Y.; Wang, Y.; Breed, D.R.; Manoharan, V.N.; Feng, L.; Hollingsworth, A.D.; Weck, M.; Pine, D.J. Colloids with valence and specific directional bonding. *Nature* **2012**, *491*, 51–55. [CrossRef] [PubMed]
35. Forge, D.; Gossuin, Y.; Roch, A.; Laurent, S.; Vander Elst, L.; Muller, R.N. Development of magnetic chromatography to sort polydisperse nanoparticles in ferrofluids. *Contrast Media Mol. Imaging* **2010**, *5*, 126–132. [CrossRef] [PubMed]

 *nanomaterials*

Article

Effect of the Heating Rate to Prevent the Generation of Iron Oxides during the Hydrothermal Synthesis of $LiFePO_4$

Francisco Ruiz-Jorge [1], Almudena Benítez [2], M. Belén García-Jarana [1], Jezabel Sánchez-Oneto [1], Juan R. Portela [1,*] and Enrique J. Martínez de la Ossa [1]

[1] Department of Chemical Engineering and Food Technology, Faculty of Sciences, International Excellence Agrifood Campus (CeiA3), University of Cadiz, 11510 Puerto Real, Spain; franciscojavier.ruiz@uca.es (F.R.-J.); belen.garcia@uca.es (M.B.G.-J.); jezabel.sanchez@uca.es (J.S.-O.); enrique.martinezdelaossa@uca.es (E.J.M.d.l.O.)

[2] Department of Inorganic Chemistry and Chemical Engineering, University Institute of Nanochemistry (IUNAN), University of Cordoba, 14071 Córdoba, Spain; q62betoa@uco.es

* Correspondence: juanramon.portela@uca.es

Abstract: Lithium-ion batteries (LIBs) have gained much interest in recent years because of the increasing energy demand and the relentless progression of climate change. About 30% of the manufacturing cost for LIBs is spent on cathode materials, and its level of development is lower than the negative electrode, separator diaphragm and electrolyte, therefore becoming the "controlling step". Numerous cathodic materials have been employed, $LiFePO_4$ being the most relevant one mainly because of its excellent performance, as well as its rated capacity (170 $mA \cdot h \cdot g^{-1}$) and practical operating voltage (3.5 V vs. Li^+/Li). Nevertheless, producing micro and nanoparticles with high purity levels, avoiding the formation of iron oxides, and reducing the operating cost are still some of the aspects still to be improved. In this work, we have applied two heating rates (slow and fast) to the same hydrothermal synthesis process with the main objective of obtaining, without any reducing agents, the purest possible $LiFePO_4$ in the shortest time and with the lowest proportion of magnetite impurities. The reagents initially used were: $FeSO_4$, H_3PO_4, and LiOH, and a crucial phenomenon has been observed in the temperature range between 130 and 150 °C, being verified with various techniques such as XRD and SEM.

Keywords: lithium iron phosphate; hydrothermal synthesis; heating rate; morphology; crystallinity and purity

Citation: Ruiz-Jorge, F.; Benítez, A.; García-Jarana, M.B.; Sánchez-Oneto, J.; Portela, J.R.; Martínez de la Ossa, E.J. Effect of the Heating Rate to Prevent the Generation of Iron Oxides during the Hydrothermal Synthesis of $LiFePO_4$. *Nanomaterials* **2021**, *11*, 2412. https://doi.org/10.3390/nano11092412

Academic Editors: Ivan Stoikov and Pavel Padnya

Received: 4 August 2021
Accepted: 14 September 2021
Published: 16 September 2021

1. Introduction

The global and increasing energy demand, and the need to replace the consequential consumption of fossil fuels because of environmental concerns, has generated a growing interest, not only in the development of renewable sources of energy, but also in the design of more advanced energy storage systems such as lithium-ion batteries (LIBs) [1], super capacitors [2], lithium sulfur batteries [3], sodium sulfur batteries [4], and redox flow batteries [5] with improved energy density and cycling performance. Nowadays, LIBs are important energy storage devices because of their high specific energy, low self-discharge, excellent cycle performance, no memory effect, and lesser environmental impact [6]. About 30% of the manufacturing budget for LIBs is spent on cathode materials, and its level of development is lower than that of the negative electrode, the diaphragm, or the electrolyte. Therefore, it is the "control step" that determines the battery performance in terms of working voltage, energy density, and rate performance [6]. Numerous cathodic materials have been employed, such as $LiMn_2O$, $Li_3V_2(PO_4)_3$, and $LiCoO_2$, but it is $LiFePO_4$ that has become the main one used because of its excellent performance, as well as its high theoretical specific capacity (170 $mA \cdot h \cdot g^{-1}$), practical operating voltage (3.5 V vs. Li^+/Li), long life cycle, superior safety, low cost, low toxicity, abundant resources, and lesser environmental impact [7].

To date, many studies have focused on the production of $LiFePO_4$ particles by different methods such as: solid state synthesis [8–10], mechanochemical activation [10,11], sol−gel synthesis [10,12], coprecipitation [10,13,14], and hydrothermal synthesis [6,10], among others [7]. Among all of them, hydrothermal synthesis has been gaining prominence since its capacity to produce extremely pure and crystalline particles using relatively low temperatures (115–400 °C) [15,16] has been demonstrated, and in relatively short reaction times (from seconds to hours) [17,18].

For a long time, the research works that have been conducted on hydrothermal synthesis have primarily focused on the improvement of the two main drawbacks that $LiFePO_4$ presents. These are its low diffusivity (10–14 $cm^2 \cdot s^{-1}$) and its low electrical conductivity (10^{-9} $S \cdot cm^{-1}$) [1]. Several studies have tried to find the way to improve lithium ion diffusivity by varying the size and morphology of its particles through the control of the different variables that have an influence on their formation process in order to shorten lithium-ion main diffusion channel ([10] channel): pH [19,20], reaction time [15,21], stirring [22], temperature [16,23], and use of surface-modifying agents and reducing agents [24–28]. Other studies have tried to improve lithium-ion electrical conductivity by coating its particles with carbon or doping them with transition metals [29–31].

In addition, another important part of the studies on hydrothermal synthesis have been based on the rapid and economic production of $LiFePO_4$ [17,32]. To accomplish an efficient production of $LiFePO_4$, continuous processes, where preheating is usually rapid, are required. However, the use of high heating rates may generate large amounts of iron oxide impurities (magnetite) and poor particle crystallization, which prevents good electrochemical performance [33,34]. In our previous work [34], we found that magnetite was generated at heating rates of 86 °C/min. This did not occur when heating rates of 5.26 °C/min were used. In addition, it was noted that the particles were less crystalline when rates of 86 °C/min were applied. One alternative approach to avoid magnetite formation consists of the use of reducing agents such as ascorbic acid, pyrrole, urea, sucrose, citric acid, etc. [25,26,35,36] which might increase the cost of the process. The main objective of this work is to reduce the overall process reaction time while obtaining a $LiFePO_4$ as pure as possible, i.e., with less magnetite impurities and without any use of reducing agents. For this purpose, we have focused on the implementation of various combinations of slow and fast heating rates in order to determine the optimal time to start the fast heating step in a process that would start with a slow heating rate. Different hydrothermal synthesis experiments where a slow heating rate from room temperature to 50, 100, 150, 150, 200, 250, and 300 °C was followed by a fast heating stage to reach 300 °C have been carried out. A particularly crucial phenomenon has been observed in the temperature range between 130 and 150 °C, where slow heating rates produce particles with less magnetite impurities and higher crystallinity.

2. Materials and Methods

2.1. Equipment and Experimental Procedures

As a first stage of the work, crystalline $LiFePO_4$ particles were synthesized in a 0.3 L commercially available stainless-steel reactor (bolted closure packless autoclave) described in detail in a previous work [37]. These $LiFePO_4$ particles would be used as a reference or base to be compared with the particles obtained, in the second part of the study (14 mL batch reactors), using the combination of slow and fast heating rates, under the same temperature conditions. The exact moment at which the first crystalline nuclei of $LiFePO_4$ were generated in the reactor was also determined, with the aim of clarifying and solidly supporting the results obtained. The reactors fill ratio was adjusted so that the heating times were as close as possible between the two reactors in order to extrapolate the data.

The procedure to generate the base $LiFePO_4$ consisted of first placing 2.0392 g of $LiOH \cdot H_2O$ and 4.5039 g of $FeSO_4 \cdot 7H_2O$ in the reactor. Then, 180 mL of deionized water and 1.248 mL of H_3PO_4 were added, the reactor was closed, and the mixture was stirred and purged with nitrogen to remove the oxygen. The selected molar ratio, in order to adjust the pH of mixture at 7, has been the same as the one used for the generation of $LiFePO_4$ particles using different configurations of heating rates in this work, being optimized and described in detail in a previous work [34] (Fe: PO_4: Li; 0.9: 1.2: 2.7). The reaction medium was then heated up to 300 °C at a rate of 4 °C/min and cooled down to room temperature without stirring. Upon completion of the hydrothermal synthesis experiments, the final solution was collected from the reactor, centrifuged to separate the $LiFePO_4$ particles located at the bottom of the supernatant, and dried by means of an oven at 80 °C. More specific details can be found in our previous work [34].

For the study on the formation of the first crystalline nuclei of $LiFePO_4$, the same procedure was used as for the generation of base $LiFePO_4$. In this case, 3 experiments were carried out, where temperature was raised up to 130, 140, and 150 °C, respectively, at a rate of 4 °C/min, and then cooled down to room temperature.

For the generation of $LiFePO_4$ particles using different configurations of slow and fast heating rates, two heating systems were combined. The first method, consisting of a sand bath (Techne model TC-8D with a power of 4 kW), had been used in our previous work [34]. For the present work, a new method with a fluidized sand bath (Techne model SBL-2 with a power of 3 kW) was added. Thus, the heating system was made up of two fluidized sand baths, two compressors that introduced air into the baths to fluidize the sand, a temperature controller, and a reactor manufactured by our research team as detailed in a previous work [34]. The experimental procedure used to fill the 14 mL volume batch reactors to perform the hydrothermal synthesis of $LiFePO_4$ from the selected reagents, and the separation of the corresponding $LiFePO_4$ microcrystals obtained had been optimized and described in detail in a previous work [34]. The procedure is highly similar to that of the base $LiFePO_4$, only that to enable agitation in small reactors, the necessary water is added in two parts.

2.2. *Reagents, Process Reactions and Products*

The $LiFePO_4$ was generated by means of the following precursors: $LiOH{\cdot}H_2O$ (99%), $FeSO_4{\cdot}7H_2O$ (98%), H_3PO_4 (85%), and deionized water. All the chemicals were supplied by Aldrich Co.

The process to generate the $LiFePO_4$, could be summarized in two stages. The first and shorter stage consists of the dissolution of the initial reagents ($FeSO_4 + H_3PO_4 + 3LiOH$) in the water, which results in the formation of the intermediated reagents (Li_3PO_4 and $Fe_3(PO_4)_2{\cdot}8H_2O$) [38]. During the second stage, the intermediate reagents, first Li_3PO_4 and finally $Fe_3(PO_4)_2{\cdot}8H_2O$, are dissolved to yield $LiFePO_4$ and Li_2SO_4 [33]. The $LiFePO_4$ particles are generated when the $Fe_3(PO_4)_2{\cdot}8H_2O$ liberates Fe^{+2} into the medium. The global equation for $LiFePO_4$ formation could be expressed as follows:

$$FeSO_4 + H_3PO_4 + 3LiOH \rightarrow LiFePO_4 + Li_2SO_4 + 3H_2O$$

When the heating rate of the reactors during the formation of $LiFePO_4$ is extremely high, large concentrations of Fe^{+2} are quickly released into the medium, which favors the generation of magnetite (Fe_3O_4) because of the Schikorr reaction [39].

Schikorr reaction:

$$2\,(Fe^{+2} \rightarrow Fe^{+3} + e^{-}) \tag{1}$$

$$2\,(H_2O + e^{-} \rightarrow 1/2\,H_2 + OH^{-}) \tag{2}$$

$$2\,Fe^{+2} + 2\,H_2O \rightarrow 2\,Fe^{+3} + H_2 + 2\,OH^{-} \tag{3}$$

The presence of the Fe^{+2} leads to:

$$3\,Fe^{+2} + 2\,H_2O \rightarrow Fe^{+2} + 2\,Fe^{+3} + H_2 + 2\,OH^{-} \tag{4}$$

Electroneutrality requires the iron cations on both sides of the equation to be counterbalanced by 6 hydroxyl anions (OH^-):

$$3\,Fe^{+2} + 6\,OH^- + 2\,H_2O \rightarrow Fe^{+2} + 2\,Fe^{+3} + H_2 + 8\,OH^- \tag{5}$$

$$3\,Fe\,(OH)_2 + 2\,H_2O \rightarrow Fe\,(OH)_2 + 2\,Fe\,(OH)_3 + H_2 \tag{6}$$

Due to the autoproteolysis of the hydroxyl anions:

$$OH^- + OH^- \rightarrow O^{-2} + H_2O \tag{7}$$

Therefore:

$$3\,Fe\,(OH)_2 + 2\,H_2O \rightarrow (FeO + 1\,H_2O) + (Fe_2O_3 + 3\,H_2O) + H_2 \tag{8}$$

$$3\,Fe\,(OH)_2 \rightarrow FeO + Fe_2O_3 + 2\,H_2O + H_2 \tag{9}$$

Finally, following the magnetite formation the following reaction takes place:

$$FeO + Fe_2O_3 \rightarrow Fe_3O_4 \tag{10}$$

Therefore, by increasing the Fe^{+2} concentration in the medium, the Schikorr reaction might accelerate.

In order to evaluate the possibility of improving the overall process, this study has focused in detailed the effect caused by different heating configurations, so that increments in the heating rate can be applied at different times during the hydrothermal synthesis. Thus, this influence from the different heating configurations has been closely analyzed to determine the critical temperature levels during the heating stage where a high heating rate can be implemented to improve the $LiFePO_4$ synthesis process, not only in terms of the crystallinity, morphology, and purity of the particles, but also with regard to the feasibility of the process.

2.3. Formation of the Base $LiFePO_4$

In order to carry out an adequate comparison of the crystallinity, purity, and morphology of the particles obtained from the subsequent experiments, we previously produced our own $LiFePO_4$ particles to be used as a reference. These will be referred to here as the "base $LiFePO_4$". Such base $LiFePO_4$ particles have to be generated under tightly controlled temperature and heating rate conditions (See Figure 1) according to the literature [15,16,18,23,24,40]. The reactor and experimental procedure have been described in Section 2.1. The base experiment was carried out under the same conditions as the rest of the experiments conducted in this work: i.e., the reaction medium was heated up to 300 °C, reached a pressure of 103 bar, was held under such conditions for 5 min, and then slowly cooled down to room temperature.

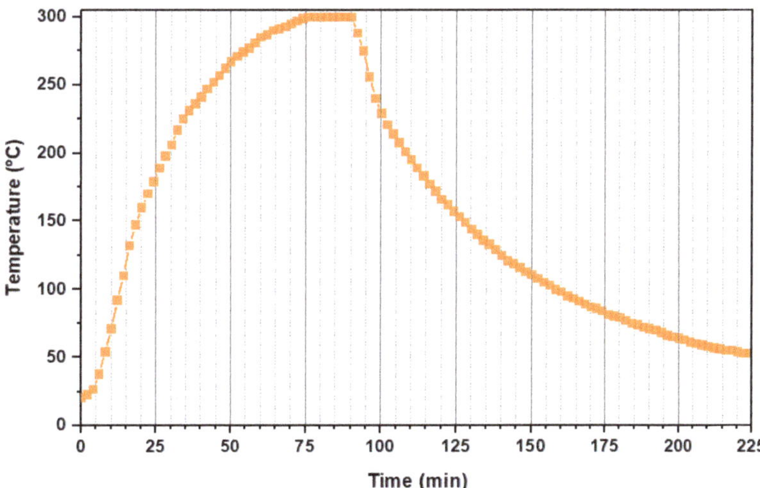

Figure 1. Temperature profile of the reaction medium during the hydrothermal synthesis of the base $LiFePO_4$.

2.4. *Study of the Formation of the First Crystalline Nuclei of* $LiFePO_4$

In order to perform an accurate analysis of the results from this study, it is necessary to determine the exact moment when the first crystalline nuclei of $LiFePO_4$ appear. Given that $LiFePO_4$ is generated when $Fe_3(PO_4)_2$-$8H_2O$ begins to dissolve, and that this depends on the length of time required to heat the reaction medium, such length of time required by our reactors should be precisely determined in order to clarify this key point regarding the purity and crystallization of $LiFePO_4$. According to the study carried out by J. Lee and A.S. Teja [16], the formation of the first crystalline nuclei would take place when the medium reaches between 120 and 190 °C, since only $Fe_3(PO_4)_2$-$8H_2O$ appeared in the samples obtained at 120 °C, while the $LiFePO_4$ particles were already visible in the samples produced at 190 °C. On the other hand, C. Min et al. [40], established that a large amount of $LiFePO_4$ nuclei are rapidly formed when the medium temperature reaches values around 124–130 °C.

Considering the temperatures that have been already studied in the bibliography, the experiments were conducted at 130, 140, and 150 °C (reaching 1.01, 6.55 and 7.58 bar, respectively). The reactor and the experimental procedures have been described in Section 2.1.

2.5. *Formation and Separation of* $LiFePO_4$ *Microcrystals*

The reactor heating procedure was as follows: First, the reactor was submerged into one of the fluidized sand baths still at room temperature (the bath heating system had not been turned on yet). Once inside, the sand bath was turned on and a slow heating rate of 4.44 °C/min was applied. When the target temperature set for each experiment was reached, the reactor was submerged into another sand bath, which had been previously heated and maintained at a constant temperature of 300 °C. Therefore, from the moment the reactor was submerged into the second sand bath until the moment the reactor reached 300 °C, the medium was subjected to a fast heating rate. Thereafter, the reactor was maintained at 300 °C for 5 min (reaction time) before being removed from the sand bath and kept in contact with the laboratory air until completely cooled down to room temperature (see Figure 2). Thus, the final reaction temperature was the same for all the experiments, but the heating time, and therefore the total time of each experiment, which comprises the heating time plus the 5 min reaction time at 300 °C, was different.

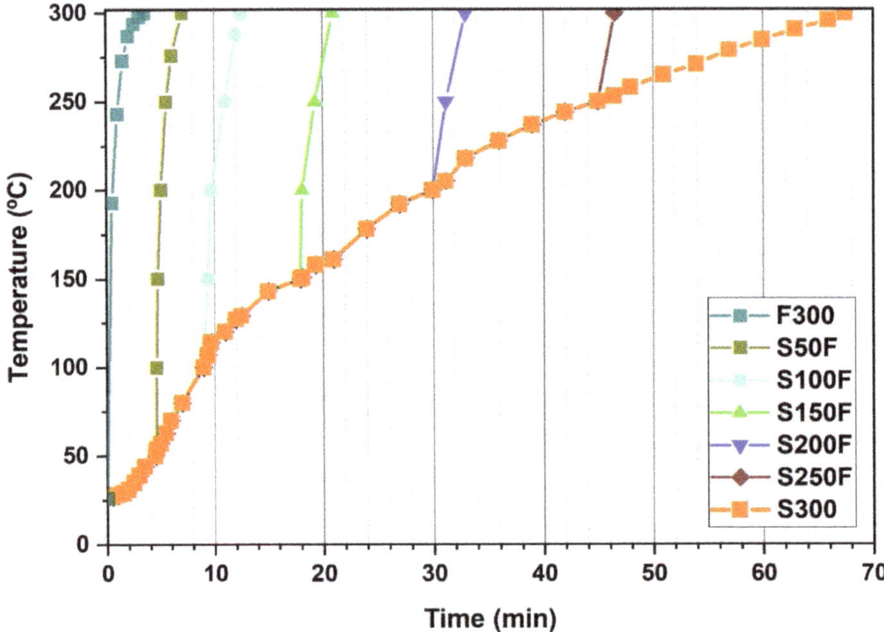

Figure 2. Temperature profile of the reaction medium during the combined slow-fast heating experiments to form $LiFePO_4$ microcrystalline particles.

Table 1 includes the operating conditions of the different hydrothermal synthesis (HS) experiments that combine slow and fast heating. In addition, in order to further consider the configuration range to be tested, two experiments with only one type of heating were carried out: (i) thus, experiment F300 was only subjected to fast heating up to 300 °C, while (ii) experiment S300 was only subjected to slow heating up to the same temperature. Figure 2 shows the evolution of the temperature profiles of the reaction medium over time for the HS experiments included in Table 1.

Table 1. Nomenclature of the samples and their corresponding experimental conditions. All the samples were heated up to 300 °C and subjected to 110 bars of pressure.

HS Experiments	Slow Heating T_o (°C)	Fast Heating T_o (°C)	Reaction Time (min)	Total Time Length of the Synthesis Process * (min)
F300	-	300	5	0 + 3.5 + 5 = 8.5
S50F	50	300	5	4.5 + 3 + 5 = 12.5
S100F	100	300	5	9 + 3 + 5 = 17
S150F	150	300	5	18 + 3 + 5 = 26
S200F	200	300	5	30 + 2.5 + 5 = 37.5
S250F	250	300	5	45 + 2 + 5 = 52
S300	300	-	5	67.5 + 0 + 5 = 72.5

* Total time length of the synthesis process: slow heating + fast heating + reaction time.

Once the hydrothermal synthesis experiments were completed, the final solution was collected from the reactor, centrifuged, and dried, following the procedure already explained in Section 2.1.

2.6. Characterization

The structural properties of $LiFePO_4$ particles were analyzed by X-ray diffraction (XRD) by means of a Bruker D8 Discover A25 diffractometer (Bruker Española S.A., Madrid, Spain) using Cu Kα radiation, Ge monochromator, and a Lynxeye detector. The patterns were registered within the 10−80° (2θ) range, according to a 140 s. step time. The lattice cell parameters, the crystallite size and the amount of impurities present in the synthesized $LiFePO_4$ were calculated using Topas software (Bruker Española S.A., Madrid, Spain) according to the full pattern matching method. The morphological properties of the samples were determined by means of a field-emission scanning electron microscope (FESEM) using a FEI-Nova Nano SEM 450 instrument (Izasa Scientific, Madrid, Spain). The presence of magnetite impurities in the synthesized $LiFePO_4$ was verified by X-ray diffraction and also by our own-built electromagnet to confirm the magnetic behavior of the samples.

3. Results and Discussion

3.1. Analysis of the Synthesized Base $LiFePO_4$

The purity and crystallinity of the synthesized base $LiFePO_4$ was determined by XRD. Figure 3 shows the XRD pattern of the $LiFePO_4$ powders synthesized in the HS base experiment, where the peaks in the different diffractograms closely match the standard $LiFePO_4$ pattern (JCPDS card no. PDF 40-1267). Moreover, no impurities (no magnetite) were detected in the diffractogram.

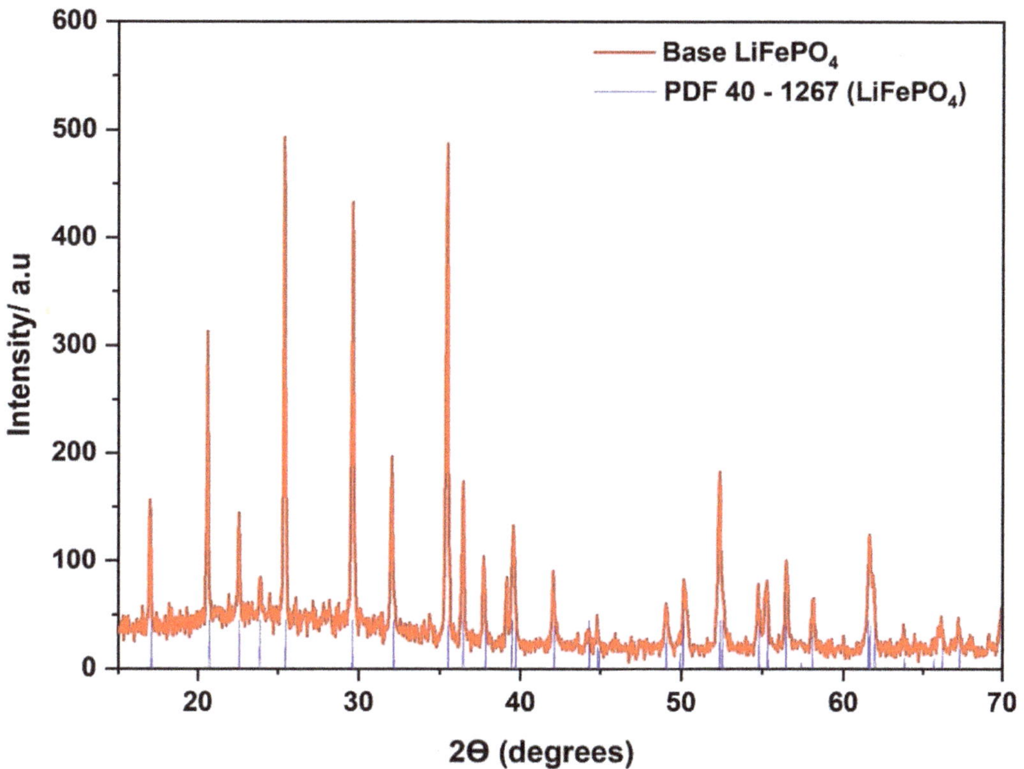

Figure 3. XRD pattern of the synthesized base $LiFePO_4$.

The refinement of the lattice parameters for the orthorhombic structure of the base $LiFePO_4$ with the *Pnma* space group provided the following values: a = 10.3328 Å, b = 6.0043 Å, and c = 4.6977 Å, which is in accordance with the literature [41].

The crystallinity of the particles obtained in the base experiment was considered optimum, and, therefore, they were suitable to be used as the reference for this study. That is, the particles obtained from the experiments, where different combinations of slow heating and fast heating was implemented, were to be mainly compared against the crystallinity of the base $LiFePO_4$.

Figure 4 displays SEM images of the base $LiFePO_4$, where flat and well-defined microparticles can be observed. Its morphology is dominated by face (010) that corresponds to diamond-shaped crystallites [42]. Thus, diamond-shaped platelets are less than 1 micron thick and their dimensions vary between 2–4 microns for the shortest diagonal and 6–8 microns for the longest one.

Figure 4. SEM micrographs of the synthesized base $LiFePO_4$ obtained from the HS base experiment (two random areas of the sample are displayed).

3.2. Study on the Formation of the First Crystalline Nuclei of $LiFePO_4$

Figure 5 shows the XRD pattern of synthesized $LiFePO_4$ powders produced at different temperatures: 130, 140, and 150 °C. As can be observed, the samples that had been synthesized at 140 and 150 °C present the characteristic peaks that can be attributed to the orthorhombic olivine structure of $LiFePO_4$. However, the sample that had been synthesized at 130 °C presented no evidence of any $LiFePO_4$ content. Although, as previously mentioned, Min et al. [34] established that a large amount of $LiFePO_4$ nuclei were formed in the 124–130 °C range. $Fe_3(PO_4)_2$-$8H_2O$ may be soluble within that temperature range albeit the heating time employed in this study may have not been long enough to allow the generation of the first crystalline nuclei of $LiFePO_4$. Therefore, even though it cannot be affirmed that the first $LiFePO_4$ crystalline nuclei are necessarily formed between 124 and 130 °C, it can be presumed that $Fe_3(PO_4)_2$-$8H_2O$ begins to dissolve at around 130 °C, so that this temperature level might be key to explain much of what happens in this process, which would make it worth a deeper discussion, as can be seen in Section 3.3.

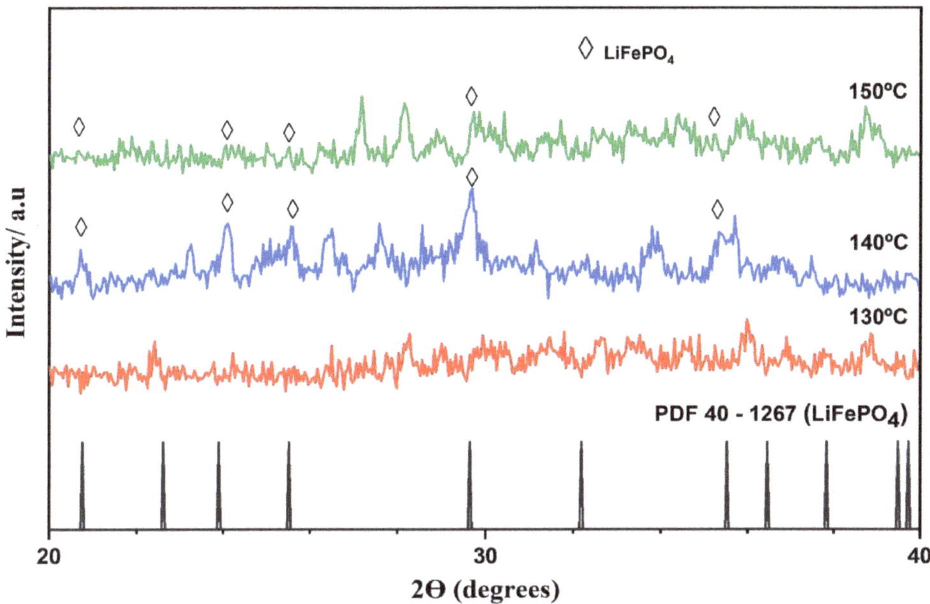

Figure 5. XRD patterns of the $LiFePO_4$ particles obtained by hydrothermal synthesis at 130, 140 and 150 °C.

3.3. Study on the Combination of Slow-Fast Heating Rates during the HS of $LiFePO_4$ Particles

3.3.1. Crystallinity and Purity of the $LiFePO_4$ Particles

The purity and crystallinity of the synthesized particles was determined by XRD. Figure 6 shows the XRD patterns of $LiFePO_4$ powders generated through the different experiments included in Table 1 and that corresponding to the base HS experiment. The peaks in all of the diffractograms closely match the reference pattern (JCPDS card no. PDF 40-1267). Therefore, the hydrothermal synthesis of $LiFePO_4$ particles has been confirmed for all the experiments that have been carried out. On the other hand, the XRD patterns revealed magnetite impurities in some of the samples, as can be seen in Figure 6.

Table 2 shows that only pure $LiFePO_4$ particles were generated in experiments S300 and S250F. On the other hand, all the particles that had been obtained from the experiments that implemented a fast heating rate before reaching 250 °C (F300, S50F, S100F, S150F, and S200F) presented magnetite impurities to a greater or lesser degree. This could be due to the dissolved form of $Fe_3(PO_4)_2{\cdot}8H_2O$. It has been already mentioned (Section 2.2) that $LiFePO_4$ is generated when $Fe_3(PO_4)_2{\cdot}8H_2O$ releases Fe^{+2} into the medium, and this occurs at around 130 °C. Therefore, the generation of magnetite in samples F300, S50F, S100F, S150F, and S200F could be due to the high concentration of Fe^{+2} in the medium generated by the rapid solubilization of $Fe_3(PO_4)_2{\cdot}8H_2O$, which in turn would allow the anaerobic oxidation of Fe^{+2} into Fe^{+3} by the water in the medium, as described by Schikorr reaction. On the other hand, the absence of magnetite in the samples S300 and S250F could be explained by the slow and gradual solubilization of most of the $Fe_3(PO_4)_2{\cdot}8H_2O$, which did not cause the medium Fe^{+2} supersaturation. It is worth noting that substantial differences could be observed between the amount of magnetite generated in the F300, S50F, and S100F samples (higher magnetite peaks) against that generated in the S150F and S200F samples (lower magnetite peaks). This lesser amount of magnetite impurities could be mainly explained by the two following factors: firstly, a slow heating rate implemented over the solubilization stage of the $Fe_3(PO_4)_2{\cdot}8H_2O$, (from 130 °C until 150 °C and from 130 °C until 200 °C; respectively) that causes the Fe^{+2} to be generated progressively in the medium, and thus the Schikorr reaction is minimized; secondly, at 150 °C, a certain amount of $LiFePO_4$ has already been generated and, consequently, there is a smaller amount of Fe^{+2} in the medium.

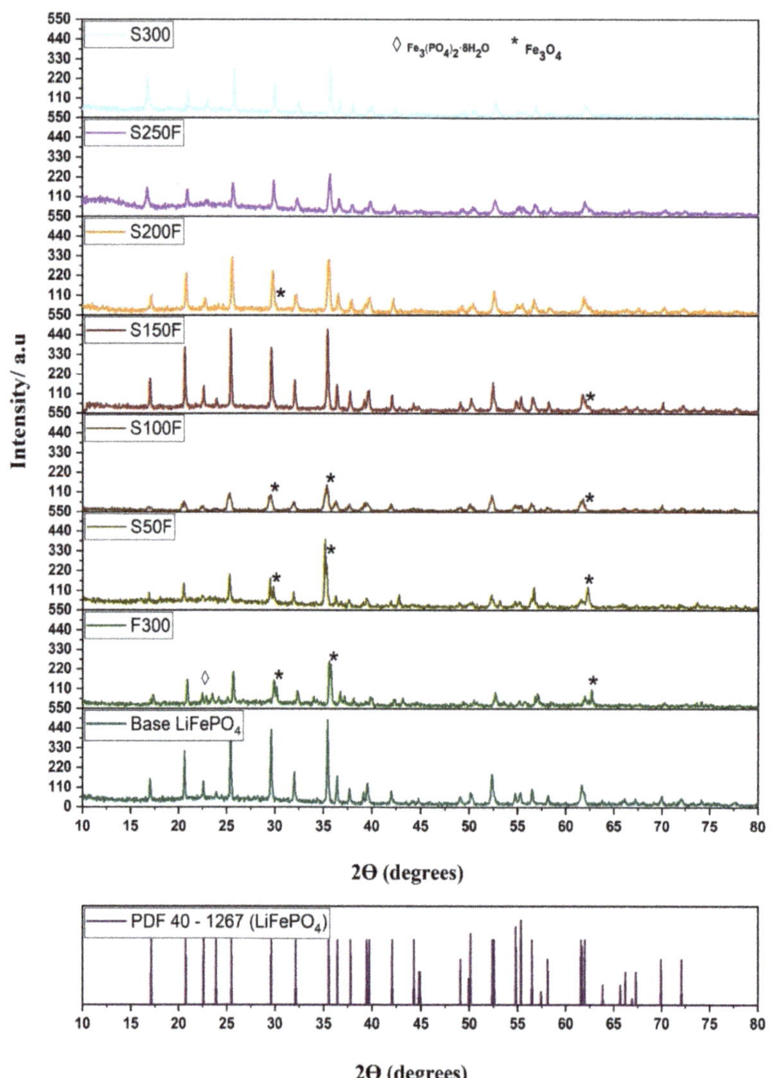

Figure 6. XRD patterns of hydrothermally synthesized $LiFePO_4$.

Table 2. Crystalline phases found in the XRD patterns of the synthesized $LiFePO_4$ particles.

HS Experiment	Crystalline Phases
F300	$LiFePO_4 + Fe_3O_4 + Fe_3(PO_4)_2{\cdot}8H_2O$
S50F	$LiFePO_4 + Fe_3O_4$
S100F	$LiFePO_4 + Fe_3O_4$
S150F	$LiFePO_4 + Fe_3O_4$
S200F	$LiFePO_4 + Fe_3O_4$
S250F	$LiFePO_4$
S300	$LiFePO_4$

Regarding crystallinity (see Figure 6), significant differences can also be observed between the F300 S50F, S100F samples and the S150F, S200F, S250F, S300 samples. Thus, the former samples were less crystalline when compared with the base $LiFePO_4$, which could be due to the rapid generation and growth rate of the $LiFePO_4$ particles as a consequence of the medium Fe^{+2} supersaturation. Consequently, this rapid particle growth would lead to a poor crystallization, i.e., a poor rearrangement of the atoms in space. On the other hand, the latter samples, which had been subjected to slow heating over 130 °C, presented a substantially improved crystallization. Particularly, the samples from experiment S150F exhibited an optimum crystallization level. In fact, the samples obtained at temperatures over 130 °C did not present any improved crystallinity but rather a poorer one was observed. According to our previous study [34], maintaining the $LiFePO_4$ particles at high temperatures (300 °C) for a long time would negatively affect their crystallinity. This could explain why the sample from experiment S150F was the most crystalline, given that the $LiFePO_4$ particles that had been formed in the 130 to 150 °C temperature range took just around 8.5 min to reach 300 °C from the moment they had been formed. This means that the length of time that they were subjected to high temperatures was relatively short (see Table 3). The high crystallinity of the S150F sample can be seen in Figure 7, where it is compared against the base sample and a close similarity can be observed.

Table 3. Approximate times to reach operation temperatures.

HS Experiments	T_i *	T_s *	T_f *
S150F	12.5	5.5	3
S200F	12.5	17.5	2.5
S250F	12.5	32.5	2
S300	12.5	55	-

* T_i: Time elapsed from the beginning of the heating of the reactor until the moment the first crystalline nuclei of $LiFePO_4$ are formed. * T_s: Time elapsed from the moment that the first crystalline nuclei of $LiFePO_4$ are formed until fast heating is implemented. * T_f: Time elapsed from the moment that fast heating is implemented and 300 °C is reached.

Therefore, an abrupt change with respect to the generation of magnetite impurities and crystallinity was observed between the S100F and S150F samples that would allow to separate them into two groups as follows: samples F300, S50F, and S100F would fall into the low crystalline particles with extremely high magnetite peaks; while samples S150F, S200F, S250F, and S300 would be classified as highly crystalline particles with low or no magnetite peaks. Figure 7 shows the XRD patterns in more detail, so that the characteristic peak of the base $LiFePO_4$ as well as those of S100F, S150F, and S200F samples can be observed. Furthermore, Figure 7 includes the diffractogram of the LaB_6 sample, which was measured under the same conditions as the base $LiFePO_4$ and the S100F, S150F, and S200F samples, so that their crystallite size would be more accurately measured after removing the instrumental broadening. The full widths at half maximum (FWHM) of the peaks (101), (301), and (311) of the synthesized samples were considered to determine the size of the crystallite (D) according to Debye-Scherrer equation:

$$D = \frac{k\,\lambda}{\beta\,\cos\theta} \tag{11}$$

where λ is the X-ray wavelength in nanometers (nm), β is the full width at half maximum of the peak in radians, θ is the scattering angle in radians, and k is a constant related to crystallite shape, at 0.9 for the Bragg reflections of $LiFePO_4$. The instrumental broadening effect on FWHM was removed using Warren's method on the assumption of a Gaussian peak [43]:

$$\beta^2 = \beta^2_{sample} - \beta^2_{instrumental} \tag{12}$$

where $\beta_{instrumental}$ is referred to LaB_6 peaks at 21.64°, 30.43°, and 37.35° associated with the planes (101), (301), and (311) of the $LiFePO_4$ samples, and correspond to the values of 0.079, 0.075, and 0.067, respectively.

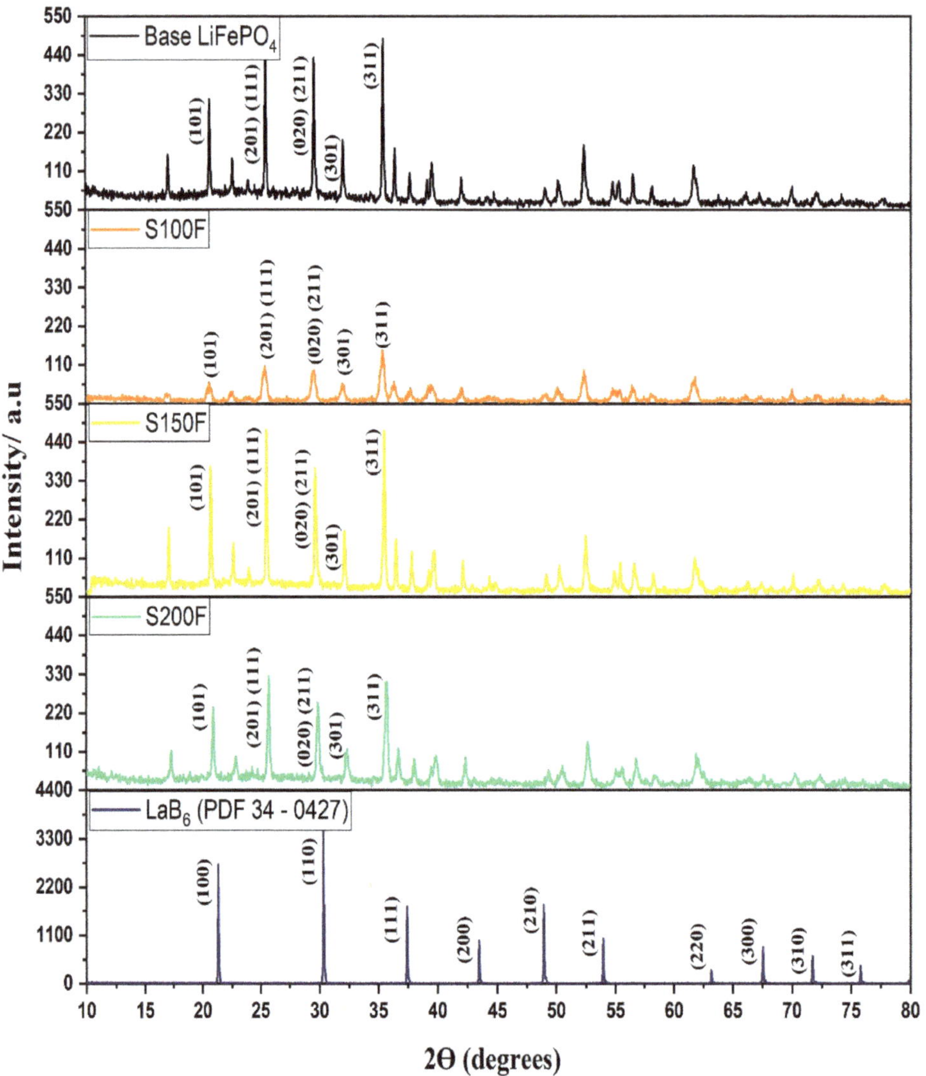

Figure 7. XRD patterns of synthesized Base $LiFePO_4$ vs. S150F. The LaB_6 sample (blue line) was used as the benchmark to determine the instrumental broadening.

Table 4 shows that the lattice parameters of the $LiFePO_4$ particles were similar for all the samples. The Scherrer crystallite size (D) was much smaller than the particle size in every case. The average size of the base $LiFePO_4$ sample was 82.5 nm, while the S150F sample was 34.26 nm, the S150F sample was 66.6 nm and the S200F sample was 39.23 nm according to Scherrer's equation. Therefore, D_{base} $LiFePO_4$ > D_{S150F} $LiFePO_4$ > DS100F $LiFePO_4$ > D_{S200F} $LiFePO_4$ due to a further growth of the crystals during the synthesis process. This claim can be confirmed by comparing the following SEM images: base $LiFePO_4$ (Figure 3), S100F, S150F, and S200F (Figure 8B–D, respectively).

Figure 8. *Cont.*

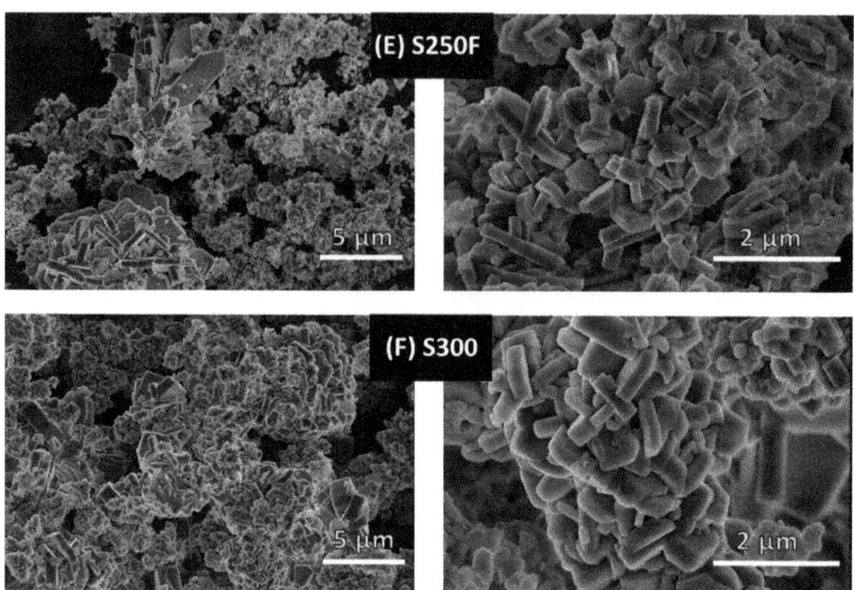

Figure 8. SEM micrographs corresponding to the following experimental samples: (**A**) S50F, (**B**) S100F, (**C**) S150F, (**D**) S200F, (**E**) S250F, (**F**) S300.

Table 4. Structural parameters derived from the XRD patterns the synthesized $LiFePO_4$ particles and Fe_3O_4 impurity content.

Samples	Planes	Peak ° (2θ)	FWHM	Crystallite Size * (nm)	Average Crystallite Size * (nm)	$LiFePO_4$ (%)	Fe_3O_4 (%)
Base $LiFePO_4$	(101)	20.643	0.110	109	82.5	99.74	0.26
	(301)	32.506	0.136	73.5			
	(311)	35.443	0.144	65.0			
S100F	(101)	20.949	0.195	45.4	34.26	82.11	17.89
	(301)	29.715	0.320	26.5			
	(311)	35.539	0.278	30.9			
S150F	(101)	20.685	0.120	88.2	66.63	96.96	3.04
	(301)	32.133	0.162	57.5			
	(311)	35.515	0.168	54.2			
S200F	(101)	20.860	0.173	52.0	39.23	97.19	2.81
	(301)	32.292	0.257	33.2			
	(311)	35.681	0.265	32.5			

* Parameter calculated considering the instrumental broadening.

In addition, the XRD patterns were fitted following the Rietveld method and the orthorhombic space group Pnma. Thus, the amount of magnetite impurities in the S100F, S150F and S200F samples (17.89%, 3.04%, and 2.81% of Fe_3O_4, respectively) were quantified and compared against that of the base $LiFePO_4$ sample (0.26% of Fe_3O_4).

Thereby, the values obtained for the samples studied confirmed the abrupt change presented by the interface of the S100F and S150F samples with regard to magnetite impurity and crystallinity, which made clear that the process requires the use of a reducing agent.

It can be, therefore, be considered that the heating rate within the 130–150 °C temperature range represents a crucial factor if highly crystalline and pure particles are to be obtained while using a smaller amount of reducing agents so that process costs can be kept under control.

3.3.2. Growth Orientation and Morphology of the $LiFePO_4$ Particles

In order to investigate the effect from different heating configurations on the morphology and size of the $LiFePO_4$ particles, all the samples were subjected to SEM analysis (see Figure 8).

In Figure 8A,B show that the reaction mediums that had not been slowly heated to a temperature of at least 130 °C generated products with widely varied morphology, where particles with fewer edges stand out (S50F, S100F). However, the samples that had been slowly heated over 130 °C (pictures C–F) presented a clear morphology mostly formed by hexagonal microparticles (S150F, S200F, S250F, and S300). In order to contrast and clearly demonstrate the crystals generated with a greater number of facets on the samples showed in Figure 8C–F, the five high peaks corresponding to the planes were selected as follows: (2 0 0), (1 0 1), (1 1 1), (0 2 0), and (3 0 1), and they were compared with the most intense peak corresponding to the (3 1 1) plane (Table 5).

Table 5. Relative intensity ratios of the diffraction peaks in the XRD patterns of the synthesized $LiFePO_4$.

Intensity Ratios	F300	S50F	S100F	S150F	S200F	S250F	S300
I (200)/(311)	0.31	0.38	0.22	0.42	0.37	0.69	0.84
I (101)/(311)	0.64	0.59	0.47	0.79	0.76	0.64	0.56
I (111)/(311)	0.82	0.78	0.81	1.01	1.06	0.64	0.56
I (020)/(311)	0.63	0.70	0.71	0.78	0.77	0.85	0.70
I (301)/(311)	0.40	0.41	0.44	0.40	0.38	0.43	0.36

The intensity ratios confirm the increasing trend that appears when the samples are generated by slow heating up to over 130 °C, by which the particles form a greater number of faces. It has, therefore, be demonstrated how the supersaturation of Fe^{+2} in the medium has a remarkable impact.

It should also be noted that that the hexagonal particles in sample S150F (see Figure 8) are evidently thinner, i.e., they have a shorter channel length [10] (see Figure 9). Given that Ceder et al. [44], had already demonstrated that lithium-ion diffusion is several orders of magnitude greater in the [10] direction than in the [1] and [101] directions, and also that M. Saiful Islam et al. [45] proved through simulations, that the energy of Li-ion migration was lower in the [10] direction than in the [1] and [101] directions (Emig [10] = 0.55 eV, Emig [1] = 2.89 eV, and Emig [101] = 3.36 eV), we can affirm that the channel length [10] is a crucial factor to determine particle functionality. The increment in length that experiments the main diffusion channel of lithium ions as the slow heating time is increased (S200F, S250F and S300) may be due to the growth pattern followed by these particles (dissolution–crystallization by Ostwald Ripening, OR) [38]. Even though the particles grow at a greater extent in other directions rather than in the [10] direction of the channel, it is a fact that they also grow in this direction over time. It should also be highlighted how as the heating time increases, size variations are also reduced. This has been corroborated by the results obtained from the different heating configurations applied in the experiments. For example, the particles in sample S150F (as can be seen in Table 3) are formed for 5.5 min under slow heating conditions, and later on, for another 3 min, more particles are formed under fast heating conditions. This difference is less pronounced for example in the S300 sample.

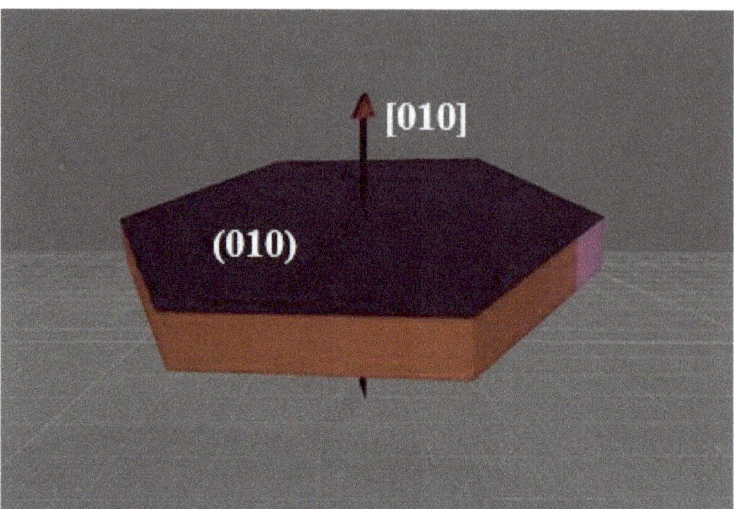

Figure 9. Direction of the particles' channel [010].

Therefore, it has been demonstrated that heating is a really essential factor at the same level as pH or lithium concentration regarding the generation of an optimal morphology, just as K. Dokko et al. [46], had already reported.

4. Conclusions

Our study has demonstrated that the configuration of the heating procedure poses a significant impact on the hydrothermal synthesis of $LiFePO_4$ microparticles. Thus, the optimal values to be applied to such heating procedure in order to achieve the desired balance between particle quality and industrial viability could be deemed as the main innovative contribution by our study.

Industrial processes generally aim at the rapid production of particles, so that both resources and operating costs can be reduced. According to the results obtained from the present study, when high heating rates are implemented (F300 samples), the resulting particles present widely varying morphology, poor crystallinity, and considerable amounts of magnetite impurities. On the contrary, the particles that were generated through slow heating exhibited the opposite characteristics, i.e., consistent morphology, good crystallinity, and lower magnetite content, even if their industrial production could present some difficulties associated with the long times that would be required to form the particles (67.5 min in the case of S300 samples).

It has also been confirmed that an abrupt change can be observed in the S100F and S150F samples interfaces, with quite significant variations in their crystallinity as well as in magnetite impurities content (17.89% and 3.04%, respectively). This fact has allowed the separation of the samples into two groups as follows: firstly, the F300, S50F and S100F samples, with low crystalline particles and extremely high magnetite peaks; and secondly the S150F, S200F, S250F, and S300 samples, containing highly crystalline particles and presenting low or no magnetite peaks. Therefore, it has been verified that the samples that were subjected to slow heating rates above the minimum $LiFePO_4$ particle formation temperature (130 °C) notably improved their crystallinity and reduced their magnetite content, with sample S150F presenting an optimum crystallinity and morphology.

It can generally be concluded that for the formation of large amounts of $LiFePO_4$ particles with good crystallinity levels, it is essential to implement low heating rates within the 130–150 °C temperature range. This would reduce the solubilisation of $Fe_3(PO_4)_2{\cdot}8H_2O$ that starts off within that temperature range and causes the medium Fe^{+2} supersaturation and its subsequent oxidization, which in turn results in a poorer crystallization of the $LiFePO_4$ particles. Nevertheless, rather long operating times are still required and the use of reducing agents, even if at lesser amounts, would be highly recommended.

In order to reduce operating times and improve industrial viability, the following procedure should be used: extremely fast heating from room temperature to 130 °C, low heating rate from 130 °C to 150 °C, and a final fast heating rate period until the desired final temperature is reached. Thanks to the actual capacity of the current industrial means to implement this fast heating rate in just a few seconds, this procedure should allow the production of optimally crystallized and rather size-consistent particles in just a few minutes. In addition, this process time could be further reduced by adding organic acids to the hydrothermal synthesis, such as J. Ni et al. [47] have already demonstrated.

In an attempt to summarize and clearly display the conclusions reached by this study, the effects from the varying parameters that have been analyzed are presented in Table 6.

Table 6. Summary of effects from the varying parameters considered in this study on the following particle characteristics: purity, crystallinity, morphology, and size.

Samples	Purity and Crystallinity	Morphology and Size	Synthesis Process Time * (min)
F300	→ Low crystallinity → Formation of impurities	Widely variable morphology	8.5
S50F	→ Low crystallinity → Formation of impurities	Widely variable morphology	12
S100F	→ Low crystallinity → Formation of impurities	Widely variable morphology	17.5
S150F	→ Extremely high crystallinity → Reduced impurity formation	→ Hexagonal based particles → Shorter lithium-ion diffusion channel. → Larger particle size distribution	26
S200F	→ Good crystallinity → Reduced impurity formation	→ Hexagonal based particles → Longer lithium-ion diffusion channel. → Decreased particle size distribution	38
S250F	→ Good crystallinity → High purity	→ Hexagonal based particles → Longer lithium-ion diffusion channel. → Decreased particle size distribution	52
S300	→ Good crystallinity → High purity	→ Hexagonal based particles → Longer lithium-ion diffusion channel. → Decreased particle size distribution	67.5

* Synthesis process time: including slow heating + fast heating + reaction time.

Author Contributions: Conceptualization, F.R.-J., A.B., M.B.G.-J., J.S.-O., J.R.P. and E.J.M.d.l.O.; methodology, F.R.-J.; software, F.R.-J. and A.B.; validation, F.R.-J., A.B., M.B.G.-J., J.S.-O., J.R.P. and E.J.M.d.l.O.; formal analysis, F.R.-J. and A.B.; investigation, F.R.-J. and A.B.; resources, F F.R.-J., A.B., M.B.G.-J., J.S.-O., J.R.P. and E E.J.M.d.l.O.; data curation, F.R.-J. and A.B.; writing original draft preparation, F.R.-J.; writing review and editing, A.B., M.B.G.-J., J.S.-O., J.R.P. and E.J.M.d.l.O.; visualization, F F.R.-J.; su-pervision, M.B.G.-J., J.S.-O., J.R.P. and E.J.M.d.l.O.; project administration, M.B.G.-J., J.S.-O., J.R.P. and E.J.M.d.l.O.; funding acquisition, M.B.G.-J., J.S.-O., J.R.P. and E.J.M.d.l.O. All authors have read and agreed to the published version of the manuscript.

Funding: This research received no external funding.

Acknowledgments: A. Benitez thanks the financial support from the European Social Fund and Junta de Andalucia. Finally, the authors wish to acknowledge the technical staff from the University Institute of Nanochemistry (IUNAN) of the University of Cordoba.

Conflicts of Interest: The authors declare no conflict of interest.

References

1. Yu, F.; Zhang, L.; Li, Y.; An, Y.; Zhu, M.; Dai, B. Mechanism studies of $LiFePO_4$ cathode material: Lithiation/delithiation process, electrochemical modification and synthetic reaction. *RSC Adv.* **2014**, *4*, 54576–54602. [CrossRef]
2. Panda, P.K.; Grigoriev, A.; Mishra, Y.K.; Ahuja, R. Progress in supercapacitors: Roles of two dimensional nanotubular materials. *Nanoscale Adv.* **2020**, *2*, 70–108. [CrossRef]
3. Benítez, A.; Morales, J.; Caballero, Á. Pistachio shell-derived carbon activated with phosphoric acid: A more efficient procedure to improve the performance of li–s batteries. *Nanomaterials* **2020**, *10*, 840. [CrossRef]
4. Xin, S.; Yin, Y.X.; Guo, Y.G.; Wan, L.J. A high-energy room-temperature sodium-sulfur battery. *Adv. Mater.* **2014**, *26*, 1261–1265. [CrossRef]
5. Perry, M.L.; Weber, A.Z. Advanced Redox-Flow Batteries: A Perspective. *J. Electrochem. Soc.* **2016**, *163*, A5064–A5067. [CrossRef]
6. Zhang, B.; Wang, S.; Li, Y.; Sun, P.; Yang, C.; Wang, D.; Liu, L. Review: Phase transition mechanism and supercritical hydrothermal synthesis of nano lithium iron phosphate. *Ceram. Int.* **2020**, *46*, 27922–27939. [CrossRef]
7. Satyavani, T.V.S.L.; Srinivas Kumar, A.; Subba Rao, P.S.V. Methods of synthesis and performance improvement of lithium iron phosphate for high rate Li-ion batteries: A review. *Eng. Sci. Technol. Int. J.* **2016**, *19*, 178–188. [CrossRef]
8. Dong, Y.Z.; Zhao, Y.M.; Chen, Y.H.; He, Z.F.; Kuang, Q. Optimized carbon-coated $LiFePO_4$ cathode material for lithium-ion batteries. *Mater. Chem. Phys.* **2009**, *115*, 245–250. [CrossRef]
9. Chen, Z.; Zhu, H.; Ji, S.; Fakir, R.; Linkov, V. Influence of carbon sources on electrochemical performances of $LiFePO_4$/C composites. *Solid State Ionics* **2008**, *179*, 1810–1815. [CrossRef]
10. Franger, S.; Le Cras, F.; Bourbon, C.; Rouault, H. Comparison between different $LiFePO_4$ synthesis routes and their influence on its physico-chemical properties. *J. Power Sources* **2003**, *119–121*, 252–257. [CrossRef]
11. Kosova, N.; Devyatkina, E. On mechanochemical preparation of materials with enhanced characteristics for lithium batteries. *Solid State Ionics* **2004**, *172*, 181–184. [CrossRef]
12. Dominko, R.; Goupil, J.M.; Bele, M.; Gaberscek, M.; Remskar, M.; Hanzel, D.; Jamnik, J. Impact of $LiFePO_4$/C composites porosity on their electrochemical performance. *J. Electrochem. Soc.* **2005**, *152*, A858. [CrossRef]
13. Park, K.S.; Son, J.T.; Chung, H.T.; Kim, S.J.; Lee, C.H.; Kim, H.G. Synthesis of $LiFePO_4$ by co-precipitation and microwave heating. *Electrochem. Commun.* **2003**, *5*, 839–842. [CrossRef]
14. Arnold, G.; Garche, J.; Hemmer, R.; Ströbele, S.; Vogler, C.; Wohlfahrt-Mehrens, M. Fine-particle lithium iron phosphate $LiFePO_4$ synthesized by a new low-cost aqueous precipitation technique. *J. Power Sources* **2003**, *119–121*, 247–251. [CrossRef]
15. Benedek, P.; Wenzler, N.; Yarema, M.; Wood, V.C. Low temperature hydrothermal synthesis of battery grade lithium iron phosphate. *RSC Adv.* **2017**, *7*, 17763–17767. [CrossRef]
16. Lee, J.; Teja, A.S. Characteristics of lithium iron phosphate ($LiFePO_4$) particles synthesized in subcritical and supercritical water. *J. Supercrit. Fluids* **2005**, *35*, 83–90. [CrossRef]
17. Xu, C.; Lee, J.; Teja, A.S. Continuous hydrothermal synthesis of lithium iron phosphate particles in subcritical and supercritical water. *J. Supercrit. Fluids* **2008**, *44*, 92–97. [CrossRef]
18. Jin, E.M.; Jin, B.; Jun, D.K.; Park, K.H.; Gu, H.B.; Kim, K.W. A study on the electrochemical characteristics of $LiFePO_4$ cathode for lithium polymer batteries by hydrothermal method. *J. Power Sources* **2008**, *178*, 801–806. [CrossRef]
19. Gu, Y.J.; Li, C.J.; Cheng, L.; Lv, P.G.; Fu, F.J.; Liu, H.Q.; Ding, J.X.; Wang, Y.M.; Chen, Y.B.; Wang, H.F.; et al. Novel synthesis of plate-like $LiFePO_4$ by hydrothermal method. *J. New Mater. Electrochem. Syst.* **2016**, *19*, 33–36. [CrossRef]
20. He, L.H.; Zhao, Z.W.; Liu, X.H.; Chen, A.L.; Si, X.F. Thermodynamics analysis of $LiFePO_4$ pecipitation from Li-Fe(II)-P-H_2O system at 298 K. *Trans. Nonferrous Met. Soc. China* **2012**, *22*, 1766–1770. [CrossRef]
21. Paolella, A.; Bertoni, G.; Hovington, P.; Feng, Z.; Flacau, R.; Prato, M.; Colombo, M.; Marras, S.; Manna, L.; Turner, S.; et al. Cation exchange mediated elimination of the Fe-antisites in the hydrothermal synthesis of $LiFePO_4$. *Nano Energy* **2015**, *16*, 256–267. [CrossRef]

22. Vediappan, K.; Guerfi, A.; Gariépy, V.; Demopoulos, G.P.; Hovington, P.; Trottier, J.; Mauger, A.; Julien, C.M.; Zaghib, K. Stirring effect in hydrothermal synthesis of nano C-$LiFePO_4$. *J. Power Sources* **2014**, *266*, 99–106. [CrossRef]
23. Wu, G.; Liu, N.; Gao, X.; Tian, X.; Zhu, Y.; Zhou, Y.; Zhu, Q. A hydrothermally synthesized $LiFePO_4$/C composite with superior low-temperature performance and cycle life. *Appl. Surf. Sci.* **2018**, *435*, 1329–1336. [CrossRef]
24. Liu, Y.; Gu, J.; Zhang, J.; Yu, F.; Wang, J.; Nie, N.; Li, W. $LiFePO_4$ nanoparticles growth with preferential (010) face modulated by Tween-80. *RSC Adv.* **2015**, *5*, 9745–9751. [CrossRef]
25. Liang, Y.P.; Li, C.C.; Chen, W.J.; Lee, J.T. Hydrothermal synthesis of lithium iron phosphate using pyrrole as an efficient reducing agent. *Electrochim. Acta* **2013**, *87*, 763–769. [CrossRef]
26. Gunawan, I.; Wagiyo, H. Synthesis of $LiFePO_4$ in the Presence of Organic Reductant by Hydrothermal Method and its Characterization. *Macromol. Symp.* **2015**, *353*, 225–230. [CrossRef]
27. Golestani, E.; Javanbakht, M.; Ghafarian-Zahmatkesh, H.; Beydaghi, H.; Ghaemi, M. Tartaric acid assisted carbonization of $LiFePO_4$ synthesized through in situ hydrothermal process in aqueous glycerol solution. *Electrochim. Acta* **2018**, *259*, 903–915. [CrossRef]
28. Qin, X.; Wang, J.; Xie, J.; Li, F.; Wen, L.; Wang, X. Hydrothermally synthesized $LiFePO_4$ crystals with enhanced electrochemical properties: Simultaneous suppression of crystal growth along [10] and antisite defect formation. *Phys. Chem. Chem. Phys.* **2012**, *14*, 2669–2677. [CrossRef]
29. Johnson, I.D.; Lübke, M.; Wu, O.Y.; Makwana, N.M.; Smales, G.J.; Islam, H.U.; Dedigama, R.Y.; Gruar, R.I.; Tighe, C.J.; Scanlon, D.O.; et al. Pilot-scale continuous synthesis of a vanadium-doped $LiFePO_4$/C nanocomposite high-rate cathodes for lithium-ion batteries. *J. Power Sources* **2016**, *302*, 410–418. [CrossRef]
30. Paolella, A.; Turner, S.; Bertoni, G.; Hovington, P.; Flacau, R.; Boyer, C.; Feng, Z.; Colombo, M.; Marras, S.; Prato, M.; et al. Accelerated Removal of Fe-Antisite Defects while Nanosizing Hydrothermal $LiFePO_4$ with Ca^{2+}. *Nano Lett.* **2016**, *16*, 2692–2697. [CrossRef]
31. Yen, H.; Rohan, R.; Chiou, C.Y.; Hsieh, C.J.; Bolloju, S.; Li, C.C.; Yang, Y.F.; Ong, C.W.; Lee, J.T. Hierarchy concomitant in situ stable iron(II)−carbon source manipulation using ferrocenecarboxylic acid for hydrothermal synthesis of $LiFePO_4$ as high-capacity battery cathode. *Electrochim. Acta* **2017**, *253*, 227–238. [CrossRef]
32. Kim, J.K. Supercritical synthesis in combination with a spray process for 3D porous microsphere lithium iron phosphate. *CrystEngComm* **2014**, *16*, 2818–2822. [CrossRef]
33. Hong, S.A.; Kim, S.J.; Chung, K.Y.; Chun, M.S.; Lee, B.G.; Kim, J. Continuous synthesis of lithium iron phosphate ($LiFePO_4$) nanoparticles in supercritical water: Effect of mixing tee. *J. Supercrit. Fluids* **2013**, *73*, 70–79. [CrossRef]
34. Ruiz-Jorge, F.; Benítez, A.; Fernández-García, S.; Sánchez-Oneto, J.; Portela, J.R. Effect of Fast Heating and Cooling in the Hydrothermal Synthesis on $LiFePO_4$ Microparticles. *Ind. Eng. Chem. Res.* **2020**, *59*, 9318–9327. [CrossRef]
35. Vasquez-Elizondo, L.J.; Rendón-Ángeles, J.C.; Matamoros-Veloza, Z.; López-Cuevas, J.; Yanagisawa, K. Urea decomposition enhancing the hydrothermal synthesis of lithium iron phosphate powders: Effect of the lithium precursor. *Adv. Powder Technol.* **2017**, *28*, 1593–1602. [CrossRef]
36. Chen, J.; Wang, S.; Whittingham, M.S. Hydrothermal synthesis of cathode materials. *J. Power Sources* **2007**, *174*, 442–448. [CrossRef]
37. Sánchez-Oneto, J.; Portela, J.R.; Nebot, E.; de la Ossa, E.M. Hydrothermal oxidation: Application to the treatment of different cutting fluid wastes. *J. Hazard. Mater.* **2007**, *144*, 639–644. [CrossRef] [PubMed]
38. Zhu, J.; Fiore, J.; Li, D.; Kinsinger, N.M.; Wang, Q.; DiMasi, E.; Guo, J.; Kisailus, D. Solvothermal synthesis, development, and performance of $LiFePO_4$ nanostructures. *Cryst. Growth Des.* **2013**, *13*, 4659–4666. [CrossRef]
39. Gerhild, T.L. *Schikorr Reaction: Iron(II) Hydroxide, Iron(II,III) Oxide*; Betascript Publishing: Beau Bassin, Mauritius, 2012.
40. Min, C.; Ou, X.; Shi, Z.; Liang, G.; Wang, L. Effects of heating rate on morphology and performance of lithium iron phosphate synthesized by hydrothermal route in organic-free solution. *Ionics* **2017**, *24*, 1285–1292. [CrossRef]
41. Ellis, B.; Kan, W.H.; Makahnouk, W.R.M.; Nazar, L.F. Synthesis of nanocrystals and morphology control of hydrothermally prepared $LiFePO_4$. *J. Mater. Chem.* **2007**, *17*, 3248–3254. [CrossRef]
42. Fisher, C.A.J.; Islam, M.S. Surface structures and crystal morphologies of $LiFePO_4$: Relevance to electrochemical behaviour. *J. Mater. Chem.* **2008**, *18*, 1209–1215. [CrossRef]
43. Amaro-Gahete, J.; Benítez, A.; Otero, R.; Esquivel, D.; Jiménez-Sanchidrián, C.; Morales, J.; Caballero, Á.; Romero-Salguero, F.J. A comparative study of particle size distribution of graphene nanosheets synthesized by an ultrasound-assisted method. *Nanomaterials* **2019**, *9*, 152. [CrossRef] [PubMed]
44. Morgan, D.; Van der Ven, A.; Ceder, G. Li Conductivity in Li_xMPO_4(M = Mn, Fe, Co, Ni) Olivine Materials. *Electrochem. Solid-State Lett.* **2004**, *7*, A30. [CrossRef]
45. Islam, M.S.; Driscoll, D.J.; Fisher, C.A.J.; Slater, P.R. Atomic-scale investigation of defects, dopants, and lithium transport in the $LiFePO_4$ olivine-type battery material. *Chem. Mater.* **2005**, *17*, 5085–5092. [CrossRef]
46. Dokko, K.; Koizumi, S.; Nakano, H.; Kanamura, K. Particle morphology, crystal orientation, and electrochemical reactivity of $LiFePO_4$ synthesized by the hydrothermal method at 443 K. *J. Mater. Chem.* **2007**, *17*, 4803. [CrossRef]
47. Ni, J.; Morishita, M.; Kawabe, Y.; Watada, M.; Takeichi, N.; Sakai, T. Hydrothermal preparation of $LiFePO_4$ nanocrystals mediated by organic acid. *J. Power Sources* **2010**, *195*, 2877–2882. [CrossRef]

Article

Synthesis, Physicochemical Characterization, and Cytotoxicity Assessment of Rh Nanoparticles with Different Morphologies-as Potential XFCT Nanoprobes

Yuyang Li, Giovanni M. Saladino, Kian Shaker, Martin Svenda, Carmen Vogt, Bertha Brodin, Hans M. Hertz and Muhammet S. Toprak *

Biomedical and X-ray Physics, Department of Applied Physics, KTH Royal Institute of Technology, SE 10691 Stockholm, Sweden; yuyangli@kth.se (Y.L.); saladino@kth.se (G.M.S.); kiansd@kth.se (K.S.); martin.svenda@biox.kth.se (M.S.); carmenma@kth.se (C.V.); berthab@kth.se (B.B.); hans.hertz@biox.kth.se (H.M.H.)

* Correspondence: toprak@kth.se; Tel.: +46-735-519-358

Received: 25 September 2020; Accepted: 23 October 2020; Published: 27 October 2020

Abstract: Morphologically controllable synthesis of Rh nanoparticles (NPs) was achieved by the use of additives during polyol synthesis. The effect of salts and surfactant additives including PVP, sodium acetate, sodium citrate, CTAB, CTAC, and potassium bromide on Rh NPs morphology was investigated. When PVP was used as the only additive, trigonal NPs were obtained. Additives containing Br^- ions (CTAB and KBr) resulted in NPs with a cubic morphology, while those with carboxyl groups (sodium citrate and acetate) formed spheroid NPs. The use of Cl^- ions (CTAC) resulted in a mixture of polygon morphologies. Cytotoxicity of these NPs was evaluated on macrophages and ovarian cancer cell lines. Membrane integrity and cellular activity are both influenced to a similar extent, for both the cell lines, with respect to the morphology of Rh NPs. The cells exposed to trigonal Rh NPs showed the highest viability, among the NP series. Particles with a mixed polygon morphology had the highest cytotoxic impact, followed by cubic and spherical NPs. The Rh NPs were further demonstrated as contrast agents for X-ray fluorescence computed tomography (XFCT) in a small-animal imaging setting. This work provides a detailed route for the synthesis, morphology control, and characterization of Rh NPs as viable contrast agents for XFCT bio-imaging.

Keywords: polyol synthesis; rhodium nanoparticles; surfactants; role of additives; morphology control; toxicity; bio-imaging; X-ray fluorescence; contrast agent; XFCT

1. Introduction

Like noble metals Pt and Pd, Rh nanomaterials have caught increasing attention due to applications in catalysis, photonics, and biosensors [1–4]. Physical and chemical properties of nanomaterials are strongly related to the size and morphology of particles. For instance, catalytic properties are associated with the active sites, which is reflected by the spatial shape and exposed crystal facets of metallic nanoparticles (NPs) [5]. In general, a high specific surface area contributes to a high catalytic activity, due to high-index crystal planes showing high binding affinity [5,6]. Likewise, nanomaterials also reveal an inherent correlation between toxicity and their surface properties within biological applications [7–10]. The specific surface-exposed groups could be the reactive sites inducing superoxide radical formation, as major reactive oxygen species (ROS) which are cytotoxic [10]. The hydrophilicity–hydrophobicity, or lipophilicity–lipophobicity, of nanomaterials functionalized by various groups also plays a key role in the toxicity and biocompatibility. Metallic nanomaterials, including Ag and Au, show biocompatibility with a strong dependence on particle size, morphology, and surface properties [11]. The toxicity of Rh NPs as a function of their particle morphology has been rarely discussed in the literature [12]. It is, therefore, scientifically important to explore Rh NPs' morphological effects in connection with their potential biological applications.

Enormous efforts have been devoted to tune the synthesis methods in order to control the morphology and surface properties of metallic NPs. Surface capping agents such as salts and surfactants have shown significant effects on directing the growth of nanocrystals. A diverse set of morphologies of Ag and Au NPs have been reported using different capping agents. For example, PVP preferentially adsorbs on and reduces the growth rate of the Ag(100) surface, while PVP passivates the Au(111) surface. Halides anions like Br^- also adsorb on the Pd(100) surface. Furthermore, citrate is able to passivate the Pd(111) surface [8]. Yu et al. [13] reported on the surface energy of noble metals Rh, Pt, Pd, and Au that is different for various nanocrystal shapes, where Rh showed the highest surface energy for the crystal planes with the same index. The difference in surface energy values is significant (about five–six times higher) when compared to Au, which makes Rh unique as a catalyst. However, if necessary, it is possible to passivate the catalytic activity with selective surface-adsorbed ligands. Furthermore, this extraordinarily high surface energy makes it rather challenging to have a good shape control for Rh [13]. There are very few studies, though, focusing on the morphology control of Rh particles. Br^- ions, for instance, were reported to passivate the (100) plane in Rh nanocrystals [14]. The morphology control of metallic Rh NPs during typical polyol synthesis was demonstrated by replacing the Rh salt precursors with organometallic Rh $Rh_2(COOCF_3)_4$, and the solvent with other polyols such as diethylene glycol, or tri-ethylene glycol [1,15].

We recently demonstrated the potential use of Rh NPs as X-ray fluorescence computed tomography (XFCT) contrast agents [16,17]. CCK-8-based cytotoxicity assays of triangular Rh NPs based on two cells lines, murine macrophages and human-derived ovarian cancer cells, showed toxicity for doses above 40 mg/L [17]. Further research is needed to understand the toxicity induced by Rh NPs. In this work, we present a systematic study on the effect of halide ions and carboxyl salts on the morphology of Rh NPs by one-pot polyol synthesis. The morphological and surface properties and cytotoxicity of morphologically different Rh NPs are studied and discussed, finally demonstrating the performance of Rh NPs in a small-animal XFCT imaging scenario.

2. Materials and Methods

2.1. Materials

Rhodium (III) chloride hydrate ($RhCl_3{\cdot}xH_2O$, Rh 38.5–45.5%), Ethylene glycol (EG, $HOCH_2CH_2OH$, >99%), Poly(vinyl pyrrolidone) (PVP, $C_2H_2N(C_6H_9NO)_nC_{13}H_{10}NS_2$, average MW = 55 kDa), Potassium bromide (KBr, >99%), Sodium Acetate trihydrate (NaAc; $CH_3COONa{\cdot}3H_2O$, >99%), Hexadecyltrimethylammonium bromide (CTAB, $CH_3(CH_2)_{15}N(Br)(CH_3)_3$, >99%), Hexadecyltrimethylammonium chloride (CTAC, $CH_3(CH_2)_{15}N(Cl)(CH_3)_3$, >99%), and Sodium citrate dihydrate (NaCit; $HOC(COONa)(CH_2COONa)_2{\cdot}2H_2O$, 99%) were obtained from Sigma Aldrich, Germany. Hydrochloric acid, Sodium hydroxide, and solvents, including acetone and ethanol, were of analytical grade and were obtained from Sigma Aldrich, Sweden. All chemicals were used without further purification.

2.2. Synthesis of Rh NPs

The synthesis was based on a modified polyol reduction process [17]. In a typical synthesis, 20 mL EG is used as solvent. The halide salts, surfactants (molar ratio of Rh precursor/additive 1:1), and PVP (4 mmol in repeating units) were dissolved in EG in a glycerol bath. A series of reactions were performed for tuning the morphology of Rh NPs, by using the following additives in combination with PVP: Na-Ac, Na-Cit, CTAB, KBr, and CTAC. Details of all the synthesized samples are given in Table 1. After the addition of 0.2 mmol $RhCl_3$, the solution was heated to 85 °C, where nucleation was initiated, monitored from the darkening of the solution's color. After 15 min, the temperature was ramped up to 115 °C, where it was kept for 2 h, followed by quenching. The NPs obtained were centrifuged and washed three times through precipitation with acetone and re-dispersion in deionized water (DIW). Thereafter, they were dispersed in DIW for further characterizations.

Table 1. Summary of Rh nanoparticles (NPs) with diverse morphologies, synthesized using various additives.

Sample Designation	Precursor	Solvent	Additive	Stabilizer	Morphology
Rh_PVP	$RhCl_3$	EG	-	PVP	triangle
Rh_PVP-KBr	$RhCl_3$	EG	KBr	PVP	cubic
Rh_PVP-CTAB	$RhCl_3$	EG	CTAB	PVP	cubic
Rh_PVP-CTAC	$RhCl_3$	EG	CTAC	PVP	polygon
Rh_PVP-NaAc	$RhCl_3$	EG	NaAc	PVP	spherical
Rh_PVP-NaCit	$RhCl_3$	EG	NaCit	PVP	spherical

2.3. Characterization Methods

The morphology and crystallinity of Rh samples were characterized by transmission electron microscopy (TEM) (JEM-2100F, 200 kV, JEOL Ltd., Tokyo, Japan). The TEM samples were prepared by transferring a ~10 µL droplet of colloidal suspension onto a carbon-coated copper grid (Ted-Pella, CA, USA), and drying for 12 h. Particle size was measured by ImageJ by counting around 200 NPs in different fields of view on several TEM micrographs. Dynamic light scattering (DLS, Malvern Nano-ZS90, Malvern, UK) was used to measure the hydrodynamic size distribution of the colloidal suspension of as-prepared Rh NPs dispersed in DI water, at pH 7. Surface charge (Zeta potential) was also measured on the same samples using the same system under AC field. Inductively coupled plasma-optical emission spectroscopy (ICP-OES) (iCAP 6000 series, Thermo Scientific, Waltham, MA, USA) was used for the determination of the elemental Rh concentration in the colloidal suspensions prior to cytotoxicity tests. Ultraviolet–visible spectroscopy (UV–Vis) (NanoPhotometer NP80, Implen, Munich, Germany) was used to measure the absorption spectroscopy of Rh samples synthesized with different additives. Fourier-transform infrared spectroscopy (FT-IR, Thermo Scientific Nicolet iS20, Stockholm, Sweden) was used to obtain FT-IR spectra in transmission mode, on KBr pellets (2 mg sample in 100 mg KBr), in the spectral range of 4000–400 cm^{-1}. Thermal gravimetric analysis (TGA) was performed on dried samples in the temperature range of 30–700 °C, using the TGA 550 system (TA instruments, Sollentuna, Sweden).

2.4. In Vitro Toxicity

Before proceeding with the in vitro toxicity analysis, Rh NP suspensions were tested for lipopolysaccharides (LPS) contamination [18] with PTS cartridges with a sensitivity of 0.005 EU/mL (Endosafe-PTS™, Charles River) following the manufacturer's instructions. All the NPs suspensions had LPS values below the maximum admissible limit of 0.1 EU/mL [19].

Toxicity tests were performed on two different cell lines which were RAW 264.7 (murine macrophages, 91062702-1VL, SigmaAldrich, Stockholm, Sweden) and SKOV-3—human-derived ovarian cancer (ATCC HTB-77, Wesel, Germany). Two different toxicity assays were used which were Cell Counting Kit-8 (CCK-8, Cat.# 96922, Sigma Aldrich, Stockholm, Sweden) and NucGreen® viability assay (NucGreen™ Dead 488 ReadyProbes™, Cat# R37109, Thermo Fisher, Stockholm, Sweden). All the tests were performed following the supplier's instructions; the absorbance (CCK8) and fluorescence (NucGreen®) measurements were performed on a Hidex Sense multi plate reader (Kem-En-Tec Nordic, Uppsala, Sweden). The viability of the cells exposed to NPs is normalized to the viability of the non-exposed cells (the negative control).

Dulbecco's modified Eagle medium (DMEM, Sigma Aldrich) containing 10% fetal bovine serum (FBS) was used as cell culture medium. As negative control, cells grown in the absence of NPs were used. The cells were about 80% confluent when the assays were performed.

The cells were split and seeded into 96-well plates (Cat. # 167008, Thermo Fisher, Stockholm, Sweden), corresponding to 30,000 to 35,000 cells per well in a 96-well plate, one day before NP exposure. The cells were exposed for 24 h to a dilution series of Rh NPs with different morphologies. The concentration of NPs tested was highest at 250 μg/mL in the first well, and further dilution series were obtained by a 2-fold dilution (as 125, 62.5, and 31.25 μg/mL) of the stock to obtain a concentration series. Only in the case of the Rh_PVP-CTAC sample, the highest concentration was 100 μg/mL in the first well, and further dilution series were obtained by a 2-fold dilution (as 50, 25, and 12.5 μg/mL) of the stock.

The CCK-8 assay, in brief, was performed by adding the substrate, 10 μL WST-8 (2-(2-methoxy-4-nitrophenyl)-3-(4-nitrophenyl)-5-(2,4-disulfophenyl)-2H-tetrazolium, monosodium salt), to 100 μL cell media/well. The plates were incubated at 37 °C, 5% CO_2 for 2 h, after which the absorbance was measured. During incubation, the substrate was reduced by the enzymes in active cells to an orange formazan dye directly proportional to the number of living cells. To avoid interference, the media were aspirated and new fresh media were added just before the initiation of the assay.

NucGreen® is a cell-impermeant stain that emits bright green (535 nm) fluorescence when bound to DNA. Cells that have lost plasma membrane integrity are stained within minutes, making NucGreen® a stain used to estimate live/dead cell ratios. For the NucGreen viability assay, 2 drops of NucGreen reagent were added per mL of medium and after 5 min incubation, the fluorescence signal was detected at 535 nm, with an excitation wavelength of 485 nm [20].

2.5. *XFCT Phantom Experiments*

The performance of the synthesized NPs as contrast agents in small-animal XFCT was investigated in an in situ imaging setting. A pipette tip (Eppendorf epTIPS, 2–200 μL) was used as an imaging target by filling it with a Rh-PVP NPs sample dispersion in DI water at a concentration of 250 μg/mL Rh. The pipette tip was then surgically inserted in a sacrificed mouse for another study (15-week-old, female, NOD.Cg-Prkdcscid Il2rgtm1Sug/JicTac, Taconic Biosciences, Lille Skensved, Denmark). The mouse was positioned inside a 60 mL tube and imaged using our in-house preclinical XFCT arrangement [21]. The choice of the pipette tip as an imaging target allowed, due to its conical shape with a decreasing cross-sectional diameter, an investigation of the smallest observable feature size at the fixed NP concentration. The inner diameter of the pipette tip ranged from 0.5 to 4 mm along its length of 35 mm.

Simultaneous XFCT and computed tomography (CT) was performed on the sacrificed mouse by acquiring 30 projection images over 180 degrees. Each projection image was acquired with steps of 200 μm and exposure time of 10 ms, resulting in a total scan time of ~1.5 min for each axial slice (of 200 μm thickness). We note that the chosen acquisition settings and the radiation dose, which is estimated to be ~25 mGy, have been demonstrated suitable for in vivo imaging in our recent study [21]. The acquired CT data were reconstructed using a standard filtered back-projection algorithm, while the acquired XFCT data were reconstructed using an in-house developed iterative algorithm. For more details on the imaging arrangement, see Ref. [21].

3. Results and Discussion

Formation of metallic NPs in polyol synthesis is rather straightforward, where metal ions are reduced by the polyalcohol solvent, in our case EG. PVP acts as a pooling agent, by forming reservoirs of metal ions. In this respect, its concentration can play an important role on the size of nucleated NPs. Parameters such as reaction temperature during nucleation and crystal growth, reaction time, solvent type, and concentration of precursors and capping agents can affect the shape of final particles. In this study, we investigated the effect of different additives (PVP, KBr, CTAB, CTAC, Na-Ac, and Na-Cit) on the NP morphology while keeping the other parameters unaltered. Details of the prepared samples are presented in Table 1.

3.1. Morphology Analysis

When all the others synthesis parameters of the reaction are kept the same, the morphological variations of NPs are dependent on the type of additives used with the anionic ligands inducing the strongest morphology variation in the synthesis of NPs. For metals with a face-centered cubic (fcc) crystal structure (including Rh), the surface energy (γ) follows the trend $\gamma(111) < \gamma(100) < \gamma(110)$ with the highest surface energy in the (110) facet. These relative energies can be modified by surface-adsorbing species exposing facets that would not normally be thermodynamically favored, introducing a degree of anisotropy into the NP geometry [22].

As expected, the morphology of the obtained Rh NPs varies greatly dependent on the additives present during the synthesis (Table 1, Figure 1). With only PVP added, predominantly triangular NPs are observed (Figure 1a), revealing anisotropy due to the adsorption of PVP on the NP surface. Br-containing additives, KBr (Figure 1b) and CTAB (Figure 1d), resulted in cubic NPs. Among the carboxyl group-containing additives, acetate (NaAc, Figure 1c) and citrate (NaCit, Figure 1e) ions led to the formation of spherical NPs, which is indicative of isotropic growth, eliminating the effect of PVP. Rh NPs synthesized by the addition of CTAC (Figure 1f) display an irregular mixed polygonic particle morphology. Cl^- ions did not show the same effect as the one observed in the presence of Br^- ions. The resultant NPs present a mixed polygon morphology, including trigonal, similar to the ones synthesized in the presence of PVP only. The mixed polygon morphology may be due to a complex formation between the additional Cl^- ions and Rh, influencing the NP growth dynamics [23].

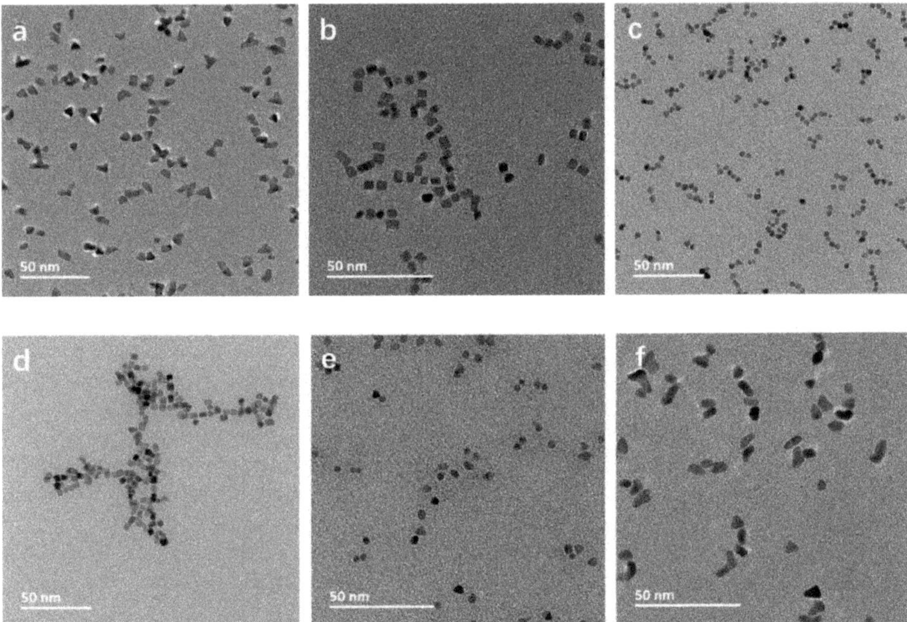

Figure 1. TEM micrographs of (**a**) Rh_PVP; (**b**) Rh_PVP-KBr; (**c**) Rh_PVP-Na-Ac; (**d**) Rh_ PVP-CTAB; (**e**) Rh_PVP-Cit; and (**f**) Rh_PVP-CTAC.

The observed morphological differences are related to the preferential adsorption of anions on certain facets with the NP growth allowed predominantly along them. One common feature observed in the HRTEM micrographs of the samples (Figure 2) is the fact that the NPs obtained in the presence of all studied anions are single crystalline, with the measured d values matching the fcc Rh (ICDD PDF card no: 03-065-2866). Lattice fringes of Rh samples synthesized in the presence of PVP (Figure 2a), Na-Ac (Figure 2c), Na-Cit (Figure 2e), and CTAC (Figure 2f) have a d value of 0.22 nm, which matches well with the (111) plane of fcc-Rh. The additives containing Br^- ions, however, showed a slightly different passivating behavior. Br- ions are reported to passivate the (100) plane in Rh nanocrystals [14]. Our observations confirmed the stabilization of the (100) surface when Br^--bearing KBr (Figure 2b) and CTAB (Figure 2d) are used as additives. In both cases, the NPs have a cubic morphology with lattice fringes of 0.19 nm corresponding to the (200) plane of Rh, proving that Br^- ions are capable of getting adsorbed on the surface of the NPs formed, even in the presence of PVP.

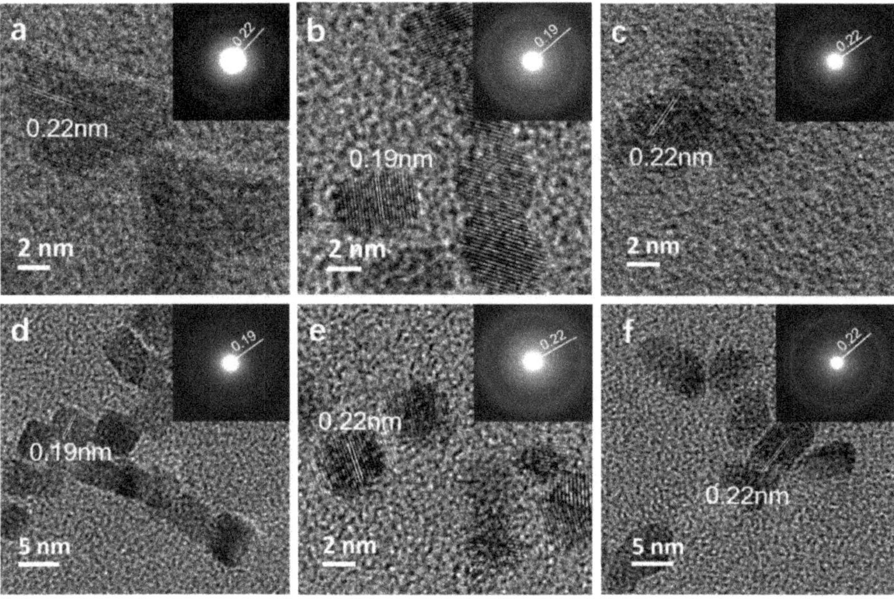

Figure 2. (**a–f**) HRTEM micrographs of Rh NPs: (**a**) Rh_PVP; (**b**) Rh_PVP-KBr; (**c**) Rh_PVP-Na-Ac; (**d**) Rh_ PVP-CTAB; (**e**) Rh_PVP-Cit; and (**f**) Rh_PVP-CTAC (the lattice spacings are indexed for fcc Rh using ICDD PDF card no: 03-065-2866).

Plasmonic NPs of Ag and Au show a size- and morphology-dependent UV–Vis absorption behavior [24]. In order to study the effect of morphology on Rh NP absorption characteristics, we performed UV–Vis analysis on the synthesized samples. UV–Vis absorption spectra indicated that the maximum absorption wavelength due to the surface plasmon resonance (SPR) of Rh varied with particles morphology, with a peak value in the range of 210–220 nm (Figure S1). The intensity of SPR absorption is the highest for Rh NPs with a cubic morphology (Rh_PVP-KBr and Rh_PVP-CTAB), with an absorption edge at 225 nm. A similar trend of absorption intensity and edge was observed for triangular NPs (Rh_PVP). Spherical NPs (Rh_PVP-Ac and Rh_PVP-Cit) display an absorption maximum at 210 nm, with an absorption edge at about 220 nm. All these samples showed a weak, broad absorption at the 270–275 nm region. Polygonic Rh NPs (Rh_PVP-CTAC) showed fluctuating absorption in the range of 200–225 nm, followed by a broad absorption centered at 275 nm, with a very high background absorption. This may be due to the polydispersity in the shapes of NPs (observed in Figure 1f) increasing the background absorption significantly.

3.2. Particle Size and Surface Chemistry Analyses

Besides the morphological effects on the Rh NPs, the dry particle size obtained from TEM micrographs, hydrodynamic size, and surface properties are also influenced by the additives used (Table 2, Figure 3). The largest NPs were formed when PVP only (7.6 nm) and PVP-CTAC (6.6 nm) are used. Bromide-containing additives, KBr and CTAB, formed NPs with an average size of 4.7–4.8 nm. Carboxylic acid functionality-bearing additives, Na-Ac and Na-Cit, resulted in NPs with an average size of 3.1–3.4 nm. The differences in NP size are ascribed to the fact that the additives clearly affect the crystal nucleation and growth kinetics in addition to directing the morphology. It may be reasonable to relate this to the notion that acetate and citrate ions allowed a fast nucleation, i.e., burst nucleation, and bromide ions constrained the growth moderately, whereas PVP only and chloride ions yielded the largest size of Rh NPs due to the impeded kinetics allowing particle ripening.

Table 2. Dry size (TEM size ± standard deviation (SD)), hydrodynamic size (DLS size, the polydispersity index (PDI)), and zeta potential of Rh NPs with different morphologies in DIW and DMEM with 10% FBS.

		In DIW		In CCM + 10%FBS	
Sample	**TEM Size; Mean ± SD (nm)**	**DLS Size [PDI] (nm)**	**Zeta Potential (mV)**	**DLS Size [PDI] (nm)**	**Zeta Potential (mV)**
Rh-PVP	7.6 ± 1.6	29.1 [0.23]	0.18	18.1 [0.49]	−9.72
Rh-PVP-KBr	4.7 ± 0.8	41.1 [0.23]	−7.74	16.5 [0.46]	−11.24
Rh-PVP-CTAB	4.8 ± 0.7	97.1 [0.19]	−11.57	37.8 [0.64]	−8.76
Rh-PVP-CTAC	6.6 ± 1.3	46.2 [0.39]	−3.59	19.7 [0.55]	−5.69
Rh-PVP-NaAc	3.4 ± 0.7	31.9 [0.41]	−9.43	17.8 [0.44]	−10.02
Rh-PVP-NaCit	3.1 ± 0.6	28.3 [0.18]	−5.05	50.7 [0.68]	−9.38

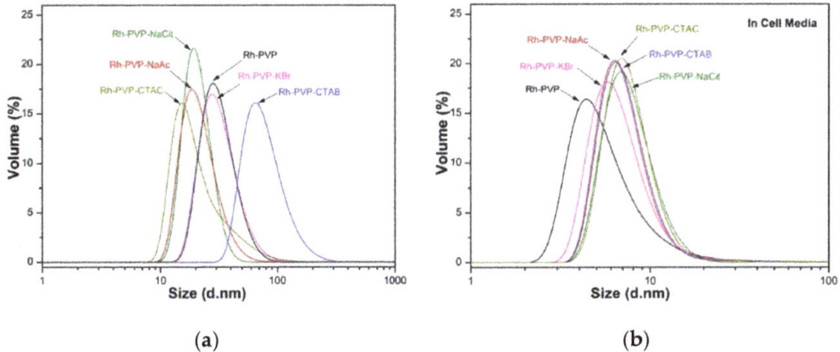

Figure 3. Hydrodynamic size distribution plots, using the DLS technique, of as-synthesized Rh NPs, with different morphologies in (**a**) DIW and (**b**) DMEM. Data presented in volume distribution.

The NPs' dispersed sizes, i.e., the hydrodynamic size, were evaluated in DIW and cell culture media-DMEM, and the results are presented graphically in Figure 3 and summarized in Table 2. Rh NPs synthesized with PVP, PVP-Cit, and PVP-Ac have an average hydrodynamic size of around 30 nm, the NPs synthesized in the presence of PVP-KBr and PVP-CTAC have an intermediate size of 40–50 nm and the size of about 100 nm was found when PVP-CTAB was used. The polydispersity index (PDI), a parameter that defines the goodness of size distribution, where a value less than 0.3 relates to a homogeneous population [25], is lower than 0.3 for all samples except the PVP-CTAC and PVP-Ac samples. As a general trend, all the Rh NPs synthesized in this work present a very low surface charge (Table 2), under (±) 30 mV. Despite the low charge density on their surface, the NPs are stable in solution probably due to the steric hindrance exerted by the polymeric chains attached to the surface. Rh_PVP NPs show near-zero surface charge at neutral pH. A common trend for all other samples is the net negative surface charge on the NPs. This correlates well with the morphological differences being controlled by the adsorbed anions on the NPs. It is important to note that the size distributions of all synthesized Rh NPs, regardless of the shape, are suitable for use in XFCT bio-imaging.

Particle size distribution is also measured upon dispersing the NPs in DMEM, containing 10% FBS. Results indicate a smaller dispersed size, except Rh_PVP-Cit, with an increase in the polydispersity index to the 0.4–0.6 level. The ionic strengths of DMEM could have helped the decrease initially observed in NP clusters, while various proteins in DMEM led to larger agglomerate formations, which are detected when scattering intensity is used for hydrodynamic NP size distribution (see ESI, Figure S2). Adsorption of proteins is also influential in the final surface charge of Rh NPs in DMEM, which are observed to have enhanced their negative charge as compared to their surface charge in DIW (Table 2).

The FT-IR spectra reveal that the surface of all Rh NPs shows dominantly PVP features with no strong indication of the presence of secondary additives (Figure 4), probably due to the easy removal of these additives during the washing steps. A common feature is the shift of the position of the C = O absorption band from 1662 for the pure PVP to around 1650 cm^{-1} for all NPs synthesized, indicating the conjugation of PVP via the carbonyl group onto the Rh NPs' surface. The quantitative determination of the organic material proportion in the samples by thermal gravimetric analysis (TGA) showed that the degradation of the organics was completed at 500 °C (ESI, Figure S3). The thermograms and a table summarizing the findings are presented in Figure S3. The Rh_PVP system showed the highest NP content of about 13 wt%, where the NP content decreased to 6 wt%, in a non-monotonic way for the other samples.

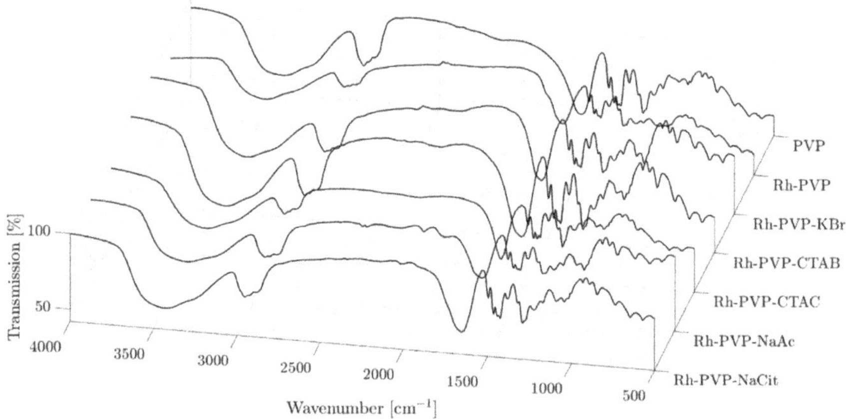

Figure 4. FT-IR spectra of pure PVP and Rh NPs; Rh_PVP, Rh_PVP-KBr, Rh_PVP-CTAB, Rh_PVP-CTAC, Rh_PVP-Ac, and Rh_PVP-Cit.

3.3. Cytotoxicity Studies

The cellular viability as an indicator for the toxicity potency of an agent can be affected in various ways. We, therefore, tested the membrane integrity (fluorescence NucGreen assay) and cellular metabolic activity (colorimetric CCK-8 assay) of cells exposed to Rh NPs with different morphologies in a concentration-dependent manner, with the highest dose of 250 µg/mL and the lowest nominal dose of 30 µg/mL for the 24-h exposure time. It is known that different cell types will react in a distinct way in the presence of external stimuli. Therefore, in this work the cytotoxicity of Rh NPs was evaluated on murine macrophages (RAW264.7) and a human ovarian cancer (SKOV-3) cell line. The RAW264.7 macrophages serve as a model for immune cells, while SKOV-3 is used as a model of tumoral cells to investigate if the exposure to NPs will induce a different level of toxicity in different cell lines. The half maximal inhibitory concentration (IC50) is specified for NPs, whenever relevant, in various toxicity assays against different cell lines.

NucGreen is staining only the cells that have lost plasma membrane integrity. The membrane integrity of RAW 264.7 is not compromised when exposed to trigonal Rh NPs (Rh_PVP). However, it is significantly compromised when exposed to cubic NPs (Rh_PVP-KBr and Rh_PVP-CTAB), at all doses. In the case of spherical NPs (Rh_PVP-NaAc (IC50: 118 μg/mL) and Rh_PVP-NaCit (IC50: 120 μg/mL)), the membrane integrity is affected in a concentration-dependent manner, the viability increasing with the decreasing concentration of NPs. Cell viability is significantly reduced for doses ≥ 125 μg/mL (ESI, Figure S4). Rh NPs with a mixed polygon morphology (Rh_PVP-CTAC), presented separately in Figure S6, due to their lower concentration reaching a maximum of 100 μg/mL, showed high viability at all the doses tested.

The exposure to triangular Rh NPs (Rh_PVP) of the RAW264.7 cell line in doses up till 125 μg/mL did not significantly decrease the cells' metabolic activity, with the viability reducing to about 50% at a dose of 250 μg/mL (Figure 5. Among the spherical NPs, Rh-Cit NPs (IC50: 56 μg/mL) reached viability beyond 60% when the dose was reduced to 60 μg/mL, while Rh-Ac NPs (IC50: 86 μg/mL) showed viability of >70% at the lowest dose of 30 μg/mL. Cubic NPs (Rh_PVP-KBr and Rh_PVP-CTAB) showed very low viability, or a high level of toxicity, at all the doses tested. NPs with a mixed polygon morphology (Rh-CTAC, Figure 6) showed a high level of toxicity in the RAW264.7 cell line.

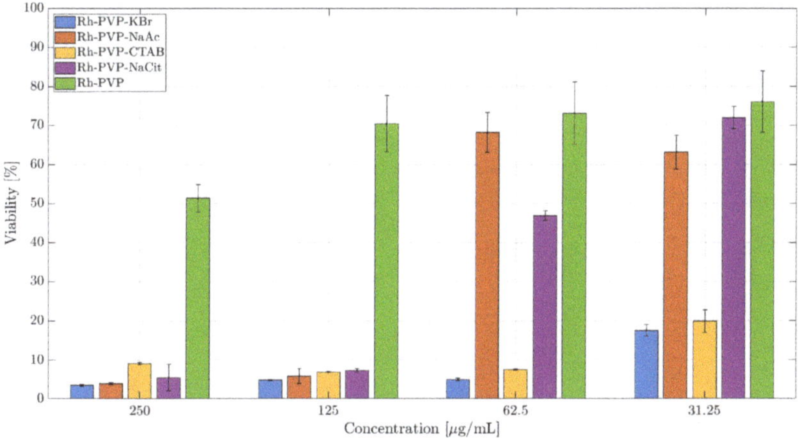

Figure 5. CCK-8 toxicity assay of Rh-NPs in the RAW 264.7 cell line after 24 h of incubation. The percentage of cell viability is calculated relative to the cells incubated in the absence of NPs (negative control) with 100% viability.

The cytotoxicity of some organic surface-coating agents used for NP synthesis has been studied in the literature, where the cationic surfactant CTAB was reported to be highly toxic [26]. In the case of Rh NPs synthesized in the presence of CTAB and CTAC, the residue of the cationic part could cause the observed negative effect. The common character of Rh NPs synthesized in the presence of KBr and CTAB is their cubic morphology, due to surface-adsorbed Br- ions, which resulted in high cytotoxicity levels. Br^- ions are known to exist in humans and animals and can reach up to almost half of the Cl^- content in red blood cells. [27] Adsorbed inorganic Br^- is, therefore, not expected to show the observed cytotoxicity in the case of KBr-assisted synthesis. Therefore, it is reasonable to ascribe the observed cytotoxicity to the cubic NP morphology. In general, SKOV-3 cells are more resistant to external agents. Indeed, the membrane integrity of SKOV-3, maintained at >80%, is not compromised significantly (Figure S5), when exposed to trigonal (Rh_PVP) and spherical NPs (Rh_PVP-NaAc and Rh_PVP-NaCit). However, in the presence of NPs with a cubic morphology (Rh_PVP-KBr (IC50:

119 μg/mL) and Rh_PVP-CTAB (IC50: 115 μg/mL)) in doses ≥ 125 μg/mL, the viability is significantly reduced. NPs with a mixed polygon morphology (Rh_PVP-CTAC) (Figure S6) showed high viability at all the doses tested, similar to the response of macrophages.

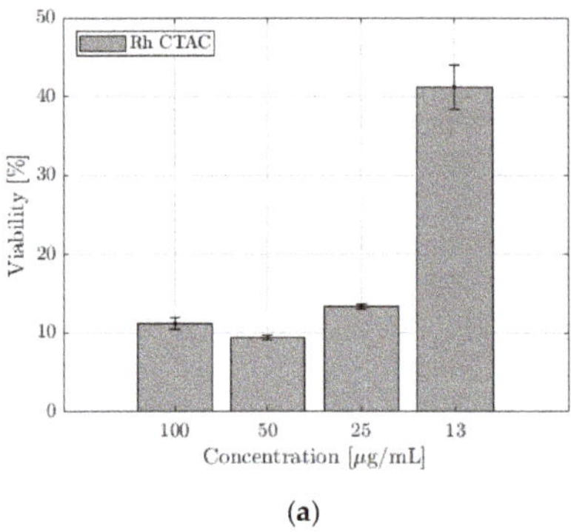

(a)

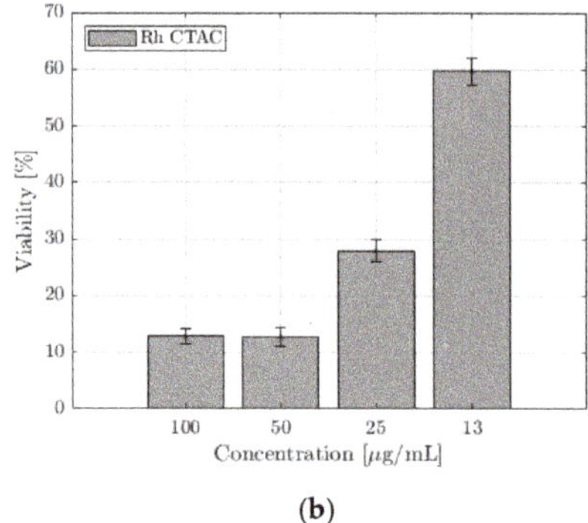

(b)

Figure 6. CCK-8 toxicity assays of Rh-CTAC NPs in the (**a**) RAW 264.7 and (**b**) SKOV-3 cell lines after 24 h of incubation. The percentage of cell viability is calculated taking negative control cells incubated in the absence of NPs with 100% viability.

A similar trend is observed in the CCK-8 assay (Figure 7) with the trigonal NPs (Rh-PVP) exerting the lowest toxicity followed by spherical NPs (RhPVP-NaCit and Rh-PVP-NaAc (IC50: 218 μg/mL)). The highest toxicity level is observed when the cells were incubated with cubic NPs (Rh_PVP-CTAB (IC50: 42 μg/mL) and Rh_PVP-KBr (IC50: 41 μg/mL)). NPs with a mixed polygon morphology (Rh_PVP-CTAC) (Figure 6b) showed a high level of toxicity in the SKOV-3 cell line at higher doses, reaching a viability of >60% at the lowest dose of 13 μg /mL NP concentration.

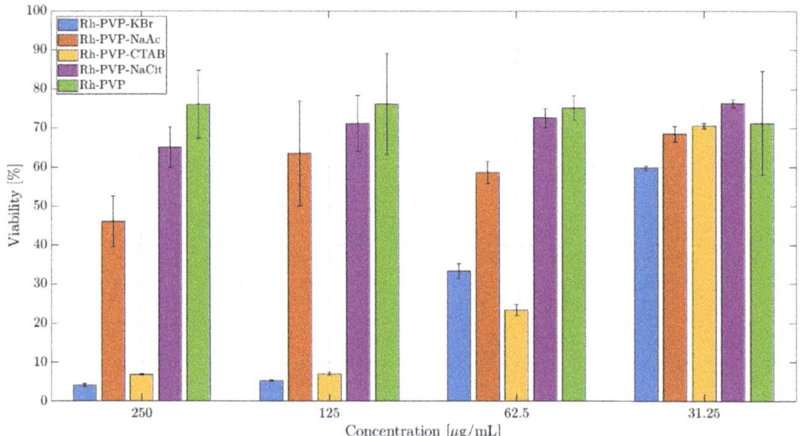

Figure 7. CCK-8 toxicity assay of Rh NPs in the SKOV-3 cell line after 24 h of incubation. The percentage of cell viability is calculated relative to the cells incubated in the absence of NPs (negative control) with 100% viability.

The differentiated cytotoxicity response of the macrophage cell line (RAW 264.7) and cancer cell line (SKOV-3) in the presence of the NPs might be explained by the different functions the macrophages and cancer cells are specialized to perform. While the macrophages should react to any external or internal aggression factors, the cancer cells are programed to adapt to any conditions in order to survive and multiply. This work showed that membrane integrity and metabolic activity, expressed as viability, are influenced to a similar extent, for both the cell lines, with respect to the morphology of Rh NPs. Particles with a mixed polygon morphology showed the highest negative impact on cellular activity, followed by cubic and spherical NPs. Trigonal NPs showed the lowest negative effect, or the highest viability, among the NP series. From the FT-IR analysis, we see no significant evidence of the presence of functional groups, especially for surfactants CTAB and CTAC. Trace amounts of these functional groups may still reside on the surface dominating PVP, or at the immediate NP surface. By looking at the significantly different response of cellular activity towards these NPs, it is reasonable to ascribe the observed effects to the NP morphology.

Similar observations were reported for noble metal NPs. For instance, gold nanorods prepared using CTAB were shown to be highly toxic to human skin cells due to the presence CTAB, not the gold nanorods themselves [28]. It was later shown that over-coating gold nanorods with polymers substantially reduced their cytotoxicity [29]. If any of the Rh NP morphologies showing a high level of toxicity prove to be more interesting candidates for biomedical applications, their surface chemistry can be further modified using polymers or other inorganic coatings, such as silica, to improve their biocompatibility, without adversely influencing the diagnostics function.

3.4. XFCT Performance

Figure 8 shows the in situ small-animal XFCT imaging experiment. Using the pipette tip (Figure 8a) as an imaging target allowed the investigation of the smallest observable feature size at the fixed NP concentration, due to its conical shape with a decreasing cross-sectional diameter. The content of the pipette tip could not be separated and visualized in the CT reconstruction, while it could be clearly visualized in the XFCT reconstruction (c.f., Figure 8b). For diameters of the pipette tip between 4 and 2.5 mm, the NP content could be reliably reconstructed, while for diameters smaller than 2.5 mm, it became increasingly difficult for the reconstruction algorithms to separate the XRF signal from background X-ray scattering. Below the 2 mm diameter mark, the reconstruction algorithms could no longer separate the XRF signal from the background. These results agree with previous simulations, which indicated that a local NP concentration of >1 mg/mL is necessary to reach sub-mm feature visualization [30].

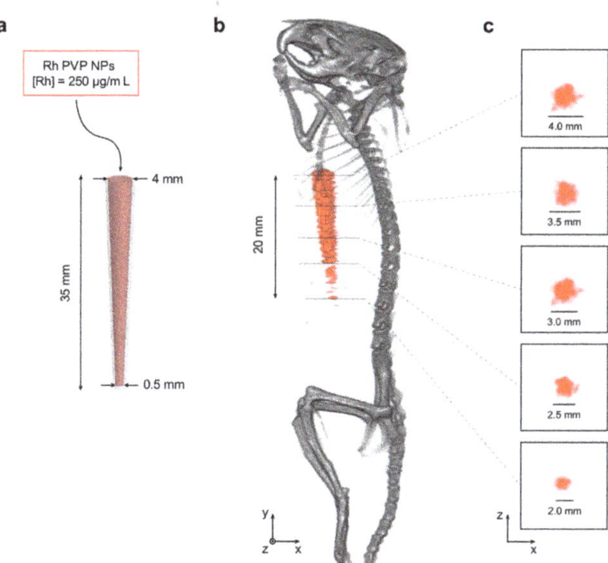

Figure 8. In situ XFCT performance of sample Rh-PVP. (**a**) A pipette tip was used as an imaging target

by filling it with Rh-PVP NPs (250 μg/mL), where the conical shape offered a target with an inner diameter ranging from 4 to 0.5 mm. (**b**) Reconstructed tomographic data visualized in 3D (grayscale: CT, red: XFCT). The portion of the pipette tip visible in the XFCT reconstruction was estimated to be ~20 mm. (**c**) Selected axial slices with 5 mm separation in y of the XFCT reconstruction, with the theoretical expected diameters at each location denoted with scale bars and numbers. Each pixel in the axial slices corresponds to 200 × 200 μm^2.

Selected cross-sections of the XFCT reconstruction are shown in Figure 8c and compared to the theoretical inner diameters of the pipette tip at the selected locations. We note that estimated cross-sectional pipette tip diameters in the XFCT reconstruction are in the range of theoretical values between 4 and 2 mm, and conclude, therefore, that few-mm features with a local NP concentration in the 100 μ g/mL range can be visualized using XFCT. In other words, this type of in situ imaging experiment offers a controlled environment to investigate relations between parameters such as feature size and observable local Rh NP concentration, which is not directly possible in in vivo scenarios.

4. Conclusions

Rh nanoparticles (NPs) with various morphologies were synthesized in the presence of a large palette of additives during the polyol synthesis process. The halide salts and surfactant additives including sodium acetate, sodium citrate, KBr, CTAB, and CTAC, in addition to PVP, showed a tuning effect on the shape of the Rh NPs' morphology. When PVP was used as the only additive, trigonal NPs were obtained. Additives with Br^- ions (CTAB and KBr) resulted in NPs with a cubic morphology, while those with carboxyl groups (sodium citrate and acetate) formed spheroid NPs. Cl^- ions (CTAC) resulted in a mixture of polygon morphologies. The observed cytotoxicity response evaluated using two different viability assays, on macrophages and ovarian cancer cell lines, shows a strong dependence on the NPs morphology. The cells exposed to trigonal Rh NPs showed the highest viability, among the NP series. Particles with a mixed polygon morphology showed the highest cytotoxicity, followed by cubic and spherical NPs, which can be correlated with a more shape-dependent response triggering. The performance of Rh NPs as contrast agents for small-animal XFCT was demonstrated in an in situ experiment, showing promise for future in vivo imaging. This work provides a detailed route for the synthesis, morphology control, and characterization of Rh NPs as viable contrast agents for XFCT bio-imaging.

Supplementary Materials: The following are available online at http://www.mdpi.com/2079-4991/10/11/2129/s1, Figure S1: UV–Vis absorption spectra of Rh NPs; Rh_PVP, Rh_PVP-KBr, Rh_PVP-Ac, Rh_PVP-CTAB, Rh_PVP-Cit, and Rh_PVP- CTAC. Figure S2: Hydrodynamic particle size distribution plots, using DLS, of as-synthesized Rh NPs, with different morphologies. Data presented in intensity distribution to show the agglomerates formed. Figure S3: TGA thermograms of Rh NPs; Rh_PVP, Rh_PVP-KBr, Rh_PVP-CTAB, Rh_PVP-CTAC, Rh_PVP-Ac, and Rh_PVP-Cit. Figure S4. NucGreen toxicity assay of Rh NPs in the RAW 264.7 cell line after 24 h of incubation. The percentage of cell viability is calculated relative to the cells incubated in the absence of NPs (negative control) with 100% viability. Figure S5. NucGreen toxicity assay of Rh NPs in the SKOV-3 cell line after 24 h of incubation. The percentage of cell viability is calculated relative to the cells incubated in the absence of NPs (negative control) with 100% viability. Figure S6. NucGreen toxicity assays of Rh-CTAC NPs in (a) RAW 264.7 and (b) SKOV-3 cell lines after 24 h of incubation. The percentage of cell viability is calculated taking negative control cells incubated in the absence of NPs with 100% viability.

Author Contributions: M.S.T. and H.M.H. conceived of the presented ideas. Conceptualization and methodology, Y.L., K.S., M.S., B.B. and M.S.T.; validation, M.S., C.V. and K.S.; formal analysis, Y.L., G.M.S.; writing—original draft preparation, Y.L.; writing—review and editing, all. M.S.T. and H.M.H. supervised the work. All authors have read and agreed to the published version of the manuscript.

Funding: This research was funded by the Wallenberg Foundation. Y.L. acknowledges the support from the Chinese Scholarship Council (CSC).

Acknowledgments: We thank Torbjörn Gräslund and Haozhong Ding (KTH Biotechnology) for fruitful discussions.

Conflicts of Interest: The authors declare no conflict of interest. The funders had no role in the design of the study; in the collection, analyses, or interpretation of data; in the writing of the manuscript, or in the decision to publish the results.

References

1. Xu, L.; Liu, D.; Chen, D.; Liu, H.; Yang, J. Size and shape controlled synthesis of rhodium nanoparticles. *Heliyon* **2019**, *5*, e01165. [CrossRef] [PubMed]
2. Lee, S.R.; Vara, M.; Hood, Z.D.; Zhao, M.; Gilroy, K.D.; Chi, M.; Xia, Y. Rhodium Decahedral Nanocrystals: Facile Synthesis, Mechanistic Insights, and Experimental Controls. *ChemNanoMat* **2018**, *4*, 66–70. [CrossRef]
3. Xie, S.; Liu, X.Y.; Xia, Y. Shape-controlled syntheses of rhodium nanocrystals for the enhancement of their catalytic properties. *Nano Res.* **2015**, *8*, 82–96. [CrossRef]
4. Cao, D.X.; Wieckowski, A.; Inukai, J.; Alonso-Vante, N. Oxygen reduction reaction on ruthenium and rhodium nanoparticles modified with selenium and sulfur. *J. Electrochem. Soc.* **2006**, *153*, A869–A874. [CrossRef]
5. Zhou, M.; Wang, H.; Vara, M.; Hood, Z.D.; Luo, M.; Yang, T.H.; Bao, S.; Chi, M.; Xiao, P.; Zhang, Y.; et al. Quantitative Analysis of the Reduction Kinetics Responsible for the One-Pot Synthesis of Pd-Pt Bimetallic Nanocrystals with Different Structures. *J. Am. Chem. Soc.* **2016**, *138*, 12263–12270. [CrossRef] [PubMed]
6. Xia, Y.; Xia, X.; Peng, H.C. Shape-Controlled Synthesis of Colloidal Metal Nanocrystals: Thermodynamic versus Kinetic Products. *J. Am. Chem. Soc.* **2015**, *137*, 7947–7966. [CrossRef]
7. Ganguly, P.; Breen, A.; Pillai, S.C. Toxicity of Nanomaterials: Exposure, Pathways, Assessment, and Recent Advances. *ACS Biomater. Sci. Eng.* **2018**, *4*, 2237–2275. [CrossRef]
8. Lewinski, N.; Colvin, V.; Drezek, R. Cytotoxicity of nanoparticles. *Small* **2008**, *4*, 26–49. [CrossRef]
9. Buzea, C.; Pacheco, I.I.; Robbie, K. Nanomaterials and nanoparticles: Sources and toxicity. *Biointerphases* **2007**, *2*, MR17–MR71. [CrossRef]
10. Nel, A.; Xia, T.; Mädler, L.; Li, N. Toxic Potential of Materials at the Nanolevel. *Science* **2006**, *311*, 622. [CrossRef]
11. Alkilany, A.M.; Murphy, C.J. Toxicity and cellular uptake of gold nanoparticles: What we have learned so far? *J. Nanopart. Res.* **2010**, *12*, 2313–2333. [CrossRef] [PubMed]
12. Kang, S.; Shin, W.; Choi, M.-H.; Ahn, M.; Kim, Y.-K.; Kim, S.; Min, D.-H.; Jang, H. Morphology-Controlled Synthesis of Rhodium Nanoparticles for Cancer Phototherapy. *ACS Nano* **2018**, *12*, 6997–7008. [CrossRef] [PubMed]
13. Yu, N.F.; Tian, N.; Zhou, Z.Y.; Huang, L.; Xiao, J.; Wen, Y.H.; Sun, S.G. Electrochemical Synthesis of Tetrahexahedral Rhodium Nanocrystals with Extraordinarily High Surface Energy and High Electrocatalytic Activity. *Angew. Chem.-Int. Ed.* **2014**, *53*, 5097–5101. [CrossRef]
14. Zhang, H.; Li, W.; Jin, M.; Zeng, J.; Yu, T.; Yang, D.; Xia, Y. Controlling the Morphology of Rhodium Nanocrystals by Manipulating the Growth Kinetics with a Syringe Pump. *Nano Lett.* **2011**, *11*, 898–903. [CrossRef] [PubMed]
15. Biacchi, A.J.; Schaak, R.E. The solvent matters: Kinetic versus thermodynamic shape control in the polyol synthesis of rhodium nanoparticles. *ACS Nano* **2011**, *5*, 8089–8099. [CrossRef]
16. Li, Y.; Shaker, K.; Larsson, J.C.; Vogt, C.; Hertz, H.M.; Toprak, M.S. A Library of Potential Nanoparticle Contrast Agents for X-Ray Fluorescence Tomography Bioimaging. *Contrast Media Mol. Imaging* **2018**, *2018*. [CrossRef] [PubMed]
17. Li, Y.; Shaker, K.; Svenda, M.; Vogt, C.; Hertz, M.H.; Toprak, S.M. Synthesis and Cytotoxicity Studies on Ru and Rh Nanoparticles as Potential X-Ray Fluorescence Computed Tomography (XFCT) Contrast Agents. *Nanomaterials* **2020**, *10*, 310. [CrossRef]
18. Dobrovolskaia, M.A.; Patri, A.K.; Zheng, J.; Clogston, J.D.; Ayub, N.; Aggarwal, P.; Neun, B.W.; Hall, J.B.; McNeil, S.E. Interaction of Colloidal Gold Nanoparticles with Human Blood: Effects on Particle Size and Analysis of Plasma Protein Binding Profiles. *Nanomed. Nanotechnol. Biol. Med.* **2009**, *5*, 106–117. [CrossRef]
19. FDA. *Guidance for Industry: Pyrogen and Endotoxins Testing: Questions and Answers*; FDA: Rockville, MD, USA, 2012.
20. Roth, B.L.; Poot, M.; Yue, S.T.; Millard, P.J. Bacterial Viability and Antibiotic Susceptibility Testing with SYTOX Green Nucleic Acid Stain. *Appl. Environ. Microbiol.* **1997**, *63*, 2421–2431. [CrossRef]

21. Shaker, K.; Vogt, C.; Katsu-Jimenez, Y.; Kuiper, R.; Andersson, K.; Li, Y.; Larsson, J.; Rodriguez-Garcia, A.; Toprak, M.; Arsenian-Henriksson, M.; et al. Longitudinal In-Vivo X-Ray Fluorescence Computed Tomography with Molybdenum Nanoparticles. *IEEE Trans. Med. Imaging* **2020**. [CrossRef]
22. Hoefelmeyer, J.D.; Niesz, K.; Somorjai, G.A.; Tilley, T.D. Radial Anisotropic Growth of Rhodium Nanoparticles. *Nano Lett.* **2005**, *5*, 435–438. [CrossRef]
23. Shlenskaya, V.I.; Efremenko, O.A.; Oleinikova, S.V.; Alimarin, I.P. Chloride Complexes of Rhodium (Iii) in Aqueous Solutions. *Bull. Acad. Sci. USSR Div. Chem. Sci.* **1969**, *18*, 1525–1527. [CrossRef]
24. Zhang, J.Z.; Noguez, C. Plasmonic Optical Properties and Applications of Metal Nanostructures. *Plasmonics* **2008**, *3*, 127–150. [CrossRef]
25. Danaei, M.; Dehghankhold, M.; Ataei, S.; Hasanzadeh Davarani, F.; Javanmard, R.; Dokhani, A.; Khorasani, S.; Mozafari, M.R. Impact of Particle Size and Polydispersity Index on the Clinical Applications of Lipidic Nanocarrier Systems. *Pharmaceutics* **2018**, *10*, 57. [CrossRef] [PubMed]
26. Zhang, Y.; Newton, B.; Lewis, E.; Fu, P.P.; Kafoury, R.; Ray, P.C.; Yu, H. Cytotoxicity of Organic Surface Coating Agents Used for Nanoparticles Synthesis and Stability. *Toxicol. Vitr.* **2015**, *29*, 762–768. [CrossRef]
27. Van Leeuwen, F.X.R.; Sangster, B.; Hildebrandt, A.G. The Toxicology of Bromide Ion. *Crit. Rev. Toxicol.* **1987**, *18*, 189–213. [CrossRef]
28. Wang, S.; Lu, W.; Tovmachenko, O.; Rai, U.S.; Yu, H.; Ray, P.C. Challenge in Understanding Size and Shape Dependent Toxicity of Gold Nanomaterials in Human Skin Keratinocytes. *Chem. Phys. Lett.* **2008**, *463*, 145–149. [CrossRef] [PubMed]
29. Alkilany, A.M.; Nagaria, P.K.; Hexel, C.R.; Shaw, T.J.; Murphy, C.J.; Wyatt, M.D. Cellular Uptake and Cytotoxicity of Gold Nanorods: Molecular Origin of Cytotoxicity and Surface Effects. *Small* **2009**, *5*, 701–708. [CrossRef]
30. Shaker, K.; Larsson, J.C.; Hertz, H.M. Quantitative Predictions in Small-Animal X-Ray Fluorescence Tomography. *Biomed. Opt. Express* **2019**, *10*, 3773. [CrossRef]

MDPI

Article

Towards Universal Stimuli-Responsive Drug Delivery Systems: Pillar[5]arenes Synthesis and Self-Assembly into Nanocontainers with Tetrazole Polymers

Dmitriy N. Shurpik [1], Lyaysan I. Makhmutova [1], Konstantin S. Usachev [2], Daut R. Islamov [3], Olga A. Mostovaya [1], Anastasia A. Nazarova [1], Valeriy N. Kizhnyaev [4] and Ivan I. Stoikov [1,*]

1 A. M. Butlerov Chemical Institute, Kazan Federal University, Kremlevskaya, 18, 420008 Kazan, Russia; dnshurpik@mail.ru (D.N.S.); lays_9393@mail.ru (L.I.M.); olga.mostovaya@mail.ru (O.A.M.); anas7tasia@gmail.com (A.A.N.)

2 Institute of Fundamental Medicine and Biology, Kazan Federal University, Kremlevskaya, 18, 420008 Kazan, Russia; k.usachev@mail.ru

3 FRC Kazan Scientific Center, Russian Academy of Sciences, Arbuzov Institute of Organic and Physical Chemistry, Arbuzov St., 8, 420088 Kazan, Russia; daut1989@mail.ru

4 Department of Theoretical and Applied Organic Chemistry and Polymerization Processes, Irkutsk State University, K. Marksa, 1, 664003 Irkutsk, Russia; kizhnyaev@chem.isu.ru

* Correspondence: ivan.stoikov@mail.ru

Citation: Shurpik, D.N.; Makhmutova, L.I.; Usachev, K.S.; Islamov, D.R.; Mostovaya, O.A.; Nazarova, A.A.; Kizhnyaev, V.N.; Stoikov, I.I. Towards Universal Stimuli-Responsive Drug Delivery Systems: Pillar[5]arenes Synthesis and Self-Assembly into Nanocontainers with Tetrazole Polymers. *Nanomaterials* **2021**, *11*, 947. https://doi.org/10.3390/nano11040947

Academic Editor: Abdelhamid Elaissari

Received: 10 March 2021
Accepted: 5 April 2021
Published: 8 April 2021

Abstract: In this work, we have proposed a novel universal stimulus-sensitive nanosized polymer system based on decasubstituted macrocyclic structures—pillar[5]arenes and tetrazole-containing polymers. Decasubstituted pillar[5]arenes containing a large, good leaving tosylate, and phthalimide groups were first synthesized and characterized. Pillar[5]arenes containing primary and tertiary amino groups, capable of interacting with tetrazole-containing polymers, were obtained with high yield by removing the tosylate and phthalimide protection. According to the fluorescence spectroscopy data, a dramatic fluorescence enhancement in the pillar[5]arene/fluorescein/polymer system was observed with decreasing pH from neutral (pH = 7) to acidic (pH = 5). This indicates the destruction of associates and the release of the dye at a pH close to 5. The presented results open a broad range of opportunities for the development of new universal stimulus-sensitive drug delivery systems containing macrocycles and nontoxic tetrazole-based polymers.

Keywords: pillar[5]arene; tetrazole; drug delivery systems; fluorescein

1. Introduction

In recent years, the pharmaceutical industry has had an increasing interest in the development and methods of introducing nanosystems in the treatment of various diseases by encapsulating drugs in biocompatible polymer matrices [1–3]. The resulting polymer-drug associates can change the pharmacokinetic properties and profile of the loaded drug after administration and provide a controlled and long-term effect of drugs on disease foci in comparison with the effect of the drug itself [1–3]. In addition, the polymer shell protects the loaded drug from premature biotransformation and can transport the drug to the focus of the disease practically without damage [1]. Water-soluble polymer systems occupy a special place among such drug delivery systems (DDSs) [4,5]. Various types of synthetic and natural polymer compositions (solid–liquid nanoparticles, liposomes, etc.) are widely studied as promising DDSs [6,7]. However, polymer systems have a number of disadvantages, namely, an extremely developed spatial structure and weak receptor properties, which complicate their controlled interaction with drugs [8,9].

The most promising DDSs are amphiphilic copolymers since their use promotes the solubilization of slightly soluble drugs [10]. It should be noted that the application of aqueous polymer systems is complicated due to the uncontrolled processes of association and aggregation of high molecular weight polymer units [11]. To address the problem, it was proposed to use various multifunctional macrocyclic compounds capable of controlled interaction with the polymer, which leads to the formation of stable associates in water [12]. Today, the most promising macrocyclic compounds are representatives of a new class of para-cyclophanes—pillar[n]arenes [13–15]. Pillar[n]arenes are macrocyclic compounds in which fragments of substituted hydroquinones are interconnected by methylene bridges [13]. Unlike other classes of macrocycles (calix[n]arenes, cyclodextrins, cucurbit[n]urils), pillar[n]arenes [16] are synthetically available and allow working in conditions (pH, water, and buffer systems) that are not suitable for other macrocycles [17–22]. Yang et al. [23] used a water-soluble pillar[5]arene to increase biocompatibility in antibacterial polymeric materials based on cationic polyaspartamide derivatives with different side-chain lengths by creating host–guest complexes to produce new antibacterial materials with pH-sensitive characteristics. Tong et al. [24] showed stimulus-responsive polymer vesicles consisting of water-soluble pillar[5]arenes with carboxylate fragments and paraquat-containing block copolymers in water. Vesicles are formed due to the formation of inclusion complexes between the pillar[5]arene and fragments of paraquats included in the structure of polymers. Additionally, Pisagatti et al. [25] developed a new type of coating based on pillar[5]arene 1 containing carboxylate fragments (Figure 1), with poly(allylamine hydrochloride) for slow release of antibiotics against Gram-positive and Gram-negative bacteria with antiadhesive and antimicrobial activity in vitro.

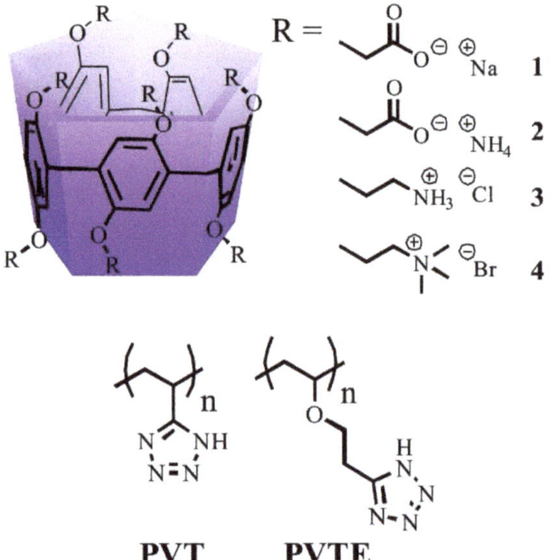

Figure 1. Structures of macrocycles **1–4** and polyvinyltetrazole (PVT) and polyvinyl (tetrazol-2-yl) ethyl ether (PVTE) polymers.

In this study, we propose a new type of universal stimulus-sensitive DDS based on decasubstituted macrocyclic structures—pillar[5]arenes and tetrazole-containing polymers. The versatility of the system lies in the principle of step-by-step supramolecular self-assembly of DDS components. The macrocyclic compound in this case will act as a link between the "protective shell"—a nontoxic water-soluble polymer and a drug. This is due to the presence of a macrocyclic cavity in pillar[5]arenes [17–22], which is involved in the formation of host-guest complexes with drugs of various structures [18,21,22]. Additionally, the introduction into the structure of macrocycles of substituents complementary to the tetrazole fragments of the polymer and sensitive to pH changes will facilitate the packing of tetrazole-containing polymers into nanosized associates [17,19].

Recently, polymer compositions based on polyvinyltetrazoles (PVTs) (Figure 1) are considered promising carriers for the formation of DDSs [26–29]. Polymers based on PVTs are easily synthesized, have a pronounced anti-inflammatory activity, promote blood clotting, accelerate wound healing [30,31]. One of the key conditions for the creation of DDSs based on water-soluble polymers is the micellization of polymer systems in a wide pH range [32–34]. DDSs based on polymer systems are of practical interest as nanocarriers for the encapsulation and controlled release of hydrophobic drugs. Such nanosized polymer systems are sensitive to the ambient pH due to the presence of charged ammonium fragments in their structure [33]. Tetrazole-containing polymer systems are ideal as components of pH-sensitive DDSs. However, it should be noted that polymers based on PVTs do not form stable nanosized aggregates in aqueous solutions [35]. In this regard, the formation of stable, stimulus-responsive nanoscale associates of PVT/pillararene capable of controlled drug release is an actual problem.

Thus, in this paper, we report the first example of the use of uncharged water-soluble derivatives of pillar[5]arene containing tertiary amino groups to obtain stimulus-responsive nanosized associates with polyvinyl (tetrazol-2-yl) ethyl ether (PVTE) and dye—fluorescein, by the method of controlled supramolecular self-assembly.

2. Materials and Methods

2.1. Characterization

^{1}H, ^{13}C NMR, and 2D NOESY, HSQC NMR spectra were obtained on a Avance-400 spectrometer (Bruker, Billerica, MA, USA) ($^{13}C\{^{1}H\}$—100 MHz and ^{1}H and 2D NOESY—400 MHz). Chemical shifts were determined against the signals of residual protons of deuterated solvent ($CDCl_3$, D_2O, CD_3OD). The concentration of sample solutions was 3–5%.

Attenuated total internal reflectance IR spectra were recorded with a Spectrum 400 (Perkin Elmer, Waltham, MA, USA) Fourier spectrometer.

Elemental analysis was performed with a Perkin Elmer 2400 Series II instrument (Perkin Elmer, Waltham, MA, USA).

Mass spectra (MALDI-TOF) were recorded on an Ultraflex III mass spectrometer (Bruker, Billerica, MA, USA) in a 4-nitroaniline matrix. Melting points were determined using a Boetius Block apparatus.

Additional control of the purity of compounds and monitoring of the reaction were carried out by thin-layer chromatography using Silica G, 200 μm plates, UV 254.

Most chemicals were purchased from Aldrich and used as received without additional purification. Organic solvents were purified in accordance with standard procedures.

2.2. Diffusion Ordered Spectroscopy (DOSY)

^{1}H diffusion ordered spectroscopy (DOSY) spectra were recorded on a Bruker Avance 400 spectrometer (Bruker, Billerica, MA, USA) at 9.4 tesla at a resonating frequency of 400.17 MHz for ^{1}H using a BBO (Bruker, Billerica, MA, USA) 5 mm gradient probe. The temperature was regulated at 298 K and no spinning was applied to the NMR tube. DOSY experiments were performed using the STE bipolar gradient pulse pair (stebpgp1s) pulse sequence with 16 scans of 16 data points collected. The maximum gradient strength produced in the z direction was 5.35 G mm^{-1}. The duration of the magnetic field pulse gradients (δ) was optimized for each diffusion time (Δ) in order to obtain a 2% residual signal with the maximum gradient strength. The values of δ and Δ were 1.800 μs and 100 ms, respectively. The pulse gradients were incremented from 2 to 95% of the maximum gradient strength in a linear ramp.

2.3. Scanning Electron Microscope (SEM)

Morphological structures of samples were observed by SEM (Carl Zeiss Auriga Cross Beam). Samples were first diluted with water to a final concentration of 1×10^{-4} g/mL, and the resulting suspension was placed on a silicon pan, which was then dried in a vacuum desiccator for 1 h. After complete drying, the samples were placed into the scanning electron microscope using a special holder for microanalysis. Analysis was held at the accelerating voltage of 80 kV.

2.4. Fluorescence Spectroscopy

Fluorescence spectra were recorded on a Fluorolog 3 luminescent spectrometer (Horiba Jobin Yvon, Longjumeau, France). The excitation wavelength was selected as 450 nm. The emission scan range was 480–700 nm. Excitation and emission slits were 1 nm. Quartz cuvettes with an optical path length of 10 mm were used. Fluorescence spectra were automatically corrected by the Fluorescence program. The spectra were recorded in water solutions with a concentration of fluorescein 1×10^{-5} M for pillar[5]arene/fluorescein systems. The obtained molar ratio of fluorescein to pillar[5]arene 7 was 1:100. The experiment was carried out at 293 K. Fluorescence spectra at different pH were fixed in a buffer for system **7** (1×10^{-5} M)/Flu (1×10^{-5} M)/PVTE (1×10^{-4} M) at 298 K. Solutions of the investigated systems were measured after incubating for an hour at room temperature.

2.5. UV–Visible Spectroscopy

UV–Vis spectra were recorded using the UV-3600 spectrometer (Shimadzu, Kyoto, Japan); the cell thickness was 1 cm, and slit width 1 nm. Deionized water with a resistivity >18.0 MΩ cm was used to prepare the solutions. Deionized water was obtained from a Millipore-Q purification system. Recording of the absorption spectra of the mixtures of PVT, PVTE, and fluorescein (Flu) with pillar[5]arenes **7–9** and **12** (1×10^{-5}–5×10^{-5} M) was carried out after mixing the solutions at 293 K. The 1×10^{-5}–5×10^{-5} M solution of the guest—PVT, PVTE (50, 75, 100, 150, 300, 400, 500, 600, 700, 900, 1200, 1500, 1800, 2100, 2400, and 2700 μL, 1×10^{-5}–5×10^{-5} M) in deionized water or phosphate buffer—was added to 300 μL of the solution of macrocycle **7–9** and **12** (1×10^{-5}–5×10^{-5} M) (in the case of Flu, the concentration of Flu was constant) in deionized water or phosphate buffer and diluted to a final volume of 3 mL with deionized water or phosphate buffer. The UV spectra of the solutions were then recorded. The association constants of complexes were calculated as described below. Three independent experiments were carried out for each series. The student's *t*-test was applied in the statistical data processing.

2.6. Dynamic Light Scattering (DLS)

2.6.1. Particle Size

Dynamic Light Scattering (DLS). The particle size was determined by the Zetasizer Nano ZS (Malvern Instruments, Malvern, UK) at 298 K. The instrument contains a 4 mW He—Ne laser operating at a wavelength of 633 nm and incorporated noninvasive backscatter (NIBS) optics. The measurements were performed at the detection angle of 173° and the software automatically determined the measurement position within the quartz cuvette. The concentration ratios of pillar[5]arenes 7–9 and 12, pillar[5]arenes **7–9** and **12**/PVT, PVTE, and Flu were 50:1, 10:1, 5:1, 2:1, 1:1, 1:2, 1:5, and 1:15, and the concentration of compounds was 1×10^{-5} M, 1×10^{-4} M, and 1×10^{-3} M. The experiments were carried out for each solution in triplicate.

2.6.2. Zeta Potentials

Zeta (ζ) potentials were measured on a Zetasizer Nano ZS from (Malvern Instruments, Malvern, UK). Samples were prepared as they were for the DLS measurements and were transferred with the syringe to the disposable folded capillary cell for measurement. The zeta potentials were measured using the Malvern M3-PALS method and averaged from three measurements.

2.7. X-ray Diffraction (XRD)

Data sets for single crystals **6**, **7**, and **11** were collected on a Rigaku XtaLab Synergy S instrument with a HyPix detector (Rigaku, Tokyo, Japan) and a PhotonJet microfocus X-ray tube (Rigaku, Tokyo, Japan) using Cu Kα (1.54184 Å) radiation at low temperature. Images were indexed and integrated using the CrysAlisPro data reduction package. Data were corrected for systematic errors and absorption using the ABSPACK module, i.e., numerical absorption correction based on Gaussian integration over a multifaceted crystal model and empirical absorption correction based on spherical harmonics according to the point group symmetry using equivalent reflections. The GRAL module was used for the analysis of systematic absences and space-group determination. The structures were solved by direct methods using SHELXT [36] and refined by the full-matrix least squares on F^2 using SHELXL [37]. Non-hydrogen atoms were refined anisotropically. The hydrogen atoms were inserted at the calculated positions and refined as riding atoms. The positions of the hydrogen atoms of methyl groups were found using rotating group refinement with idealized tetrahedral angles. The figures were generated using Mercury 4.1 [38] program. To refine the model against the measured data of **11** the SQUEEZE as implemented in PLATON was used to model the disordered solvent in the big voids of the structure. SQUEEZE identified a void with a volume of 2190 Å^3 containing the equivalent of 768 electrons. This would correspond to about 3[$CHCl_3$ + C_2H_3N]. Crystal data, data collection, and structure refinement details are summarized in Table S2, Supplementary Materials.

2.8. Synthesis of O-Substituted Hydroquinones and PVT and PVTE Polymers

Compound **5**, **10** were synthesized according to the literature procedures [39,40].

Polymers PVT and PVTE were synthesized according to the literature procedures [35].

Polymer sample PVT (M = 7 × 10^4) used in this study was synthesized via the azidation of polyacrylonitrile, which was synthesized through the 2,2′-azodi(isobutyronitrile) (AIBN)-initiated polymerization of 5-Vinyl-1H-tetrazole (VT) in acetonitrile at 60 °C.

We used a PVTE polymer sample (M = 5 × 10^4) synthesized by cyanoethylation of polyvinyl alcohol with acrylonitrile, followed by azidation of nitrile groups.

2.9. Synthesis

2.9.1. General Procedure for the Synthesis of Compounds 6 and 11

Paraformaldehyde (PFA) 0.17 g (5.9 mmol) and 50 mL of dichloroethane were placed in a round bottom flask equipped with a magnetic stirrer. Then, (1.9 mmol) of compound **5** (or **10**) and 0.17 mL (1.9 mmol) of trifluoromethanesulfonic acid were added. The reaction was carried out at 0 °C for 30 min. Then, the temperature was gone up to 45 °C and stirred for 1 h. Then, the reaction mixture was quenched with 5% $NaHCO_3$ solution (50 mL). The organic layer was separated on a separatory funnel and washed with distilled water (2 × 30 mL) and then evaporated under reduced pressure. The resulting light brown reaction mixture was purified by reprecipitation from isopropyl alcohol. The reaction mixture was dissolved in acetonitrile (20 mL) and slowly poured into isopropyl alcohol (40 mL). The formed precipitate was separated by centrifugation and dried at room temperature.

*4,8,14,18,23,26,28,31,32,35-Deca-(4-methylbenzylsulfonate-1-ethoxy)-pillar[5]arene (**6**).*

Light yellow crystalline substance. Yield 0.74 g (85%). ^{1}H NMR (400 MHz, $CDCl_3$): δ 2.27 (s., 30H, -CH_3), 3.60 (s., 10H, -CH_2-), 4.02–4.11 (m., 10H, -O-CH_2-$\underline{CH_2}$-O-), 4.21–4.29 (m., 10H, -O-CH_2-$\underline{CH_2}$-O-), 4.29–4.38 (m., 10H, -O-$\underline{CH_2}$-CH_2-O-), 4.40–4.53 (m., 10H, -O-$\underline{CH_2}$-CH_2-O-), 6.78 (s, 10H, Ar$\underline{H}$), 7.05 (d., J = 7.9 Hz, 20H, $\underline{Ar_{Ts}}CH_3$), 7.71 (d., J = 7.9 Hz, 20H, $\underline{Ar_{Ts}}CH_3$). ^{13}C NMR (100 MHz, $CDCl_3$): δ 21.62, 28.96, 66.01, 69.85, 114.50, 127.98, 128.55, 130.01, 132.28, 144.98, 149.49. IR, ν/cm^{-1}: 3065 (ArH), 2927, 2877 (-CH_2-, CH_3), 1597, 1496 (-C=C-), 1353, 1172 (-S=O), 917 (-S-O). MS (MALDI-TOF) calc. [M^+] m/z = 2591.5, found [M^+] m/z = 2591.7. Found (%): C, 56.12; H, 5.25; S, 10.93. Calc. for $C_{125}H_{130}O_{40}S_{10}$. (%): C, 57.90; H, 5.05; S 12.36.

*4,8,14,18,23,26,28,31,32,35-Deca-[(isoindoline-1,3-dione)propoxy]-pillar[5]arene (**11**).*

Light yellow crystalline substance. Yield 0.70 g (74%). ^{1}H NMR (400 MHz, $CDCl_3$): δ 2.47–2.63 (m., 10H, -O-CH_2-$\underline{CH_2}$-CH_2-N), 4.10 (s., 10H, -CH_2-), 4.22–4.32 (m., 10H, -O-CH_2-CH_2-$\underline{CH_2}$-N), 4.32–4.50 (m., 20H, -O-$\underline{CH_2}$-CH_2-CH_2-N), 7.26 (s., 10H, Ar$\underline{H}$), 7.97–8.21 (Ar_{Pht}H). ^{13}C NMR (100 MHz, $CDCl_3$): δ 29.26, 29.43, 35.81, 66.35, 115.28, 123.25, 128.53, 132.40, 133.84, 149.95, 168.31. IR, ν/cm^{-1}: 3029 (ArH), 2941, 2878 (-CH_2-, CH_3), 1703 (-C=O), 1203 (Ar-O-CH_2-). MS (MALDI-TOF) calc. [M$^+$] *m/z* = 2481.8, found [M+H$^+$] *m/z* = 2484.7. Found (%): C, 68.96; H, 4.46; N, 4.84. Calc. for $C_{145}H_{120}N_{10}O_{30}$. (%): C, 70.15; H, 4.87; N, 5.64.

2.9.2. General Procedure for the Synthesis of Compounds **7–9** and **12**

Compound **6** (0.2 g, 0.07 mmol) and 1 mL amine (pyrrolidine, piperidine, morpholine) were placed in a round-bottom flask with a magnetic stirrer and a reflux condenser. The reaction mixture was boiled for 48 h. Then, the dark brown reaction mixture was poured into diethyl ether (10 mL). The formed precipitate was collected by filtration on a Schott filter. Then, the precipitate was dissolved in dichloromethane (20 mL) and washed with distilled water (3 × 20 mL). The organic layer was separated and evaporated under reduced pressure. The target products were obtained by recrystallization from isopropyl alcohol.

*4,8,14,18,23,26,28,31,32,35-Deca-[2-(pyrrolidin-1-yl)ethoxy]-pillar[5]arene (**7**).*

White crystalline substance. Yield 0.11 g (87%). ^{1}H NMR (400 MHz, $CDCl_3$): δ 1.70–1.90 (m., 40H, -N-$(CH_2)_2$-), 2.55–2.77 (m., 40H, -N-$(CH_2)_2$-), 2.87–3.05 (m., 20H, -O-CH_2-$\underline{CH_2}$-), 3.75 (s., 10H, -CH_2-), 3.93–4.16 (m., 20H, -O-$\underline{CH_2}$-CH_2-), 6.85 (s., 10H, Ar$\underline{H}$). ^{13}C NMR (100 MHz, $CDCl_3$): δ 22.88, 28.97, 54.57, 54.95, 66.96, 117.26, 124.41, 149.66. IR, ν/cm^{-1}: 3029 (ArH), 2954, 2882 (-CH_2-), 2607, 2483 (-N-$(CH_2)_2$-), 1204 (Ar-O-CH_2-). MS (MALDI-TOF) calc. [M$^+$] *m/z* = 1582.1, found [M+H$^+$] *m/z* = 1583.2. Found (%): C, 71.20; H, 7.98; N, 7.74. Calc. for $C_{95}H_{140}N_{10}O_{10}$. (%): C, 72.12; H, 8.92; N, 8.85.

*4,8,14,18,23,26,28,31,32,35-Deca-[2-(piperidin-1-yl)ethoxy]-pillar[5]arene (**8**).*

Light yellow crystalline substance. Yield 0.10 g (78%). ^{1}H NMR (400 MHz, $CDCl_3$): δ 1.33–1.54 (m., 20H, -CH_2-), 1.54–1.70 (m., 40H, -N-$(CH_2)_2$-$(\underline{CH_2})_2$-), 2.50–2.70 (m., 40H, -N-$(\underline{CH_2})_2$-$(CH_2)_2$-), 2.70–2.91 (m., 20H, -O-CH_2-$\underline{CH_2}$-), 3.73 (s., 10H, -CH_2-), 3.90–4.14 (m., 20H, -O-$\underline{CH_2}$-CH_2-), 6.83 (s., 10H, Ar$\underline{H}$). ^{13}C NMR (100 MHz, $CDCl_3$): δ 24.34, 26.15, 29.48, 58.68, 66.81, 115.19, 128.53, 149.91. IR, ν/cm^{-1}: 2929, 2851 (-CH_2-), 2781, 2748 (-N-$(CH_2)_2$-), 1203 (Ar-O-CH_2-). MS (MALDI-TOF) calc. [M$^+$] *m/z* = 1722.2, found [M+H$^+$] *m/z* = 1723.1. Found (%): C, 72.54; H, 8.93; N, 7.86. Calc. for $C_{105}H_{160}N_{10}O_{10}$. (%): C, 73.22; H, 9.36; N, 8.13.

*4,8,14,18,23,26,28,31,32,35-Deca-(2-morpholinoethoxy)-pillar[5]arene (**9**).*

Light yellow crystalline substance. Yield 0.09 g (72%). ^{1}H NMR (400 MHz, $CDCl_3$): δ 2.60 (d., 40H, J = 4.4 Hz, -N-$(\underline{CH_2})_2$-$(CH_2)_2$-O-), 2.75–2.90 (m., 20H, -O-CH_2-$\underline{CH_2}$-), 3.69–3.77 (m., 50H, -N-$(CH_2)_2$-$(\underline{CH_2})_2$-O-, -CH_2-), 3.92–4.14 (m., 20H, -O-$\underline{CH_2}$-CH_2-), 6.84 (s, 10H, Ar$\underline{H}$). ^{13}C NMR (100 MHz, $CDCl_3$): δ 29.59, 54.09, 54.33, 58.28, 66.76, 67.03, 115.73, 128.95, 150.03. IR, ν/cm^{-1}: 2935, 2891 (-CH_2-), 2800 (-N-$(CH_2)_2$-), 1113 (Ar-O-CH_2-). MS (MALDI-TOF) calc. [M$^+$] *m/z* = 1742.0, found [M+H$^+$] *m/z* = 1743.0. Found (%): C, 66.35; H, 9.25; N, 8.57. Calc. for $C_{95}H_{140}N_{10}O_{20}$. (%): C, 65.49; H, 8.10; N, 8.04.

4,8,14,18,23,26,28,31,32,35-Deca-(aminopropyloxy)-pillar[5]arene (**12**).

Macrocycle **11** (0.3 g, 0.12 mmol), hydrazine hydrate (65% aqueous solution, 0.53 mL, 7.2 mmol), and 10 mL of methanol were placed in a round bottom flask with a magnetic stirrer and a reflux condenser. The reaction mixture was boiled for 48 h. Then, the reaction mixture was cooled to room temperature. The formed precipitate was filtered off on a Schott filter. The crude product was washed on the filter with methanol excess. Then, the resulting product was dissolved in 10% HCl (15 mL). The resulting cloudy solution was filtered on a Schott filter. Next, an ammonia solution (15%) was added to the filtrate to pH 10. The precipitated amine **12** was filtered off on a Schott filter. The resulting white crystalline substance was dried under vacuum. Yield 0.12 g (85%). ^{1}H NMR (400 MHz, D_2O): δ 1.90–2.06 (m., 20H, -O-CH_2-$\underline{CH_2}$-CH_2-NH_2), 3.15 (t., 20H, J = 7.3 Hz, -O-CH_2-CH_2-$\underline{CH_2}$-NH_2), 3.78–3.90 (m., 30H, -O-$\underline{CH_2}$-CH_2-CH_2-NH_2, -CH_2-), 6.76 (s, 10H, Ar$\underline{H}$). ^{13}C NMR (100 MHz,

$CDCl_3$): δ 26.78, 29.50, 37.31, 67.14, 116.61, 129.78, 150.05. IR, ν/cm^{-1}: 3298 (-NH_2), 2924, 2864 (-CH_2-), 1199 (Ar-O-CH_2-). MS (MALDI-TOF) calc. [M^+] m/z = 1180.8, found [M+H^+] m/z = 1181.6. Found (%): C, 65.93; H, 7.45; N, 10.84. Calc. for $C_{65}H_{100}N_{10}O_{10}$. (%): C, 66.07; H, 8.53; N, 11.85.

Detailed information on physical–chemical characterization is presented in Supplementary Materials (SM).

3. Results

3.1. Synthesis of Pillar[5]arene Derivatives

The two most studied tetrazole-containing polymers were selected as objects of this study, namely, poly-5-vinyltetrazole (PVT) and polyvinyl (tetrazol-5-yl) ethyl ether (PVTE) (Figure 1), which have relatively low toxicity (from LD_{50} = 166 mg/kg for p-5VT to 1340 mg/kg for IPT-VPD copolymer) [41]. Interest in tetrazole-containing polymers is growing due to the fact that the tetrazole fragment in the polymer structure is a pharmacophore group. Today, there are 43 drugs containing 1H- or 2H-tetrazole, 23 of them are approved by the FDA [42]. Moreover, 5-substituted tetrazole is of particular interest because it has a mobile proton. This makes 5-substituted tetrazoles strong NH acids (pKa = 4.5–4.9 depending on the substituent). Therefore, 5-substituted tetrazoles have been widely used in recent years as an isosteric analog of the COOH group [42–44]. Similar to polymers containing carboxylate groups, polymers containing tetrazole moieties (PVT and PVTE) are capable of ionizing at physiological pH (7.4). However, it has been shown [43] that tetrazolate anions are almost 10 times more lipophilic than carboxylate anions, which is an important factor in the passage of the polymer composition through cell membranes. In addition, the polymers used in DDSs and containing carboxylate groups (*N*-isopropylacrylamide (NIPAM), poly(alkyl acrylic acid)s, modified poly(glycidol)s (PGs)) do not penetrate into the cell due to the submicron size, but they are fixed on the surface [44]. Delivery of the drug into the cell is carried out by loosening the cell membrane with subsequent release of the drug [44]. Loosening of the cell membrane during drug transport can be detrimental to the cell. The interaction between water-soluble polymers and lipid membranes is mainly governed by the hydrophobic/hydrophilic balance and the polymer architecture [45]. Charged polymers (polyelectrolytes), such as polycarboxylates or polyamines, can adsorb on oppositely charged lipid bilayers, which can lead to membrane deformations. This manifests itself in an increase in the curvature of the membrane, which can lead to sealing or destruction of the membrane in the case of cationic polymers [45,46]. Membrane damage occurs through hole formation, and some cationic polymers/oligomers can move across the cell membrane and bind to intracellular DNA or RNA to prevent intracellular synthesis [46]. In previous works, we showed [47] that polymers based on tetrazole in the presence of pillar[5]arene are capable of forming nanoscale associates. In [47], uncharged pillar[5]arenes containing hydroxyl groups were used. We have shown that the macrocyclic platform plays a key role in the formation of supramolecular associates. The resulting nanosized associates turned out to be insensitive to pH changes. Therefore, in this work, it was decided to use pillar[5]arenes containing primary and tertiary amino groups capable of protonation in an acidic medium.

In this regard, we hypothesize that the use of water-soluble derivatives of the pillar[5]arene containing amino groups in the presence of tetrazole-containing polymers will contribute to the formation of nanoscale associates sensitive to changes in the pH environment. Additionally, the presence of a free electron-donating macrocyclic cavity makes it possible to use substituted pillar[5]arenes as molecular containers. To confirm this hypothesis, it was necessary to solve a number of problems, i.e., (1) to synthesize macrocycles with functional groups complementary to tetrazole fragments and capable of interacting with the polymer, (2) to study the interaction of synthesized macrocycles with polymers (PVT and PVTE), and (3) to study the interaction of macrocycles and polymers with the model drug. The key task of the study is to establish the patterns of drug release when the system is exposed to a certain stimulus. The following functional groups of pillar[5]arene can act as groups interacting with the polymer: -NH (aliphatic or aromatic), -OH (aliphatic, phenolic, aromatic), and carbonyl (esters, amides, ketones, etc.) groups [42–44]. Analysis of the literature has shown that four main pillar[5]arenes **1–4** (Figure 1) are the most studied as components of self-assembling systems [48–53]. The presence of acidic protons in the tetrazole fragments of PVT and PVTE (Figure 1) excludes the use of carboxylate groups (compounds **1**, **2** in Figure 1) in the pillar[5]arene structure for further study of the interaction with polymers. For this reason, we chose macrocycles **3** and **4**. However, the method for their preparation requires the use of an expensive palladium catalyst and gaseous compounds (hydrogen and trimethylamine) [52,53].

In order to develop an alternative approach to the production of pillar[5]arenes containing primary and tertiary amino acids, the synthesis of the pillar[5]arene platform was carried out by macrocyclization of functionalized 1,4-dialkoxybenzenes: *p*-bis[2-(phthalimido)ethyloxy]benzene and 2,2′-bis(tosylethyl)-hydroquinone. The cyclization of O-substituted hydroquinones in the presence of paraformaldehyde is currently the main method for the preparation of substituted pillar[n]arenes. However, the yield of the anticipated pillararenes modified with functional groups is quite low [16]. Therefore, the development of additional methods of macrocyclization makes it possible to significantly expand the number of introduced functional groups into the pillar[5]arene structure.

The starting bis[2-(phthalimido)ethyloxy]benzene **5** and 2,2′-bis(tosylethyl)-hydroquinone **10** were obtained according to published methods [39]. At the next stage, we selected the conditions for macrocyclization (see SM, Figures S1–S25). Three main solvents were selected in the synthesis of pillar[n]arenes: $CH_2Cl\text{-}CH_2Cl$, $CHCl_3$, CH_2Cl_2 [13–16]. Additionally, catalysts (Lewis acids) were varied. A number of main catalysts were selected, namely, $BF_3 \times Et_2O$, $AlBr_3$, CF_3COOH, and CF_3SO_3H. These catalysts are most often used in the synthesis of pillar[5]arenes [13–16]. Paraform and trioxane were chosen as a reagent, allowing the introduction of methylene bridge into the macrocycle. The reaction time varied from 30 min to 24 h. The reaction temperature range was from 0 °C to 85 °C (Scheme 1) (see SM, Table S1).

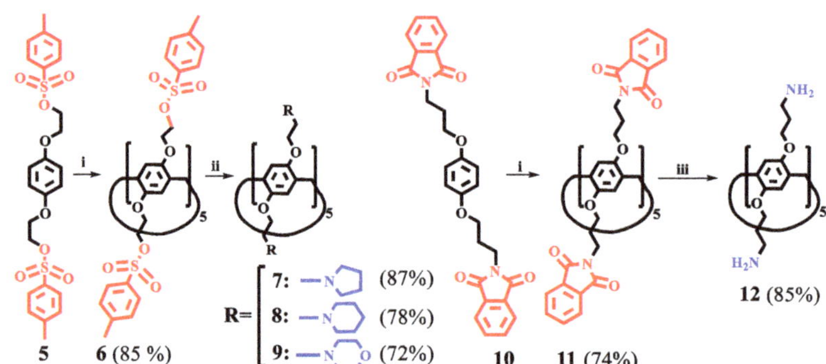

Scheme 1. Synthesis of macrocycles **6–9** and **11**, **12**. Reagents and conditions: **i**—CF_3SO_3H, $(CH_2O)_n$, $CH_2Cl\text{-}CH_2Cl$, 0–45 °C, 1.5 h; **ii**—amine (pyrrolidine, piperidine, morpholine), 100 °C, 48 h; and **iii**—N_2H_2, CH_3OH, 48 h.

The change in the nature of the solvent showed that the reaction of **5** and **10** with paraformaldehyde or trioxane in anhydrous CH_2Cl_2 in the presence of Lewis acids ($BF_3 \times Et_2O$, $AlBr_3$, CF_3COOH, CF_3SO_3H) does not lead to the formation of desired products **6** and **11**. In the case of $CHCl_3$, macrocyclization proceeded only with 2,2′-bis(tosylethyl)-hydroquinone in the presence of paraformaldehyde and CF_3SO_3H. Analysis of ^1H NMR spectra showed the presence of a mixture of products—pillar[5]arene, pillar[6]arene, and polymer (see SM, Table S1). Unfortunately, it was not possible to separate the reaction mixture using column chromatography. The use of $CH_2Cl\text{-}CH_2Cl$ as a solvent and CF_3SO_3H as catalyst made it possible to obtain target macrocycles **6** and **11** (Scheme 1). The use of $BF_3 \times Et_2O$ as catalyst also led to the production of **6** (64%) and **11** (53%). However, a large amount of polymer was formed during the reaction, which required laborious column chromatography. The formation of target products was not observed in the case of using catalysts $AlBr_3$ and CF_3COOH. Changing the paraformaldehyde to trioxane in the $BF_3 \times Et_2O/CH_2Cl\text{-}CH_2Cl$ and $CF_3SO_3H/CH_2Cl\text{-}CH_2Cl$ systems also led to **6** and **11**, but the yield of the target products did not exceed 20%.

Thus, the use of model compounds **5** and **10** in the presence of paraformaldehyde and CF_3SO_3H leads to the formation of target macrocycles **6** and **11** in 85% and 74% yields, respectively (Scheme 1). The temperature regime in all performed reactions varied from 0–85 °C. However, the best results were achieved when the reaction was carried out in the temperature range from 0 °C to 45 °C for an hour and a half for both macrocycles. It should be noted that the formation of non-macrocyclic polymer is almost minimized when the reaction goes in the selected temperature range and use CF_3SO_3H as a catalyst. The target products were isolated by reprecipitation from isopropyl alcohol. The technique developed by us allows producing large quantities of target macrocycles without losing the yield (see SM, Figures S1–S8).

The spatial structure of the resulting products **6** and **11** (see SM, Table S2.) was fully confirmed using structural analysis by single-crystal XRD (Figure 2). The crystals were grown in both cases from a mixture of solvents $CHCl_3\text{-}CH_3CN$. Syngony of product **6** is monoclinic, group symmetry is C2/c (Figure 2a,b). As can be seen from Figure 2c, the formation of a supramolecular polymer in the crystalline state is observed in the case of compound **6** by including two tosylate fragments in the cavity of a single macrocycle.

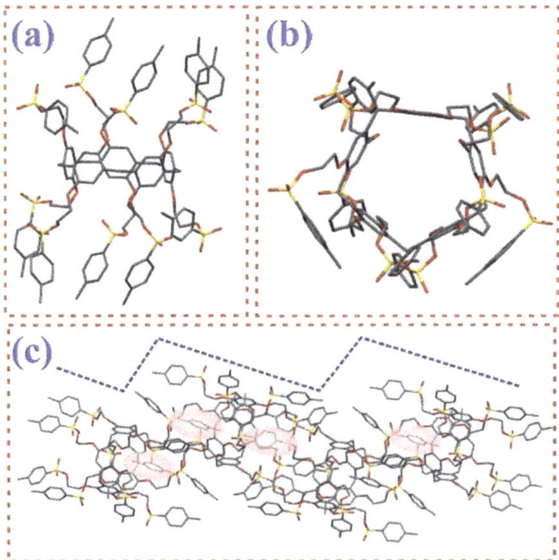

Figure 2. (**a**) Side view of macrocycle **6**, (**b**) top view of macrocycle **6**, and (**c**) packaging of macrocycle **6** in a crystal, demonstrating the formation of pseudorotaxane structures.

The type of crystal lattice is triclinic, and the symmetry group is *P*-1 in the case of macrocycle **11** (Figure 3). Unlike macrocycle **6**, compound **11** does not form a supramolecular polymer in the crystalline state. All phthalimide fragments **11** (Figure 3a,b) are deployed in the "propeller blade" form outward from the cavity of the macrocycle **11**. This arrangement can be explained by intermolecular π–π-stacking interactions between phthalimide fragments of neighboring macrocycles in structural motive **11**. Structural analysis by single-crystal XRD showed the presence of pseudocavity (Figure 3c) in the crystal structure of

11, which is formed between two macrocycles bound by stacking phthalimide fragments (Figure 3c, orange balls). It is interesting to note that the absence of substituent fragments in the cavity of the macrocyclic platform **7** (Figure 3c, green balls) makes it possible to use the resulting structures as nonporous adaptive crystals [54–58].

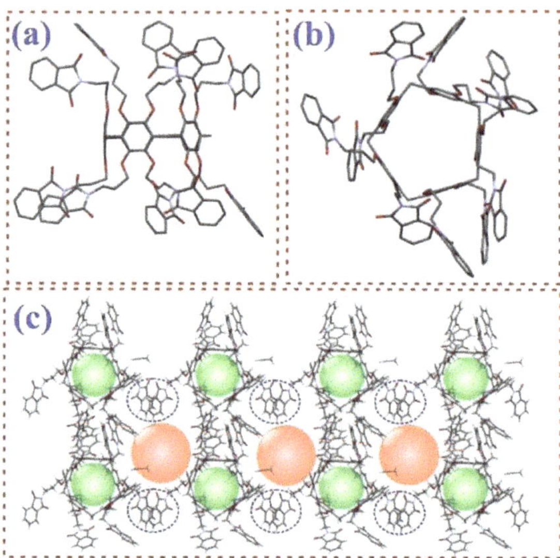

Figure 3. (**a**) Side view of macrocycle **11**, (**b**) top view of the macrocycle **11**, and (**c**) packing of macrocycle **11** in a crystal, showing π–π-stacking interactions between phthalimide fragments, macrocycles and the formation of corresponding pseudocavities (orange balls), and the presence of a free macrocyclic cavity (green balls).

Then, macrocycle **6** was involved in the reaction with secondary amines—pyrrolidine, piperidine, and morpholine. The reaction was performed in excess of amine at 100 °C (Scheme 1). The reaction time was 48 h. The yield of macrocycles **7–9** was from 72% to 87% (see SM, Figures S9–S20). In order to obtain a primary amine, macrocycle **11** was involved in a reaction with hydrazine hydrate in methanol at the boiling point of the solvent (Scheme 1). The reaction time was 48 h. Decaamine **12** precipitated from the reaction mixture was filtered off and purified by conversion into the hydrochloride form. The yield of macrocycle **12** was 85%. The structures of the obtained products **7–9** and **12** were fully confirmed by a complex of physical methods: ^{1}H and ^{13}C one-dimensional NMR, IR spectroscopy, and mass spectrometry, and the compositions were confirmed by elemental analysis data (see SM, Figures S9–S20). The structure of product **7** was finally established by XRD (Figure 4). The type of crystal lattice is triclinic, symmetry groups are *P*-1 (Figure 4a). As can be seen from Figure 4b, the crystal structure of macrocycle **7** is a hollow tube in which the macrocycle molecules are arranged one after the other. It should be noted that such packing of macrocycles into hollow tubes (Figure 4b, yellow balls) makes these compounds promising for the development of crystalline organic materials based on macrocycles [56–58].

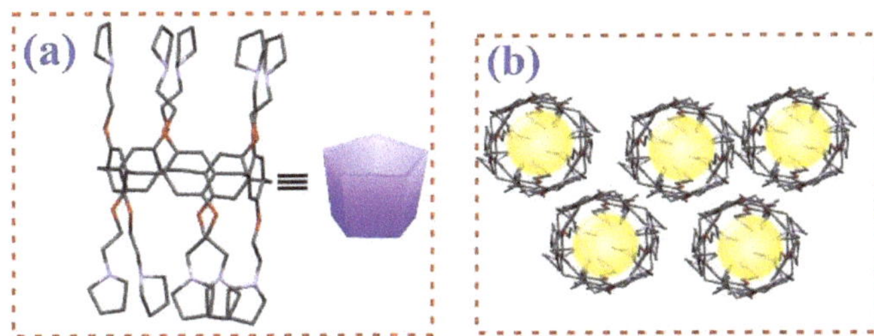

Figure 4. (**a**) Side view of macrocycle **7** and (**b**) packing of macrocycle **7** in the crystal, showing the formation of molecular channels in the crystal, with a free macrocyclic cavity (yellow balls).

3.2. Study of the Interaction of Pillar[5]arenes with PVT and PVTE Polymers

At the next stage of the study, the ability of macrocycles **7–9** and **12** to interact with PVT, PVTE, and fluorescein dye in ethanol and water was studied using UV–Vis, ^{1}H, 2D NMR, fluorescence spectroscopy, and dynamic light scattering (DLS). Unfortunately, changes in the UV–Vis spectra of compounds **7–9** and **12** in the presence of PVT and PVTE turned out to be insignificant in water and buffer solutions (pH = 5–9). It was found that the spectral changes were significant only during the interaction of macrocycles **7** and **9** with PVTE in ethanol, which made it possible to establish the quantitative characteristics of the binding (see SM, Figures S27 and S28). The association constant was determined by spectrophotometric titration, in which the concentration of PVTE was changed, and the concentration of **7** (1×10^{-5} M) remained constant. The results were processed using BindFit [59–61] and fitted to a 1:1 binding model (see SM, Figures S27 and S28). To confirm the proposed stoichiometry, the titration data were also processed using a binding model with a host:guest ratio of 1:2. However, in this case, the constants were determined with much greater uncertainty (see SM, Figures S27 and S28). Calculated association constant (Ka) was for **9**/PVTE = 1898.15 and for **7**/PVTE—Ka = 2993.95. The interaction of macrocycle **12** with PVT and PVTE is accompanied by the formation of a precipitate. The calculation of the association constant by UV-Vis spectroscopy is impossible. The spectral pattern, in this case, is complicated by light scattering, which is manifested in a significant rise of the baseline.

We chose 2D ^{1}H-^{1}H NOESY and 2D DOSY NMR spectroscopy methods to confirm the formation of a complex between macrocycles **7** and **9** with PVTE and its structure. (Figure 5). However, an analysis of the experimental data for mixtures **9**/PVTE and **7**/PVTE, obtained using ^{1}H NMR spectroscopy, did not allow determining the nature of the interaction since the polymer signals were too broadened. In this regard, to confirm the interaction of pillar[5]arenes with polymer fragments, the corresponding monomer, 5-vinyltetrazole (5-VT) was chosen. Thus, in the 2D ^{1}H-^{1}H NOESY NMR spectrum of the **7**/5-VT (5×10^{-3} M) associate in CD_3OD (Figure 5a), cross peaks were observed between the protons of the 5-VT $\mathbf{H^3}$ double bond and the protons of the pyrrolidine fragment $\mathbf{H^g}$ and $\mathbf{H^f}$ of macrocycle **7** (Figure 5c). The formation of the **7**/5-VT complex was additionally confirmed by 2D DOSY NMR spectroscopy (Figure 5b). Diffusion coefficients of **7**, 5-VT and **7**/5-VT at 298 K (1×10^{-3} M) were determined. The DOSY NMR spectrum of the **7**/5-VT system showed the presence of signals from a complex lying on one straight line (Figure 5b), with a diffusion coefficient (D = 3.36×10^{-10} m^2 s^{-1}), which is significantly lower than the self-diffusion coefficient of macrocycle **7** (D = 4.95×10^{-10} m^2 s^{-1}) and 5-VT (D = 7.44×10^{-10} m^2 s^{-1}) under the same conditions [21]. The results obtained unambiguously indicate the formation of the **7**/5-VT complex. The absence of cross peaks between 5-VT and aromatic protons of pillar[5]arene 7 indicates that the monomer is not in the macrocyclic cavity but in the pseudocavity of the macrocycle formed by fragments of substituents [62–64].

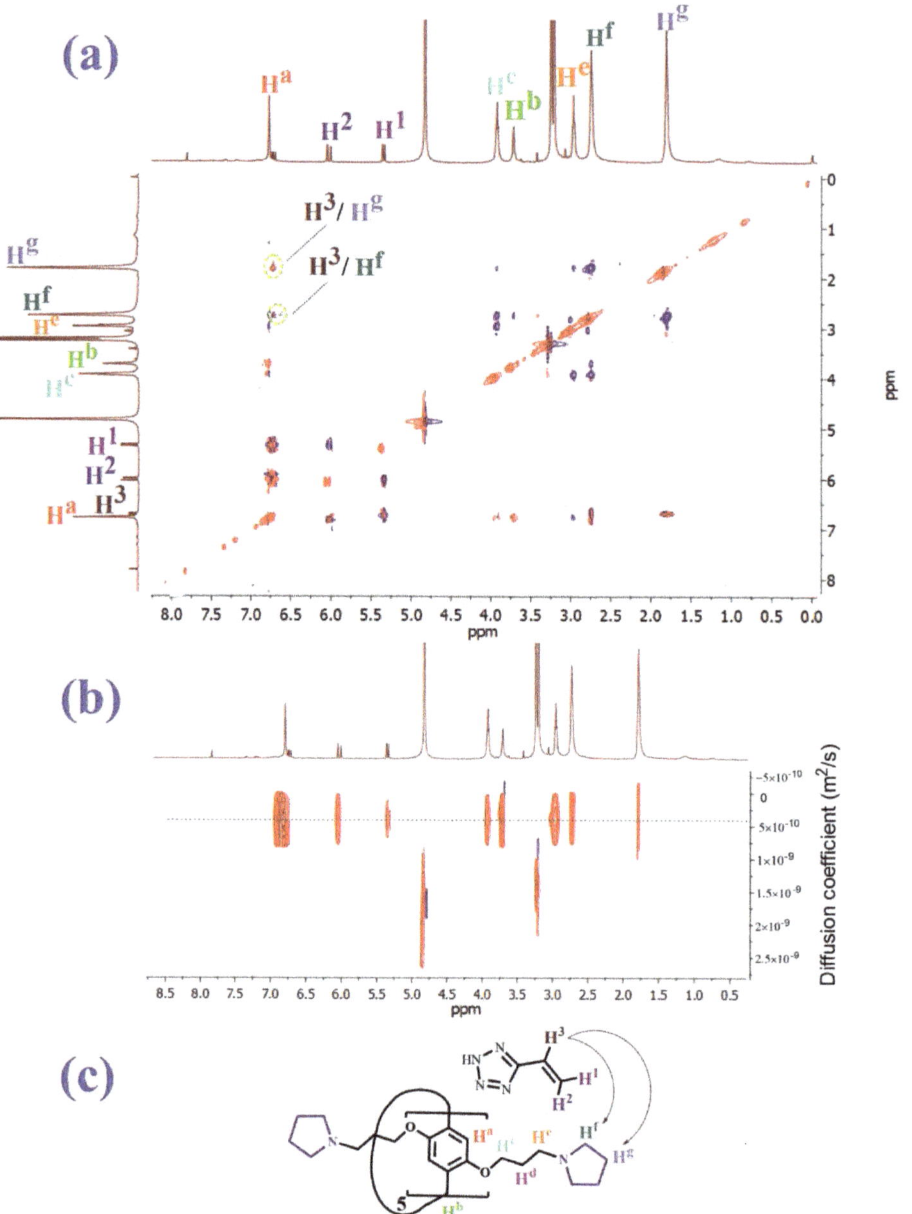

Figure 5. (**a**) the 2D ^1H—^1H NOESY NMR spectrum of the **7**/5—VT complex (1:1, 5×10^{-3} M) in CD3OD at 25 °C; (**b**) 2D DOSY NMR **7**/5—VT (1:1, 5×10^{-3} M) spectrum; and (**c**) structures **7** and 5—VT with the indication of close protons.

3.3. *Study of the Interaction of Pillar[5]arenes with Fluorescein*

The presence of a macrocyclic cavity in the pillar[5]arene 7 which is "not occupied" by a guest in the 7/5-VT system can facilitate the incorporation of an additional guest molecule into the pillararene cavity. To confirm this hypothesis, we studied the interaction of macrocycles **7** and **9** with a dye—fluorescein (Flu). Fluorescein was chosen as a convenient model compound. The choice of a specific drug for research is difficult since it is of interest to create universal DDS. Fluorescein is a well-studied compound with a developed spatial structure similar to a number of commercially used drugs: [65,66] quinoline derivatives (mepacrine, amodiaquine, etc.), which have antibacterial action, phenothiazine antipsychotics (aminazine, propazine, etc.), antineoplastic antibiotics (pixantrone, epirubicinum, etc.). A distinctive feature of fluorescein is the presence of its own fluorescence, which makes it possible to use sensitive fluorescent methods for establishing complexation processes.

As a result, the ability of macrocycles **7** and **9** to interact with fluorescein was studied using UV–Vis and fluorescence spectroscopy methods. Pillar[5]arenes 7 and 9 have one absorption maximum in ethanol with λmax at 292 nm according to UV–Vis spectroscopy data. Fluorescein solution absorbs at λmax = 276 nm, 453 nm, and 480 nm. Therefore, to study the interaction of macrocycles 7 and 9 with fluorescein, we chose the spectral range 350–600 nm, in which there is no absorption of pillar[5]arenes **7** and **9**. It turned out that the interaction of macrocycle **7** and **9** (1×10^{-5} M) with fluorescein leads to a hyperchromic effect, and the absorption band undergoes a redshift (see SM, Figure S26). The association constant and the stoichiometry of the associate were determined by the spectrophotometric titration data, in which the concentration of macrocycles **7** and **9** varied at a constant concentration of fluorescein (1×10^{-5} M).

The processing of the results was based on the analysis of binding isotherms, for which the BindFit [59–61] application was used and fitted to a 1:1 binding model (see SM, Figures S29–S31). The association constant of pillar[5]arenes 7 and 9 with fluorescein was 4730 M^{-1} and 2699 M^{-1}, respectively. Additionally, the stoichiometry of the complex was confirmed by titration data processed by the binding model at the ratio host-guest = 1:2 and 2:1. However, in this case, the constants are determined with great uncertainty. The binding constant of **7**/Flu in ethanol was additionally calculated by the method of fluorescence spectroscopy from analysis of binding isotherms. The dye in ethanol has maximum emission at 520 nm. The addition of pillararene **7** to it leads to the flare-up of fluorescence without a shift in the emission maximum. The association constant calculated using the 1:1 binding model in BindFit for this system is 10614.

At the next stage of the study, the self-association of macrocycles **7** and **9**, and the association with fluorescein and PVTE were studied using the dynamic light scattering (DLS) method. A self-association study of macrocycles **7** and **9** was carried out in water and ethanol using the DLS method (SM Table S3). It turned out that pillar[5]arenes 7 and 9 do not form stable self-associates in ethanol over the entire studied concentration range (1×10^{-3}–1×10^{-5} M). Further, the systems of macrocycle **7**/Flu and **9**/Flu were studied at the 1:1, 1:2, 2:1 ratios in the concentration range (1×10^{-3}–1×10^{-5} M) (see SM, Table S3). However, stable associates are formed only when Flu is added to a solution of macrocyclic **7** in ethanol. The minimum values of the polydispersity index (PDI = 0.16) were recorded for the ratio **7**/Flu = 1:1 (1×10^{-5} M) with an average hydrodynamic diameter of 155 nm in ethanol (see SM Table S3). An increase in the concentration of components **7**/Fly = 1:1 (1×10^{-4}–1×10^{-3} M) leads to increasing the average hydrodynamic diameter of the formed associates. The largest value of the average hydrodynamic diameter of the **7**/Flu system at 1:1 ratio (D = 428 nm) is observed at 1×10^{-3} M concentration with PDI = 0.36. The **9**/Flu system at 1:1, 1:2, 2:1 ratios and in the studied concentration range (1×10^{-3}–1×10^{-5} M) does not form stable associates in ethanol. It should be noted that both systems **7**/Flu and **9**/Flu do not form stable associates in aqueous and aqueous–alcoholic (H_2O/C_2H_5OH = 100/1) solutions (1×10^{-3}–1×10^{-5} M).

Then, the association of the **7**/PVTE and **9**/PVTE systems in ethanol and water was studied. PVTE does not form stable associates over the studied concentration range (1×10^{-3}–1×10^{-5} M). The association of the **7**/PVTE and **9**/PVTE systems was studied at the 50:1, 10:1, 5:1, 2:1, 1:1, 1:2, 1:5, and 1:15 ratios in the concentration range (1×10^{-3}–1×10^{-5} M). It is interesting that only in the case of macrocycle **7** (1×10^{-4} M) in the presence of

PVTE (1×10^{-5} M) (**7**/PVTE = 10:1), monodisperse (PDI = 0.23) stable associates **7**/PVTE with an average hydrodynamic diameter 116 nm are formed (see SM, Table S3). There is a dramatic increase in the mean hydrodynamic diameter and PDI with increasing PVTE concentration. Both systems **7**/PVTE and **9**/PVTE do not form stable associates in aqueous and aqueous–alcoholic (H_2O/C_2H_5OH = 100/1) solutions (1×10^{-3}–1×10^{-5} M) (see SM, Table S3).

3.4. Study of Self-Assembly of a Three-Component System *7/Flu/PVTE*

It should be noted that despite the presence of the interaction for **9**/Flu and **9**/PVTE confirmed by UV–Vis spectroscopy, the formation of associates **9**/Flu and **9**/PVTE by the DLS method was not detected. However, in the case of the **7**/Flu and **7**/PVTE systems, the formation of stable associates occured, which was confirmed by UV–Vis method, fluorescence spectroscopy, and DLS. The results obtained indicate the possibility of the formation of ternary self-assembling systems **7**/Flu/PVTE, in which the pillar[5]arene acts as a molecule capsule for the drug compound, and the polymer acts as an outer protective shell (Figure 6a).

To confirm this hypothesis, it is necessary to check the interaction between PVTE and Flu. The UV–Vis spectroscopy and the dynamic light scattering methods showed no changes in the spectra of the PVTE/Flu system, and the formation of PVTE/Flu associates in the 50:1,10:1, 5:1, 2:1, 1:1, 1:2, 1:5, and 1:15 ratios (1×10^{-3}–1×10^{-5} M) (see SM, Table S3). Next, the interaction of the **7**/Flu = 1:1 system with the PVTE polymer in the ratios **7**/Flu/PVTE = 1:1:0.1, 1:1:1, 1:1:5, 1:1:10 was studied by the method of DLS (1×10^{-3}–1×10^{-5} M) in ethanol and aqueous–alcoholic solution (H_2O/C_2H_5OH = 100/1). It turned out that the addition of small amounts of PVTE (1×10^{-6}–5×10^{-6} M) to the system **7**/Flu = 1:1 (1×10^{-5} M) does not lead to a change in the average hydrodynamic diameter of the system itself **7**/Flu = 1:1 (Figure 6a, SM, Table S3), whereas adding a 10-fold excess of PVTE (1×10^{-4} M) stabilizes the system ($\zeta = -12.81$ mV) **7**/Flu/PVTE (see SM, Table S3). The hydrodynamic diameter **7**/Flu/PVTE = 1:1:10 decreases to 48 nm with PDI = 0.16. Interestingly, the tendency for the formation of nano-sized associates **7**/Flu/PVTE = 1:1:10 persists, with a slight increase in the diameter particle and average PDI system, when the concentration of **7**/Flu = 1:1 (1×10^{-3}–1×10^{-4} M) is increasing (see SM, Table S3).

As noted above, 7/Flu and 7/PVTE separately do not form associates in the H_2O/C_2H_5OH = 100/1 system. However, the nanoscale associates remain ($\zeta = -34.12$ mV), when a solution of the ternary system 7/Flu/PVTE = 1:1:10 in ethanol is transferred into water (Figure 6a, SM, Figure S35, Table S3). Attempts to form the 7/Flu/PVTE ternary system directly in H_2O/C_2H_5OH resulted in a polydisperse system. The ability to form nanostructured 7/Flu/PVTE associates in the H_2O/C_2H_5OH system was investigated by scanning electron microscopy (SEM). According to SEM data (Figure 6, SM, Figures S36–S41), PVTE polymer (Figure 6b, SM, Figures S36–S41) forms micron-sized dendritic aggregates. The 7/Flu/PVTE system (Figure 6c,d, SM, Figures S36–S41) is characterized by spherical associates with an average diameter of 66 nm. It should be noted that the 7/Flu system in H_2O/C_2H_5OH represents formless aggregates of submicron sizes (Figure 6e, SM, Figures S36–S41). Thus, the preliminary self-assembly of the 7/Flu/PVTE system in ethanol followed by its transfer to water opens up the possibility of forming nanostructured associates with an incorporated drug. However, drug release processes are also of interest. One convenient way to detect the release of a potential drug is to track spectral changes when a nanostructured system is exposed to certain stimuli. Among these stimuli, one of the most promising is the response of the system to a change in pH. This is due to the pH values in late endosomes and/or lysosomes in tumor cells that differ from those in healthy tissues and equal to approximately 5.0 [67,68]. Thus, the methods of DLS and fluorescence spectroscopy were used to study the behavior of the 7/Flu/PVTE (1:1:10) system at different pH. System 7/Flu/PVTE was prepared in buffers at pH 2–9 by adding an alcohol solution of the system to the buffer in the ratio C_2H_5OH/buffer = 1/100.

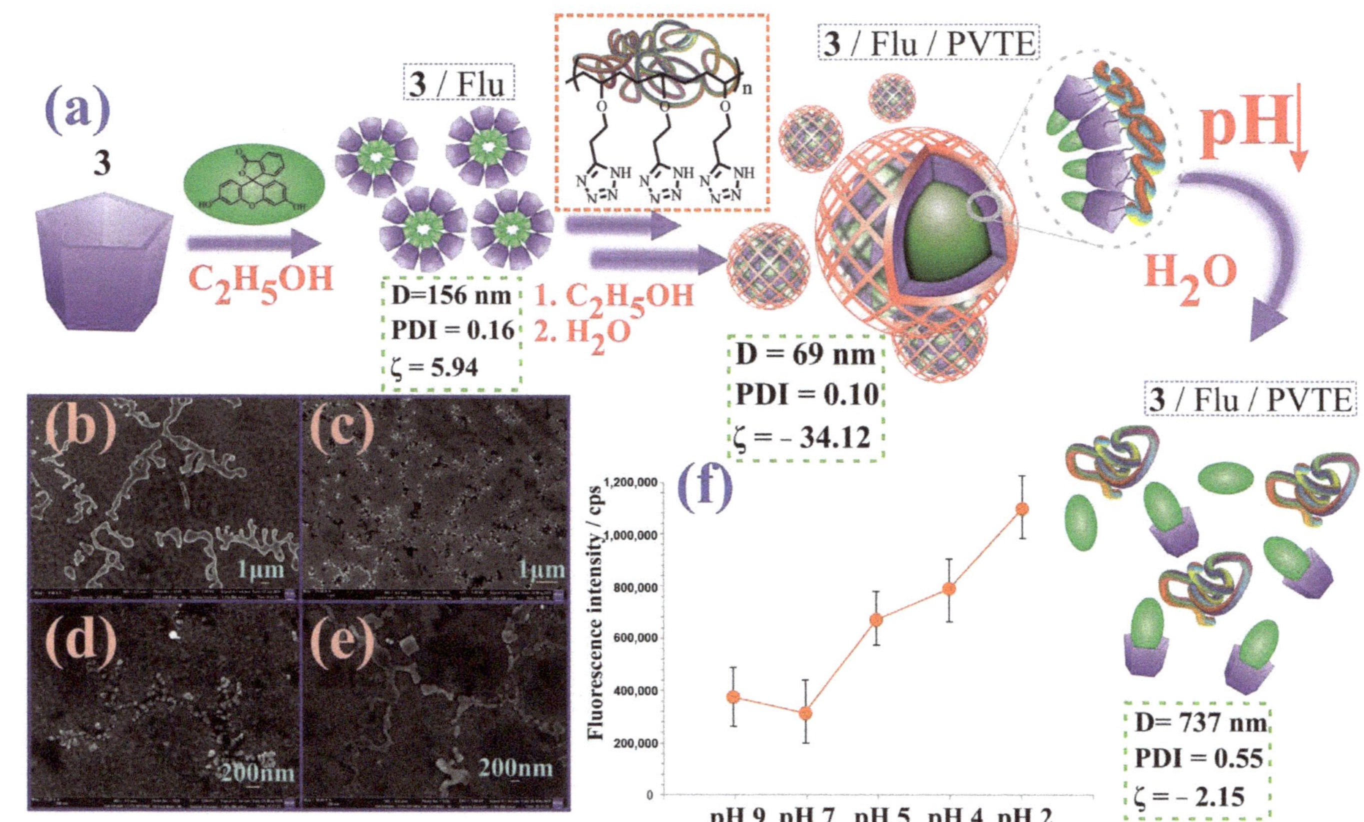

Figure 6. (**a**) The sketch represents the **7**/Flu associates, the formation of the **7**/Flu/PVTE triple associates, and the **7**/Flu/PVTE responses to the action of an external stimulus (pH change); scanning electron microscope (SEM) image of units: (**b**) PVTE (1×10^{-5} M); (**c**,**d**) associates **7** (1×10^{-5} M)/Flu (1×10^{-5} M)/PVTE (1×10^{-4} M); (**e**) associates **7**/Flu (1×10^{-5} M) in the H_2O/C_2H_5OH system; and (**f**) change in fluorescence Flu in the **7**/Flu/PVTE system when changing pH (9–2).

The **3**/Flu/PVTE (1:1:10) system in a buffer at pH 9–7 (see SM, Figure S35) remains stable ($\zeta = -31.22$ mV—pH 9 and $\zeta = -33.14$ mV—pH 7) and monodisperse (D = 83 nm—pH 9 and D = 84—pH 7) with a polydispersity index (PDI) = 0.12–0.15, according to the DLS results. The retention of encapsulated Flu in the **7**/Flu/PVTE system is confirmed by changes in the fluorescence spectra (Figure 6f). Thus, the position of the emission maximum does not change and a slight flare-up of fluorescence is observed at λ_{max} = 512 nm in the fluorescence spectra. There is a dramatic increase in PDI (see SM, Figure S42) and particle diameter **7**/Flu/PVTE in going from pH 7 to pH 5–2. The **7**/Flu/PVTE system becomes polydisperse in the pH range 5–4. The highest values of PDI (0.55) and average hydrodynamic diameter (D = 737 nm) are at pH 2 (Figure 6a, see SM, Figure S35). According to the data of fluorescence spectroscopy, from pH 7 to pH 5–2, a sharp flare-up of fluorescence is observed, which indicates a change in the environment of the dye at a pH close to **7** (Figure 6f). At the same time, it is well known that the quantum yield of fluorescein upon protonation in an acidic medium is minimal [69]. Thus, it is obvious that even at a low pH value, Flu is still isolated from the solvent [70], which is possible when it is inside pillararene. Probably, its destruction occurs with the release of the polymer and the dye encapsulated in pillararene as a result of protonation of the outer shell of the ternary complex consisting of PVTE.

4. Conclusions

Thus, for the first time, a new type of universal stimulus-sensitive system DDS based on decasubstituted macrocyclic structures—pillar[5]arenes and tetrazole-containing polymers—was demonstrated. Decasubstituted pillar[5]arenes 6 and 11, containing easily leaving tosylate and phthalimide fragments, were synthesized. Removal of tosylate and phthalimide protections yielded pillar[5]arenes 7–9 and 12, containing primary and tertiary amino groups in high yield. The ability of pillar[5]arenes 7 and 9 containing tertiary amine groups to interact with the tetrazole-containing polymer PVTE and the Flu dye was shown by UV–Vis and fluorescence spectroscopy. The DLS method showed the absence of associates in the 9/Flu and 9/PVTE systems. However, in the case of 7/Flu systems (1:1, C7/Flu = 1 × 10−5 M, D = 155 nm, PDI = 0.16) and 7/PVTE (10:1, C7 = 1 × 10−4 M, CPVTE = 1 × 10−5 M, D = 116 nm, PDI = 0.23), stable associates are formed. The possibility of forming a triple self-assembly system 7/Flu/PVTE was shown. The average hydrodynamic diameter 7/Flu/PVTE decreases to 48 nm with PDI = 0.16. The behavior of the 7/Flu/PVTE (1:1:10) system was studied at different pH by DLS and fluorescence spectroscopy. It was shown that the 7/Flu/PVTE system becomes polydisperse in going from pH 7 to pH 5-2. The results obtained open a broad range of possibilities for the development of new universal stimulus-responsive DDS containing nontoxic tetrazole-based polymers.

Supplementary Materials: The following are available online at https://www.mdpi.com/article/10.3390/nano11040947/s1, Figure S1: 1H NMR spectrum of 4,8,14,18,23,26,28,31,32,35-deca-(4-methylbenzylsulfonate-1-ethoxy)-pillar[5]arene (6), CDCl3, 298 K, 400 MHz, Figure S2: 13C NMR spectrum of 4,8,14,18,23,26,28,31,32,35-deca-(4-methylbenzylsulfonate-1-ethoxy)-pillar[5]arene (6), CDCl3, 298 K, 100 MHz, Figure S3: Mass spectrum (MALDI-TOF, 4-nitroaniline matrix) of 4,8,14,18,23,26,28,31,32,35-deca-(4-methylbenzylsulfonate-1-ethoxy)-pillar[5]arene (6), Figure S4: IR spectrum of 4,8,14,18,23,26,28,31,32,35-deca-(4-methylbenzylsulfonate-1-ethoxy)-pillar[5]arene (6), Figure S5: 1H NMR spectrum of 4,8,14,18,23,26,28,31,32,35-deca-[(isoindoline-1,3-dione)propoxy]-pillar[5]arene (11), CDCl3, 298 K, 400 MHz, Figure S6: 13C NMR spectrum of 4,8,14,18,23,26,28,31,32,35-deca-[(isoindoline-1,3-dione)propoxy]-pillar[5]arene (11), CDCl3, 298 K, 100 MHz, Figure S7: IR spectrum of 4,8,14,18,23,26,28,31,32,35-deca-[(isoindoline-1,3-dione)propoxy]-pillar[5]arene (11), Figure S8: Mass spectrum (MALDI-TOF, 4-nitroaniline matrix) of 4,8,14,18,23,26,28,31,32,35-deca-[(isoindoline-1,3-dione)propoxy]-pillar[5]arene (11), Figure S9: 1H NMR spectrum of 4,8,14,18,23,26,28,31,32,35-deca-[2-(pyrrolidin-1-yl)ethoxy]-pillar[5]arene (7), CDCl3, 298 K, 400 MHz, Figure S10: 13C NMR spectrum of 4,8,14,18,23,26,28,31,32,35-deca-[2-(pyrrolidin-1-yl)ethoxy]-pillar[5]arene (7), CDCl3, 298 K, 100 MHz, Figure S11: Mass spectrum (MALDI-TOF, 4-nitroaniline matrix) of 4,8,14,18,23,26,28,31,32,

35-deca-[2-(pyrrolidin-1-yl)ethoxy]-pillar[5]arene (7), Figure S12: IR spectrum of 4,8,14,18,23,26,28,31, 32,35-deca-[2-(pyrrolidin-1-yl)ethoxy]-pillar[5]arene (7), Figure S13: 1H NMR spectrum of 4,8,14,18,23, 26,28,31,32,35-deca-[2-(piperidin-1-yl)ethoxy]-pillar[5]arene (8), CDCl3, 298 K, 400 MHz, Figure S14: 13C NMR spectrum of 4,8,14,18,23,26,28,31,32,35-deca-[2-(piperidin-1-yl)ethoxy]-pillar[5]arene (8), CDCl3, 298 K, 100 MHz, Figure S15: Mass spectrum (MALDI-TOF, 4-nitroaniline matrix) of 4,8,14,18,23,26,28,31,32,35-deca-[2-(piperidin-1-yl)ethoxy]-pillar[5]arene (8), Figure S16: IR spectrum of 4,8,14,18,23,26,28,31,32,35-deca-[2-(piperidin-1-yl)ethoxy]-pillar[5]arene (8), Figure S17: 1H NMR spectrum of 4,8,14,18,23,26,28,31,32,35-deca-(2-morpholinoethoxy)-pillar[5]arene (9), CDCl3, 298 K, 400 MHz, Figure S18: 13C NMR spectrum of 4,8,14,18,23,26,28,31,32,35-deca-(2-morpholinoethoxy)-pillar[5]arene (9), CDCl3, 298 K, 100 MHz, Figure S19: Mass spectrum (MALDI-TOF, 4-nitroaniline matrix) of 4,8,14,18,23,26,28,31,32,35-deca-(2-morpholinoethoxy)-pillar[5]arene (9), Figure S20: IR spectrum of 4,8,14,18,23,26,28,31,32,35-deca-(2-morpholinoethoxy)-pillar[5]arene (9), Figure S21: 1H NMR spectrum of 4,8,14,18,23,26,28,31,32,35-deca-(aminopropyloxy)-pillar[5]arene (12), D2O, 298 K, 400 MHz, Figure S22:13C NMR spectrum of 4,8,14,18,23,26,28,31,32,35-deca-(aminopropyloxy)-pillar[5]arene (12), D2O, 298 K, 100 MHz, Figure S23: Mass spectrum (MALDI-TOF, 4-nitroaniline matrix) of 4,8,14,18,23,26,28,31,32,35-deca-(aminopropyloxy)-pillar[5]arene (12), Figure S24. IR spectrum of 4,8,14,18,23,26,28,31,32,35-deca-(aminopropyloxy)-pillar[5]arene (12), Figure S25: 1H-13C HSQC NMR spectrum of 4,8,14,18,23,26,28,31,32,35-deca-(4-methylbenzylsulfonate-1-ethoxy)-pillar[5]arene (6), CDCl3, 298 K, 400 MHz, Figure S26: UV spectra and Bindfit (Fit data to 1:1, 1:2 and 2:1 Host-Guest equilibria), Figure S27: Screenshots taken from the summary window of the website supramolecular.org. This screenshots shows the raw data for UV–Vis titration of 7 with PVTE, the data fitted to 1:1 binding model (A), 1:2 binding model (B) and 2:1 binding model (C), Figure S28: Screenshots taken from the summary window of the website supramolecular.org. This screenshots shows the raw data for UV–Vis titration of 9 with PVTE, the data fitted to 1:1 binding model (A), 1:2 binding model (B) and 2:1 binding model (C), Figure S29: Screenshots taken from the summary window of the website supramolecular.org. This screenshots shows the raw data for UV–Vis titration of 7 with fluorescein, the data fitted to 1:1 binding model (A), 1:2 binding model (B) and 2:1 binding model (C), Figure S30: Screenshots taken from the summary window of the website supramolecular.org. This screenshots shows the raw data for UV–Vis titration of 9 with fluorescein, the data fitted to 1:1 binding model (A), 1:2 binding model (B) and 2:1 binding model (C), Figure S31: Screenshots taken from the summary window of the website supramolecular.org. This screenshots shows the raw data for fluorescence titration of 7 with fluorescein, the data fitted to 1:1 binding model (A), 1:2 binding model (B) and 2:1 binding model (C), Figure S32: Size distribution of the particles by intensity for PVTE ($1 \times 10{-}5$ M) in ethanol, Figure S33: Size distribution of the particles by intensity for 7 ($1 \times 10{-}5$ M) in ethanol, Figure S34: Size distribution of the particles by intensity for 7 ($1 \times 10{-}4$ M)/PVTE ($1 \times 10{-}5$ M) (10:1) in ethanol, Figure S35. Size distribution of the particles by intensity for 7/Flu, 7/Flu/PVTE in ethanol and buffer at different pH, Figure S36: SEM image of silicon substrate, Figure S37: SEM image of 7/Flu/PVTE ($1 \times 10{-}5$ M) after the solvent (H2O/EtOH) evaporation, Figure S38: SEM image of 7/Flu/PVTE ($1 \times 10{-}5$ M) after the solvent (H2O/EtOH) evaporation, Figure S39: SEM image of 7/Flu/PVTE ($1 \times 10{-}5$ M) after the solvent (H2O/EtOH) evaporation, Figure S40: SEM image of 7/Flu ($1 \times 10{-}5$ M) after the solvent (H2O/EtOH) evaporation, Figure S41: SEM image of PVTE ($1 \times 10{-}5$ M) after the solvent (H2O/EtOH) evaporation, Figure S42: Fluorescence spectra of 7/Flu/PVTE ($1 \times 10{-}5$ M) in buffer at different pH (9-2), Table S1: Reaction conditions for the synthesis of target macrocycles 6 and 11 from starting compounds 5 and 10, respectively, Table S2: Crystal data and structure refinement for 6, 7 and 11, Table S3: Dynamic light scattering.

Author Contributions: Conceptualization, I.I.S.; writing—original draft preparation, D.N.S. and L.I.M.; data curation, K.S.U., D.R.I. and O.A.M.; writing—review and editing, A.A.N. and V.N.K. All authors have read and agreed to the published version of the manuscript.

Funding: This study was supported by Russian Science Foundation (No.20-73-00161). The investigation of the spatial structure of the compounds by NMR spectroscopy was carried out within the framework of the grant of the President of the Russian Federation for state support of leading scientific schools of the Russian Federation (NSh-2499.2020.3).

Data Availability Statement: Data is available upon the reasonable request from the corresponding author.

Acknowledgments: The authors are grateful to the staff of the Spectral-Analytical Center of Shared Facilities for Study of Structure, Composition, and Properties of Substances and Materials of the A.E. Arbuzov' Institute of Organic and Physical Chemistry of the Kazan Scientific Center of the Russian Academy of Sciences for their research and assistance in the discussion of the results.

Conflicts of Interest: The authors declare no conflict of interest.

References

1. Marasini, N.; Haque, S.; Kaminskas, L.M. Polymer-drug conjugates as inhalable drug delivery systems: A review. *Curr. Opin. Colloid Interface Sci.* **2017**, *31*, 18–29. [CrossRef]
2. Grossen, P.; Witzigmann, D.; Sieber, S.; Huwyler, J. PEG-PCL-based nanomedicines: A biodegradable drug delivery system and its application. *J. Control. Release* **2017**, *260*, 46–60. [CrossRef]
3. Ji, X.; Ahmed, M.; Long, L.; Khashab, N.M.; Huang, F.; Sessler, J.L. Adhesive supramolecular polymeric materials constructed from macrocycle-based host–guest interactions. *Chem. Soc. Rev.* **2019**, *48*, 2682–2697. [CrossRef]
4. Masood, F. Polymeric nanoparticles for targeted drug delivery system for cancer therapy. *Mater. Sci. Eng. C* **2016**, *60*, 569–578. [CrossRef] [PubMed]
5. Wang, Z.; Deng, X.; Ding, J.; Zhou, W.; Zheng, X.; Tang, G. Mechanisms of drug release in pH-sensitive micelles for tumour targeted drug delivery system: A review. *Int. J. Pharm.* **2018**, *535*, 253–260. [CrossRef] [PubMed]
6. George, A.; Shah, P.A.; Shrivastav, P.S. Natural biodegradable polymers based nano-formulations for drug delivery: A review. *Int. J. Pharm.* **2019**, *561*, 244–264. [CrossRef] [PubMed]
7. Kakkar, A.; Traverso, G.; Farokhzad, O.C.; Weissleder, R.; Langer, R. Evolution of macromolecular complexity in drug delivery systems. *Nat. Rev. Chem.* **2017**, *1*, 1–17. [CrossRef]
8. Dong, P.; Rakesh, K.; Manukumar, H.; Mohammed, Y.H.E.; Karthik, C.; Sumathi, S.; Mallu, P.; Qin, H.-L. Innovative nano-carriers in anticancer drug delivery-a comprehensive review. *Bioorg. Chem.* **2019**, *85*, 325–336. [CrossRef]
9. Raza, A.; Sime, F.B.; Cabot, P.J.; Maqbool, F.; Roberts, J.A.; Falconer, J.R. Solid nanoparticles for oral antimicrobial drug delivery: A review. *Drug Discov. Today* **2019**, *24*, 858–866. [CrossRef] [PubMed]
10. Jacob, J.; Haponiuk, J.T.; Thomas, S.; Gopi, S. Biopolymer based nanomaterials in drug delivery systems: A review. *Mater. Today Chem.* **2018**, *9*, 43–55. [CrossRef]
11. Hossen, S.; Hossain, M.K.; Basher, M.; Mia, M.; Rahman, M.; Uddin, M.J. Smart nanocarrier-based drug delivery systems for cancer therapy and toxicity studies: A review. *J. Adv. Res.* **2019**, *15*, 1–18. [CrossRef] [PubMed]
12. Chen, Y.; Sun, S.; Lu, D.; Shi, Y.; Yao, Y. Water-soluble supramolecular polymers constructed by macrocycle-based host-guest interactions. *Chin. Chem. Lett.* **2019**, *30*, 37–43. [CrossRef]
13. Fa, S.; Kakuta, T.; Yamagishi, T.A.; Ogoshi, T. One-, Two-, and Three-Dimensional Supramolecular Assemblies Based on Tubular and Regular Polygonal Structures of Pillar[n]arenes. *CCS Chem.* **2019**, *1*, 50–63. [CrossRef]
14. Lou, X.Y.; Li, Y.P.; Yang, Y.W. Gated Materials: Installing Macrocyclic Arenes-Based Supramolecular Nanovalves on Porous Nanomaterials for Controlled Cargo Release. *Biotechnol. J.* **2019**, *14*, 1800354–1800363. [CrossRef] [PubMed]
15. Zhu, H.; Shangguan, L.; Shi, B.; Yu, G.; Huang, F. Recent progress in macrocyclic amphiphiles and macrocyclic host-based supra-amphiphiles. *Mater. Chem. Front.* **2018**, *2*, 2152–2174. [CrossRef]
16. Cragg, P.J. Pillar[n]arenes at the Chemistry-Biology Interface. *Isr. J. Chem.* **2018**, *58*, 1194–1208. [CrossRef]
17. Smolko, V.; Shurpik, D.; Porfireva, A.; Evtugyn, G.; Stoikov, I.; Hianik, T. Electrochemical Aptasensor Based on Poly(Neutral Red) and Carboxylated Pillar[5]arene for Sensitive Determination of Aflatoxin M1. *Electroanalysis* **2018**, *30*, 486–496. [CrossRef]
18. Shurpik, D.N.; Padnya, P.L.; Basimova, L.T.; Evtugin, V.G.; Plemenkov, V.V.; Stoikov, I.I. Synthesis of new decasubstituted pillar[5]arenes containing glycine fragments and their interactions with Bismarck brown Y. *Mendeleev Commun.* **2015**, *25*, 432–434. [CrossRef]
19. Stoikova, E.E.; Sorvin, M.I.; Shurpik, D.N.; Budnikov, H.C.; Stoikov, I.I.; Evtugyn, G.A. Solid-Contact Potentiometric Sensor Based on Polyaniline and Unsubstituted Pillar[5]Arene. *Electroanalysis.* **2014**, *27*, 440–449. [CrossRef]
20. Yakimova, L.S.; Shurpik, D.N.; Makhmutova, A.R.; Stoikov, I.I. Pillar[5]arenes Bearing Amide and Carboxylic Groups as Synthetic Receptors for Alkali Metal Ions. *Macroheterocycles* **2017**, *10*, 226–232. [CrossRef]
21. Shurpik, D.N.; Sevastyanov, D.A.; Zelenikhin, P.V.; Subakaeva, E.V.; Evtugyn, V.G.; Osin, Y.N.; Cragg, P.J.; Stoikov, I.I. Hydrazides of glycine-containing decasubstituted pillar[5]arenes: Synthesis and encapsulation of Floxuridine. *Tetrahedron Lett.* **2018**, *59*, 4410–4415. [CrossRef]
22. Shurpik, D.N.; Mostovaya, O.A.; Sevastyanov, D.A.; Lenina, O.A.; Sapunova, A.S.; Voloshina, A.D.; Petrov, K.A.; Kovyazina, I.V.; Cragg, P.J.; Stoikov, I.I. Supramolecular neuromuscular blocker inhibition by a pillar[5]arene through aqueous inclusion of rocuronium bromide. *Org. Biomol. Chem.* **2019**, *17*, 9951–9959. [CrossRef] [PubMed]
23. Yan, S.; Chen, S.; Gou, X.; Yang, J.; An, J.; Jin, X.; Yang, Y.; Chen, L.; Gao, H. Biodegradable Supramolecular Materials Based on Cationic Polyaspartamides and Pillar[5]arene for Targeting Gram-Positive Bacteria and Mitigating Antimicrobial Resistance. *Adv. Funct. Mater.* **2019**, *29*, 1904683–1904694. [CrossRef]
24. Zhong, J.; Tang, Q.; Ju, Y.; Lin, Y.; Bai, X.; Zhou, J.; Luo, H.; Lei, Z.; Tong, Z. Redox and pH responsive polymeric vesicles constructed from a water-soluble pillar[5]arene and a paraquat-containing block copolymer for rate-tunable controlled release. *J. Biomater. Sci. Polym. Ed.* **2019**, *30*, 202–214. [CrossRef]

25. Barbera, L.; De Plano, L.M.; Franco, D.; Gattuso, G.; Guglielmino, S.P.P.; Lando, G.; Notti, A.; Parisi, M.F.; Pisagatti, I. Antiadhesive and antibacterial properties of pillar[5]arene-based multilayers. *Chem. Commun.* **2018**, *54*, 10203–10206. [CrossRef]
26. Zhao, H.; Qu, Z.R.; Ye, H.Y.; Xiong, R.G. ChemInform Abstract: In situ Hydrothermal Synthesis of Tetrazole Coordination Polymers with Interesting Physical Properties. *Chem. Soc. Rev.* **2008**, *39*, 84–100. [CrossRef] [PubMed]
27. Gaponik, P.N.; Ivashkevich, O.A.; Karavai, V.P.; Lesnikovich, A.I.; Chernavina, N.I.; Sukhanov, G.T.; Gareev, G.A. Polymers and copolymers based on vinyl tetrazoles, 1. Synthesis of poly(5-vinyl tetrazole) by polymer-analogous conversion of polyacrylonitrile. *Angew. Makromol. Chem.* **1994**, *219*, 77–88. [CrossRef]
28. Zakerzadeh, E.; Salehi, R.; Mahkam, M. Smart tetrazole-based antibacterial nanoparticles as multifunctional drug carriers for cancer combination therapy. *Drug Dev. Ind. Pharm.* **2017**, *43*, 1963–1977. [CrossRef] [PubMed]
29. Piradashvili, K.; Simon, J.; Paßlick, D.; Höhner, J.R.; Mailänder, V.; Wurm, F.R.; Landfester, K. Fully degradable protein nanocarriers by orthogonal photoclick tetrazole-ene chemistry for the encapsulation and release. *Nanoscale Horiz.* **2017**, *2*, 297–302. [CrossRef]
30. Kizhnyaev, V.N.; Vereshchagin, L.I. Vinyltetrazoles: Synthesis and properties. *Russ. Chem. Rev.* **2003**, *72*, 143–164. [CrossRef]
31. Pu, H.; Luo, H.; Wan, D. Synthesis and properties of amphoteric copolymer of 5-vinyltetrazole and vinylbenzyl phos-phonic acid. *J. Polym. Sci. A Polym. Chem.* **2013**, *51*, 3486–3493. [CrossRef]
32. Atanase, L.I.; Riess, G. Micellization of pH-stimulable poly (2-vinylpyridine)-b-poly (ethylene oxide) copolymers and their complexation with anionic surfactants. *J. Colloid Interface Sci.* **2013**, *395*, 190–197. [CrossRef] [PubMed]
33. Iurciuc-Tincu, C.E.; Cretan, M.S.; Purcar, V.; Popa, M.; Daraba, O.M.; Atanase, L.I.; Ochiuz, L. Drug delivery system based on pH-sensitive biocompatible poly (2-vinyl pyridine)-b-poly (ethylene oxide) nanomicelles loaded with curcumin and 5-fluorouracil. *Polymers* **2020**, *12*, 1450. [CrossRef]
34. Atanase, L.I.; Lerch, J.P.; Caprarescu, S.; Tincu, C.E.I.; Riess, G. Micellization of pH-sensitive poly(butadiene)-block -poly(2 vinylpyridine)-block -poly(ethylene oxide) triblock copolymers: Complex formation with anionic surfactants. *J. Appl. Polym. Sci.* **2017**, *134*, 45313–45321. [CrossRef]
35. Kizhnyaev, V.N.; Petrova, T.L.; Pokatilov, F.A.; Smirnov, V.I. Synthesis of crosslinked poly(5-vinyltetrazole) and properties of hydrogels formed on its basis. *Polym. Sci. Ser. B* **2011**, *53*, 626–633. [CrossRef]
36. Sheldrick, G.M. SHELXT—Integrated space-group and crystal-structure determination. *Acta Crystallogr.* **2015**, *71*, 3–8. [CrossRef]
37. Sheldrick, G.M. A short history of SHELX. *Acta Crystallogr.* **2008**, *64*, 112–122. [CrossRef]
38. Macrae, C.F.; Edgington, P.R.; McCabe, P.; Pidcock, E.; Shields, G.P.; Taylor, R.; Towler, M.; Streek, J.V.D. Mercury: Visualization and analysis of crystal structures. *J. Appl. Crystallogr.* **2006**, *39*, 453–457. [CrossRef]
39. Wong, W.; Curiel, D.; Cowley, A.R.; Beer, P.D. Dinuclear zinc(ii) dithiocarbamate macrocycles: Ditopic receptors for a variety of guest molecules. *Dalton Trans.* **2005**, *2*, 359–364. [CrossRef]
40. Tsukamoto, T.; Sasahara, R.; Muranaka, A.; Miura, Y.; Suzuki, Y.; Kimura, M.; Miyagawa, S.; Kawasaki, T.; Kobayashi, N.; Uchiyama, M.; et al. Synthesis of a Chiral [2]Rotaxane: Induction of a Helical Structure through Double Threading. *Org. Lett.* **2018**, *20*, 4745–4748. [CrossRef]
41. Salamone, J.C. *Concise Polymeric Materials Encyclopedia*; CRC Press: Boca Raton, FL, USA, 1998.
42. Neochoritis, C.G.; Zhao, T.; Dömling, A. Tetrazoles via Multicomponent Reactions. *Chem. Rev.* **2019**, *119*, 1970–2042. [CrossRef] [PubMed]
43. Kraus, J. Isosterism and molecular modification in drug design: Tetrazole analogue of GABA: Effects on enzymes of the γ-aminobutyrate system. *Pharmacol. Res. Commun.* **1983**, *15*, 183–189. [CrossRef]
44. Felber, A.E.; Dufresne, M.H.; Leroux, J.C. pH-sensitive vesicles, polymeric micelles, and nanospheres prepared with polycarboxylates. *Adv. Drug Deliv. Rev.* **2012**, *64*, 979–992. [CrossRef] [PubMed]
45. Schulz, M.; Olubummo, A.; Binder, W.H. Beyond the lipid-bilayer: Interaction of polymers and nanoparticles with membranes. *Soft Matter.* **2012**, *8*, 4849–4864. [CrossRef]
46. Fu, L.; Wan, M.; Zhang, S.; Gao, L.; Fang, W. Polymyxin B Loosens Lipopolysaccharide Bilayer but Stiffens Phospholipid Bilayer. *Biophys. J.* **2020**, *118*, 138–150. [CrossRef]
47. Shurpik, D.N.; Nazarova, A.A.; Makhmutova, L.I.; Kizhnyaev, V.N.; Stoikov, I.I. Uncharged water-soluble amide de-rivatives of pillar[5]arene: Synthesis and supramolecular self-assembly with tetrazole-containing polymers. *Russ. Chem. Bull.* **2020**, *69*, 97–104. [CrossRef]
48. Dasgupta, S.; Mukherjee, P.S. Carboxylatopillar[n]arenes: A versatile class of water soluble synthetic receptors. *Org. Biomol. Chem.* **2016**, *15*, 762–772. [CrossRef]
49. Zhang, H.; Liu, Z.; Zhao, Y. Pillararene-based self-assembled amphiphiles. *Chem. Soc. Rev.* **2018**, *47*, 5491–5528. [CrossRef] [PubMed]
50. Ogoshi, T.; Kakuta, T.; Yamagishi, T.A. Separation and detection of meta-and ortho-substituted benzene isomers by water-soluble pillar[5]arene. *Angew. Chem. Int. Ed.* **2019**, *131*, 2219–2229. [CrossRef]
51. Feng, W.; Jin, M.; Yang, K.; Pei, Y.; Pei, Z. Supramolecular delivery systems based on pillararenes. *Chem. Commun.* **2018**, *54*, 13626–13640. [CrossRef]
52. Hu, X.B.; Chen, L.; Si, W.; Yu, Y.; Hou, J.L. Pillar[5]arene decaamine: Synthesis, encapsulation of very long linear diacids and formation of ion pair-stopped [2]rotaxanes. *Chem. Commun.* **2011**, *47*, 4694–4696. [CrossRef]

53. Ma, Y.; Ji, X.; Xiang, F.; Chi, X.; Han, C.; He, J.; Abliz, Z.; Chen, W.; Huang, F. A cationic water-soluble pillar[5]arene: Synthesis and host–guest complexation with sodium 1-octanesulfonate. *Chem. Commun.* **2011**, *47*, 12340–12342. [CrossRef]
54. Zhou, Y.; Jie, K.; Zhao, R.; Huang, F. Cis-Trans Selectivity of Haloalkene Isomers in Nonporous Adaptive Pillararene Crystals. *J. Am. Chem. Soc.* **2019**, *141*, 11847–11851. [CrossRef]
55. Sheng, X.; Li, E.; Zhou, Y.; Zhao, R.; Zhu, W.; Huang, F. Separation of 2-Chloropyridine/3-Chloropyridine by Nonporous Adaptive Crystals of Pillararenes with Different Substituents and Cavity Sizes. *J. Am. Chem. Soc.* **2020**, *142*, 6360–6364. [CrossRef]
56. Zhou, Y.; Jie, K.; Zhao, R.; Li, E.; Huang, F. Highly Selective Removal of Trace Isomers by Nonporous Adaptive Pillararene Crystals for Chlorobutane Purification. *J. Am. Chem. Soc.* **2020**, *142*, 6957–6961. [CrossRef]
57. Hosseini, M.W. Molecular Tectonics: From Simple Tectons to Complex Molecular Networks. *Acc. Chem. Res.* **2005**, *38*, 313–323. [CrossRef]
58. Hosseini, M.W.; De Cian, A. Crystal engineering: Molecular networks based on inclusion phenomena. *Chem. Commun.* **1998**, *7*, 727–734. [CrossRef]
59. Thordarson, P. Determining association constants from titration experiments in supramolecular chemistry. *Chem. Soc. Rev.* **2010**, *40*, 1305–1323. [CrossRef] [PubMed]
60. Hibbert, D.B.; Thordarson, P. The death of the Job plot, transparency, open science and online tools, uncertainty estimation methods and other developments in supramolecular chemistry data analysis. *Chem. Commun.* **2016**, *52*, 12792–12805. [CrossRef]
61. Bindfit v0.5 (Open Data Fit, 2016). Available online: http://supramolecular.org/bindfit/ (accessed on 12 January 2021).
62. Panneerselvam, M.; Kumar, M.D.; Jaccob, M.; Solomon, R.V.; Panneesrselvam, M. Computational Unravelling of the Role of Alkyl Groups on the Host-Guest Complexation of Pillar[5]arenes with Neutral Dihalobutanes. *ChemistrySelect* **2018**, *3*, 1321–1334. [CrossRef]
63. Sengupta, A.; Singh, M.; Sundarajan, M.; Yuan, L.; Fang, Y.; Yuan, X.; Feng, W. Understanding the extraction and complexation of thorium using structurally modified CMPO functionalized pillar[5]arenes in ionic liquid: Experimental and theoretical investigations. *Inorg. Chem. Commun.* **2017**, *75*, 33–36. [CrossRef]
64. Chen, L.; Cai, Y.; Feng, W.; Yuan, L. Pillararenes as macrocyclic hosts: A rising star in metal ion separation. *Chem. Commun.* **2019**, *55*, 7883–7898. [CrossRef] [PubMed]
65. Fisher, K.A.; Huddersman, K.D.; Taylor, M.J. Comparison of micro-and mesoporous inorganic materials in the uptake and release of the drug model fluorescein and its analogues. *Chem. Eur. J.* **2003**, *9*, 5873–5878. [CrossRef]
66. Jamwal, H.S.; Chauhan, G.S. Designing Silica-Based Hybrid Polymers and Their Application in the Loading and Release of Fluorescein as a Model Drug and Diagnostic Agent. *Adv. Polym. Technol.* **2016**, *37*, 411–418. [CrossRef]
67. Ju, C.; Mo, R.; Xue, J.; Zhang, L.; Zhao, Z.; Xue, L.; Ping, Q.; Zhang, C. Sequential Intra-Intercellular Nanoparticle Delivery System for Deep Tumor Penetration. *Angew. Chem.* **2014**, *126*, 6367–6372. [CrossRef]
68. Bae, Y.; Fukushima, S.; Harada, A.; Kataoka, K. Design of environment-sensitive supramolecular assemblies for intracellular drug delivery: Polymeric micelles that are responsive to intracellular pH change. *Angew. Chem. Int. Ed. Engl.* **2003**, *115*, 4788–4791. [CrossRef]
69. Sjöback, R.; Nygren, J.; Kubista, M. Absorption and fluorescence properties of fluorescein. *Spectrochim. Acta Part A Mol. Biomol. Spectrosc.* **1995**, *51*, L7–L21. [CrossRef]
70. Lakowicz, J.R.; Masters, B.R. *Principles of Fluorescence Spectroscopy*, 3rd ed.; Kluwer-Plenum: New York, NY, USA, 2008.

Article

New Amphiphilic Imidazolium/Benzimidazolium Calix[4]arene Derivatives: Synthesis, Aggregation Behavior and Decoration of DPPC Vesicles for Suzuki Coupling in Aqueous Media

Vladimir Burilov [1,*], Ramilya Garipova [1], Elsa Sultanova [1], Diana Mironova [1], Ilya Grigoryev [1], Svetlana Solovieva [2] and Igor Antipin [1]

1 Kazan Federal University, 18 Kremlevskaya st., 420008 Kazan, Russia; aukhadieva.ramilya@yandex.ru (R.G.); elsultanova123@gmail.com (E.S.); mir_din@mail.ru (D.M.); os18sir@yandex.ru (I.G.); iantipin54@yandex.ru (I.A.)

2 A.E.Arbuzov Institute of Organic & Physical Chemistry, 8 Arbuzov str., 420088 Kazan, Russia; evgersol@yandex.ru

* Correspondence: ultrav@bk.ru; Tel.: +7-843-2337344

Received: 11 May 2020; Accepted: 8 June 2020; Published: 10 June 2020

Abstract: In this study, new types of amphiphilic calix[4]arene derivatives bearing *N*-alkyl/aryl imidazolium/benzimidazolium fragments were designed and synthesized by two step transformation: Regioselective Blanc chloromethylation of distal-di-O-butyl calix[4]arene and subsequent interaction with *N*-Substituted imidazole/benzimidazole. Critical aggregation concentration (CAC) values were estimated using pyrene fluorescent probe. Obtained macrocycles were found to form submicron particles with electrokinetic potential +44–+57 mV in aqueous solution. For the first time it was found that amphiphilic calixarene causes the fast transformation of 1,2-dipalmitoyl-sn-glycero-3-phosphocholine (DPPC) multilamellar vesicles into unilamellar ones and leads to the ordering of the lipid in membranes at the molar calixarene/DPPC ratio more than 0.07. In situ complexes of calixarene aggregates with $Pd(OAc)_2$ were found to be active in Suzuki–Miyaura coupling of 1-bromo-4-nitrobenzene with phenylboronic acid in water. It was shown that bulky *N*-substituents of heterocycle decrease the catalytic activity of the aggregates. These result can be assigned to the inhibition effect of Pd(II) complex in situ formation by bulky substituents located on the aggregate surface. Embedding of the most active palladium *N*-heterocyclic carbene (NHC) complex with methylimidazolium headgroups into DPPC vesicles enhances its catalytic activity in Suzuki–Miyaura coupling.

Keywords: calix[4]arene; NHC complex; Suzuki–Miyaura coupling; DPPC vesicles

1. Introduction

Effective approaches to the new carbon–carbon and carbon–heteroatom bond formation are attracting a lot of attention among chemists [1,2]. One of the most important tools is cross-coupling reactions, available to synthetic chemists in their quest to create artificial (or reproduce natural) organic scaffolds [3]. In recent decades, a special place in the cross-coupling catalysis has been occupied by palladium complexes with *N*-heterocyclic carbene (NHC) ligands [4–6]. Exponential growth of interest to palladium NHC complexes is quite expected: (i) NHC complexes can be easily synthesized both ex situ and in situ in the reaction medium from palladium salt and appropriate *N*-heterocycles in the presence of base; (ii) the electron-donor and steric properties of the ligand can be easily regulated by substituents on the heterocycle core; (iii) synthesis of precursors (salts of *N*-heterocycles) can be carried out easily through *N*-alkylation of appropriate *N*-heterocycles (imidazole derivatives are used most often). However, the most attractive property of NHC complexes is their stability to moisture and air oxygen, thus allowing to perform specific catalytic reactions in aqueous media without inert atmosphere [7,8]. The application of water as solvent brings down the overall cost and significantly reduces the health risks due to its non-toxicity and nonflammability, unlike most organic solvents [9]. However, not all organic substances are water-soluble and to solve this problem the surfactants are usually used. Thus, NHC complexes with amphiphilic properties allow to combine both micellar and metallocomplex types of catalysis in aqueous media [10,11]. So, the design of new amphiphilic NHC complexes is very promising for enhancing the cross-coupling reactions' efficiency.

Recently, much attention has been paid to NHC complexes on the calixarene platform [12]. Embedding of NHC units on the calixarene scaffold can serve as a tool to change the environment of central metal due to steric hindrance and electronic effects. A calixarene cavity can also impact on catalytic transformations [13]. The unique structure of calix[4]arenes as well as the variety of stereoisomeric forms and the facile functionalization of both upper and lower rims allows to create amphiphilic structures with spatial separation of lipophilic and hydrophilic domains, that is why they have been widely used to construct macrocyclic amphiphiles [14–17].

Herein we report the synthesis, aggregation behavior, embedding into DPPC vesicles and catalytic activity of their in situ complexes with $Pd(OAc)_2$ in Suzuki–Miyaura coupling of new amphiphilic calix[4]arene derivatives containing two imidazolium/benzimidazolium headgroups and two lipophilic fragments on the upper and lower rim, respectively.

2. Materials and Methods

2.1. Characterisation Methods

TLC was performed on Merck UV 254 plates with Vilber Lourmat VL-6.LC UV lamp (254 nm) control. Elemental analysis of synthesized compounds was done on the PerkinElmer PE 2400 CHNS/O Elemental Analyzer. NMR spectra were recorded on Bruker Avance 400 Nanobay with signals from residual protons of deuterated solvents ($CDCl_3$ or DMSO-d6) as the internal standard. MALDI mass spectra were measured on UltraFlex III TOF/TOF with PNA matrix, laser Nd:YAG, $\lambda = 355$ nm. The IR spectra were recorded on a Bruker Vector-22 spectrometer. Samples were prepared as thin films, obtained from chloroform solutions dried on the surface of the KBr tablet. The melting points were measured using the Stuart SMP10.

2.2. Reagents

All reagents were purchased from either Acros or Sigma-Aldrich and used without further purification. Solvents were purified according to standard methods [18]. 1-*Iso*propylimidazole [19], 1-(2,6-diisopropylphenyl)imidazole [20], 1-mesitylimidazole [20], 5,11,17,23-tetra-*tert*-butyl-25,26,27,28-tetrahydroxycalix[4]arene [21], 25,26,27,28-tetrahydroxycalix[4]arene [22], 25,27-dihydroxy-26,28-dibutoxycalix[4]arene [23] were synthesized according to published methods.

11,23-Bis(chloromethyl)-25,27-dihydroxy-26,28-dibutoxycalix[4]arene (**4**) 5,11,17,23-tetra-H-25,27-dihydroxy-26,28-dibutoxycalix[4]arene (0.67 g (1.22 mmol)) and 0.55 g (18.36 mmol) of paraform were suspended in 10 mL of glacial acetic acid and then bubbled with HCl gas for 2 h. Reaction was stopped by adding 30 mL of water. The crude product was filtered off, washed by 10% acetic acid (3 × 20 mL), water (3 × 20 mL) and dried in vacuo. The final product was obtained as a white solid (0.75 g, 97%). Melting point: 188 °C (decomp). ^{1}H NMR (400 MHz, $CDCl_3$, 25 °C) δ_H ppm: 1.09 (6H, t, CH_3, $J = 7.3$ Hz,), 1.72–1.83 (4H, m, -CH_2-), 1.98–2.08 (4H, m, -CH_2-), 3.38 (4H, d, -CH_2-, $J = 12.9$ Hz,), 4.00 (4H, t, -CH_2O-, $J = 6.35$ Hz), 4.29 (4H, d, -CH_2-, $J = 12.9$ Hz,), 4.51 (4H, s, CH_2Cl), 6.78 (2H, t, HAr, $J = 7.6$ Hz), 6.93 (4H, d, HAr, $J = 7.6$ Hz), 7.08 (4H, s, HAr), 8.48 (2H, s, OH). $^{13}C\{^{1}H\}$ NMR: (100.6 MHz, $CDCl_3$, 25 °C) δ_c ppm: 14.22, 19.52, 31.52, 32.36, 47.08, 125.55, 127.87, 128.46, 129.12, 129.18, 133.25.

IR (KBr) ν_{max} cm^{-1}: 2931 (-CH_2-), 2959 (-CH_3), 3318 (OH). MALDI-TOF (*m/z*): 561 [M-2HCl+H]$^+$. Elemental analysis calculated (calcd) for $C_{38}H_{42}Cl_2O_4$: C, 72.03; H, 6.68. found: C, 72.09; H, 6.72.

General procedure for synthesis of compounds ***5–9****.*

N-substituted imidazole (benzimidazole) (0.8 mmol) and calixarene 4 (0.16 mmol) were dissolved in 3 mL of dry acetonitrile. Resulting mixture was refluxed during 6 h, and then solvent was evaporated in vacuo to give light-yellow oil. Diethyl ether was added to the resulting oil, and the precipitate formed was filtered and dried in a desiccator.

*11,23-Bis[(3-methyl-1H-imidazolium-1-yl)methyl]-25,27-dihydroxy-26,28-dibutoxycalix[4]arene dichloride (**5**)* White solid, 0.10 g, 85%. Melting point: 185 °C (decomp). ^{1}H NMR (400 MHz, DMSO-d_6, 25 °C) δ_H ppm: 1.06 (6H, t, CH_3, J = 7.1 Hz), 1.83–1.71 (4H, m, -CH_2-), 1.88–2.03 (4H, m, -CH_2-), 3.44 (4H, d, Ar-CH_2-Ar, J = 12.8 Hz), 3.85 (6H, s, CH_3Im), 3.97 (4H, t, -CH_2O-, J = 6.4 Hz), 4.16 (4H, d, Ar-CH_2-Ar, J = 12.8 Hz), 5.19 (4H, s, Ar-CH_2N), 6.80 (2H, t, HAr, J = 7.3 Hz), 7.05 (4H, d, HAr, J = 7.4 Hz), 7.28 (4H, c, HAr), 7.69 (2H, brs, Im), 7.75 (2H, brs, Im), 8.70 (2H, s, OH), 9.18 (2H, s, N-CH-N). ^{13}C{^{1}H} NMR: (100.6 MHz, $CDCl_3$, 25 °C) δ_c ppm: 14.21, 19.47, 31.25, 32.31, 34.49, 52.51, 122.39, 122.96, 124.78, 125.48, 128.93, 129.27, 129.52, 133.47, 137.19, 152.24, 154.21. IR (KBr) ν_{max} cm^{-1}: 1434 (-CH_3), 2932 (-CH_2-), 2959 (-CH_3), 3376 (OH), 1560 (C = N). MALDI-TOF (*m/z*): 644 [M-$C_4H_6N_2$-2HCl+H]$^+$. Elemental analysis calcd for $C_{46}H_{54}Cl_2N_4O_4$: C, 69.25; H, 6.82; N, 7.02, found: 69.32; H, 6.87; N, 6.98.

11,23-Bis[(3-isopropyl-1H-imidazolium-1-yl)methyl]-25,27-dihydroxy-26,28-dibutoxycalix[4]arene dichloride (6) White solid, 0.10 g, 82%. Melting point: 174 °C (decomp). ^{1}H NMR (400 MHz, DMSO-d_6, 25 °C) δ_H ppm: 1.06 (6H, t, CH_3, J = 7.1 Hz), 1.47 (12H, d, CH-CH_3, J = 6.5 Hz), 1.71–1.84 (4H, m, -CH_2-), 1.90–2.02 (4H, m, -CH_2-), 3.44 (4H, d, -CH_2-, J = 12.7 Hz), 3.97 (4H, t, CH_2O, J = 6.0 Hz), 4.16 (4H, d, Ar-CH_2-Ar, J = 12.7 Hz), 4.69–4.59 (2H, sept, CH_3-CH-CH_3, J = 6.4 Hz), 5.18 (4H, s, Ar-CH_2N), 6.76 (2H, t, HAr, J = 7.5 Hz), 7.04 (4H, d, HAr, J = 7.5 Hz), 7.29 (4H, s, HAr), 7.78 (2H, brs, CH-Im), 7.89 (2H, brs, CH-Im), 8.70 (2H, s, OH), 9.37 (2H, s, N-CH-N). ^{13}C{^{1}H} NMR: (100.6 MHz, $CDCl_3$, 25 °C) δ_c ppm: 14.18, 19.44, 22.60, 31.30, 32.26, 53.05, 119.70, 121.88, 123.78, 125.52, 129.10, 129.28, 129.52, 129.79 133.27, 136.01, 152.07, 154.48. IR (KBr) ν_{max} cm^{-1}: 2934 (-CH_2-), 2960 (-CH_3), 3318 (OH), 1554 (C = N). MALDI-TOF (*m/z*): 672 [M-$C_6H_{10}N_2$-2HCl+H]$^+$. Elemental analysis calcd for $C_{50}H_{62}Cl_2N_4O_4$: C, 70.32; H, 7.32; N, 6.56, found: C, 70.37; H, 7.38; N, 6.51.

11,23-Bis[(3-(2,6-diisopropylphenyl)-1H-imidazolium-1-yl)methyl]-25,27-dihydroxy-26,28-dibutoxycalix[4]arene dichloride (7) White solid, 0.14 g, 82%. Melting point: 175 °C (decomp). ^{1}H NMR (400 MHz, DMSO-d_6, 25 °C) δ_H ppm: 1.07 (6H, t, CH_3, J = 7.4 Hz), 1.13 (12H, d, CH_3, J = 6.4 Hz), 1.17 (12H, d, CH_3, J = 6.5 Hz), 1.72–1.83 (4H, m, -CH_2-), 1.91–2.02 (4H, m, -CH_2-), 2.18–2.30 (4H, m, CH-Im), 3.43 (4H, d, Ar-CH_2-Ar, J = 12.6 Hz), 3.99 (4H, s, CH_2O), 4.20 (4H, d, Ar-CH_2-Ar, J = 12.6), 5.36 (4H, s, Ar-CH_2N), 6.77 (2H, t, HAr, J = 7.3 Hz), 7.02 (4H, d, HAr, J = 7.4 Hz), 7.29 (4H, s, HAr), 7.47 (4H, , HAr-Im, J = 7.7), 7.64 (2H, , HAr-Im), 8.06 (2H, s, CH-Im), 8.09 (2H, s, CH-Im), 8.71 (2H, s, OH), 9.84 (2H, s, N-CH-N). ^{13}C{^{1}H} NMR: (100.6 MHz, $CDCl_3$, 25 °C) δ_c ppm: 14.21, 19.47, 24.21, 24.38, 28.78, 31.34, 32.31, 53.63, 122.61, 123.38, 123.62, 124.03, 124.75, 125.53, 129.23, 129.80, 131.96, 133.08, 138.33, 145.34, 151.99, 154.59. IR (KBr) ν_{max} cm^{-1}: 2934 (-CH_2-), 2960 (-CH_3), 3318 (OH), 1544 (C = N), 1484 (C = C). MALDI-TOF (*m/z*): 790 [M-$C_{15}H_{20}N_2$-2HCl+H]$^+$. Elemental analysis calcd for $C_{68}H_{82}Cl_2N_4O_4$: C, 74.91; H, 7.58; N, 5.14, found: C, 74.96; H, 7.56; N, 5.11.

11,23-Bis[(3-(mesityl)-1H-imidazolium-1-yl)methyl]-25,27-dihydroxy-26,28-dibutoxycalix[4]arene dichloride (8) White solid, 0.15 g, 85%. Melting point: 172 °C (decomp). ^{1}H NMR (400 MHz, DMSO-d_6, 25 °C) δ_H ppm: 1.07 (6H, t, CH_3, J = 7.2 Hz), 1.72–1.83 (4H, m, CH_2), 1.90–2.07 (16H, m, CH_3Im, CH_2), 2.33 (6H, s, CH_3Im), 3.45 (4H, d, Ar-CH_2-Ar, J = 12.8 Hz), 3.97 (4H, t, CH_2O, J = 5.8 Hz), 4.18 (4H, d, Ar-CH_2-Ar, J = 12.7 Hz), 5.31 (4H, s, Ar-CH_2-N), 6.75 (2H, t, HAr, J = 7.5 Hz), 7.03 (4H, d, HAr, J = 7.5 Hz), 7.16 (4H, s, HAr), 7.33 (4H, s, HAr-Im), 7.88 (2H, s, CHIm), 8.00 (2H, s, CHIm), 8.72 (2H, s, OH), 9.62 (2H, s, N-CH-N). $^{13}C\{^1H\}$ NMR: (100.6 MHz, $CDCl_3$,25 °C) δ_c ppm: 14.22, 17.70, 19.49, 21.19, 31.37, 32.33, 53.61, 122.25, 122.97, 123.70, 125.60, 129.21, 129.28, 129.79, 129.95, 130.89, 133.06, 134.24, 138.31, 141.31, 151.98, 154.60. IR (KBr) ν_{max} cm^{-1}: 1553 (C = N), 2933 (CH_2), 2959 (CH_3), 3318 (OH). MALDI-TOF (*m/z*): 748 $[M\text{-}C_{12}H_{14}N_2\text{-}2HCl+H]^+$, 562 $[M\text{-}2C_{12}H_{14}N_2\text{-}2HCl+H]^+$. Elemental analysis calcd for $C_{62}H_{70}Cl_2N_4O_4$: C, 74.01; H, 7.01; N, 5.57, found: C, 74.07; H, 7.05; N, 5.59.

11,23-Bis[(3-methyl-1H-benzimidazolium-1-yl)methyl]-25,27-dihydroxy-26,28-dibutoxycalix[4]arene dichloride (9) White solid, 0.08 g, 56%. Melting point: 167 °C (decomp). ^1H NMR (400 MHz, DMSO-d_6, 25 °C) δ_H ppm: 1.04 (6H, t, CH_3, J = 6.7 Hz), 1.70–1.80 (4H, m, CH_2), 1.88–1.98 (4H, m, CH_2), 3.43 (4H, d, Ar-CH_2-Ar, J = 12.8 Hz), 3.95 (4H, brt, CH_2O), 4.06–4.18 (10H, m, Ar-CH_2-Ar, CH_3Im), 5.53 (4H, s, Ar-CH_2-N), 6.76 (2H, t, HAr, J = 7.2 Hz), 7.01 (4H, d, HAr, J = 7.4 Hz), 7.39 (4H, s, HAr), 7.68 (4H. brs, HAr-Im), 8.02 (4H, m, HAr-Im), 8.67 (2H, s, OH), 9.88 (2H, s, N-CH-N). $^{13}C\{^1H\}$ NMR: (100.6 MHz, $CDCl_3$, 25 °C) δ_c ppm: 14.18, 19.44, 31.28, 32.28, 32.57, 51.13, 112.62, 114.01, 123.66, 125.51, 127.15, 129.06, 129.53, 130.92, 131.95, 133.30, 143.02, 152.02, 154.31. IR (KBr) ν_{max} cm^{-1}: 1485 (C = C), 1567 (C = N), 2933 (CH_2), 2959 (CH_3), 3334 (OH). MALDI-TOF (*m/z*): 861 $[M\text{-}Cl]^+$, 694 $[M\text{-}C_8H_8N_2\text{-}2HCl+H]^+$. Elemental analysis calcd for $C_{54}H_{58}Cl_2N_4O_4$: C, 72.23; H, 6.51; N, 6.24, found: C, 72.29; H, 6.58; N, 6.27.

2.3. Dynamic Light Scattering and Zeta-Potential

Dynamic light scattering (DLS) experiments and zeta-potential measurements were carried out on the Zetasizer Nano ZS instrument (Malvern Instruments, Worcestershire, UK) with 4 mW 633 nm He–Ne laser light source and the light scattering angle of 173°. The data were treated with DTS software (Dispersion Technology Software 5.00). The solutions were filtered through a 0.8 μM filter before the measurements to remove dust. The experiments were carried out in the disposable plastic cells DTS 0012 (size) or in the disposable folded capillary cells DTS 1070 (zeta potential) (Sigma–Aldrich, St. Louis, MO, USA) at 298 K with at least three experiments for each system. Statistical data treatment was done using t-Student coefficient and the particle size determination error was <2%. The prepared samples were ultra-sonicated within 30 min at 25 °C before measurements.

2.4. Critical Aggregation Concentration Determination

Critical aggregation concentration (CAC) values were measured using pyrene fluorescent probe and calculated from the dependence of the intensity ratio of the first (373 nm) and third (384 nm) bands in the emission spectrum of pyrene vs. calixarene concentration. Fluorescence experiments with pyrene were performed in 10.0 mm quartz cuvettes and recorded on a Fluorolog FL-221 spectrofluorimeter (HORIBA Jobin Yvon, Edison Township, NJ, USA) in the range of 350 to 430 nm and excitation wavelength 335 nm with 2.5 nm slit. All studies were conducted in buffered aqueous solution (TRIS buffer, pH 7.4) at 298 K.

2.5. Vesicles Preparation

DPPC lipid films were formed from chloroform solutions, dried at 85 °C, and left under reduced pressure for a minimum of 2 h to remove all traces of organic solvent. Turbid MLV-suspensions were prepared by adding 1.15 mL of water Milli-Q water to the films and heating for 1 h at 65 °C. SUV-dispersions were obtained by extrusion through 100 nm-pore size polycarbonate membranes with a Mini-Extruder from Avanti. Concentration of DPPC stock dispersion was typically 1 mM. For binary system with calixarene 43, 65 or 98 μM of **5** were added to the chloroform solutions during preparation the lipid films.

2.6. Turbidity Measurements

The dependence of optical density at wavelengths of 400 nm on temperature was recorded using a Shimadzu UV-2600 spectrophotometer equipped with a Shimadzu TCC-100 thermostat. The temperature was varied in the range from 25 to 55 °C, the heating rate was 0.1 °C/min. The obtained plots were mathematically treated using Van't-Hoff's two-state model. [24] In accordance with this model the main temperature of phase transition of the lipid bilayer corresponds to the inflection point of the respective turbidity plot.

2.7. Gas Chromatography Mass Spectrometry

Gas chromatography mass spectrometry was performed on a GCMS-QP2010 Ultra gas chromatography mass spectrometer (Shimadzu, Kyoto, Japan) equipped with an HP-5MS column (the internal diameter was 0.25 μm and the length was 30 m). The parameters were as follows: Helium A was the carrier gas, the temperature of an injector was 250 °C, the flow rate through the column was 2 mL/min, the thermostat temperature program was a gradient temperature increase from 70 to 250 °C with a step of 10 °C/min. The range of scanned masses was m/z 35 ÷ 400. The internal standard method using dodecane was used for the quantitative analysis of 4-nitro-1,1'-biphenyl.

2.8. Suzuki–Miyaura Coupling

Reactions were performed in 2 mL Pyrex vials in IKA heating block with vigorous stirring. Aqueous dispersion (1 mL) containing 50 μM $Pd(OAc)_2$, **5–9** (0.098, 0.033, 0.060, 0.080 or 0.18 mM for **5–9**, respectively) or DPPC-**5** vesicles was filled with 10 mM 1-bromo-4-nitrobenzene, 11 mM phenylboronic acid, 1 mM of dodecane and 30 mM K_2CO_3 (30 mM). The mixture was degassed with Ar for 10 min by piercing the septum of the vial with two needles for supplying and discharging gas. The mixture was stirred at 70 °C. Two aliquots (0.1 mL) for GCMS analysis were taken from the reaction mixture after 1 and 4 h and then extracted with chloroform (3 × 0.5 mL).

3. Results and Discussion

3.1. Synthesis of Imidazolium/Benzimidazolium Calix[4]arene Derivatives

As it was previously shown by Schatz group, tetra-O-alkyl calixarene derivatives bearing at the distal positions of upper rim two [(3-*N*-substituted imidazolium-1-yl)methyl] fragments formed stable *N*-heterocyclic carbenes and corresponding cis/trans Pd(II) complexes which effectively catalyzed Suzuki–Miyaura coupling [25,26]. This structural motive of NHC precursors with two imidazolium fragments on the upper rim seems to be very attractive for designing self-organizing systems for different catalytic transformations in water. The preparation of calixarene imidazolium salts was based on the tetra-O-propyl substituted dichloromethyl calix[4]arene as precursor. However, its synthesis is rather complicated and included at least seven stages starting from parent *p*-*tert*-butyl calix[4]arene (Scheme 1) [23,27].

To design amphiphilic calixarene imidazolium salts we suggest a more synthetically accessible precursor containing two (instead of four) O-alkyl groups on the macrocycle lower rim (**3**, Scheme 1). In this case the number of stages for the target product is halved (4 vs. 8). The application of distal di-O-alkyl calixarene derivative **3** (Scheme 1) allows to introduce selectively two chloromethyl substituents on the upper rim in one step due to significantly different reactivity of *p*-positions of aromatic rings containing OH and OAlk groups [28]. Taking into account that tetra-O-alkyl substituted bis-imidazolium calixarene has a very low solubility in water [29], a decrease of the number of alkyl groups on the calixarene lower rim (2 vs. 4) gives another very important advantage for the applications in aqueous solutions: Better solubility in water.

Parent *p*-*tert*-butyl calix[4]arene **1** was de-*tert*-butylated using the earlier described procedure [22]. Obtained *p*-H-calix[4]arene **2** was treated by butyl bromide under microwave irradiation and O,O-dialkyl derivative **3** was formed. Then it was involved in the Blanc reaction to give dichloromethyl macrocycle **4** with nearly quantitative yield. The structure of **4** was established by NMR ^{1}H and ^{13}C spectroscopy as well as IR spectroscopy and MALDI-TOF mass spectrometry (see Supplementary Materials). The composition was determined by the element analysis. A set of signals which are characteristic for distal chloromethylated calixarene was found in the NMR ^{1}H spectra of **4**: A singlet of chloromethylene protons at 4.51 ppm, signals of aromatic calixarene protons as singlet, doublet and triplet at 7.08, 6.93 and 6.78 ppm, respectively.

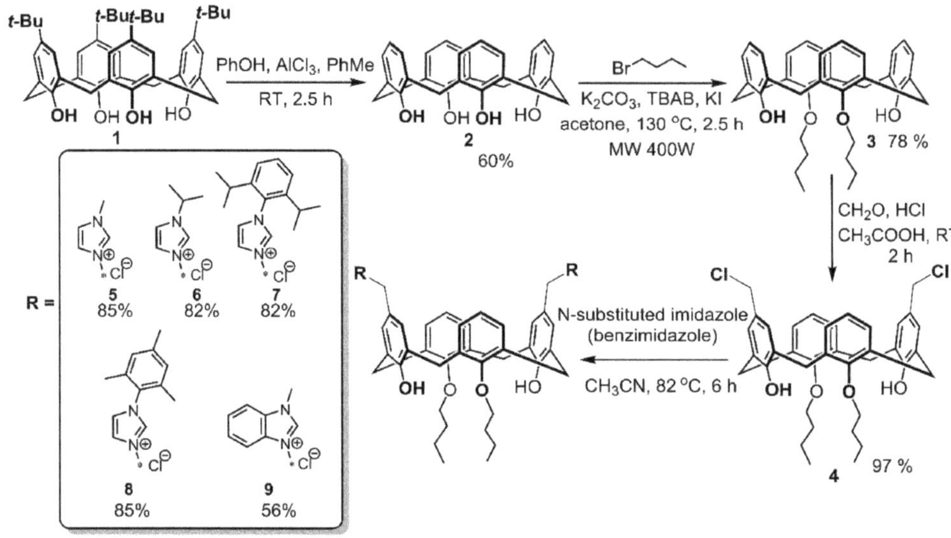

Scheme 1. Synthetic pathway for imidazolium/benzimidazolium salts **5–9**.

Chloromethylated calixarene **4** was functionalized by imidazolium/benzimidazolium moieties. Macrocycle **4** was heated with five-fold excess of appropriate *N*-substituted imidazole/benzimidazole in acetonitrile for 6 h. Corresponding imidazolium salts **5–8** were obtained with high yields (82–85%). The lower yield of benzimidazolium salt **9** (56%) can be attributed to its slightly better solubility in diethyl ether, which was used to precipitate compound compared to that for imidazolium salts **5–8**. The structures of **5–9** were well-established with the same set of methods used for **4**. In all synthesized salts the singlet of the methylene spacer between macrocycle and heterocycle cores is downshifted to $\Delta\delta$ ~0.7–1 ppm in comparison to **4** due to the effect of a positive charge of imidazolium/benzimidazolium fragments and the location of the methylene spacer in the deshielding region of the imidazolium/benzimidazolium molecular ring current. Moreover, *N*-substituted imidazolium/benzimidazolium fragments appeared as a standard set of signals: Two broad singlets of heterocycle core at 7.7–8.0 and 7.8–8.1 ppm and one singlet of NCHN protons at 9.2–9.9 ppm. Interestingly, molecular ion peaks with expulsion of one imidazolium/benzimidazolium fragment and two molecules of HCl were observed in MALDI-TOF spectra of salts **5–9** due to easy formation of *p*-quinone methide structures [30]. The stereoisomeric form of **5–9** can be easily identified using the Gutsche's "1H NMR $\Delta\delta$" rule [31]: Chemical shift separation of $ArCH_2Ar$ protons $\Delta\delta \approx 0.7$–1.0 is indicative of a *syn* orientation of the two pertinent aromatic rings, typical of the *cone* stereoisomeric form. Indeed, in all NMR spectra of **5–9** $\Delta\delta$ of signals of $ArCH_2Ar$ protons are 0.72–0.76 ppm indicating thus the *cone* stereoisomeric form. Additionally, the stereoisomeric form of macrocycle **5** was established using two-dimensional ^{1}H-^{1}H NOESY NMR spectroscopy (Figure 1).

The presence of a cross peak between equatorial $ArCH_2Ar$ protons with protons of hydroxyl group (δ = 4.24 and 8.78 ppm) as well as cross peaks between axial $ArCH_2Ar$ protons and aryl protons (δ = 3.44 and 7.23, 7.01 ppm) clearly indicates the *cone* stereoisomeric form of **5**.

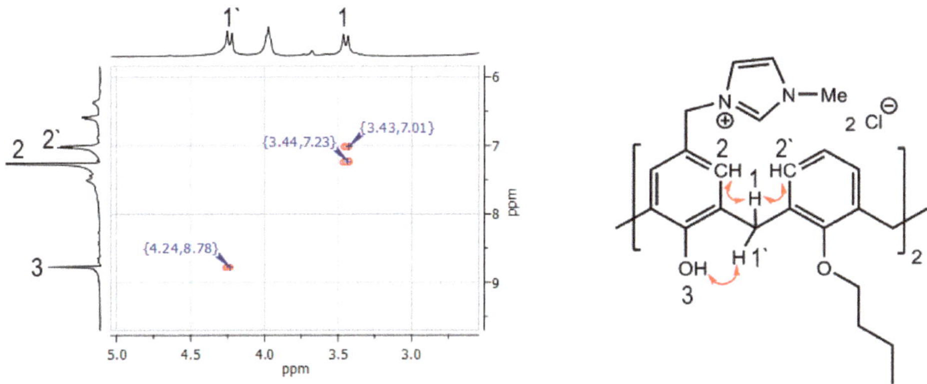

Figure 1. Fragment of the ^1H-^1H NOESY NMR spectra of **5** ($CDCl_3$).

3.2. *Aggregation Behavior of 5–9 in Aqueous Solutions*

The aggregates formation in the aqueous solutions is typical behavior for calix[4]arenes containing positively charged headgroups and alkyl moieties on the upper and lower rim, respectively [32,33]. To estimate amphiphilic properties of calixarenes **5–9** their critical aggregation concentration (CAC) values were measured using a pyrene fluorescent probe.

The intensity ratio changes of the first and third peaks of pyrene (so called polarity index [34]) (Figure 2) indicate a decrease in the pyrene polarity environment caused by its solubilization in the hydrophobic core of the calixarene aggregates. Calculated CAC values are given in Table 1.

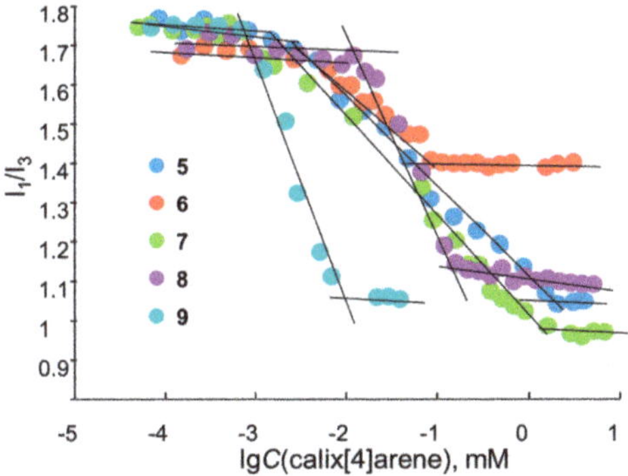

Figure 2. Dependence of I_1/I_3 pyrene ratio vs. lg(C) of **5–9**, C (pyrene) = 1 μM in ultrapure water, 25 °C.

Table 1. Critical aggregation concentration (CAC) values, dynamic light scattering (DLS) and electrophoretic (ELS) data of aggregates formed by macrocycles **5–9**.

Calixarene *	CAC, μM	d, nm	PDI	ζ, mV
5	65	340 ± 13	0.426 ± 0.081	+57 ± 2
6	45	351 ± 51	0.477 ± 0.042	+45 ± 1
7	40	420 ± 5	0.473 ± 0.055	+52 ± 1
8	53	390 ± 2	0.513 ± 0.030	+44 ± 1
9	120	410 ± 18	0.416 ± 0.070	+54 ± 0.5

* Pyrene concentration is 1 μM, C(**5–9**) = 1.5× CAC μM.

It was found that, in the case of imidazolium-containing calixarenes (**5–8**), CAC values were not sensitive to the *N*-substituents of the heterocycle. Neither nature (alkyl or aryl) or size (steric hindrance) of *N*-substituents had no effect on the CAC values (50 ± 10 μM). This can be explained by the fact that *N*-substituents do not participate in the aggregate hydrophobic core formation because they are constrained to orientate into the aqueous phase for steric reasons. On other hand the aromatic ring condensed at 4,5-positions of imidazole (benzimidazolium calixarene **9**) increased the CAC values by more than double. Such a difference in CAC values between imidazole and benzimidazole derivatives may be associated with the delocalization of the positive charge on the entire aromatic benzimidazole system, while in the case of imidazolium compounds it concentrates in the imidazole cycle [35]. Thus, in the case of benzimidazole, the charged groups occupy a larger volume, which leads to an increase of electrostatic repulsion between molecules during the aggregation process, and, as a result, increases CAC.

Additionally, the study of aggregation properties was carried out using dynamic (DLS) and electrophoretic (ELS) light scattering methods (Table 1) at a concentration above corresponding CAC of macrocycles (1.5× CAC μM). All macrocycles formed submicron particles in aqueous solution with a hydrodynamic diameter in the range of 340–420 nm. Electrokinetic potential of the aggregates corresponds to the positive charge of the imidazolium/benzimidazolium headgroups and is about +44–+57 mV that refers them to the stable colloids. However, a rather high polydispersity index was indicated on a wide size distribution of aggregates. This is clearly seen from the DLS curves (Supplementary Materials).

3.3. *Complexes of 5–9 with Pd(II) Obtained In Situ in Model Suzuki–Miyaura Coupling*

NHC-Pd complexes are well-known catalysts [36] in Suzuki–Miyaura coupling [1]. Obviously, many catalytic systems have been applied for this reaction including in situ systems, generated from appropriate imidazolium salts and $Pd(OAc)_2$ [37]. The attempts to isolate and characterize corresponding NHC complexes were unsuccessful because they decomposed easily during column chromatography. For this reason, catalytic activity of in situ prepared complexes of amphiphilic calixarenes **5–9** and $Pd(OAc)_2$ was investigated in Suzuki–Miyaura coupling of 1-bromo-4-nitrobenzene with phenylboronic acid in water (Figure 3). Reactions were performed in the presence of 0.5 mol% of $Pd(OAc)_2$ and K_2CO_3 as a base at 70 °C. The concentration of calixarenes **5–9** was one and a half times greater than their CAC values to form aggregates in the solution for effective substrate solubilization. Conversion of haloarenes and 4-nitro-1,1′-biphenyl formation selectivity were evaluated using GCMS with dodecane as the internal standard.

It is important that selectivity of 4-nitro-1,1′-biphenyl formation was very high (96–99%) for all investigated catalytic systems. Calixarene containing catalytic systems showed a higher activity than pure $Pd(OAc)_2$ (Figure 3). The influence of alkyl/aryl substituent in N-position on catalytic activity in the series of calixarenes **5–9** was evaluated. Compound **5** with methyl substituent was found to be the most active—conversion of 1-bromo-4-nitrobenzene was achieved at 80% and 98% after 1 and 4 h, respectively. Compounds with more bulky isopropyl (**6**), mesityl (**7**) and 2,6-diisopropylphenyl (**8**) substituents in the N-position of the imidazolium unit demonstrated lower conversion of 1-bromo-4-nitrobenzene: 50–60% and 75–85% after 1 and 4 h, respectively. The benefits of using calixarenes **5–9** with $Pd(OAc)_2$ are more apparent when comparing turnover number (TON) and turnover frequency (TOF) values (Table 2). According to this data, calixarenes give at least a 2.5-fold increase in TON and TOF, and macrocycle **5** shows better results and gives a four-fold increase.

These results are not consistent with the usual effect of the bulky substituents on the catalytic activity of pre-prepared NHC complexes: Usually bulky groups promote the last reductive elimination step of Suzuki–Miyaura coupling as well as the formation of active Pd(0) species through favorable repulsion of halides from the coordination sphere of Pd(II) [38,39]. However, it can be assumed that catalytically active species in the case of less bulky NHC forms faster. In confirmation of this, recently we showed that the reaction time of bis-imidazolium thiacalix[4]arenes with $Pd(OAc)_2$ to give

NHC–Pd(II) complexes increases from 28 to 60 h upon going from methyl to the more bulky mesityl and 2,6-diisopropylphenyl substituents in the N-position of the imidazole [40]. Thus, lower rates of the NHC–Pd(II) complex in situ formation could be a reason for their lower activity in Suzuki–Miyaura coupling. The lower activity of benzimidazole derivative **9** can be attributed to a decrease in the relative stability of the in situ complex, which correlates with the basicity of NHC ligands and is decreased going from *N*-methylimidazole to *N*-methylbenzimidazole [41].

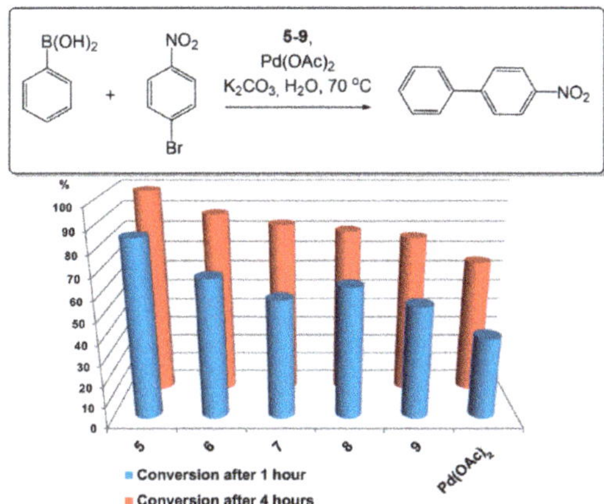

Figure 3. Conversion of 1-bromo-4-nitrobenzene after 1 or 4 h of reaction upon reaction with phenylboronic acid. *C*(1-bromo-4-nitrobenzene) = 10 mM, *C*(phenylboronic acid) = 11 mM, *C*(K_2CO_3) = 30 mM, *C*($Pd(OAc)_2$) = 50 μM, *C*(**5–9**) = 1.5× CAC μM, H_2O, 70 °C.

Table 2. (TON) and turnover frequency (TOF) values for coupling of 1-bromo-4-nitrobenzene with phenylboronic acid using different catalytic systems.

Catalyst *	TON	TOF 10^{-2} s^{-1}
$Pd(OAc)_2$	22	1.8
5+$Pd(OAc)_2$	89	7.4
6+$Pd(OAc)_2$	67	5.6
7+$Pd(OAc)_2$	56	4.6
8+$Pd(OAc)_2$	62	5.2
9+$Pd(OAc)_2$	53	4.4

* Values were calculated using conversion after 20 min of reaction, C(1-bromo-4-nitrobenzene) = 10 mM, C(phenylboronic acid) = 11 mM, C(K_2CO_3) = 30 mM, C($Pd(OAc)_2$) = 50 μM, C(**5–9**) = 1.5× CAC μM, H_2O, 70 °C.

3.4. Embedding of 5 into DPPC Vesicles and Their Catalytic Activity in Suzuki–Miyaura Coupling

An alternative approach to the application of amphiphilic compounds containing receptor groups for the catalytic systems design is their embedding into vesicles and liposomes, which have a well-defined size/shape and can act as microreactors. This is a well-known and very promising way to increase the reaction rates and to solubilize hydrophobic reagents [42,43]. So, metallocomplex and micellar catalysis can be combined in one catalytic system.

Recently, phospholipid vesicles were successfully used as both effective stabilizers for Pd(0) nanoparticles and microreactors for Suzuki–Miyaura coupling in aqueous media [44]. To the best of our knowledge, vesicles doped by amphiphilic NHC metal complexes and their application for the catalysis are not presented in literature.

In our investigation DPPC (1,2-dipalmitoyl-sn-glycero-3-phosphocholine) liposomes were chosen as a typical membrane mimicking system, which is characterized by close to zero zeta potential and a hydrodynamic diameter of ca. 100 nm. As the amphiphilic guest molecule, macrocycle **5** was used because it demonstrated the best catalytic activity among synthesized compounds. The vesicles modified by amphiphilic calixarene were obtained by the film-hydration method [40] from a mixture of DPPC and calixarene **5** chloroform solutions (from 0.04 up to 0.10 calixarene/DPPC molar ratio). Size, polydispersity index and electrokinetic potential of mixed phospholipid vesicles before and after extrusion are presented in Table 3.

Table 3. DLS and ELS data of aggregates formed by DPPC and calixarene **5**.

System	Calixarene/DPPC Molar Ratio	d, nm		PDI		ζ, mV
		Before Extrusion	After Extrusion	Before Extrusion	After Extrusion	
DPPC	0	600 ± 63	106 ± 2	0.780 ± 0.140	0.078 ± 0.006	-
DPPC+ **5**	0.04	130 ± 5	64 ± 5	0.703 ± 0.049	0.269 ± 0.074	+22 ± 1
DPPC+ **5**	0.07	60 ± 1	51 ± 2	0.331 ± 0.015	0.239 ± 0.015	+36 ± 3
DPPC+ **5**	0.1	79 ± 1	63 ± 1	0.365 ± 0.011	0.285 ± 0.025	+35 ± 5

It can be seen that the size of the DPPC multilamellar vesicles (MLVs) before extrusion is 600 ± 63 nm with high PDI values. The extrusion process leads to the transformation of MLVs into unilamellar ones (SUVs) with a narrow size distribution (106 ± 2 nm, PDI 0.078). For the first time it was found that addition of amphiphilic calixarene leads to the same effect. DPPC multilamellar vesicles were easily destroyed and transformed into unilamellar ones in the presence of macrocycle **5**. Moreover, amphiphile concentration plays a major role in this process. At the small calixarene/DPPC molar ratio (0.04) the aggregates size almost decreased five times. At the higher ratio, the size and PDI values of obtained aggregates before and after extrusion became practically equal. Evidently, the embedding of a positively charged macrocycle into the multilamellar vesicles leads to an "explosion" of huge particles due to Coulomb repulsion between charged monolayers. Of course, the size and shape of the amphiphile may play a key role as well. The commonality and reasons of such behavior are now under investigation. Nevertheless, it can be concluded from DLS data that calixarene **5** was embedded into DPPC vesicles, thus forming mixed nanosized vesicles, since both calixarene and DPPC separately form larger aggregates.

The gel-to-liquid crystalline phase transition, named the main phase transition, is an informative parameter that is sensitive to guest molecules. In particular, a change of the main phase transition temperature, the so-called melting point T_m, indicates that a perturbation of the lipid organization occurs, and so indicates the incorporation of guest molecules into the lipid bilayer. The T_m value can be monitored by the turbidimetry technique, a well-known method of detection of this parameter with high reproducibility [24]. Turbidity curves and T_m values for DPPC+**5** systems are presented on Figure S9 (Supplementary Materials).

For a single DPPC bilayer Tm is ca. 41 °C [45], while the introduction of surfactants may change this value. Usually a decrease of T_m indicates that the macrocyclic amphiphile disorders the lipid bilayer due to the interaction with lipids and promotes the gel-to-liquid phase transition [46]. In our case T_m values did not change (~40.9 °C) up to a molar calixarene/DPPC ratio of 0.07, after which the increase of calixarene concentration promotes the ordering of the lipid in membranes (Figure 4), which is in line with an increase in the T_m value (43.3 °C). An increase in the temperature of the main phase transition, correlated with the stabilizing effect on the lipid packing, was demonstrated earlier [47] and was explained in terms of an increase in the distribution coefficient value for amphiphile in the gel phase in comparison with the distribution coefficient in the liquid phase.

Preliminarily, to estimate the catalytic effect of amphiphilic calixarene embedding into the vesicles, a blank experiment was carried out. It was found that the catalytic activity of Pd(II) in the presence of DPPC vesicles increases by 20% (Figure 4), which may be assigned to both the stabilization of catalytically active palladium species by DPPC phosphonate groups and/or the micellar catalytic effect due to solubilization and concentration of reagents inside the vesicle.

Embedding of calixarene **5** into the vesicles at the small calixarene/DPPC molar ratio (up to 0.07) gave practically quantitative reaction yield after 4 h. Surprisingly, further growth of calixarene concentration in the reaction mixture led to the decrease in the yield. However, according to the turbidity data, an ordering of the lipid in membranes at a larger calixarene/DPPC molar ratio was observed and the lower catalytic activity of such vesicles can be explained by the fact that the better packing of hydrophobic tails in a lipid bilayer prevents free diffusion of reagents inside vesicle. Moreover, the significant effect of the calixarene/DPPC molar ratio on the differences of 1-bromo-4-nitrobenzene conversion for 1 and 4 h indicates diffusion reaction control (Figure 5).

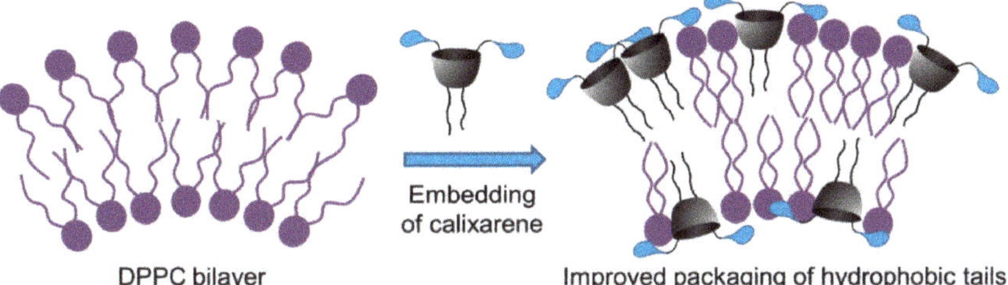

Figure 4. Schematic representation of improved packaging of hydrophobic tales in mixed DPPC–calixarene **5** vesicles.

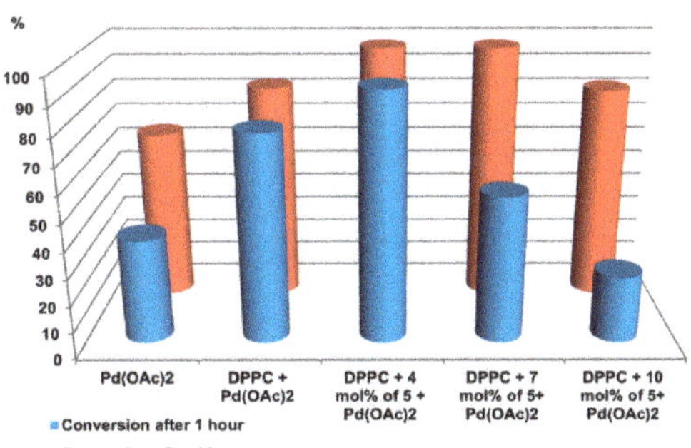

Figure 5. Conversion of 1-bromo-4-nitrobenzene after 1 or 4 h of reaction. C(1-bromo-4-nitrobenzene) = 10 mM, C(phenylboronic acid) = 11 mM, C(K_2CO_3) = 30 mM, C(DPPC) = 1 M, C(**5** mixed with DPPC) = 43, 65 and 98 μM, C($Pd(OAc)_2$) = 50 μM, H_2O, 70 °C.

4. Conclusions

In this study, new types of amphiphilic calix[4]arene derivatives bearing N-alkyl/aryl imidazolium/benzimidazolium fragments were synthesized according to a two step scheme including regioselective chloromethylation of distal di-O-butyl calix[4]arene and subsequent interaction with *N*-substituted imidazole/benzimidazole. Obtained amphiphilic macrocycles are formed submicron particles with electrokinetic potential +44–+57 mV in aqueous solution. For the first time it was found that amphiphilic calixarene causes the fast transformation of DPPC multilamellar vesicles into unilamellar ones and leads to the ordering of the lipid into membranes at a molar calixarene/DPPC ratio more than 0.07.

In situ complexes of calixarene aggregates with $Pd(OAc)_2$ were found to be active in Suzuki–Miyaura coupling of 1-bromo-4-nitrobenzene with phenylboronic acid in water. It was shown that the bulky *N*-substituents of a heterocycle decrease the catalytic activity of the aggregates. These results are not consistent with the usual influence of the bulky substituents on the catalytic activity of pre-prepared NHC complexes and can be assigned to the inhibition effect of Pd(II) complex formation by bulky substituents located on the aggregate surface.

An alternative approach to the catalytic application of amphiphilic receptor molecules based on their embedding into liposome membrane was investigated. Amphiphilic calixarene embedding into DPPC vesicles enhances the catalytic activity in Suzuki–Miyaura coupling.

Supplementary Materials: The following are available online at http://www.mdpi.com/2079-4991/10/6/1143/s1, Figures S1–S6: NMR ^{1}H, ^{13}C and MALDI-TOF spectra of new compounds, Figure S7: DLS size graph of aggregates formed by **5–9**, Figure S8: DLS size graph of aggregates formed DPPC or DPPC mixed with **5**, Figure S9: T_m plots of DPPC and DPPC-5 vesicles, Scheme S1: Known synthetic pathway for NHC-precursors made by Schatz group.

Author Contributions: Conceptualization, V.B., I.A. and S.S.; methodology, V.B.; validation, V.B., I.A. and S.S.; investigation, R.G., I.G., E.S. and D.M.; data curation, V.B.; writing—original draft preparation, V.B.; writing—review and editing, I.A., S.S.; visualization, V.B.; supervision V.B. and I.A.; project administration, V.B. and I.A.; funding acquisition, V.B. All authors have read and agreed to the published version of the manuscript.

Funding: This research was funded by Russian Science Foundation, grant number 18-73-10033.

Conflicts of Interest: The authors declare no conflict of interest.

References

1. Beletskaya, I.; Alonso, F.; Tyurin, V. The Suzuki-Miyaura reaction after the Nobel prize. *Coord. Chem. Rev.* **2019**, *385*, 137–173. [CrossRef]
2. Chinchilla, R.; Najera, C. The Sonogashira reaction: A booming methodology in synthetic organic chemistry. *Chem. Rev.* **2007**, *107*, 73874–73922. [CrossRef] [PubMed]
3. Corbet, J.-P.; Mignani, G.R. Selected patented cross-coupling reaction technologies. *Chem. Rev.* **2006**, *106*, 2651–2710. [CrossRef] [PubMed]
4. Nolan, S.P.; Scott, N.M. *N-Heterocyclic Carbenes in Synthesis*, 1st ed.; Wiley-VCH: Weinheim, Germany, 2006.
5. Peris, E. Smart N-Heterocyclic Carbene Ligands in Catalysis. *Chem. Rev.* **2018**, *118*, 9988–10031. [CrossRef] [PubMed]
6. Fortman, G.C.; Nolan, S.P. N-Heterocyclic carbene (NHC) ligands and palladium in homogeneous cross-coupling catalysis: A perfect union. *Chem. Soc. Rev.* **2011**, *40*, 5151–5169. [CrossRef] [PubMed]
7. De, S.; Udvardy, A.; Czegeni, C.E.; Joo, F. Poly-*N*-heterocyclic carbene complexes with applications in aqueous media. *Coord. Chem. Rev.* **2019**, *400*, 213038. [CrossRef]
8. Levin, E.; Ivry, E.; Diesendruck, C.E.; Lemcoff, N.G. Water in N-Heterocyclic Carbene-Assisted Catalysis. *Chem. Rev.* **2015**, *115*, 4607–4692. [CrossRef] [PubMed]
9. Anastas, P.T.; Kirchhoff, M.M. Origins, Current Status, and Future Challenges of Green Chemistry. *Acc. Chem. Res.* **2002**, *35*, 686–694. [CrossRef] [PubMed]
10. Sorella, G.L.; Strukul, G.; Scarso, A. Recent advances in catalysis in micellar media. *Green Chem.* **2015**, *17*, 644–683. [CrossRef]
11. Donner, A.; Hagedorn, K.; Mattes, L.; Drechsler, M.; Polarz, S. Hybrid Surfactants with *N*-Heterocyclic Carbene Heads as a Multifunctional Platform for Interfacial Catalysis. *Chem. Eur. J.* **2017**, *23*, 18129–18133. [CrossRef] [PubMed]
12. Yang, J.; Liu, J.; Wang, Y.; Wang, J. Synthesis, structure and catalysis/applications of *N*-heterocyclic carbene based on macrocycles. *J. Incl. Phenom. Macrocycl. Chem.* **2018**, *90*, 15–37. [CrossRef]
13. Raynal, M.; Ballester, P.; Vidal-Ferran, A.; van Leeuwen, P.W.N.M. Supramolecular catalysis. Part 2: Artificial enzyme mimics. *Chem. Soc. Rev.* **2014**, *43*, 1734–1787. [CrossRef] [PubMed]
14. Solovieva, S.E.; Burilov, V.A.; Antipin, I.S. Thiacalix[4]arene's Lower Rim Derivatives: Synthesis and Supramolecular Properties. *Macroheterocycles* **2017**, *10*, 134–146. [CrossRef]
15. Helttunen, K.; Shahgaldian, P. Self-assembly of amphiphilic calixarenes and resorcinarenes in water. *New J. Chem.* **2010**, *34*, 2704–2714. [CrossRef]
16. Burilov, V.A.; Valiyakhmetova, A.M.; Mironova, D.A.; Sultanova, E.D.; Evtugyn, V.G.; Osin, Y.N.; Katsyuba, S.A.; Burganov, T.I.; Solovieva, S.E.; Antipin, I.S. Novel amphiphilic conjugates of: P-tert-butylthiacalix[4]arene with 10,12-pentacosadiynoic acid in 1,3-alternate stereoisomeric form synthesis and chromatic properties in the presence of metal ions. *New J. Chem.* **2018**, *42*, 2942–2951. [CrossRef]
17. Yakimova, L.S.; Padnya, P.L.; Kunafina, A.F.; Nugmanova, A.R.; Stoikov, I.I. Sulfobetaine derivatives of thiacalix[4]arene: Synthesis and supramolecular self-assembly of submicron aggregates with Ag^I cations. *Mendeleev Commun.* **2019**, *29*, 86–88. [CrossRef]

18. Armarego, W.L.F.; Chai, C. *Purification of Laboratory Chemicals*, 6th ed.; Elsevier: New York, NY, USA, 2009.
19. Kumar, N.; Jain, R. Convenient syntheses of bulky group containing imidazolium ionic liquids. *J. Heterocycl. Chem.* **2012**, *49*, 370–374. [CrossRef]
20. Jingping, L.; Jingbo, C.; Jingfeng, Z.; Yuanhong, Z.; Liang, L.; Hongbin, Z. A modified procedure for the synthesis of 1-arylimidazoles. *Synthesis* **2003**, *17*, 2661–2666.
21. Gutsche, D.C.; Dhawan, B.; No, K.H.; Muthukrishnan, R. Calixarenes. 4. The synthesis, characterization, and properties of the calixarenes from p-tert-butylphenol. *J. Am. Chem. Soc.* **1981**, *103*, 3782–3792. [CrossRef]
22. Gutsche, D.C.; Levine, J.A. Calixarenes. 6. Synthesis of a functionalizable calix[4]arene in a conformationally rigid cone conformation. *J. Am. Chem. Soc.* **1982**, *104*, 2653–2655. [CrossRef]
23. Frank, M.; Maas, G.; Schatz, J. Calix[4]arene-Supported N-Heterocyclic Carbene Ligands as Catalysts for Suzuki Cross-Coupling Reactions of Chlorotoluene. *Eur. J. Org. Chem.* **2004**, *3*, 607–613. [CrossRef]
24. Eker, F.; Durmus, H.O.; Akinoglu, B.G.; Severcan, F. Application of turbidity technique on peptide-lipid and drug-lipid interactions. *J. Mol. Struct.* **1999**, *482–483*, 693–697. [CrossRef]
25. Fahlbusch, T.; Frank, M.; Maas, G.; Schatz, J. N-Heterocyclic Carbene Complexes of Mercury, Silver, Iridium, Platinum, Ruthenium, and Palladium Based on the Calix[4]arene Skeleton. *Organometallics* **2009**, *28*, 6183–6193. [CrossRef]
26. Fahlbusch, T.; Frank, M.; Schatz, J. The Suzuki Coupling of Aryl Chlorides in Aqueous Media Catalyzed by in situ Generated Calix[4]arene-Based *N*-Heterocyclic Carbene Ligands. *Eur. J. Org. Chem.* **2006**, *10*, 2378–2383.
27. Larsen, M.; Jørgensen, M. Selective Halogen-Lithium Exchange Reaction of Bromine-Substituted 25,26,27,28-Tetrapropoxycalix[4]arene. *J. Org. Chem.* **1996**, *61*, 6651–6655. [CrossRef] [PubMed]
28. Huang, Z.-T.; Wang, G.-Q.; Yang, L.-M.; Lou, Y.-X. The Selective Chloromethylation of 25, 27-Dihydroxy-26, 28-Dimethoxycalix{4}arene and Nucleophilic Substitution Therefrom. *Synth. Commun.* **1995**, *25*, 1109–1118. [CrossRef]
29. Rehm, M.; Frank, M.; Schatz, J. Water-soluble calixarenes—Self-aggregation and complexation of noncharged aromatic guests in buffered aqueous solution. *Tetrahedron Lett.* **2009**, *50*, 93–96. [CrossRef]
30. Regnouf-de-Vains, J.-B.; Berthalon, S.; Lamartine, R. Electrospray mass spectrometric evidence of calixarene p-quinone methide formation. *J. Mass Spectrom.* **1998**, *33*, 968–970. [CrossRef]
31. Gutsche, C.D. *Calixarenes;* Royal Society of Chemistry: Cambridge, UK, 1989.
32. Rodik, R.V.; Anthony, A.-S.; Kalchenko, V.I.; Mely, Y.; Klymchenko, A.S. Cationic amphiphilic calixarenes to compact DNA into small nanoparticles for gene delivery. *New J. Chem.* **2015**, *39*, 1654–1664. [CrossRef]
33. Burilov, V.A.; Mironova, D.A.; Ibragimova, R.R.; Evtugyn, V.G.; Osin, Y.N.; Solovieva, S.E.; Antipin, I.S. Imidazolium *p*-tert-Butylthiacalix[4]arene Amphiphiles—Aggregation in Water Solutions and Binding with Adenosine 5-Triphosphate Dipotassium Salt. *BioNanoScience* **2018**, *8*, 337–343. [CrossRef]
34. Aguiar, J.; Carpena, P.; Molina-Bolıvar, J.A.; Carnero Ruiz, C. On the determination of the critical micelle concentration by the pyrene 1:3 ratio method. *J. Colloid Interface Sci.* **2003**, *258*, 116–122. [CrossRef]
35. Rozengart, E.; Basova, N. Ammonium Compounds with Localized and Delocalized Charge as Reversible Inhibitors of Cholinesterases of Different Origin. *J. Evol. Biochem. Physiol.* **2001**, *37*, 604–610. [CrossRef]
36. Marion, N.; Nolan, S.P. Well-Defined N-Heterocyclic Carbenes–Palladium(II) Precatalysts for Cross-Coupling Reactions. *Acc. Chem. Res.* **2008**, *41*, 1440–1449. [CrossRef] [PubMed]
37. Akkoç, S.; Gök, Y.; Özer lhan, l.; Kayser, V. In situ Generation of Efficient Palladium N-heterocyclic Carbene Catalysts Using Benzimidazolium Salts for the Suzuki-Miyaura Cross-coupling Reaction. *Curr. Org. Synth.* **2016**, *13*, 761–766. [CrossRef]
38. Froese, R.D.J.; Lombardi, C.; Pompeo, M.; Rucker, R.P.; Organ, M.G. Designing Pd–N-Heterocyclic Carbene Complexes for High Reactivity and Selectivity for Cross-Coupling Applications. *Acc. Chem. Res.* **2017**, *50*, 2244–2253. [CrossRef] [PubMed]
39. Szilvási, T.; Veszprémi, T. Internal Catalytic Effect of Bulky NHC Ligands in Suzuki–Miyaura Cross-Coupling Reaction. *ACS Catal.* **2013**, *3*, 1984–1991. [CrossRef]
40. Burilov, V.; Gafiatullin, B.; Mironova, D.; Sultanova, E.; Evtugyn, V.; Osin, Y.; Islamov, D.; Usachev, K.; Solovieva, S.; Antipin, I. Amphiphilic Pd (II)-NHC complexes on 1,3-alternate *p*-tertbutylthiacalix[4]arene platform: Synthesis and catalytic activities in coupling and hydrogenation reactions. *Eur. J. Org. Chem.* **2020**. [CrossRef]

41. Kostyukovich, A.Y.; Tsedilin, A.M.; Sushchenko, E.D.; Eremin, D.B.; Kashin, A.S.; Topchiy, M.A.; Asachenko, A.F.; Nechaev, M.S.; Ananikov, V.P. In situ transformations of Pd/NHC complexes with N-heterocyclic carbene ligands of different nature into colloidal Pd nanoparticles. *Inorg. Chem. Front.* **2018**, *6*, 482–492. [CrossRef]
42. Lasic, D.D.; Barenholz, Y. *Handbook of Nonmedical Applications of Liposomes*, 1st ed.; CRC Press Inc.: Boca Raton, FL, USA, 1996.
43. Gruber, B.; König, B. Self-Assembled Vesicles with Functionalized Membranes. *Chem. Eur. J.* **2012**, *19*, 438–448. [CrossRef] [PubMed]
44. Duss, M.; Vallooran, J.J.; Salvati Manni, L.; Kieliger, N.; Handschin, S.; Mezzenga, R.; Jessen, H.J.; Landau, E.M. Lipidic Mesophase-embedded Palladium Nanoparticles: Synthesis and Tunable Catalysts in Suzuki-Miyaura Cross Coupling Reactions. *Langmuir* **2019**, *35*, 120–127. [CrossRef] [PubMed]
45. Inoue, T.; Muraoka, Y.; Fukushima, K.; Shimozawa, R. Interaction of surfactants with vesicle membrane of dipalmitoylphosphatidylcholine: Fluorescence depolarization study. *Chem. Phys. Lipids* **1988**, *46*, 107–115. [CrossRef]
46. Batna, A.; Spiteller, G. Oxidation of furan fatty acids by soybean lipoxygenase-1 in the presence of linoleic acid. *Chem. Phys. Lipids* **1994**, *70*, 179–185. [CrossRef]
47. Samarkina, D.A.; Gabdrakhmanov, D.R.; Lukashenko, S.S.; Khamatgalimov, A.R.; Kovalenko, V.; Zakharova, L.Y. Cationic amphiphiles bearing imidazole fragment: From aggregation properties to potential in biotechnologies. *Colloids Surf. A Physicochem. Eng. Asp.* **2017**, *529*, 990–997. [CrossRef]

Article

Electrochemical DNA Sensor Based on the Copolymer of Proflavine and Azure B for Doxorubicin Determination

Anna Porfireva [1] and Gennady Evtugyn [1,2,*]

1 A.M. Butlerov' Chemistry Institute of Kazan Federal University, 18 Kremlevskaya Street, 420008 Kazan, Russia; porfireva-a@inbox.ru

2 Analytical Chemistry Department of Chemical Technology Institute of Ural Federal University, 19 Mira Street, 620002 Ekaterinburg, Russia

* Correspondence: Gennady.Evtugyn@kpfu.ru

Received: 6 April 2020; Accepted: 8 May 2020; Published: 10 May 2020

Abstract: A DNA sensor has been developed for the determination of doxorubicin by consecutive electropolymerization of an equimolar mixture of Azure B and proflavine and adsorption of native DNA from salmon sperm on a polymer film. Electrochemical investigation showed a difference in the behavior of individual drugs polymerized and their mixture. The use of the copolymer offered some advantages, i.e., a higher roughness of the surface, a wider range of the pH sensitivity of the response, a denser and more robust film, etc. The formation of the polymer film and its redox properties were studied using scanning electron microscopy and electrochemical impedance spectroscopy. For the doxorubicin determination, its solution was mixed with DNA and applied on the polymer surface. After that, charge transfer resistance was assessed in the presence of $[Fe(CN)_6]^{3-/4-}$ as the redox probe. Its value regularly grew with the doxorubicin concentration in the range from 0.03 to 10 nM (limit of detection 0.01 nM). The DNA sensor was tested on the doxorubicin preparations and spiked samples mimicking blood serum. The recovery was found to be 98–106%. The DNA sensor developed can find application for the determination of drug residues in blood and for the pharmacokinetics studies.

Keywords: electropolymerization; poly(Azure B); poly(proflavine); DNA sensor; doxorubicin determination; electrochemical impedance spectroscopy

1. Introduction

Electropolymerization is an advanced tool of modern electrochemistry that is frequently used for the modification of the electrodes in the assembly of electrochromic devices and electrochemical biosensors [1,2]. In these reactions initiated by primary electron transfer, oligomeric products are formed and deposited on the electrode as a dense uniform film, whose properties are derived from those of the monomers [3–5]. In most cases, electropolymerized products play role of heterogeneous mediators of the electron transfer or of the matrix wiring the bioreceptors and the nanomaterials (metals, carbon nanoparticles) [6–8] and providing the immobilization of biochemical components [9,10]. They offer many advantages over conventional modifiers used for the same purpose. Among them, one-step synthesis, controlled electrochemical activity, simple modification with various functional groups, and quantification of the electrodeposition by the current or the charge transferred are mentioned. All the polymers obtained by the electrolysis are divided into three groups, i.e., electroconductive, electrochemically active, and inactive polymers. Polyaniline [11,12], polypyrrole [13], polythiophene [14], and their derivatives are mostly used in electrochemical sensors and biosensors as electroconductive polymers.

Polymeric forms of the phenazine and phenothiazine dyes have found application among electrochemically active polymers [15–17] and some functionalized polyphenols as inactive coatings.

Methylene blue and Neutral red in polymeric form have been described in the assembly of the electrochemical sensors for the determination of many small organic molecules interesting for medicine, pharmacy, and environmental monitoring, e.g., catechin [18], nevirapine [19], paracetamol [20], vanillomandelic and homovanillic acids [21], catechol and hydroquinone [22], and ascorbic acid [23]. Methylene blue has found a broad application as a redox probe and diffusionally-free indicator in DNA- and aptasensors [24,25]. Neutral red was also implemented in the biosensor assembly for mediation of the electron transfer [26–28]. Although they show satisfactory characteristics of the electron transfer, their application can be limited by some drawbacks, e.g., very high non-specific adsorption of the Methylene blue and low selectivity of mediation by the Neutral red.

Recently, other derivatives of the phenazine dyes (Azure A, Azure B) and proflavine able to change their electrochemical properties in the presence of the proteins and DNA have been introduced in the electroanalysis [29–32]. Thus, proflavine, an acridine dye, was successfully used for the detection of the DNA hybridization events due to the ability to intercalate double-stranded DNA helix and to influence the electron exchange conditions on the electrode interface [33]. Additionally, it was used for the assessment of the DNA melting point changed due to the ligand binding [34]. Azure B was electropolymerized on the Pt and glassy carbon electrode (GCE) in acidic media [30] and polycrystalline Au [35] showing reversible redox behavior both in monomeric and polymeric form. The only examples of the application of the polymeric proflavine and Azure B in the assembly of the DNA sensors were done in our previous works [29,31], where rather low efficiency of the electropolymerization referred to the low solubility of the monomers in neutral media.

In this work, a mixed copolymer of the proflavine and Azure B has been obtained for the first time and applied for the doxorubicin determination. It was shown that the use of the copolymer changed the mechanism of the signal formation and the permeability of the surface layer became more important than the electrostatic interactions in the doxorubicin determination. The DNA sensor developed made it possible to detect 0.01 nM doxorubicin with a satisfactory recovery demonstrated in artificial blood plasma.

2. Materials and Methods

2.1. Reagents

Azure B (3-(dimethylamino)-7-(methylamino)phenothiazin-5-ium chloride, 97%), proflavine hydrochloride (3,6-Diamino-10-methylacridinium chloride, 95%), doxorubicin hydrochloride ((8S,10S)-6,8,10,11-tetrahydroxy-8-(hydroxyacetyl)-1-methoxy-7,8,9,10-tetrahydrotetracene-5,12-dione, 97%), low molecular double-stranded DNA from salmon sperm (average mol. mass 4.6 kDa [36]) HEPES (4-(2-hydroxyethyl)-1-piperazineethanesulfonic acid), hydroquinone, and bovine serum albumin (BSA) were purchased from Sigma-Aldrich, Dortmund, Germany (https://www.sigmaaldrich.com/catalog). Other reagents were of analytical grade. Deionized Millipore® water (Simplicity® water purification system, Merck-Millipore, Molsheim, France, https://www.merckmillipore.com/) was used for the preparation of working solutions. The pH dependence of the polymer coating properties was monitored using Britton–Robinson buffer consisting of 0.04 M H_3PO_4, 0.04 M H_3BO_3, 0.04 M CH_3COOH, 0.05 M Na_2SO_4). The DNA stock solutions (1 or 10 mg/mL) were prepared in 0.1 M HEPES containing 0.03 M NaCl, pH 7.0. Electropolymerization was performed in 0.025 M phosphate buffer (PB) containing 0.1 M KCl. Ringer–Locke's solution was used for mimicking influence of blood electrolytes. It contained 0.45 g NaCl, 0.021 g KCl, 0.016 g $CaCl_2 \cdot 2H_2O$, 0.005 g $NaHCO_3$, 0.075 g of glucose, 0.015 g of $MgSO_4$, and 0.025 g of $NaH_2PO_4 \cdot 2H_2O$ per 50 mL of water [37]. Doxorubicin preparations LANS® ("LANS-Verofarm", Belgorod, Russia, https://products.veropharm.ru/en) and TEVA ("Teva Pharmaceutical Industries", Petah Tikva, Israel, https://tevapharm.com) were purchased in the local pharmacy market.

2.2. Apparatus

Voltammetric and impedimetric measurements were conducted at ambient temperature with the potentiostat-galvanostat AUTOLAB PGSTAT 302N (Metrohm Autolab b.v., Utrecht, The Netherlands, https://www.metrohm.com/en/products/electrochemistry) equipped with the FRA2 module. The electropolymerization and the DNA deposition were performed using the GCE (1.67 mm^2) as working electrode, Pt wire as auxiliary electrode and Ag/AgCl/3 M KCl as reference electrode (Metrohm Autolab b.v.). The electrochemical impedance (EIS) spectra were recorded at the equilibrium potential with the amplitude of applied sine potential of 5 mV and the frequency varied in the range from 100 kHz to 0.04 Hz in the presence of 0.01 M $K_3[Fe(CN)_6]$ and 0.01 M $K_4[Fe(CN)_6]$. The equilibrium potential was calculated as the half-sum of the cathodic and anodic peak potentials of the $[Fe(CN)_6]^{3-/4-}$ peak pair. The impedance parameters were determined by fitting with the equivalent circuit $(R_{ct}C)_1(R_{ct}C)_2$, where R_{ct} is the charge transfer resistance and C the constant-phase element representing the non-ideal capacitance behavior. Indices *1* and *2* correspond to the outer and inner interfaces (electrode–film and film–solution). Equivalent circuit fitting was performed with the NOVA software (Metrohm Autolab b.v.).

The pH measurements were performed with the EXPERT-001-1 digital pH meter-ionometer (Econix-Expert Ltd, Moscow, Russia, http://ionomer.ru/index.php?lang=english).

Scanning electron microscopy (SEM) images were recorded on a field emission scanning electron microscope Merlin™ (Zeiss, Jena, Germany, https://www.zeiss.com/microscopy/int/products/scanning-electron-microscopes.html). The films obtained by electropolymerization were preliminarily coated with the Au/Pd layer in vacuum by a T150ES sputter coater (Quorum Technologies Ltd, Laughton, United Kingdom, https://www.quorumtech.com).

2.3. Electropolymerization of Azure B and Proflavine and DNA Sensor Preparation

The GCE was first mechanically polished to a mirror-like surface and cleaned with acetone and sulfuric acid. Then, it was immersed in the solution obtained by mixing 4.6 mL of 0.025 M PB, 385 µL of Azure B (1.03 mg/mL), and 22 µL of proflavine (15 mg/mL). Resulting concentrations of the dyes were equal to 0.25 mM of each. Then, 20 cycles of the potential in the range from −0.4 to 1.2 V were run with the scan rate of 100 mV/s. After that, the electrode was rinsed with deionized water. After drying, the electrode was fixed upside down and 2 µL of DNA solution (1 mg/mL) in 0.1 M HEPES containing 0.03 M NaCl, pH = 7.0, were spread on the working area. After drying, the electrode was washed several times with the PB and deionized water and used for electrochemical measurements.

2.4. Doxorubicin Determination

The doxorubicin solution was mixed with the 10 mg/mL DNA solution in the 9:1 ratio (*v*:*v*). Then, 2 µL of the mixture were placed on the working surface of the GCE covered with the Azure B–proflavine copolymer and dried at ambient temperature. The electrode was washed with deionized water to remove unbound reactants and transferred to the cell containing 4.5 mL of 0.025 M PB with 0.1 M KCl, pH = 7.0, and 0.5 mL of 0.1 M equimolar mixture of $K_3[Fe(CN)_6]$ and $K_4[Fe(CN)_6]$. After magnetic stirring, the EIS spectrum was recorded and the Nyquist diagram plotted to determine the charge transfer resistance as a measure of the doxorubicin concentration.

3. Results

3.1. Copolymerization of Azure B and Electrochemical Properties of the Polymerization Product

The cyclic voltammograms obtained in a multiple cycling of the potential in the mixture of monomeric Azure B and proflavine are presented in Figure 1. Most changes on the voltammograms have been finished to the 20th cycle. Thus, in the following experiments, the electropolymerization was performed using this number of potential cycles.

At the first scan, high anodic peak (a_1 on voltammogram) appears at −0.13 V. In the second cycle, it was decreased about twice and shifted to less negative potential (−0.04 V). Meanwhile, the second anodic peak (a_2) appeared at −0.20 V. In the following scans, anodic peak (a_1) disappeared and the second one was shifted to 0.35 V and insignificantly decreased in height.

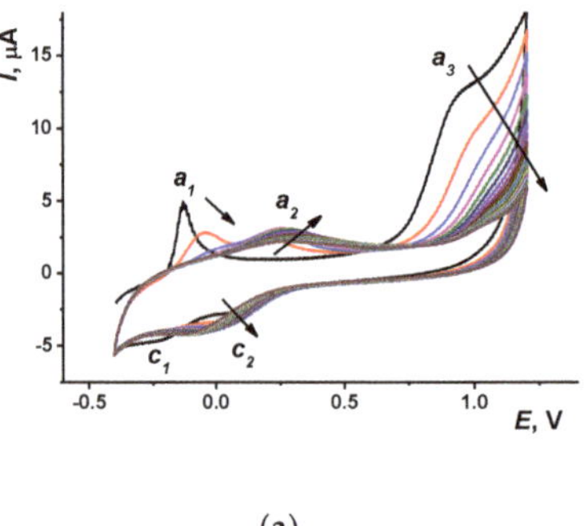

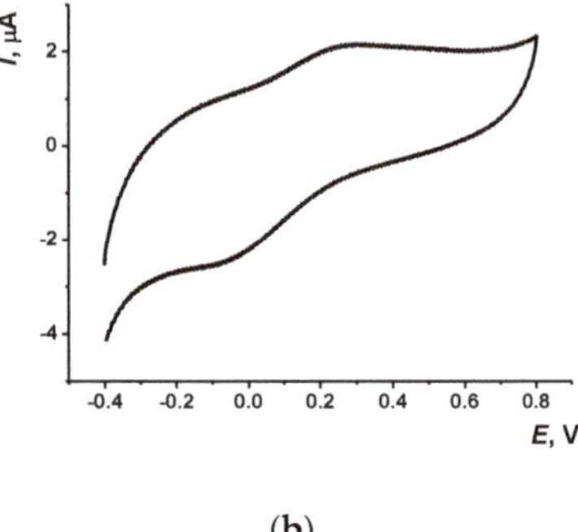

(**a**) (**b**)

Figure 1. (**a**) Multiple cyclic voltammograms recorded on glassy carbon electrode in 0.025 M PB, pH = 7.0, containing 0.03 M NaCl, 0.25 mM Azure B, and 0.25 mM proflavine, scan rate 100 mV/s (20 cycles). Arrows indicate direction of the changes observed with increasing number of the cycles. (**b**) Cyclic voltammograms recorded with glassy carbon electrode covered with Azure B–proflavine copolymer (20 cycles) in 0.025 M PB with no monomers.

In the opposite direction of the potential scan, changes in the peaks recorded in the same potential range were quite different. On the first scan, two small cathodic peaks appeared on voltammogram, from which one (c_1) derived from the peak a_1 and second one (c_2) was probably related to the anodic peak a_2 though it appeared on the next cycles. Additionally, these two pairs of reversible peaks, a high irreversible anodic peak was observed at the potentials higher than 0.93 V. This peak decreased in height and shifted to about 1.0 V to the 20th cycle. As for other phenothiazine dyes, this peak can be attributed to the formation of di-cation initiating the polymerization of the monomeric dye [15]. Indeed, if the cyclic voltammograms were recorded in the range from −0.3 to 0.5 V, no changes in the position and height of reversible peaks were observed indicating no polymerization of the monomers. Moreover, transfer of the electrode in a fresh PB with no monomers resulted in sharp decrease of the above peaks down to the background currents typical for supporting electrolyte. Thus, the adsorption of monomeric dyes on bare GCE is rather weak and reversible and cannot result in reproducible peaks on appropriate voltammograms. After four cycles of the potential, two peaks remained on the voltammogram (E_{pc} = −0.07 V and E_{pa} = −0.27 V), in which the height slowly increased in the following cycling of the potential.

It is interesting to compare the peaks with those of individual dyes in similar experimental conditions. Previously we reported about electropolymerization of Azure B and proflavine taken alone in the same range of potentials. Azure B was oxidized in 0.1 M HEPES, pH = 6.9, with formation of the broaden peak pair at −0.11 and −0.09 V and of one irreversible anodic peak at the potentials higher than 0.8 V [29]. However, the Azure B peak pair at low potentials remained on voltammograms during the whole period of potential cycling (20 cycles) whereas the anodic irreversible peak was decreased with the number of cycles similarly to that observed here for the mixture of dyes. Regarding proflavine [31], it formed quasi-reversible pair of the peaks with the peak potential difference (0.22 and −0.13 V) higher than that discussed above for the Azure B polymer and the Azure B–proflavine copolymer. In this pair, cathodic peak was slowly increased with the number of potential cycles and anodic one was quite stable. The behavior of irreversible anodic peak initiating polymerization was similar to that reported for individual dyes polymerized in similar conditions.

The comparison of the voltammograms obtained earlier for individual dyes and that, in their mixture, made it possible to conclude that the polymerization starts with the Azure B, which anodic peaks dominated in first 2–3 cycles. The implementation of proflavine became significant in the next cycles and resulted in general simplification of the voltammograms and in the shift of the peaks in the low potential range to higher potentials. The coverage of the surface with the polymer was mainly finished after the fourth cycle. The following deposition of the polymer increased the thickness of the layer and suppressed the access of both monomers to the electrode surface. As a result, the electron exchange became less effective and the equilibrium potential was slightly shifted to its higher value. The involvement of both monomers is confirmed by appropriate peak potentials and existence of two peaks referred to individual products at first cycles of the potential.

The resulting film is rather stable and exerts redox activity being transferred in a fresh PB with no monomers (Figure 1b). On this voltammogram, the anodic peak (0.28 V) is closer to that of the proflavine and the cathodic peak (−0.08 V) to that of poly(Azure B). The voltammograms of all three reactants, i.e., Azure B, proflavine, and their mixture, are rather close to each other so that variation in the molar ratio of the dyes in the mixture did not result in quantitative changes of appropriate voltammograms. The ratio used in this work corresponded to the maximal difference in the peaks related to individual dyes and their mixture. The only conclusion was that lower quantities of Azure B decreased the relative height of the peak a_1 at first scan. This confirms the attribution of the peak to this monomer.

The stability of the coating was confirmed in the series of voltammetric measurements with individual sensors in the PB with no monomers. The range of the potential scan (from −0.4 to 0.8 V) excluded polymerization of the monomers that could be entrapped in the surface film. Contrary to electropolymerization, the runs of the potential were separated by a certain period of switched-off voltage (5 min.), in which the solution was magnetically stirred. The anodic peak current was stabilized after the 2–3 cycle, whereas the cathodic current with no stirring was constant starting from the first run (Figure 2b).

In addition to the peak currents, the charge passed was determined by integration of the current–voltage curves. As could be seen from the Figure 2c, the charge passed was stabilized to the third measurement with no stirring. However, magnetic stirring resulted in a regular increase of the charge during the whole measurement series. This might be due to the elimination of small particles of the coating that are weakly attached to the surface and leave them between the measurements. This would increase the specific surface of the electrode interface. The charge passed also tends to increase with the scan rate to the limit similar to that reached in consecutive potential cycling (compare Figure 2c,d). At a low scan rate, oxidized forms of the dyes preferably determine the charge passed, which is negative and corresponds to a predominantly cathodic process of the oxidized dyes reduction. This coincides well with the fact that the dyes are commonly present in the oxidized (salt) form in aqueous solutions containing dissolved oxygen. However, at a scan rate higher than 40 mV/s, the charge passed becomes positive indicating the prevalence of the anodic reactions over the cathodic ones. This can result from the formation of rather dense film and low rate of electron exchange in the polymer layer. In such conditions, changes in the overall charge passed are less compensated for by the movements of the negatively-charged counter ions and, hence, less affect the charge passed in the cathodic branch of the voltammogram.

The electrode modified with the Azure B–proflavine copolymer was tested after stabilization at various scan rates to determine the nature of the rate-limiting stage. In the bilogarithmic plots, the slope of the linear graph was found in-between 0.5 (surface confined reactions) and 1.0 (diffusion limitation) [38] in the range from 10 to 500 mV/s. Appropriate regression equations are presented in Equations (1) and (2). Here, I_{pa} and I_{pc} are the anodic and cathodic peak currents, respectively, and ν is the potential scan rate.

$$\text{Anodic peak current: } \log(I_{pa}, \mu\text{A}) = (0.77 \pm 0.01) + (0.67 \pm 0.01) \times \log(\nu, \text{mV/s}),\ R^2 = 0.9961,\ n = 14 \quad (1)$$

$$\text{Cathodic peak current: } \log(I_{pc}, \mu\text{A}) = (1.04 \pm 0.01) + (0.64 \pm 0.01) \times \log(\nu, \text{mV/s}),\ R^2 = 0.9946,\ n = 14 \quad (2)$$

The dependence of the peak currents on the scan rate was investigated in the absence of the monomers in solution so that diffusional transfer of redox species was impossible. In these conditions, the decrease of the slope against the value typical for the surface reactions (1.0) can be referred to the influence of the counter ions transfer mentioned above or to the slow electron exchange between the oxidized and reduced monomer items within the layer. Similar behavior was found earlier for the polymeric Neutral red [39]. The transfer coefficient α = 0.83 was determined using Laviron's theory from the dependence of the peak potential on the scan rate (Equation (3)) [40,41]:

$$E_{pc} = E^{0'} + \frac{RT}{\alpha nF} \ln \frac{RTk_{et}}{\alpha nF} - \frac{RT}{\alpha nF} \ln v \tag{3}$$

Here, $E^{0'}$ is the formal redox potential, α is the transfer coefficient, n is the number of the electrons transferred, F is the Faraday constant, R is the universal gas constant, T is the temperature, K, and k_{el} is the heterogeneous constant of the electron transfer.

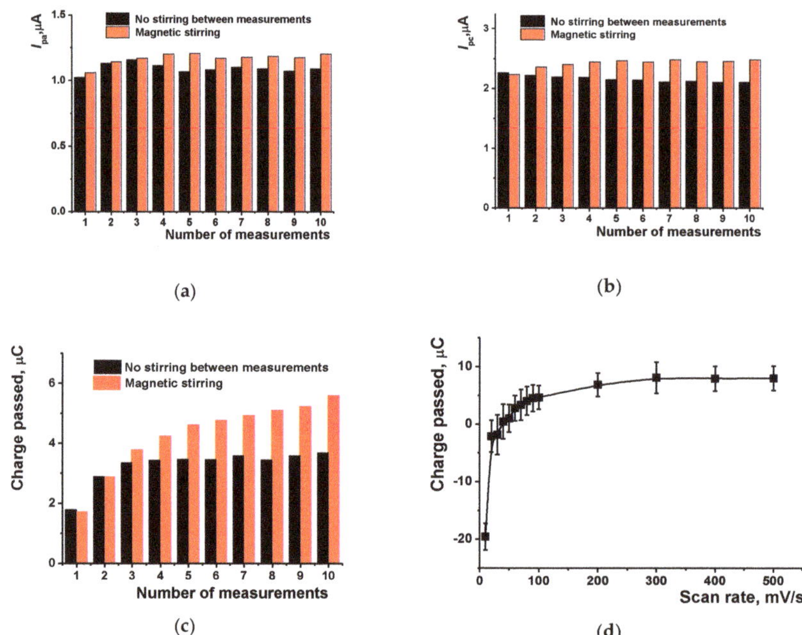

Figure 2. The dependency of cathodic (**a**), anodic (**b**) peak currents and charge passed (**c**) on the number of consecutive measurements separated by magnetic stirring of solution or by its equalization in open circuit mode. 0.025 M PB, pH = 7.0, scan rate 100 mV/s; (**d**) The dependence of the charge passed in the whole potential cycle between −0.4 and 0.8 V on the scan rate.

3.2. The Comparison of the Redox Properties of Poly(Azure B), Poly(proflavine) and Their Copolymer

Although the behavior of the dyes used was very similar to each other, there is a difference, which can be related to the intrinsic processes of the electron exchange and counter ions transfer. To simplify the consideration, some of the changes observed are summarized in the Table 1 based on this work and our previous investigations performed in similar experimental conditions.

Among other properties, the pH dependence of the equilibrium potential is most sensitive to the monomer mixing. All three coatings mentioned in Table 1 exert a linear response toward pH according to the Nernst equation with the slopes of 29 mV/pH for poly(Azure B) and 59 mV/pH for the poly(proflavine) that correspond to the transfer of one H^+ ion, two electrons, and two H^+, two electrons, respectively. In case of the copolymer, the pH dependence is more complicated. Cathodic peak potential shows the pH dependence similar to that of Azure B (slope 53 mV/pH), whereas the anodic dependence formally corresponds to the transfer of 1.5 H^+ ions per electron. Appropriate slope (43 mV/pH) is in between those obtained for the poly(Azure B) and poly(proflavine). In the same dependencies involving equilibrium potential corresponded to the half-sum of peak potentials, the copolymer showed maximal range of linearity (pH from 2.0 to 9.0), whereas the poly(proflavine) changed this potential only in basic media (pH from 6.0 to 9.0) and that of poly(Azure B) in weakly acidic conditions (pH from 3.0 to 6.0). It can be concluded that the copolymer combines the pH sensitivity areas of individual dyes and exerts better reversibility of electron/H^+ exchange. The nominal transfer of 1.5 H^+ per electron can be explained by simultaneous redox reactions of both monomers resulting in averaging of the stoichiometry of equilibrium.

Table 1. The comparison of the properties of electropolymerized coatings of the phenazine dyes obtained by multiple cycling of the potential.

Property	Poly(Azure B) [29]	Poly(proflavine)	Copolymer [31]
Stability of redox parameters	The currents regularly decrease in consecutive measurements	Stable	Stable to 2–3 cycle
logI_p-logv slope	Oxidation: 1.1 (polymer), 0.83 (monomer)	Oxidation: 0.81	Oxidation: 0.67
	Reduction: 0.96 (monomer)	Reduction: 0.64	Reduction: 0.64
Transmission coefficient	0.56 (polymer)	0.57	0.83

3.3. DNA Deposition on the Copolymer of Azure B and Proflavine

Polyelectrolyte complexes of the redox active polymers and DNA have been successfully applied for the detection of specific biochemical interactions of the DNA influencing redox equilibria of the support [3]. For this purpose, polyaniline [42], poly(Neutral red) [27,43], poly(Methylene blue) [43], poly(Methylene Green) [43], poly(Azure B) [29], and poly(proflavine) [31] have been used. In this work, the DNA solution was applied on the working surface of the electrode covered with the copolymer and either dried or left capped with plastic tube preventing drying for a certain time. In voltammetric experiments, the contact with the DNA resulted in 10–15% decrease of the oxidation peak current irrespective of the DNA quantities and application protocol. However, incubation resulted in much higher deviation of the signal. For this reason, drying DNA solution was used in other experiments described. Taking into account low sensitivity of the voltammetry to such changes of the surface layer, the deposition of DNA was confirmed by SEM and EIS.

3.3.1. SEM Monitoring of the Surface Layer Assembling

Figure 3 represents the morphology of the electrode surface on the stages of the electropolymerization and the DNA casting.

Polished glassy carbon showed smoothen surface with randomly positioned mechanical scratches. Deposition of the copolymer resulted in the formation of rough surface covered with the angular fragments. Their main fraction (60%) has the size of 40–55 nm. In comparison with individual dyes, which electropolymerization was studied earlier, the surface morphology is similar to that of proflavine. The latter one also forms aggregates covered with roundish particles [29], whereas poly(Azure B) had a uniform structure with elongated parallel inclusions [31].

Deposition of the DNA onto the copolymer surface changed its morphology. First, angular fragments distinguishable on the surface folds, disappear. Instead of them, well-defined roundish DNA aggregates with diameters of about 25–30 nm fill the hollows and the surface of the film. They are much better discernible than those described on the poly(proflavine) film and are about twice smaller.

(a) (b) (c)

Figure 3. SEM microimages of bare glassy carbon electrode (**a**), of the copolymer of Azure B and proflavine prior to (**b**) and after DNA adsorption (**c**).

Disappearance of the fragments visible on the copolymer film prior to the DNA application can be referred to their leaching from the surface to the solution. Negatively charged DNA molecules promote such a leaching for the weakly attached particles due to neutralization of their positive charge by the negatively-charged phosphate groups of the DNA skeleton. Different aggregation of the DNA molecules adsorbed on the surface of the poly(proflavine) and of the copolymer of Azure B and proflavine can be explained by the difference in the specific surface charge and in the hydrophobicity of both coatings.

3.3.2. EIS Measurements

EIS is a powerful tool of electrochemistry that is used for investigation of the charge transfer and for the monitoring of the interactions influencing the efficiency of electron exchange on the electrode interface. The Nyquist diagrams were obtained using 10 mM $[Fe(CN)_6]^{3-/4-}$ as redox probe at the mid-point potential calculated as the half-sum of the peak potentials (Figure 4). Here, R is the charge transfer resistance and C constant phase element, indices *1* and *2* correspond to the internal and external interface of the electrode-polymer layer. Roughness coefficient n specifying a non-ideal capacitance behavior of the constant phase element was found to be 0.78–0.90. Figure 3 represents the morphology of the electrode surface on the stages of the electropolymerization/DNA casting.

The Nyquist diagram contains two semi-circles corresponding to the electrode–film (smaller one) and film–solution (larger one) interfaces. Most changes observed during the surface layer assembling related to the charge transfer resistance of the outer interface. This coincides well with the accessibility of this interface to the reactant addition and hence higher sensitivity to various stages of the layer assembling in comparison with the inner interface electrode—polymer, which remains about constant within the whole series of measurements.

Electropolymerization of the Azure B and proflavine increased the R_2 value from 4 ± 1 to 130 ± 5 kΩ. Such changes can be attributed either to lesser electrostatic attraction of the oppositely-charged redox probe and cationic surface or to the denser coating (lower diffusion coefficient) of the ferricyanide ions penetrating the film. The increase in the number of the potential cycles and anodic polarization at the stage of the electrode modification did not significantly alter this value. This makes it possible to conclude that slower diffusion appeared to be more important than electrostatic interactions. The following application of the DNA molecules increased the R_2 value to 170 ± 3 kΩ. It is interesting to note that the treatment of the DNA with doxorubicin, an anthracycline dye intercalating the DNA helix, increased this value to 215 ± 8 kΩ. The capacitance remains about constant during the DNA application and doxorubicin introduction. Its small value (about 1 μF) confirms a low charge separation on the electrode interface and the suggestion about the predominant influence of the diffusion factors on the EIS parameters. Doxorubicin molecules partially compensate for the negative charge of the phosphate residues in the DNA backbone and make weaker electrostatic interactions between them and the $[Fe(CN)_6]^{3-/4-}$ ions of the redox probe.

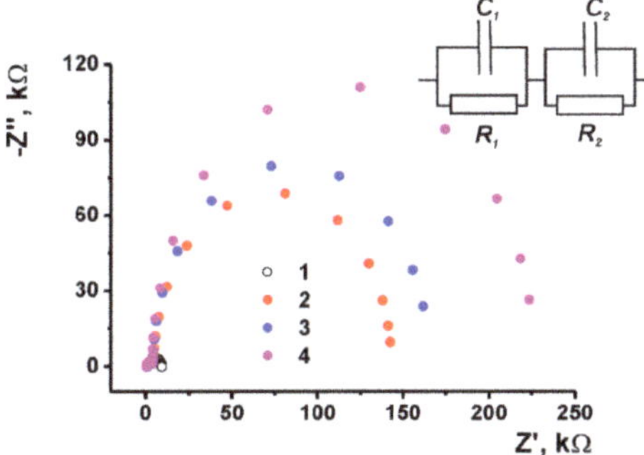

Figure 4. The Nyquist diagram obtained with the glassy carbon electrode (1) covered with poly(Azure B–proflavine) prior to (2) and after (3) application of DNA (2 μL of 1.0 mg/mL solution in 0.1 M HEPES + 0.03 M NaCl, pH = 7.0) or DNA-1.0 nM doxorubicin aliquot (4). Measurements in 0.025 M PB + 0.1 M KCl, pH = 7.0, in the presence of 10 mM $[Fe(CN)_6]^{3-/4-}$. Inset: equivalent circuit, *C*—constant phase element, *R*—charge transfer resistance.

Thus, the modification protocol used provides the deposition of the polymeric film and adsorption of the DNA molecules, which affect the behavior of the sensor due to the variation of the permeability of the surface layer for small ions of redox probe.

3.4. Determination of Doxorubicin

3.4.1. Doxorubicin Oxidation on Glassy Carbon Prior to and After the Electropolymerization Stage

In accordance with the literature data [44,45], hydroquinone and benzoquinone units of doxorubicin are involved in electrode reaction with formation of separated signals. The largest one is commonly observed at negative potentials (from −0.5 to 0.65 V depending on the electrode material) and another one, much lower, at positive potential (0.4–0.6 V). This information coincides well with the results obtained on bare GCE (Figure 5).

With no modifier, an irreversible cathodic peak at −0.45 V and one irreversible anodic peak at 0.55 V appear on the voltammogram. They are regularly increasing with the doxorubicin concentration. However, after the electrode modification, doxorubicin does not affect the signals of the underlying copolymer within the concentration range of five orders of magnitude (Figure 5b). Probably, the molecules of doxorubicin cannot reach the electrode or compete with the electron exchange chain within the surface layer. This might result from hydrophobicity of the analyte or steric limitation of its adsorption on the polymer layer.

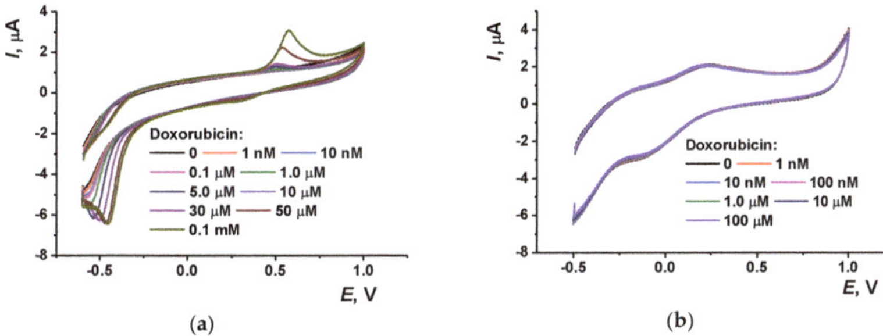

Figure 5. Cyclic voltammograms of doxorubicin recorded on glassy carbon, bare (**a**) and modified with copolymer of Azure B and proflavine (**b**). 0.025 M PB + 0.1 M KCl, pH = 7.0, scan rate 100 mV/s.

3.4.2. Determination of Doxorubicin with DNA Sensor Based on Copolymer of Azure B and Proflavine

The investigations were continued in the EIS mode, which is more sensitive than cyclic voltammetry to the surface-confined reactions. Doxorubicin intercalates the DNA helix with inclusion between the pairs of nucleobases. This reaction changes the volume and conformation of the DNA molecules and partially compensates for the negative charge of the DNA and its ability for electrostatic attractions with the positively-charged moieties. However, incubation of the electrode covered with the copolymer Azure B–proflavine and adsorbed DNA did not result in an increase of the charge transfer resistance. Contrary to that, the R_2 value decreased to the level typical for the copolymer prior to the DNA application. This might be due to the lesser repulsion of the $[Fe(CN)_6]^{3-/4-}$. For this reason, the following experiments were performed with preliminary treatment of the DNA with the doxorubicin solution performed prior to the DNA application on the electrode surface. After 20 min. of incubation, the mixture was applied on the electrode covered with the copolymer. After drying, the electrode was carefully washed with deionized water and then immersed in the working PB for EIS measurements. As was mentioned previously, all the changes of the surface layer affected mostly the impedimetric parameters on outer interface. The charge transfer resistance linearly increased with the doxorubicin concentration in the range from 10 nM to 0.03 nM. Figure 6 shows appropriate changes of the Nyquist diagram.

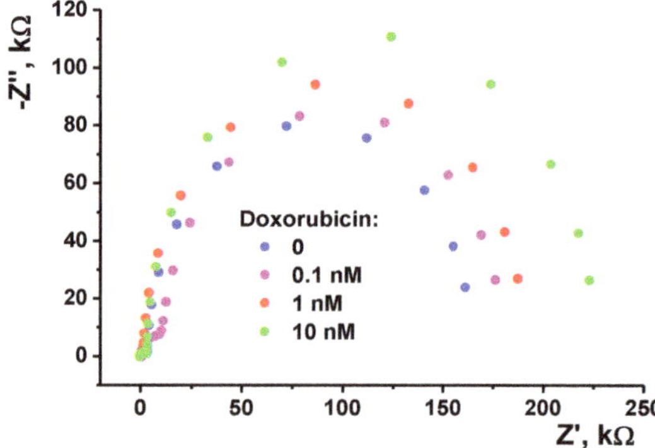

Figure 6. The Nyquist diagram obtained on the glassy carbon electrode covered with poly(Azure B–proflavine) and DNA previously mixed with doxorubicin solution (2 μL of the mixture containing 1.0 mg/mL DNA in 0.1 M HEPES + 0.03 M NaCl, pH = 7.0). Measurements in 0.025 M PB + 0.1 M KCl, pH = 7.0, in the presence of 10 mM $[Fe(CN)_6]^{3-/4-}$.

Thus, the sensitivity of the EIS parameters toward doxorubicin was much higher than that of voltammetry. The linear range of the concentrations determined was 3×10^{-11}–1×10^{-8} M (calibration Equation (4)):

$$R_{ct}, \text{k}\Omega = (380 \pm 12) + (21 \pm 1) \log(c, \text{M}), R^2 = 0.9673, n = 5 \quad (4)$$

In the same concentration range, the capacitance of the outer interface increases with the doxorubicin concentration from 1.3 to 1.9 μF and remains constant after reaching 1 nM concentration. The limit of detection (LOD) corresponded to the S/N = 3 criterion was found to be 1×10^{-11} M. These characteristics are similar or better than those of other electrochemical sensors and DNA sensors described for doxorubicin determination. The comparison of the characteristics of such sensors is summarized in Table 2.

Table 2. Analytical characteristics of the determination of doxorubicin with electrochemical sensors and DNA sensors.

Surface Layer/Electrode	Concentration Range	LOD, nM	Ref
Electrochemical sensors			
Carbon nanotubes	20–500 nM	-	[46]
Ionic liquid/ZnO in carbon paste	0.07–5000 µM	9.0	[47]
Mesoporous carbon nanospheres/reduced graphene oxide	10 nM–10 µM	1.5	[44]
Basal plane pyrographite	0.01–1 µM	10	[45]
Fe_2Ni@Au/reduced graphene oxide	5.5–9.2 µM	1460	[48]
Electrochemical DNA sensors			
Poly(Azure B)/DNA	0.1 µM–0.1 nM	0.07	[29]
Poly(proflavine)/DNA	1 nM–0.1 µM	0.3	[31]
Polyaniline/DNA	1.0 pM–1 mM	0.0006	[49]
Carbon nanotubes/polylysine/DNA	2.5 nM–0.25 µM	1.0	[50]
Poly(Neutral red)/DNA	0.01–100 µM	0.1	[51]
Poly(Azure B–proflavine)	0.03–10 nM	0.01	This work

The only DNA sensor that showed higher sensitivity was comprised of the polyaniline with the physically adsorbed DNA. The advantage in its characteristics can be referred to a higher level of the redox activity of the polymer and higher sensitivity of its redox properties toward microenvironment than those of polyphenothiazines. Meanwhile, the electropolymerization of aniline requires strongly acidic media and is sensitive to the presence of the oxygen that can partially suppress the electroconductivity of the polymer formed. For this reason, only freshly distilled aniline is used in the electropolymerization. The incubation performed at pH 3.0 can also negatively affect the results if complex media is analyzed.

For the selectivity assessment, the impedimetric signal of 1.0 nM doxorubicin was measured in the presence of 10 µM bovine serum albumin and 1.0 µM sulfamethoxazole. The variation of the charge transfer resistance averaged for six individual DNA sensors did not exceed 6%. This coincides well with the results obtained earlier for the polyaniline based DNA sensors with impedimetric and voltammetric signals [3,49].

The results obtained make it possible to conclude that impedimetric measurements allow for the sensitive determination of doxorubicin in the sub-nanomolar range of its concentrations.

3.4.3. Measurement Precision and DNA Sensor Lifetime

Sensor-to-sensor repeatability calculated from the EIS data was equal to 5.5% (six individual sensors, 1.0 nM doxorubicin). Each sensor was used only once because of irregular changes in the sensitivity observed in the next attempts of the signal determination. The GCE covered with the copolymer of the Azure B and proflavine can be stored in dry conditions at 4 °C for at least six months. In the following DNA application and doxorubicin signal measurement, the deviation tends to increase to 10% toward the end of the storage period.

3.4.4. Real Sample Analysis

The DNA sensor developed was tested in the determination of the doxorubicin content in two medications, i.e., Doxorubicin-TEVA® and Doxorubicin-LANS® ((lyophilizates for intravascular injection solutions). In both cases, the medications were dissolved first in deionized water and then in working buffer solution and mixed in 9:1 ratio with 10.0 mg/mL DNA to final nominal concentration of 1.0 mg/mL (see Section 2.4 for details). Then the mixture was applied on the electrode surface and dried. Unbounded components were carefully removed with deionized water and PB and the EIS measurements were performed as described above. The recovery was assessed using the calibration TEVA, plot of the drug obtained in buffered media with standard drug solutions. For the doxorubicin-the recovery was equal to 99 ± 7% (six measurements) and that for the doxorubicin-LANS 102 ± 10%. It should be mentioned that both preparations contained stabilizers (lactose and mannite, respectively).

It is probably that the low activity of the polymer coating to oxidizable species showed, for doxorubicin itself, suppressed its influence on the biosensor signal.

In a similar manner, the influence of serum proteins (for the example of bovine serum protein, see Section 3.4.3) and plasma electrolytes (Ringer–Locke's solution) was estimated with 1 nM doxorubicin solution. The recoveries of 106 ± 8% and 102 ± 12% were found. Certainly, more attention should be paid to the assessment of the factors that affect the biosensor response with real blood samples. However, on this stage of "proof-of-concept" it can be concluded that the DNA sensor developed can find application for preliminary assessment of the concentration of doxorubicin in patients' serum after further validation.

4. Discussion

The electropolymerization of an equimolar mixture of the Azure B and proflavine made it possible to modify the electrode with a layer that provides both reliable adsorption of the native DNA and determination of its interaction with anthracycline dye. Contrary to similar protocols previously elaborated for individual dyes, the mixed composition offered some advantages, i.e., a denser and durable film with a rough surface capable of electron exchange and DNA implementation. The size distribution of the polymer and DNA adsorbed was different from the polymeric dyes described elsewhere by lower size of the DNA aggregates and their narrow distribution. The comparison of the cyclic voltammograms and the SEM images with those of poly(Azure B) and poly(proflavine) showed the participation of both components in the electron exchange and related H^+ transfer.

Doxorubicin is one of the anthracycline antibiotics active against solid tumors and hematological malignancies [52]. Its application is limited by rather high cardiotoxicity [53]. For this reason, its monitoring in the biological fluids and preparations is extremely important in chemotherapy for individual dose establishment. Additionally, doxorubicin has found a broad application as a standard intercalator in investigations of electrochemical DNA sensors. In addition to intercalation it is involved in the generation of reactive oxygen species that damage DNA and produce 8-oxoguanine as an indicator of oxidative damage [54]. For this reason, new systems of doxorubicin determination based on polymer-DNA films exerting own redox activity are important for both pharmaceutical applications and DNA biosensors progress.

For EIS measurements of DNA sensors, an $[Fe(CN)_6]^{3-/4-}$ redox probe is mostly used. Due to the negative charge and repulsion from phosphate residues of the DNA backbone, the impedimetric signal becomes sensitive to any biochemical interactions that take place on the electrode interface with DNA molecules. Two main mechanisms are mostly considered to explain changes in the charge transfer resistance of these biosensors contacted with target analytes, i.e., (i) changes in electrostatic interactions caused by shielding phosphate groups of the DNA, and (ii) changes in the permeability of the surface coating for small ions due to denser packing and deposition of non-conductive molecules interacting with DNA [54].

Considering the influence of doxorubicin on EIS parameters, one of the unexpected results was that electrostatic interactions mentioned played less significant role in the signal generation than usually for other DNA sensors based on the redox active polymers. Variation of the permeability of the surface layer for small ions (ferricyanide as redox probe) is most important factor explaining the performance of the biosensor. Changes in the DNA aggregation on the polymer film initiated by an intercalator could also suppress the transfer of the redox probe to the electrode but SEM data did not allow quantifying changes in the aggregation on the images.

Although the copolymer synthesized showed reversible redox behavior, its activity was found to be too small for effective participation in the electron transfer to the diffusionally-free reactants. The attempts to determine doxorubicin by its mediated oxidation on the modified electrode showed only minor variation of the current. However, this low activity can be considered as an advantage if the biosensor is applied for drug testing. Indeed, medications contain antioxidants to stabilize drugs and increase the drug storage period. In this work, no influence of such stabilizers was found for two different species (mannite and lactose). This makes biosensor also attractive for the drug residues detection in biological liquids with minimal sample treatment. Summarizing the specific properties of the DNA sensor, one could suppose, hydrophobicity of the electrode interface can be critical factor affecting the biosensor behavior and its high sensitivity toward doxorubicin.

Author Contributions: Conceptualization: G.E.; investigation: A.P.; writing—original draft preparation: G.E.; writing—review and editing: G.E.; supervision: G.E.; project administration: A.P. All authors have read and agreed to the published version of the manuscript.

Funding: This research was funded by Russian Science Foundation, grant no. 17-73-20024.

Acknowledgments: SEM images were obtained on the equipment of the Interdisciplinary Center for Analytical Microscopy of Kazan Federal University by V.V. Vorobev.

Conflicts of Interest: The authors declare no conflict of interest.

References

1. Palma-Cando, A.; Rendón-Enríquez, I.; Tausch, M.; Scherf, U. Thin functional polymer films by electropolymerization. *Nanomaterials* **2019**, *9*, 1125. [CrossRef]
2. Ates, M. A review study of (bio)sensor systems based on conducting polymers. *Mater. Sci. Eng. C* **2013**, *33*, 1853–1859. [CrossRef] [PubMed]
3. Evtugyn, G.; Hianik, T. Electrochemical DNA sensors and aptasensors based on electropolymerized materials and polyelectrolyte complexes. *TrAC Trends Anal. Chem.* **2016**, *79*, 168–178. [CrossRef]
4. Berkes, B.B.; Bandarenka, A.S.; Inzelt, G. Electropolymerization: Further insight into the formation of conducting polyindole thin films. *J. Phys. Chem. C* **2015**, *119*, 1996–2003. [CrossRef]
5. Inzelt, G. Rise and rise of conducting polymers. *J. Solid State Electrochem.* **2011**, *15*, 1711–1718. [CrossRef]
6. Omar, F.S.; Duraisamy, N.; Ramesh, K.R.S. Conducting polymer and its composite materials based electrochemical sensor for nicotinamide adenine dinucleotide (NADH). *Biosens. Bioelectron.* **2016**, *79*, 763–775. [CrossRef]
7. Runsewe, D.; Betancourt, T.; Irvin, J.A. Biomedical application of electroactive polymers in electrochemical sensors: A review. *Materials* **2019**, *12*, 2629. [CrossRef]
8. Tomczykowa, M.; Plonska-Brzezinska, M.E. Conducting polymers, hydrogels and their composites: Preparation, properties and bioapplications. *Polymers* **2019**, *11*, 350. [CrossRef]
9. Moro, G.; De Vael, K.; Moretto, L.M. Challenges in the electrochemical (bio)sensing of nonelectroactive food and environmental contaminants. *Curr. Opin. Electrochem.* **2019**, *16*, 57–65. [CrossRef]
10. Prajapati, D.G.; Kandasubramanian, B. Progress in the development of intrinsically conducting polymer composites as biosensors. *Macromol. Chem. Phys.* **2019**, *220*, 1800561. [CrossRef]
11. Singh, P.; Shukla, S.K. Advances in polyaniline-based nanocomposites. *J. Mater. Sci.* **2020**, *55*, 1331–1365. [CrossRef]
12. Zare, E.N.; Makvandi, P.; Ashtari, B.; Rossi, F.; Motahari, A.; Perale, G. Progress in conductive polyaniline-based nanocomposites for biomedical applications: A review. *J. Med. Chem.* **2020**, *63*, 1–22. [CrossRef] [PubMed]
13. Jaina, R.; Jadon, N.; Pawaiya, A. Polypyrrole based next generation electrochemical sensors and biosensors: A review. *TrAC Trends Anal. Chem.* **2017**, *97*, 363–373. [CrossRef]
14. Hui, Y.; Bian, C.; Xia, S.; Tong, J.; Wang, J. Synthesis and electrochemical sensing application of poly(3,4-ethylenedioxythiophene)-based materials: A review. *Anal. Chim. Acta* **2018**, *1022*, 1–19. [CrossRef]
15. Karyakin, A.A.; Karyakina, E.E.; Schmidt, H.-L. Electropolymerized azines: A new group of electroactive polymers. *Electroanalysis* **1999**, *11*, 149–155. [CrossRef]
16. Barsan, M.M.; Ghica, E.M.; Brett, C.M.A. Electrochemical sensors and biosensors based on redox polymer/carbon nanotube modified electrodes: A review. *Anal. Chim. Acta* **2015**, *881*, 1–23. [CrossRef]
17. Topçu, E.; Alanyalıoğlu, M. Electrochemical formation of poly(thionine) thin films: The effect of amine group on the polymeric film formation of phenothiazine dyes. *J. Appl. Polym. Sci.* **2014**, *131*. [CrossRef]
18. Manasa, G.; Mascarenhas, R.J.; Satpati, A.K.; D'Souza, O.J.; Dhason, A. Facile preparation of poly(methylene blue) modified carbon paste electrode for the detection and quantification of catechin. *Mater. Sci. Eng. C* **2017**, *73*, 552–561. [CrossRef]

19. Gholivand, M.B.; Ahmadi, E.; Haseli, M. A novel voltammetric sensor for nevirapine, based on modified graphite electrode by MWCNs/poly(methylene blue)/gold nanoparticle. *Anal. Biochem.* **2017**, 552–561. [CrossRef]
20. Devi, C.L.; Narayanan, S.S. Poly(amido amine) dendrimer/silver nanoparticles/multi-walled carbon nanotubes/poly (neutral red)-modified electrode for electrochemical determination of paracetamol. *Ionics* **2019**, *25*, 2323–2335. [CrossRef]
21. Baluchová, S.; Barek, J.; Tomé, L.I.N.; Brett, C.M.A.; Schwarzová-Pecková, K. Vanillylmandelic and homovanillic acid: Electroanalysis at non-modified and polymer-modified carbon-based electrodes. *J. Electroanal. Chem.* **2018**, *821*, 22–32. [CrossRef]
22. Kumar, T.S.S.; Swamy, B.E.K. Modification of carbon paste electrode by electrochemical polymerization of neutral red and its catalytic capability towards the simultaneous determination of catechol and hydroquinone: A voltammetric study. *J. Electroanal. Chem.* **2017**, *804*, 78–86. [CrossRef]
23. Liu, L.; Zhai, J.; Zhu, C.; Han, L.; Ren, W.; Dong, S. One-step synthesis of functional pNR/rGO composite as a building block for enhanced ascorbic acid biosensing. *Anal. Chim. Acta* **2017**, *981*, 34–40. [CrossRef] [PubMed]
24. Ihmels, H.; Otto, D. Intercalation of organic dye molecules into double-stranded DNA—General principles and recent developments. In *Supermolecular Dye Chemistry. Topics in Current Chemistry*; Würthner, F., Ed.; Springer: Berlin/Heidelberg, Germany, 2005; Volume 258, pp. 161–204. [CrossRef]
25. Han, X.; Yu, Z.; Li, F.; Shi, W.; Fu, C.; Yan, H.; Zhan, G. Two kanamycin electrochemical aptamer-based sensors using different signal transduction mechanisms: A comparison of electrochemical behavior and sensing performance. *Bioelectrochemistry* **2019**, *129*, 270–277. [CrossRef]
26. Attar, A.; Ghica, M.E.; Amine, A.; Brett, C.M.A. Poly(neutral red) based hydrogen peroxide biosensor for chromium determination by inhibition measurements. *J. Hazard. Mater.* **2014**, *279*, 348–355. [CrossRef]
27. Kuzin, Y.; Kappo, D.; Porfireva, A.; Shurpik, D.; Stoikov, I.; Evtugyn, G.; Hianik, T. Electrochemical DNA sensor based on carbon black—Poly(Neutral red) composite for detection of oxidative DNA damage. *Sensors* **2018**, *18*, 3489. [CrossRef]
28. Zhang, Y.; Yuan, R.; Chai, Y.; Xiang, Y.; Hong, C.; Ran, X. An amperometric hydrogen peroxide biosensor based on the immobilization of HRP on multi-walled carbon nanotubes/electro-copolymerized nano-Pt-poly(neutral red) composite membrane. *Biochem. Eng. J.* **2010**, *51*, 102–109. [CrossRef]
29. Porfireva, A.; Vorobev, V.; Babkina, S.; Evtugyn, G. Electrochemical sensor based on poly(Azure B)-DNA composite for doxorubicin determination. *Sensors* **2019**, *19*, 2085. [CrossRef]
30. Shen-Tu, C.; Liu, Z.; Kong, Y.; Yao, C.; Tao, Y. Electrochemical synthesis and properties of poly(azure B). *J. Electrochem. Soc.* **2013**, *160*, G83–G87. [CrossRef]
31. Porfireva, A.V.; Goida, A.I.; Rogov, A.M.; Evtugyn, G.A. Impedimetric DNA sensor based on poly(proflavine) for determination of anthracycline drugs. *Electroanalysis* **2020**, *32*, 827–834. [CrossRef]
32. Chen, C.; Gan, Z.; Xu, C.; Lu, L.; Liu, Y.; Gao, Y. Electrosynthesis of poly(aniline-co-azure B) for aqueous rechargeable zinc-conducting polymer batteries. *Electrochim. Acta* **2017**, *252*, 226–234. [CrossRef]
33. Grützke, S.; Abdali, S.; Schuhmann, W.; Gebala, M. Detection of DNA hybridization using electrochemical impedance spectroscopy and surface enhanced Raman scattering. *Electrochem. Commun.* **2012**, *19*, 59–62. [CrossRef]
34. Robinson, S.M.; Shen, Z.; Askim, J.R.; Montgomery, C.B.; Sintim, H.O.; Semancik, S. Ligand-based stability changes in duplex DNA measured with a microscale electrochemical platform. *Biosensors* **2019**, *9*, 54. [CrossRef] [PubMed]
35. Bayındır, O.; Alanyalıoğlu, M. Formation mechanism of polymeric thin films of Azure B on gold electrodes. *ChemistrySelect* **2018**, *3*, 2167–2173. [CrossRef]
36. Nishida, Y.; Domura, R.; Sakai, R.; Okamoto, M.; Arakawa, S.; Ishiki, R.; Salick, M.R.; Turng, L.-S. Fabrication of PLLA/HA composite scaffolds modified by DNA. *Polymer* **2015**, *56*, 73–81. [CrossRef]
37. Hongpaisan, J.; Roomans, G.M. Retaining ionic concentrations during in vitro storage of tissue for microanalytical studies. *J. Microsc.* **1999**, *193*, 257–267. [CrossRef]
38. Bard, A.J.; Faulkner, L.R. *Electrochemical Methods. Fundamentals and Applications*; J. Willey & Sons: New York, NY, USA, 1980.

39. Evtugyn, G.A.; Kostyleva, V.B.; Porfireva, A.V.; Savelieva, V.A.; Evtugyn, V.G.; Sitdikov, R.R.; Stoikov, I.I.; Antipin, I.S.; Hianik, T. Label-free aptasensor for thrombin determination based on the nanostructured phenazine mediator. *Talanta* **2012**, *102*, 156–163. [CrossRef]
40. Guidelli, R.; Compton, R.G.; Feliu, J.M.; Gileadi, E.; Lipkowski, J.; Schmickler, W.; Trasatti, S. Defining the transfer coefficient in electrochemistry: An assessment (IUPAC Technical Report). *Pure Appl. Chem.* **2014**, *86*, 245–258. [CrossRef]
41. Laviron, E. General expression of the linear potential sweep voltammogram in the case of diffusionless electrochemical systems. *J. Electroanal. Chem.* **1979**, *101*, 19–28. [CrossRef]
42. Chen, Y.; Li, Y.; Yang, Y.; Wu, F.; Cao, J.; Bai, K. A polyaniline-reduced graphene oxide nanocomposite as a redox nanoprobe in a voltammetric DNA biosensor for Mycobacterium tuberculosis. *Microchim. Acta* **2017**, *184*, 1801–1808. [CrossRef]
43. Kuzin, Y.; Ivanov, A.; Evtugyn, G.; Hianik, T. Voltammetric detection of oxidative DNA damage based on interactions between polymeric dyes and DNA. *Electroanalysis* **2016**, *28*, 2956–2964. [CrossRef]
44. Liu, J.; Bo, X.; Zhou, M.; Guo, L. A nanocomposite prepared from metal-free mesoporous carbon nanospheres and graphene oxide for voltammetric determination of doxorubicin. *Microchim. Acta* **2019**, *186*, 639. [CrossRef] [PubMed]
45. Vacek, J.; Havran, L.; Fojta, M. Ex situ voltammetry and chronopotentiometry of doxorubicin at a pyrolytic graphite electrode: Redox and catalytic properties and analytical applications. *Electroanalysis* **2009**, *21*, 21399–22144. [CrossRef]
46. Jiang, H.; Wang, X.-M. Highly sensitive detection of daunorubicin based on carbon nanotubes–drug supramolecular interaction. *Electrochem. Commun.* **2009**, *11*, 126–129. [CrossRef]
47. Alavi-Tabari, S.A.R.; Khalilzadeh, M.A.; Karimi-Maleh, H. Simultaneous determination of doxorubicin and dasatinib as two breast anticancer drugs uses an amplified sensor with ionic liquid and ZnO nanoparticle. *J. Electroanal. Chem.* **2018**, *811*, 84–88. [CrossRef]
48. Ilkhani, H.; Hughes, T.; Li, J.; Zhong, C.J.; Hepel, H. Nanostructured SERS-electrochemical biosensors for testing of anticancer drug interactions with DNA. *Biosens. Bioelectron.* **2016**, *80*, 257–264. [CrossRef]
49. Kulikova, T.; Porfireva, A.; Evtugyn, G.; Hianik, T. Electrochemical DNA sensors with layered polyaniline-DNA coating for detection of specific DNA interactions. *Sensors* **2019**, *19*, 469. [CrossRef] [PubMed]
50. Peng, A.; Xu, H.; Luo, C.; Ding, H. Application of a disposable doxorubicin sensor for direct determination of clinical drug concentration in patient blood. *Int. J. Electrochem. Sci.* **2016**, *11*, 6266–6278. [CrossRef]
51. Evtugyn, A.; Porfireva, A.; Stepanova, V.; Budnikov, H. Electrochemical biosensors based on native DNA and nanosized mediator for the detection of anthracycline preparations. *Electroanalysis* **2015**, *27*, 629–637. [CrossRef]
52. Hortobágyi, G.N. Anthracyclines in the treatment of cancer. An overview. *Drugs* **1997**, *54*, 1–7. [CrossRef]
53. Li, M.; Russo, M.; Pirozzi, F.; Tocchetti, C.G.; Ghigo, A. Autophagy and cancer therapy cardiotoxicity: From molecular mechanisms to therapeutic opportunities. *Biochim. Biophys. Acta* **2020**, *1867*, 118493. [CrossRef] [PubMed]
54. Oliveira Brett, A.M.; Piedade, J.A.P.; Serrano, S.H.P. Electrochemical oxidation of 8-oxoguanine. *Electroanalysis* **2000**, *12*, 969–973. [CrossRef]

nanomaterials

MDPI

Review

Nanostructured Ceria: Biomolecular Templates and (Bio)applications

Petr Rozhin [1], Michele Melchionna [1,2,*], Paolo Fornasiero [1,2,3] and Silvia Marchesan [1,2,*]

[1] Chemical and Pharmaceutical Sciences Department, University of Trieste, 34127 Trieste, Italy; petr.rozhin@phd.units.it (P.R.); pfornasiero@units.it (P.F.)
[2] Unit of Trieste, INSTM, 34127 Trieste, Italy
[3] Istituto di Chimica dei Composti Organometallici, Consiglio Nazionale delle Ricerche (ICCOM-CNR), 34127 Trieste, Italy
* Correspondence: melchionnam@units.it (M.M.); smarchesan@units.it (S.M.); Tel.: +39-040-5583923 (M.M. & S.M.)

Abstract: Ceria (CeO_2) nanostructures are well-known in catalysis for energy and environmental preservation and remediation. Recently, they have also been gaining momentum for biological applications in virtue of their unique redox properties that make them antioxidant or pro-oxidant, depending on the experimental conditions and ceria nanomorphology. In particular, interest has grown in the use of biotemplates to exert control over ceria morphology and reactivity. However, only a handful of reports exist on the use of specific biomolecules to template ceria nucleation and growth into defined nanostructures. This review focusses on the latest advancements in the area of biomolecular templates for ceria nanostructures and existing opportunities for their (bio)applications.

Keywords: ceria; nanoparticles; nanorods; nanosheets; nanozyme; biomolecule; template; catalysis; anti-oxidant; oxygen radicals

Citation: Rozhin, P.; Melchionna, M.; Fornasiero, P.; Marchesan, S. Nanostructured Ceria: Biomolecular Templates and (Bio)applications. *Nanomaterials* **2021**, *11*, 2259. https://doi.org/10.3390/nano11092259

Academic Editors: Ivan Stoikov and Pavel Padnya

Received: 8 August 2021
Accepted: 30 August 2021
Published: 31 August 2021

1. Introduction

Ceria is among the most studied metal oxides and it has attracted researchers' interest for its ability to capture, store, and release oxygen, and has been widely applied to "clean-air" catalytic conversion technologies, and, more generally, in catalysis [1,2]. Nanosized ceria can mimic a variety of enzymatic activities, earning the name of "nanozyme", and it can catalyze chemical reactions with potential biomedical applications, for instance for sensing or reducing oxidative stress in pathological conditions [3].

Ceria nanomorphology is an important factor affecting its reactivity, as described further below. It is thus not surprising that the search has been very active for suitable templates (Figure 1) to exert control over ceria nucleation and growth, and the topic of artificial or bio-based, hard or soft, templates for ceria has been recently reviewed [4]. In addition, microbial cultures of *Bacillus subtilis* were successfully used as bioreactors for the conversion of cerium (III) nitrate to ceria nanoparticles (NPs) [5]. The use of plant extracts for the green synthesis of nanoceria exploiting naturally occurring redox-active agents, such as polyphenols, is well-established [6]. Various plant parts have been used to this end, such as seeds [7–9], nut shells [10], leaves [11–14], flowers [15–17], bean sprouts [18], fruits [19,20], and kapok fibers (lignin) [21]. The use of anisotropic biotemplates, such as cotton, can be used to reproduce their morphology in ceria, e.g., microfibers [22], although the concept is typically applied at the microscale rather than nanoscale. Since green protocols for the preparation of nanosized ceria, along with its biological applications, were reviewed in 2017 [23], this work will focus on the latest developments in this area since then. It is worth noting that the vast majority of biotemplates, as discussed in the literature until now, have been obtained using top-down approaches, such as grinding or extraction, or through the use of microscale structures, usually consisting of mixtures of compounds. By contrast, the opportunities offered by pure biomolecules and their folding or self-assembly in bottom-up approaches have yet to be deeply explored, and to our knowledge, they have not been reviewed until now based on the molecular classes they belong to (i.e., carbohydrates, proteins, nucleic acids, etc.). This is particularly important also considering the catalytic activities of nanosized ceria that are responsible for the observed biological effects, and how they are affected upon binding to biomolecules. In this work, we gather the detailed knowledge pertaining to the interactions between specific biomolecules and ceria, both for the formation of nanoceria and for its (bio)catalytic activity. Furthermore, we discuss relevant examples described in the last five years to outline potential applications for these materials, with an emphasis on the biological ones to maximize the benefits offered by using biomolecular templates to attain nanosized ceria.

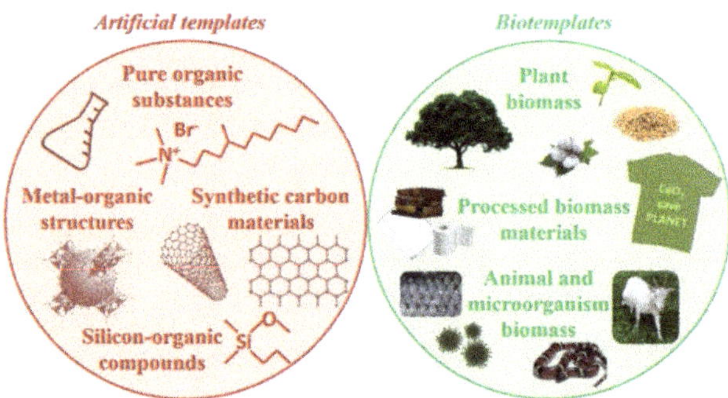

Figure 1. The search towards either artificial or bio-based templates to exert control over the growth of ceria has been very active. Reproduced from [4].

2. Ceria Nanomorphology-Reactivity Relationship

The morphology of ceria nanostructures can exert an important influence on its properties and functions, and it is thus a key parameter to consider. Consequently, in the template-assisted synthetic protocol, the choice of the specific template is of great relevance. Ceria is typically formed from cerium (III) salts (such as nitrate or chloride precursors), and the nature of the anion was shown to be critical for the morphological control, for example leading to NPs or nanorods [24].

A comparative study of cubic, rodlike, and polyhedral ceria geometries (Figure 2) on the catalytic conversion of glycerol into biorenewable methanol revealed that cubic particles displayed low catalytic activity. The low activity of the cubic NPs was ascribed to reduced surface area, relatively high acidity, and exclusive exposure of the (100) facet in the cubic geometry. This facet was more prone to hydroxylation under the reaction conditions, resulting in being detrimental for the investigated catalytic conversion [25]. Therefore, the nanocube morphology is a disadvantage for the catalytic production of methanol from glycerol. Different ceria nanomorphologies revealed varying levels of oxidase-like activity after interacting with DNA [26]. In particular, NPs and nanocubes demonstrated increased oxidase-mimetic activity upon binding DNA, whilst the opposite was true for nanorods [26]. Therefore, the former two geometries would be advantageous whenever DNA-binding could be a useful trigger to increase oxidase-like catalytic activity, for instance in biosensing. Conversely, a nanorod morphology would be disadvantageous for the same type of application. The structure–activity relationship of ceria nanorods, nanocubes, and nanooctahedra was studied for the generation of hydroxyl radicals through the catalytic decomposition of hydrogen peroxide. The reactivity was found to be highest for nanorods, followed by nanocubes and nanooctahedra. This trend was rationalized in terms of atomic defects, the percentage of surface Ce (III) ions, and the average coordination number of oxygen anions surrounding each cerium cation [27]. Therefore, nanorod or nanocube morphology could be advantageous to selectively trigger ROS-induced cell damage in pathological environments whereby there are higher levels of hydrogen peroxide, such as during inflammation.

Figure 2. Transmission electron microscopy (TEM) images of ceria nanoparticles (NPs) with different morphologies: (**a**,**b**) cubic, (**c**,**d**) rodlike, and (**e**,**f**) polyhedral. Reproduced from [25].

Morphology maps revealed that the redox performance, particle size, and surface roughness could be optimized by engineering the oxygen vacancies' density, which was influenced by the concentration of the precipitant/oxidant used during NP formation. These vacancies can be positioned at the surface, subsurface, or bulk regions, and it is the subsurface vacancies that are responsible for the main redox activity. These features can have important implications in the biological performance of nanosized ceria. For instance, ceria NPs exhibit antioxidant and cytoprotective effects at physiological pH 7.4. Conversely, at the acidic pH 6.4 that is typical of the tumor microenvironment, ceria NPs are oxidant and exert cytotoxic effects. The relative cytotoxicity thus depends on ceria nanomorphology with increasing levels in the order nanocubes < nanorods < truncated nanooctahedra [28]. This study suggested that in the case of osteosarcoma cells, truncated nanooctahedra was the ideal morphology to induce selective cytotoxicity.

A combination of experiments and in silico studies were used to design structures for nanoceria that maximize its catalytic activity. Polyhedral and nanocube morphologies expose active (100) surfaces (Figure 3), which should contain oxygen vacancies and surface hydroxyl groups. However, it was found that phosphate anions can strongly bind to (100) surfaces, inhibiting the oxygen capture and release, hence poisoning the ceria nanozyme. By contrast, the phosphate interaction with (111) surfaces is weaker, therefore these surfaces protect the ceria nanostructure against passivation [29].

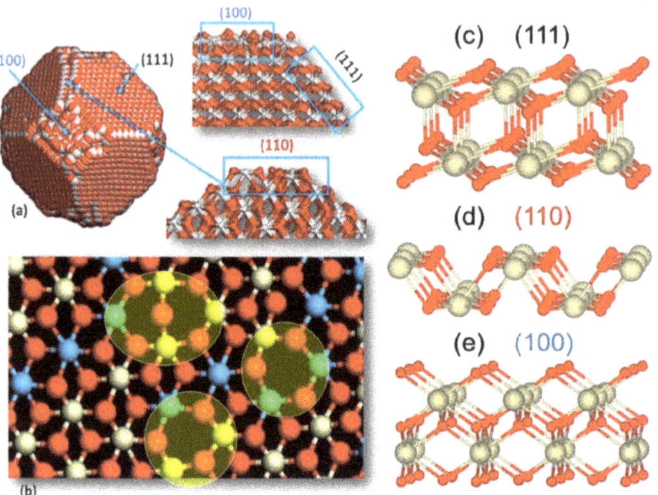

Figure 3. (**a**) Structure of ceria nanoparticle (NP) with (111), (110), and (100) surfaces. (**b**) View of one of the (111) surfaces on reduced nanoceria reveals oxygen vacancies (yellow ovals). The structures of defect-free (**c**) (111), (**d**) (110), and (**e**) (100) surfaces. Ce (IV) is white, Ce (III) is blue, and oxygen is red. Reprinted with permission from [29], Copyright © 2021, American Chemical Society.

3. Biomolecular Templates for Ceria Nanomaterials

Biotemplates have been attracting increasing interest to exert control over ceria morphology, although typically they are used for microscale (rather than nanoscale) assembly. They are often obtained through top-down approaches (e.g., grinding of biomass) and consist of mixtures of diverse molecules. A handful of studies instead focused on the use of specific biomolecules belonging to different chemical classes to template nanostructured ceria, as summarized in Table 1. These naturally derived molecules predominantly feature oxygen-bearing functional groups (e.g., carbohydrates, catechols, nucleic acids, proteins). Ligands on the surface of ceria NPs can have key effects on its antioxidant properties and even reverse it [30], therefore their choice should be carefully evaluated.

Table 1. Biomolecular templates used for ceria nanostructure definition.

Biotemplate Class	Biomolecule	Ceria Nanomorphology	Crystallite Size (nm)	Ceria NP Size (nm)	Application	Reference
Carbohydrates	Alginate	Spherical	3–5	40–200	Antioxidant	[31]
	Cellulose	Nanoparticles	8	7–10	Catalysis	[32]
	Chitosan	Spherical	8	24	Bioimaging	[33]
	Cyclodextrin	Nanoparticles	n.a.	61	Antioxidant	[34]
	Starch	Irregular	7–8	7–13	Catalysis	[35]
Catechols	PDA [1] NPs	Spherical	10	180	Catalysis	[36]
	rGO@PDA [1,2,3]	Nanosheets	n.a.	3–4	Biosensing	[37]
	Gallate	Nanoflowers	8–13	n.a.	Detection	[38]
Carboxylic acids	Citric acid	Nanocrystals	11–35	n.a.	Catalysis	[39]
Phosphates	DNA	Nanocrystals	5 ± 1	5 ± 1	Antioxidant	[40]
	DNA	Nanocrystals	6 ± 2	6–18 nm	Optoelectronics	[41]
	DNA	Nanocrystals	n.a.	50–400	Catalysis	[42]
	Phytic acid	Nanosheets	n.a.	n.a.	Flame retardant	[43]
Proteins	Albumin	Nanoparticles	n.a.	15	Antioxidant	[44]
	Albumin	Spherical, Nanochains	2	2–100	Catalysis	[45]
	Apoferritin	Nanocrystals	5.0 ± 0.7	5.0 ± 0.7	Catalysis	[46]
	Ferritin	Spherical	n.a.	7	Catalysis	[47]
	Silicatein	Nanocrystals	<3	2.56 ± 0.38	Catalysis	[48]

[1] PDA = polydopamine. [2] rGO = reduced graphene oxide. [3] @ denotes core@shell structure.

3.1. Carbohydrates

Considering the affinity of ceria for oxygen, the choice for a suitable template for ceria nucleation often falls on biomolecules that are rich in hydroxyl groups, such as carbohydrates and catechols. For instance, *Aloe vera* gel [49] and xantham gum [50] have been used as a source of polysaccharides to template nanosized ceria. Cellulose is a popular choice, although typically in the form of biomass or plant parts [21], as opposed to the purified polymer, despite the fact that the defined composition of the latter is more promising to attain finer control over homogeneously sized NPs. A recent report used microcrystalline cellulose in a sol-gel protocol to produce 8-nm sized ceria nanocrystals with a very narrow size distribution of the resulting NPs [32]. Similarly, starch templated 7–8 nm-sized ceria nanocrystals that, depending on experimental conditions, formed 7–13 nm-sized NPs [35]. In another study, alginate was used both as a precursor and a template, by providing a ceria-alginate gel, whose thermal decomposition produced spherical ceria NPs with a size < 5 nm and presence of functional groups, whose spectroscopic signatures were ascribed to carbonate and carboxylates [31]. Chitosan is another polysaccharide that was used as a template and capping agent for ceria NPs using a sol-gel method [33]. However, it is worth noting that when alginate and chitosan were used to coat ceria NPs, their stability was negatively affected, as they showed a tendency to agglomerate and sediment, and their antioxidant activity was altered [51].

3.2. Catechols

Catechols are naturally occuring polyphenolic compounds that are widely known for their crosslinking and metal chelating abilities, as well as their redox chemistry [52,53]. One compound of this class that can be extracted from green tea and is attracting attention for its anti-oxidant properties is epigallocatechin gallate [54]. In a recent study, this template was successfully employed to form Eu-doped ceria flowers (Figure 4) composed of a few-nm-sized crystallites, and they have been applied for the luminescent detection of latent fingerprints [38].

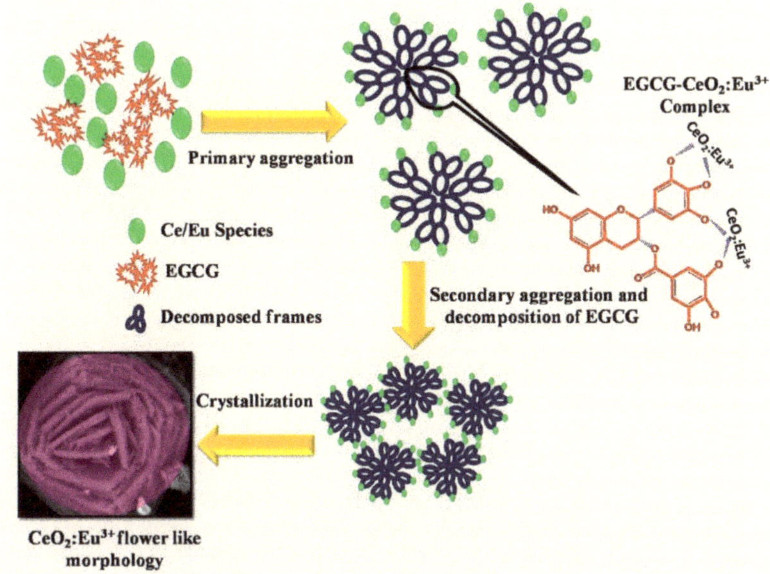

Figure 4. Epigallocatechin gallate (EGCG) biomolecular template for the nucleation and growth of ceria nanocrystallits that further assemble into a flower morphology. Reprinted from [38], Copyright © 2021, with permission from Elsevier.

Polydopamine (PDA) is another bioderived catechol that has gained widespread popularity in materials science, especially for its adhesiveness [55], which has proved effective in the generation of composites [56] and functional nanomaterials [57,58] with uses spanning from catalysis to theranostics [59]. The catechol groups of PDA NPs served to reduce first gold (III) to gold (0) onto the surface of the NPs, and then to anchor cerium (III) for the formation of ceria that found photocatalytic applications [36]. PDA was also used to coat reduced graphene oxide (rGO) and template ceria nanosheet formation [37].

3.3. Carboxylic Acids

Among carboxylic acids, the biocompatible and low-cost citric acid is possibly the most widely used capping agent and reductant for the preparation of metallic nanoparticles [60], and even beyond, to yield fluorescent biomaterials [61]. In the synthesis of nanosized ceria, the carboxylic moiety of citric acid can react with metal ions and form metallic citrate, with subsequent addition of ethanol leading to a gel. Interestingly, ceria crystallite size can be tailored depending on the calcination temperature used to form the metal oxide [39].

3.4. Phosphates and Nucleic Acids

Ceria is well-known for its affinity to phosphate groups, as discussed further below in the relevant applications. Among the phosphate-containing biomolecules, nucleic acids are an obvious choice for the templating of nanoceria. DNA has been used as a biotemplate and capping agent for the nucleation of ceria nanocrystals since the major groove of the DNA double-helix was hypothesized to be appropriate both in size and chemical composition to nucleate ceria NPs. For instance, 5 nm-sized crystals with enhanced stability against agglomeration could be formed upon DNA-assisted CeO_2 NP synthesis [40]. Indeed, it had been shown that nanoceria can adsorb the phosphate groups of DNA on its surface in a sequence-independent manner (Figure 5), although this interaction can lead to the inhibition of nanoceria oxidase-like activity [62]. However, the oxidase-like activity of nanoceria was shown to be enhanced to different extents in the presence of various nucleoside triphosphates (NTPs), with GTP exerting the highest effect, followed by ATP [63]. This effect was ascribed to the coupling of the oxidative reaction with NTP hydrolysis, catalyzed by the nanoceria phosphatase-like activity [63].

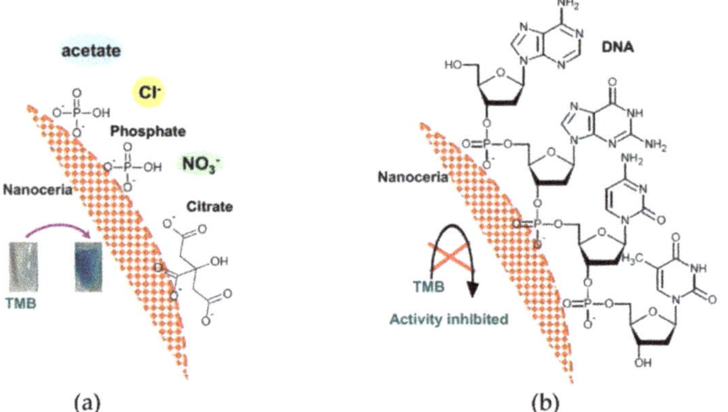

Figure 5. (**a**) Ceria binds with phosphate and citrate with higher affinity than acetate, chloride, or nitrate. In all these buffers, its oxidase-like activity is not affected, as it is able to convert 3,3′,5,5′-tetramethylbenzidine (TMB) into a blue product. (**b**) Ceria binds DNA through the phosphate groups independently from the DNA sequence and DNA adsorption impairs its oxidase-like activity. Reprinted with permission from [62], Copyright © 2021, American Chemical Society.

3.5. Proteins

Proteins have been far less used as templates for ceria NPs with respect to polysaccharides. Typically, proteins such as albumin are studied for their interactions with pre-formed ceria NPs, relevant to the formation of a biocorona that enhances NP colloidal stability [64,65] and can affect the mechanism of cellular uptake [66]. In particular, the adsorption of amino acids with carboxylic acid groups on their sidechains has been studied [67].

One work used albumin as a biomolecular template for the nucleation and growth of ceria spherical NPs, and their association into nanochains (Figure 6). Albumin displays disulfide bridges that could be reduced to thiols, thus promoting the concomitant oxidation of cerium (III) in the nitrate precursor to cerium (IV) oxide. The variation of experimental conditions, in particular the temperature used for the nucleation of the NPs, allowed them to fine-tune their final morphology, with cooling at 4 °C favoring 40 nm-long nanochains and heating at 80 °C promoting the formation of homogeneous 3.7 $\pm$ 0.7 nm spherical NPs [45].

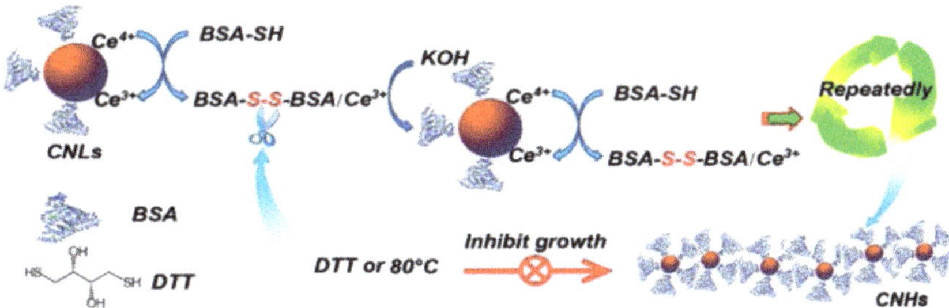

Figure 6. Proposed mechanism for the biomolecular templating effect of albumin on ceria NPs nucleation and growth into nanochains. Reprinted with permission from [45]. Copyright © 2021, American Chemical Society.

In another study, bovine serum albumin was biomineralized with ceria to produce Ce-doped carbonaceous NPs that were envisaged for antioxidant therapy [44]. Glycine was also investigated as a very simple biotemplate to form ceria via a hydrothermal route. Varying the amino acid concentration led to differing morphologies of ceria microparticles, while the addition of ethanol as a co-solvent yielded homogeneously sized spherical NPs [68].

Apoferritin is a ubiquitous protein for iron storage as ferric hydroxide NPs, with a distinctive container morphology with an inner cavity of 7 nm that was successfully used to nucleate ceria nanocrystals of 5 nm [46]. A superlattice was engineered from ferritin cages with inner cavities of 7 nm and charged surfaces (Figure 7) to nucleate the growth of ceria NPs, leading to oxidase-like and peroxidase-like catalytically active crystals [47,69].

Finally, silicateins are a class of proteins that has been widely applied for biomineralization, and although many researchers have attempted to identify the optimal peptide sequence for this purpose, usually the target is the preparation of silica NPs [70]. A remarkable, bioinspired approach exploited a mutated silicatein for optimal expression as a recombinant protein in *E. coli* was successfully used for templating the nucleation of ceria nanocrystals < 3 nm in size [48].

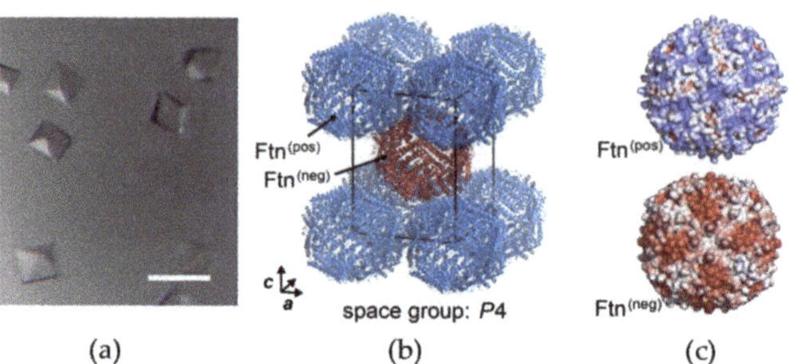

Figure 7. (**a**) Optical micrograph of crystals of engineered ferritin proteins to display a cationic (Ftn(pos)) or anionic (Ftn(neg)) surface, scale bar 200 μm. (**b**) Molecular structure of binary crystals. One unit cell of the tetragonal lattice is shown, protein backbone depicted in cartoon representation. (**c**) Electrostatic potential (red, −5 kT/e; blue, +5 kT/e) of Ftn(pos) and Ftn(neg), viewed along the 4-fold axis. Adapted with permission from [69]. Copyright © 2021, American Chemical Society.

4. (Bio)applications

Nanosized ceria finds a large variety of applications [71], and extensive reviews exist on the topic, especially relevant to catalysis [1] for energy [72,73] and environmental preservation [74] and remediation [75,76]. Therefore, here we will focus on the latest advancements described in the most recent years, with an emphasis on biological applications, for which the use of biomolecular templates for ceria nanostructures' assembly is particularly relevant.

4.1. Nanocarrier for Therapeutics

Nanomaterials are highly promising to innovate in medicine, thanks to their unique properties that arise from working at the nanoscale [77]. A fascinating opportunity is to use ceria NPs as nanovectors, as shown with a nanoemulsion obtained from lemon and corn oils that was envisaged for drug delivery [78]. Biomimetic lipids successfully yielded a vector for ceria NPs to cross the blood–brain barrier. Once internalized by neurons, they acted both as neuroprotective and pro-neurogenic agents, as demonstrated using co-culture systems [79]. NP size is a discriminating parameter for the cell uptake mechanism. NPs with a diameter as small as 3–5 nm can passively cross the membranes, a process relevant for the delivery of therapeutic biomolecules that could be otherwise negatively affected by the harsh chemical environments of endosomal pathways [80]. A fluorescent assay was developed to study ceria NP uptake by cells [81].

Ceria nanorods were shown to be able to deliver RNA interference (RNAi) to treat atherosclerosis, through silencing of the mTOR gene that controls autophagy and lipid metabolism. The use of a targeting peptide allowed for selective penetration into pathological plaques, PEGylation extended the nanocarrier's circulation time, while the anisotropic morphology facilitated endosomal escape whilst ensuring the "on-demand" release of the RNAi cargo through competitive coordination of cytosolic hydrogen peroxide for gene therapy [82]. MicroRNAs are another type of therapeutic biologics that were envisaged to be delivered with ceria NPS as carriers to target directly the site of interest, such as the lung, and avoid systemic distribution [83].

4.2. Phosphoproteomics and Phosphatase-like Nanozymes

The area of phosphoproteomics is advancing at a rapid pace, in order to gain a better understanding of the biochemical profiling of several pathologies, especially for their early diagnosis [84]. However, phosphate groups are labile, making the detection of phosphopeptides quite a challenge, for which metal oxide affinity chromatography (MOAC) has offered promising solutions [85]. Amongst the various metal oxides, titania (TiO_2) is widely applied, and the morphology of the nanocomposites has been revealed to be critical for the resulting MOAC performance [86]. However, ceria has been studied for the enrichment and detection of low-abundance phosphopeptides too [87,88]. A combination of its MOAC sorbent ability with its peroxidase-mimicking activity allowed the development of a colorimetric assay for phosphoproteins' detection, since the nanozyme activity was reduced upon adsorption of the biomolecular target [89].

A multiplexed quantitative matrix-assisted laser desorption/ionization mass spectrometry (MALDI MS) approach was developed to simultaneously assess the activity and inhibition of multiple protein kinases, which are an important class of cancer biomarkers (Figure 8). In particular, phosphorylated peptides that act as substrates for the kinases of interest were captured and dephosphorylated by ceria, allowing for enhanced detection. The method was successfully applied to Abl and Src, the kinases that are involved in chronic myeloid leukemia [90].

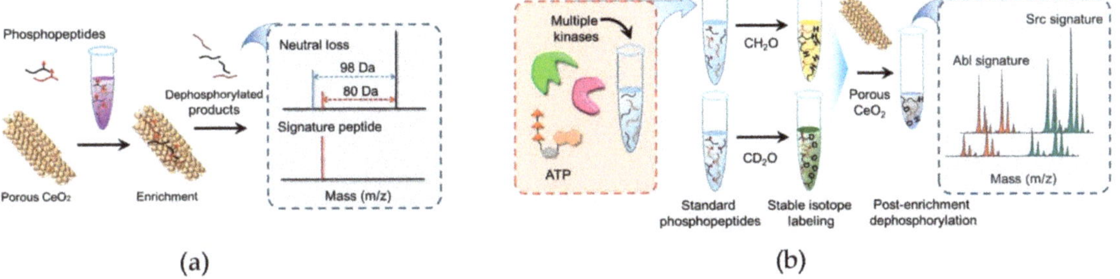

Figure 8. Multiplexed mass spectrometry approach for kinase biomarkers. (**a**) Ceria is used as MOAC sorbent and phosphorylase-mimic for phosphopeptide detection. (**b**) Ceria allows for the enhanced detection of Abl and Src kinases. Adapted with permission from [90]. Copyright © 2021, American Chemical Society.

Indeed, the ability of ceria to adsorb phosphate can be used to design phosphatase-mimics and applied to the conversion of phosphate prodrugs into active therapeutics (Figure 9), including chemotherapeutics for advanced cancer therapy [91]. It also proved to be efficient in the dephosphorylation of thiamine pyrophosphate first to thiamine monophosphate, and then to thiamine, thus producing the free vitamin form [92].

Besides, this nanozyme activity can be applied to the decomposition of pollutants, such as phosphate-bearing nerve agents [93,94]. Phosphatase-like activity is extensively studied to induce DNA cleavage, with potential applications in DNA repair, gene editing, and biosensing, as recently reviewed [95]. The interaction of polyphosphates bearing various phosphate units with nanoceria was studied ad it was shown that their esterification significantly reduces affinity for ceria [96]. Phosphate efficiently displaced DNA from ceria, contrarily to phosphite and hypophosphite. This observation could be used to screen for differing phosphorus species or oxidizing agents [97].

Ceria NP morphology was shown to affect its nanozyme activity, probed as phosphatase. In particular, ceria nanofibers were more active than commercial nanoceria and nanopolyhedra, while nanocubes displayed negligible activity. The results were correlated with the higher amount of Ce (IV) in the nanofibers that could bind hydroxide and phosphate to catalyze hydrolysis. This also meant that the catalytic activity was dependent on the buffer used, being completely quenched in phosphate buffer, and preserved in TRIS, glycine, or HEPES buffers. Furthermore, the nanozyme could adsorb enzymes or antibodies and was envisaged as a protein vehicle [98]. However, another study demonstrated that through the appropriate formulation of ceria NPs, it is possible to avoid phosphate-induced inhibition of catalytic activity [99].

Ceria NPs catalyzed the hydrolysis of 3′,5′-cyclic adenosine monophosphate (cAMP), which is a second messenger, involved in a plethora of signal transduction pathways. Importantly, the catalysis was only slightly affected by the pH, and highly specific to ceria, as opposed to other lanthanide oxides (i.e., La_2O_3, Pr_6O_{11}, and Nd_2O_3). The unusual phosphatase mimicry arose from an interplay between properly positioned Ce (III) and Ce(IV) cations, as well as cerium-activated hydroxyl moieties [100].

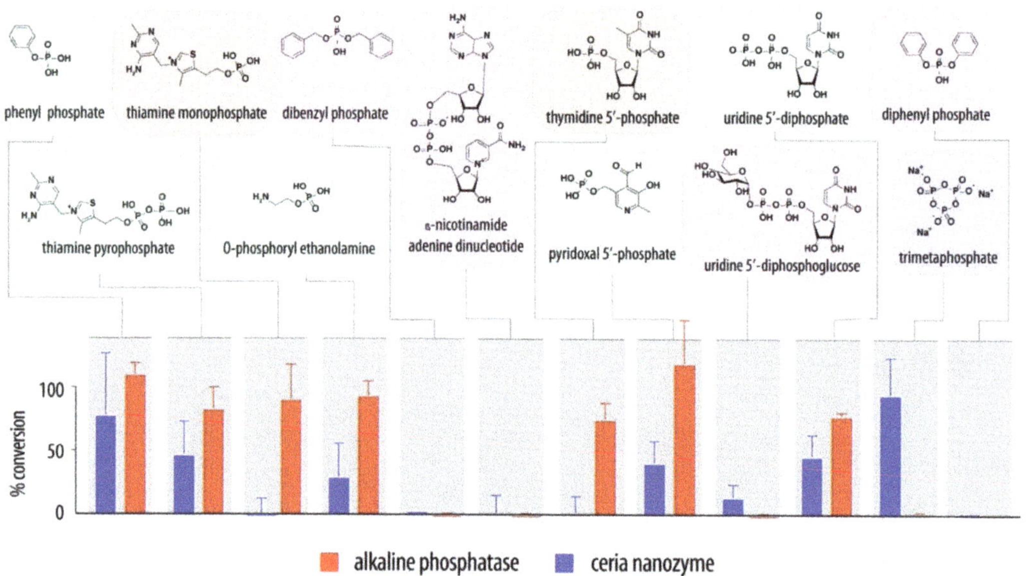

Figure 9. Substrate scope of ceria nanozyme, in comparison with alkaline phosphatase, for the hydrolysis of organic phosphates in a study for the delivery of phosphate prodrugs. Reproduced with permission from [91], Copyright © 2021, American Chemical Society.

4.3. Photocatalysis and Catalysis

Ceria is a well-known material for photocatalytic processes, such as dye and drug photocatalytic degradation for environmental remediation [16,38,101]. To this end, ceria has been combined with ferrihydrites for the photo-Fenton degradation of tetracycline or other model pollutants (i.e., tetrabromobisphenol A, Rhodamine B, and 2,4-dichlorophenol) via the generation of reactive hydroxyl radicals [102]. Ceria NPs have been applied to the catalytic ozonation of phenol too, as a model pollutant [32].

Ceria NPs were coated with polyacrylic acid for the subsequent grafting of different amino acids. It was found that the use of Phe led to the best performance as chiral catalysts for the stereoselective oxidation of DOPA, a drug used in Parkinson's disease, to dopachrome (Figure 10). Interestingly, while the use of enantiomers inverted the stereoselectivity as expected, it was found that the use of Phe as a chiral agent led to opposite stereoselectivity relative to His. Higher stereoselectivity was obtained for D-amino acids and it was rationalized in terms of their ability to engage in π–π interactions and H-bonding with the substrate [103].

Another interesting field of application is DNA repair. UV-induced damage includes the formation of cyclobutane pyrimidine dimers, which can be dissociated by photolyase enzymes, or nanosized ceria as their mimicry [104]. Protection from DNA damage was also reported through the ability of ceria to absorb ionizing radiation [105], combined with its antioxidant, as studied on irradiated cells [106].

Ceria NPs were used as supports for (2,2,6,6-tetramethylpiperidin-1-yl)oxy (TEMPO) free radical catalysts for the selective oxidation of primary alcohols of carbohydrates, and they demonstrated good stability and better performance than homogeneous catalysts over six consecutive runs [107].

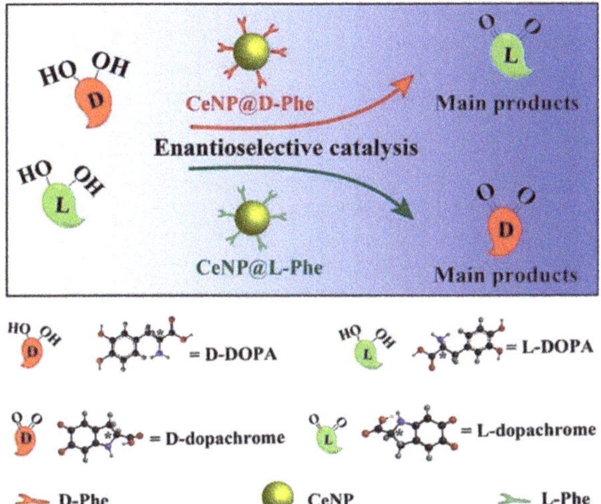

Figure 10. Stereoselective catalytic oxidation of DOPA by phenylalanine (Phe)-modified ceria NPs. The catalyst with D-Phe oxidized L-DOPA into L-dopachrome more effectively. Conversely, the catalyst with L-Phe oxidized D-DOPA into D-dopachrome more effectively. Reproduced with permission from [103], Copyright © 2021 Wiley-VCH Verlag GmbH & Co. KGaA, Weinheim.

The ability of nanosized ceria to promote the formation of radical oxygen species has been exploited in photocatalytic water oxidation, too [15,108]. For instance, a porphyrin photosensitizer was embedded in ceria nanotubes so that the electronic communication between the two components allowed for the enhanced production of molecular oxygen over the competing hydrogen peroxide generation from the partial water oxidation [109]. Nanosized ceria was also reported to enhance the performance of microbial fuel cells by creating oxygen reservoirs for the oxygen reduction reaction (ORR), while direct involvement of the cerium redox couple in the catalysis was only marginal [110].

4.4. Reactive Oxygen Species (ROS) Mitigation

4.4.1. Mechanisms of Nanozyme Activity Pertaining to ROS Mitigation

Nanostructured ceria can mimic the activity of several enzymes, including superoxide dismutase (SOD) and catalase (CAT), which have been the subject of intense mechanistic investigations. SOD catalyzes the conversion of $O_2^{\cdot-}$ to H_2O_2, which then undergoes catalytic dismutation by the CAT into water and molecular oxygen. Interestingly, in the past, the SOD and CAT mimetic activity had been proposed to occur via direct electron transfer from $O_2^{\cdot-}$ or H_2O_2 to ceria with concomitant redox cycling between Ce (III) and Ce (IV) of surface sites by drawing an analogy with those of the natural enzyme. However, recent studies demonstrated that the redox potential of the Ce (III)/Ce (IV) couple is unfavorable for such a mechanism [111]. In the case of SOD-mimicking activity, the catalytic cycle requires surface defective sites and consists of two key steps, similarly to the SOD-mimicry by noble metals: (i) $HO_2^{\cdot}$ chemisorption onto the ceria surface and (ii) O_2 and H_2O_2 generation. In the case of CAT mimicry, the mechanism is markedly different from that of noble metals, the latter occurring only with the aid of a pre-adsorbed OH group. In this case, the two key steps are (i) H_2O_2 oxidation by the CeO_2 (111) surface, with the generation of O_2 and the reduced H_2-CeO_2 (111) surface; and (ii) the subsequent reaction between another H_2O_2 molecule and H_2-CeO_2 (111), producing two water molecules [111].

Ceria nanomorphology also influences the nanozyme activity towards being CAT-like or SOD-like. In particular, (111)/(100) nanopolyhedra with a high concentration of Ce^{4+} ions promoted catalase mimicry. Conversely, (100) nano/submicron cubes and (111)/(100) nanorods that grew in the (110) longitudinal direction, both with high Ce(III) levels, enhanced SOD mimicry [112]. Interestingly, changing the temperature of preparation of Ce NPs allowed the modulation of their mimicking properties of multiple enzymes, namely superoxide dismutase (SOD), catalase (CAT), oxidase (OXD), peroxidase (POD), alkaline phosphatase (ALP) enzymes, as well as their ability to scavenge the 2,2-diphenyl-1-picrylhydrazyl (DPPH) free radical. In particular, the NPs with the highest level of antioxidant activity provided cells with cytoprotection against aging- and H_2O_2-induced oxidative damage; contrarily, those with pro-oxidant activity were proposed as candidates to induce cancer cell death [113]. The POD-like activity can be boosted by doping the CeO_2 with several first-row transition metals (i.e., Mn, Fe, Co, Ni, and Cu), with maximal effect noted for Mn, followed by Co [114]. The POD mimicry by metal and metal oxide NPs has been recently reviewed [115]. The ceria precursor concentration represents an additional parameter to be carefully assessed, as it is a knob for controlling the particle size and the amount of Ce (III) sites at the surface, which also impact the type of nanozyme activity (oxidase-like or antioxidant) [116]. Atomic layer deposition (ALD) proved to be a promising approach to modulate the Ce (III)/Ce (IV) ratio through the adjustment of film thickness, analogously to what is observed for ceria NPs with differing diameter [117]. The combination of ceria with gold in core-shell NPs allowed for multi-enzyme mimicry that exhibited POD, CAT, and SOD activity that could be controlled through pH adjustment [118].

4.4.2. ROS Mitigation for the Treatment of Cancer and Chemotherapy's Consequences

There are several pathological conditions that could be alleviated through ROS mitigation, and first and foremost, cancer is an obvious target. Furthermore, ROS can induce acute kidney injury in patients that receive chemotherapy. Ceria NPs can catalyze the dismutation of $\cdot O_2-$ into H_2O_2 at both neutral and acidic pH values, as well as decomposing H_2O_2 under neutral, but not acidic, pH. Therefore, ceria NPs exerted a general cytoprotection on cells that were challenged by oxidizing chemotherapeutics at a neutral pH, without compromising their anti-tumor activity at an acidic pH, which is a common feature of tumor microenvironments (Figure 11) [119]. Ceria NPs demonstrated a promising performance to partially revert the cellular mechanisms involved in tumor progression. They increased overall survival in vivo in rats with hepatocellular carcinoma, and following ex vivo uptake, perfused human livers and human hepatocytes. In vivo, ceria accumulated mainly in the liver, where it reduced inflammation and proliferation, and increased liver apoptotic activity [120].

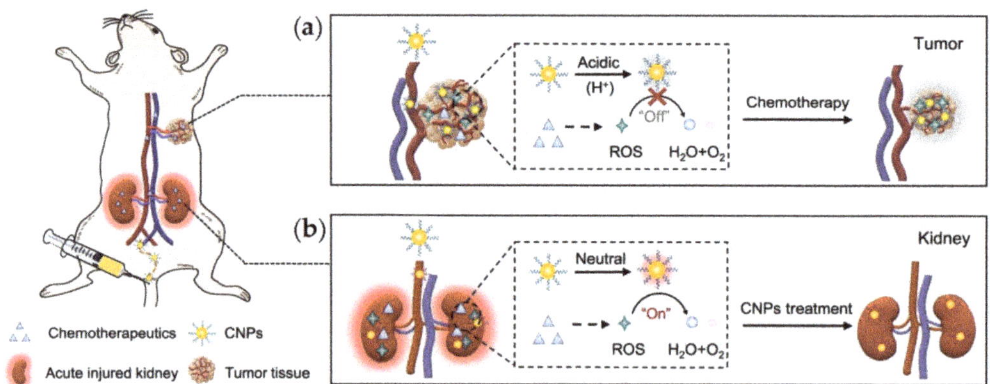

Figure 11. Schematic representation of pH-dependent redox activity of ceria NP as tested in vivo. (**a**) In the acidic tumor microenvironment, ceria NPs ROS scavenging ability is switched off and does not interfere with oxidizing chemotherapeutics; (**b**) in the neutral pH of the kidneys, ceria NPs scavenge ROS and exert cytoprotection. Reproduced from [119], under a Creative Commons license http://creativecommons.org/licenses/by/4.0/.

Ceria NPs have also been inserted into the mesopores of dendritic silica-coated bismuth sulfide nanorods to impede their aggregation. The nanocomposite was proposed for the photothermal cancer therapy, through dual enzyme-activity mimicry. In particular, ceria was used to induce oxygen radicals-mediated cancer cell death, while the nanorods endowed the material with photothermal energy conversion ability using near-infrared light absorption of the bismuth sulfide nanorods. Furthermore, bismuth enabled high-contrast CT imaging [121]. Ceria NPs have also been proposed for the treatment of colon cancer, as it was noted that they can exert higher cytotoxicity in cancer cells relative to healthy cells through the generation of ROS and subsequent triggering of apoptosis [122]. In pancreatic cancer cells, ceria NPs acted as sensitizers to radiation therapy through activation of c–Jun kinase that promoted apoptosis, whilst protecting normal tissues from radiotherapy adverse effects [123].

4.4.3. ROS Mitigation for Neurodegenerative Disorders

In Parkinson's disease, ceria can selectively scavenge mitochondrial, intracellular, and extracellular ROS [124]. For Alzheimer's, ceria has been combined with metal–organic framework (MOF) nanoparticles to attain multiple functions. They consisted of the co-delivery of small interfering RNA (siSOX9) and retinoic acid, as well as nanozyme activity mimicking superoxide dismutase (SOD) and catalase (CAT) enzymes based on the electron transfer between cerium (III) and cerium (IV). In vitro they promoted neural stem cells' differentiation to neurons and alleviated oxidative stress, leading to reduced cell death and increased neurite length. In vivo they ameliorated cognitive impairment in an Alzheimer's mouse model [125]. Finally, an artificial nanozyme, consisting of a ceria/polyoxometalate hybrid, displayed both proteolytic and SOD activities, so that it degraded amyloid β aggregates through oxidative damage and reduced intracellular ROS. Furthermore, it promoted PC12 cell proliferation, demonstrated an ability to cross the blood−brain barrier, and inhibited amyloid β-induced microglial cell activation. In addition, in vivo studies revealed a promising biocompatibility profile that paved the way to the development of multifunctional artificial nanozymes to treat neurological disorders [126].

4.4.4. ROS Mitigation to Treat the Liver and the Kidneys

Ceria NPs' ability to accumulate in the mononuclear phagocyte system (e.g., liver, spleen, and kidneys) can be used in an advantageous way for selective targeting of this nanosized therapy in areas of localized inflammation that require antioxidant treatment [127]. In particular, ceria NPs have been proposed to treat liver diseases thanks to their ROS and reactive nitrogen species (RNS) scavenging activity, as recently reviewed [128]. Another disease model where they have been studied is the non-alcohol-dependent, fatty liver, which is associated with impairment and inflammation of liver tissue, and which was alleviated in a rat model by ceria [129]. Treatment of hypoxia-induced acute kidney injury had been proposed for ceria combined with zirconia for enhanced radical scavenging activity [130].

4.4.5. ROS Mitigation for Osteoporosis

Additionally, ROS have been implicated in osteoporosis, as they can induce osteoblast and osteocyte apoptosis, thus promoting osteoclastogenesis, which is the formation of bone-resorbing cells. The ability of ceria NPs to scavenge ROS translated into antioxidant and osteogenic properties that were thus proposed for the treatment of osteoporosis [131]. This type of activity could be combined with bone tissue regeneration as discussed further below.

4.4.6. ROS Mitigation for Inflammatory Diseases and Immune System Regulation

Their ROS-scavenging activity was also proposed to treat rheumatoid arthritis [132], and to adjuvate in the treatment of traumas whereby blood transfusions are not an option, and nanocarriers to deliver hemoglobin are being studied as a possible alternative [133]. Ceria NPs have been stabilized with cyclodextrin that coated the NPs and encapsulated dithranol for the combined therapy of psoriasis, and they showed effective mitigation of ROS-induced damage in vitro and in vivo in a mouse model [34].

Cytotoxic CD8+ T cells (CTLs) play a key role to control intracellular pathogens as well as cancer. Their treatment with nanosized ceria led to higher production of cytokines, including interleukin-2 (IL-2) and tumor necrosis factor-a (TNF-a), higher release of effector molecules, higher killing activity, and stronger viral clearance capacity in vivo. Mechanistically, the treatment inhibited ROS production, and therefore promoted the activity of NF-KB signaling, overall promoting the cytotoxic activity of CTL cells [134].

ROS mitigation by ceria has been shown to limit the inflammatory response of overactivated microglia in a model of inflammation [135]. This effect could be beneficial to treat neuropathic pain, which is a chronic pathology that is caused by injury or dysfunction in the nervous system. An effective strategy to treat pain hypersensitivity employed microglia-targeting ceria-zirconia NPs that had been decorated with microglia-specific antibodies, enabling the rapid and effective inhibition of microglial activation, as demonstrated in a spinal nerve transection-induced neuropathic pain mouse model, proving the potent analgesic effect of the NPs [136]. Finally, lipid-coated magnetic silica NPs doped with ceria were designed for theranostics, as they allowed for MRI and mitigation of the inflammatory response of macrophages after being engulfed by these cells in a rodent model of intracerebral hemorrhage [137].

4.5. Reactive Nitrogen Species (RNS) Mitigation

Nitrogen stress can be caused by nitrogen excess or deficiency, and it is rather common in agriculture, where it can lead to impairment of plant growth. When tested in hydroponic rice stressed with altered nitrogen levels, ceria NPs mitigated the oxidative damage through the regulation of antioxidant enzymes, proline, and phytohormone levels. Interestingly, in the case of nitrogen deficiency, ceria NPs' dissolution, and consequent uptake and accumulation by the plant, was increased, and it was ascribed to altered root biomass and increased presence of carboxyl compounds in the root exudates. However, at high concentrations, ceria NPs inhibited plant growth in the absence of stress in the hydroponic culture. Therefore, ceria NPs can exert positive effects but their concentration needs to be monitored carefully [138]. Furthermore, it has been postulated that nanosized ceria may interfere with plant–bacteria interactions at the root level, as demonstrated in a study on nitrogen-fixing symbiotic organisms [139].

4.6. Sensing

4.6.1. Ceria as Nanozymes for Sensing

Ceria NPs find many applications in sensors' development, as recently reviewed [140]. In sensing, ceria NPs have been proposed as mimetics of various oxidizing enzymes, since they display increased stability relative to biological catalysts. To this end, they have been employed as glucose-oxidase mimics [141], but also peroxidase mimics in Enzyme-Linked Immunosorbent Assay (ELISA), as recently reviewed [142]. Ceria-titania mesoporous nanosheets' SOD mimicry has been used also for the electrochemical detection of $\cdot O_2-$ for biosensing [37].

Ceria's ability to mimic peroxidase, catalase, and oxidase enzymes [143] allows for the detection of hydrogen peroxide [144–147]. Hydrogen peroxide is produced by cancer cells and can be detected by ceria NPs in combination with $CuCo_2O_4$ nanosheets and carbon nanotubes for increased sensitivity [148]. Hydrogen peroxide is also produced by activated macrophages in atherosclerotic plaques, for which a ceria nanowire sensor was developed displaying fluorescent DNA for the competitive binding with hydrogen peroxide, as well as folic acid and CD36 antibody for cell targeting and imaging in vivo [149].

4.6.2. Sensing for Drinking Water and Food Safety

Ceria NPs have been envisaged for the quality assessment of both drinking water and food. In the first type of application, ceria NPs have been used to develop an electrochemical sensor for arsenic (V) in polluted waters using DNA. Ceria coordinates the phosphate of the nucleic acid, albeit with lower affinity relative to arsenic (V), so the latter can displace the DNA from ceria NPs and generate an electrochemical signal [150,151]. The ability of ceria NPs to bind free DNA was also exploited to adsorb antibiotic-resistance genes from tap water to target this issue of public health concern [152].

With regards to the molecular target for detection, ceria NPs have been proposed to sense hypoxanthin, which is a product of nucleotide degradation in fish and meat that can be used to monitor food freshness [153]. In this case, the NPs were immobilized with xanthine oxidase onto the surface of silanized paper, which was placed in contact with fish extract, so that hypoxanthine could be converted into xanthine first, and then uric acid, with concomitant generation of hydrogen peroxide to form molecular oxygen. The peroxide could then be reduced by ceria NPs that in turn get oxidized to colored Ce (IV) in a colorimetric paper-based sensor for monitoring food quality (Figure 12) [153].

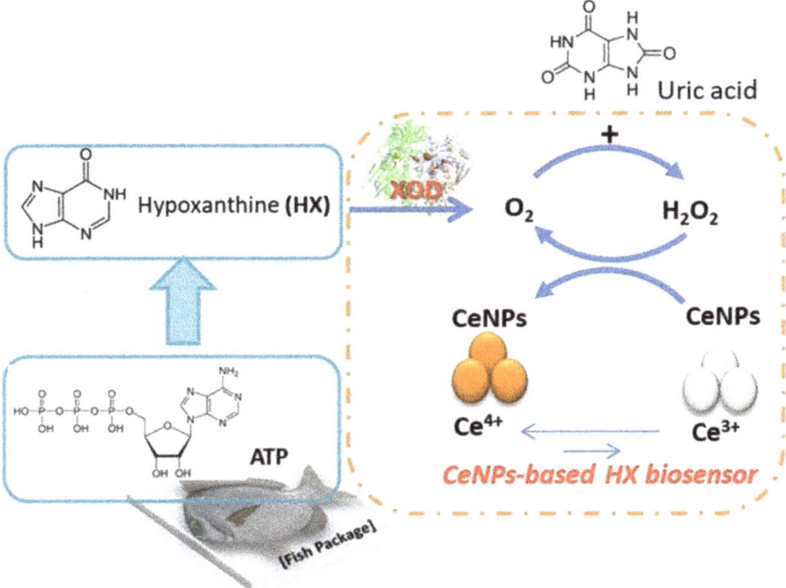

Figure 12. Ceria NPs (CeNPs) and xanthine oxidase (XOD) for the biosensing of hypoxanthine (HX) to monitor fish freshness. Reprinted from [153], Copyright © 2021, with permission from Elsevier.

Cerium oxide nanorods were used to develop an electrochemical DNA biosensor to detect *Salmonella* for food safety applications [154,155]. In addition, DNA from *E. coli* could be detected using a ceria-based biosensor that exploited its oxidase-mimicking activity [156]. Ceria was also used in combination with graphene quantum dots to develop a sensor to detect ochratoxin A, with relevance to food safety [157].

4.6.3. Sensing for the Detection of Disease Biomarkers

Ceria hollow NPs have also been proposed for sensing useful biomolecules for health monitoring. These not only included sensors for glucose [21,146,158–160], whose levels require careful monitoring for diabetic patients, but also the C-reactive protein [161], which is a biomarker for acute inflammation. Other sensing targets were the amyloid beta protein [162,163] and dopamine [164,165], both of which are relevant to neurodegenerative processes, and, in particular, Alzheimer's and Parkinson's disease, respectively. Ceria NPs were used to develop sensors for lactate, whose levels can be diagnostic of various diseases [166]. ATP [167], microRNAs [168,169], and DNA [170] were successfully detected too using ceria, for cancer diagnostics [171] or forensics applications [172]. Sensors were developed to monitor the presence of drugs, such as sulfonamide [173] and omeprazole [174]. Additional targets were a hypertension biomarker consisting of the epithelial sodium channel [175], and TNF-alpha for the early screening of neonatal diseases [176]. Ceria NPs were combined with iron oxide for the binding of carboxylic acid groups of a specific antibody that recognizes the carbohydrate antigen 19-9 as a cancer marker. In this case, the NPs were embedded in a mesoporous carbon matrix to develop an electrochemical sensor based on impedance variations for cancer detection [177].

4.6.4. DNA Sensors

Ceria NPs' affinity for phosphate groups and nucleic acids has been exploited in numerous DNA sensors. In one example, ceria NPs were anchored on gold nanorods that were combined with quantum dot-derivatized zinc oxide nanoflowers. The system was deposited on a paper working electrode to provide a photoelectric layer that was used to label one end of assistant DNA, for its subsequent hybridization with a capture probe to yield a triple helix. Incubation with a mixture of 4-chloro-1-naphthol and hydrogen peroxide resulted in peroxidase mimicry, leading to precipitation of the reaction product on the electrode and quenching of the signal. Conversely, the presence of the DNA triple helix led to the release of the gold nanorods-ceria NP system, thus removing their quenching effect, and leading to a photocurrent signal. Consequent migration of the nanozyme onto another area allowed for the conversion of a chromogenic substrate into a colored product for colorimetric visual detection [178].

In another example, ceria nanorods were coated with iridium nanorods because of their propensity to adsorb oxygen species, which could be beneficial for the peroxidase-like activity of ceria NPs. The nanomaterials were combined with an aptamer that recognizes a protein overexpressed on the surface of cancer cells for their detection. Sensitivity of the system was enhanced through the inclusion of DNA walker technology. DNA walker strands display RNAse-like activity and cleave the RNA of chimeric DNA/RNA oligonucleotides (D-RNA). The name comes from the fact that the strand displacement reaction causes the DNA walker strands to "walk forward" along the adjacent D-RNA, thus continuously cleaving it and amplifying the signal through the release of many probes [179].

Ceria NPs also display glucose-oxidase biomimicking ability, as they catalyze the oxidation of glucose to gluconic acid, with an activity that is more pronounced for smaller NPs with higher surface area. This property was exploited for the development of a sensor that could detect DNA products amplified through PCR. The amplified nucleic acids bound to ceria and led to NP aggregation, thus impeding the nanozyme activity. Therefore, in the presence of glucose, its levels could be related to the amplified DNA [141]. In another example, CuMn-CeO_2 functionalized with luminol was successfully applied to develop an electrochemiluminescent biosensor to specifically detect Group B *Streptococci* from clinical vaginal and anal swabs [180].

4.7. Medical Implants

Ceria has been studied as an additive for well-known materials used in medical implants in virtue of its antioxidant and anti-inflammatory properties and to favor the direct osseointegration at the biointerface. This concept was applied to titania alloys [181], and nanotubes, and an increased level of early hydroxyapatite precipitation was noted from simulated body fluid [182]. Furthermore, ceria-stabilized zirconia has been proposed as a promising biomedical implant material. In particular, TEM imaging revealed zirconia nanocrystals directly bound to osteoblastic cell-precipitated hydroxyapatite crystals at the lattice-fringe scale, without any pretreatment of the substrate surface [183]. These results suggested the possibility for nanoscale direct osseointegration with bone in vivo (Figure 13) to improve the outcome of dental implants, as opposed to titanium surfaces that typically get coated by cell-secreted proteins [183].

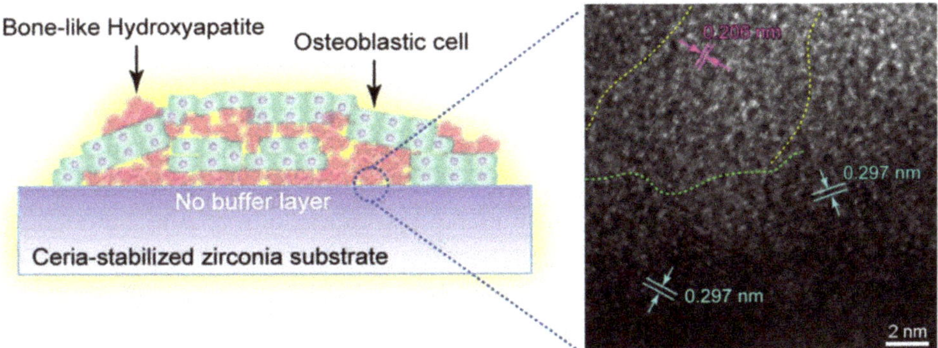

Figure 13. (**a**) Schematic representation of osteoblasti-cell (green) deposited hydroxyapatite (pink) into a ceria-zirconia substrate. (**b**) HR-TEM micrograph of the interface (green-dotted line) between the hydroxyapatite nanofibers (d = 0.206 nm of (400) face, magenta) and the zirconia nanocrystal (d = 0.297 nm of (101) face, cyan) at the nanoscale. Reproduced from [183], Copyright © 2021, with permission from Elsevier.

The valence state of cerium is an important factor for the performance of ceria coatings on medical implants. Plasma spraying allows one to fine-tune the Ce (III)/Ce (IV) ratio in the resulting materials as shown on titanium devices. Interestingly, macrophages' adhesion was restricted by a higher level of Ce (III) and promoted by an abundance of Ce (IV), which also resulted in an M2-type response associated with healing, as opposed to the M1 pro-inflammatory response, which was ascribed to the catalase-like activity of ceria that mitigates ROS [184].

4.8. Antimicrobial Activity

Ceria NPs exerted antimicrobial activity that was proposed for wastewater treatment [185], but also for wound-healing applications [14]. It can control biofilm formation and quorum sensing [186]. The antibacterial activity exerted by nanoceria was shown to be more effective on Gram-negative than Gram-positive strains [17]. Ceria demonstrated the ability to mimic haloperoxidase, and this property was applied to develop antimicrobials. In particular, a bioinspired strategy was used, since marine organisms rely on biohalogenation as a defense strategy against bacterial colonization, since it interferes with bacterial communication. Bismuth substitution enhanced the enzyme-like activity of ceria NPs three-fold, and the incorporation of the nanosystem into polyethersulfone beads, which are typical constituents of water filter membrane supports, led to decreased adhesion of the Gram-negative soil bacterium *Pseudomonas aeruginosa* and of *Phaeobacter gallaeciensis*, a primary bacterial colonizer in marine biofilms [187].

Streptococcus mutans forms biofilms that contribute to dental caries in the presence of fermentable carbohydrates and constitutes a target for therapeutic intervention. Ceria NPs reduced bacterial adhesion by 40%, and planktonic growth and dispersal assays supported a non-bactericidal mode of biofilm inhibition [188]. A detailed study of the bio–nano interface demonstrated that rod-like ceria NPs could be reduced by *Bacillus subtilis* under planktonic conditions, so that the Ce (III) ions adjacent to the surface oxygen vacancies would be chelated by the adsorption sites present on the bacterial cell wall. The bacterial biosorption of the dissolved Ce (III) ions unveiled a new mechanism for the toxicity of ceria NPs [189]. Ceria was also combined with zinc oxide onto halloysite nanotubes to reduce NP agglomeration and improve interfacial reactions between the nanocomposite and bacterial cells. The synergistic effects of the different components led to superior antibacterial activity of the nanocomposite against *Escherichia coli* [190].

Finally, the nitric oxide (NO) donor, *S*-nitroso-*N*-acetylpenicillamine (SNAP), was administered together with ceria NPs to induce synergistic antimicrobial effects against *Staphylococcus aureus*, *Escherichia coli*, and *Candida albicans*, as model organisms for Gram-positive bacteria, Gram-negative bacteria, and fungi, respectively. The best results were obtained with equimolar solutions of 3 mM for each agent and provided a promising outlook for the future development of broad-spectrum antimicrobials [191].

4.9. Tissue Engineering

The use of nanosized ceria for tissue engineering has been recently reviewed [192]. In particular, the ROS mitigation ability and positive interaction with hydroxyapatite nucleation render nanoceria highly promising for the regeneration of bone tissue. To this end, it was combined with an alginate-gelatin scaffold that demonstrated enhanced cell attachment and proliferation as well as the promotion of mesenchymal stem cell osteogenic differentiation [193], which was also shown for ceria NPs alone [194]. Pulsed laser deposition was applied to produce pyramid-shaped nanosized ceria for bone tissue regeneration. Modulation of the NP nuclei density allowed for the fine-tuning of the hydrophilic character of the material. The nanosized ceria induced a reorganization of the cell cytoskeleton in both osteosarcoma and osteoprogenitor cells, with the former showing elongated cell bodies and the latter increased adhesion, which might be beneficial for osteogenic differentiation [195].

Ceria NPs have been used to attain nanocomposites with hydroxyapathite for bone tissue regeneration [12]. Ceria NPs accelerated the formation of new bone and boosted endochondral ossification-based bone regeneration in both a subcutaneous ectopic osteogenesis model and a mouse model of critically sized bone defects. In particular, ceria NPs significantly promoted endochondral ossification-based bone regeneration by ensuring sufficient hypertrophic differentiation [196].

Ceria NPs have been applied to dental pulp regeneration. They were embedded as an insoluble antioxidant in a mineral trioxide aggregate that served as a biomaterial scaffold. Ceria accelerated odontoblastic differentiation of dental pulp stem cells via ROS downregulation, with minimal influence on the physico-mechanical properties of the scaffold [197]. ROS mitigation by nanoceria was also used for wound healing using a miRNA-loaded hydrogel to promote angiogenesis in the oxidative environment of diabetic wounds [198].

4.10. Energy Applications

Microalgae are an attractive source of biomass, and, potentially, biofuels. In particular, hydrothermal liquefaction is one of the main processes used to degrade lignocellulosic biomass into a crude bio-oil in hot compressed water. To this end, ceria NPs with a crystallite size of 6 nm were successfully applied as catalysts for the production of bio-oil from Spirulina platensis, in an attempt to exploit ceria's ability to bind to oxygen-bearing organic compounds and catalyze their decomposition into biofuel [199]. Another study demonstrated efficient hydrothermal liquefaction of Scenedesmus obliquus microalgae into biofuel by using ceria nanorods decorated with nickel and activated carbon [200]. The same type of process was also catalyzed by ceria using rice straw as a starting material to produce bio-oil [201]. Ceria has been applied also as a catalyst for the pyrolysis of cellulose, which is the major constituent of lignocellulosic biomass, for the production of high-value bio-oil [202]. Carbon monoxide oxidation is another key reaction catalyzed by ceria nanocrystals [48], as well as carbon dioxide conversion into dimethyl carbonate [203].

4.11. Optoelectronics and Bioimaging

With human progress, we have witnessed an exponential increase in optoelectronic devices that require illumination. To this end, light-emitting diodes (LEDs) are highly attractive for their low-energy consumption, extended working lifetime, high efficacy, and environmental friendliness. In particular, lanthanides are promising to generate phosphors with chromatic tunability. Therefore, praseodymium was used to dope ceria nanopowders generated using *Aloe vera* as a fuel for a green combustion process, and yielded an orange-red light emitter [49]. Europium has been combined with ceria too, to yield red phosphors for optoelectronics [38].

Ytterbium and thulium were used to dope ceria NPs to exploit their near-infrared light absorptivity and up-conversion ability for deep-tissue imaging. Lanthanides are highly attractive for biomedical imaging thanks to their low toxicity, high stability, and minimized background auto-fluorescence. Furthermore, during the up-conversion, part of the absorbed light is converted into heat, thus upconverting NPs can serve as photothermal agents, potentially useful for concomitant sterilization during biomedical imaging. However, in vitro studies on *E. coli* as a bacterial model demonstrated a modest 44% sterilization efficacy when the doped NPs were irradiated at 980 nm [204]. Finally, ceria NPs have been embedded into electrospun chitosan nanofibers to endow them with fluorescence at 520 nm upon excitation at 430 nm for bioimaging [205].

4.12. Amyloidosis Inhibition

Amyloid-associated disorders constitute a large and diverse class of pathological states of growing societal concern, as the increasing life expectancy is leading to an increasing aged population, and consequent occurrence of amyloidoses. Amongst the many potential treatments, NPs have attracted growing interest as inhibitors of amyloid aggregation, including those formed by peptides [206], carbon nanostructures [207], gold [208], and also ceria. In particular, ceria NPs were shown to bind to the α-synuclein monomer and extend the lag phase time of amyloid fibril formation, and the resulting aggregates are relatively less toxic than those formed in the absence of ceria [209]. It will be interesting in the future to assess whether interesting new properties can arise from the use of other nanomorphologies, since it was shown that neurons' interaction is favored when in contact on anisotropic, elongated nanomaterials (e.g., tubes, rods, fibers) [210].

Using a different approach, magnetite-nanoparticle assemblies were coated by nanosized ceria and further functionalized with anti-Aβ antibodies for body cleansing from amyloids. In this case, extracorporeal treatment of blood allowed for the selective capture of Aβ into the NP assemblies (Figure 14), which were separated form blood by using magnetic fields, and ceria allowed for the reduction of the oxidative stress generated by the concomitant immune response to the treatment [211].

4.13. Toxicological Studies

The widespread use of ceria in catalytic filters by the automobile industry has raised some concerns over potential undesired effects arising from its release in the environment. Therefore, the vast majority of toxicological studies have been focused either on aquatic or agriculture-relevant organisms. In the first group, studies have analyzed toxicity on diatoms [212], algae [213], clams [214], mussels [215], oysters [216], and zebrafish [217]. For instance, ceria NPs have been coated with polysaccharides (levan and pullulan) or a monosaccharide (glucose) and their toxicity was evaluated on three aquatic organisms: The bacterium *Vibrio fischeri*, the crustacean *Daphnia magna*, and zebrafish *Danio rerio*. The last one is a common model for toxicological studies on embryo development and revealed no adverse effects for either coated or uncoated NPs. The coating reduced the toxicity on bacteria and crustaceans, although it increased the bioaccumulation in the latter organism, for which respiration levels also appeared altered. However, no adverse effects were noted for NP concentrations up to 200 mg per liter [218]. Ceria NPs were also studied for their effects on river biofilm communities and were showed to exert selection pressure that ultimately altered the composition of the microbial community [219].

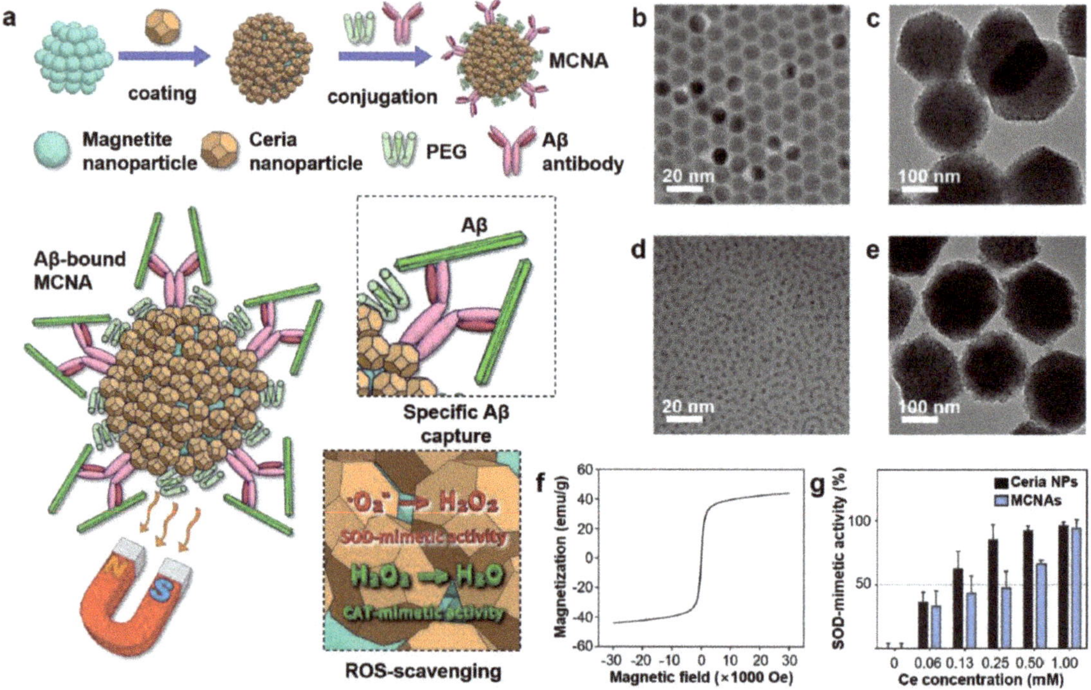

Figure 14. Magnetite/ceria NP assemblies (MCNAs). (**a**) Schematic preparation of MCNAs. (**b–e**) TEM images of magnetite NPs (**b**), their assemblies (**c**), ceria NPs (**d**), and MCNAs (**e**). (**f**) Magnetization curve of MCNAs. (**g**) SOD-mimetic activity of MCNAs. Reproduced with permission from [211], Copyright © 2021 WILEY-VCH Verlag GmbH & Co. KGaA, Weinheim.

In the second group, nanosized ceria's toxicity has been evaluated on agricultural crops [220], the tomato plant [221], hydroponic cucumber plants [222], and so on. Ceria NPs and released Ce (III) ions were also evaluated for their different mechanisms of inducing toxicity, in a study on plant growth and physiological and nutritional parameters relevant to applications in agriculture [223]. Ceria nanostructures demonstrated better biocompatibility towards plant tissues than analogous silver nanostructures [224]; however, transcriptome changes were noted in plants after exposure to nanoceria [225,226]. Ceria NPs' effects were also noted on soil bacterial community composition [227] and rhizosphere biocommunities and were studied for their accumulation in soybeans [228]. An interesting study on ceria NPs applied to the leaves of common bean plants to evaluate adverse effects showed that both plant growth and development were not negatively affected, and no variation in the nutritional quality of the pods was noted other than mineral contents. However, dose-dependent oxidative damage occurred in the leaves, giving scope for further investigations on the biochemical effects of plants' exposure to ceria NPs [229].

Positive effects of ceria NPs were instead noticed on tomato plant growth and metabolism up to 100 mg/L concentrations. However, higher levels (>200 mg/L) were detrimental to the growth and metabolism of the test plant and severe oxidative stress, with significant reduction in pigment, increased lipid peroxidation, electrolyte leakage, and H_2O_2 content. The activities of antioxidant enzymes were significantly upregulated [13]. This result may be attributed to SOD mimetic activity of nanoceria while the toxicity above the optimum concentration was probably attributed to the biotransformation of NPs and the high sensitivity of test seedlings exposed to the released Ce^{3+} ions. Further research is therefore required to gain deeper insights in understanding the uptake, accumulation, translocation, biotransformation, and toxicity of nanoceria in food crops, as well as the subsequent impact of possible transmission up the food chain.

Given the well-established use of ceria in car exhausts for the abatement of toxic gases, studies assessed the possibility of using ceria as diesel additives to reduce the emission of toxic compounds (such as benzo(a)pyrene) from the incomplete combustion of fossil fuels, and to increase fuel economy. As mentioned above, the potential release of the NPs in the environment required an accurate evaluation of the toxicity of nano-sized ceria, either alone or in combination with benzo(a)pyrene on reproductive cells revealing DNA damage [230]. This result was confirmed on sperm cells, where DNA damage [231] and oxidative stress due to ROS generation was found [232]. A report by Sundararajan et al. highlighted that *Drosophila* insects, in various developmental stages, were not liable to any significant toxicity in third instar larvae and adult flies over one month of continuous oral administration of ceria NPs up to 1 mM doses [233].

A route of potential exposure to nanosized ceria for humans is inhalation, therefore this type of exposure and potential pulmonary toxicity has been the subject of many investigations, in various models. Ceria NPs of heterogeneous size distribution ranging from a few nm to over 100 nm in diameter demonstrated no cytotoxicity on cell cultures exposed to NP aerosols [234]. The pulmonary and systemic effects of nanoceria exposure in mice were studied through repeated transnasal instillation, in order to assess the potential health risks posed by airborne nanoceria. In this study, 7-nm and 25-nm ceria NPs were used as representative models, for the most common ceria-NP fuel additives. Lung damage was manifest, with consequent penetration of nanoceria through the air–blood barrier, leading to NP distribution to other organs, especially the liver and the spleen. Interestingly, nanoceria could also reach the central nervous system through the olfactory nerve. Overall, the systemic accumulation triggered lipid peroxidation in multiple organs, with the smaller NPs inducing more severe pulmonary damage, albeit similar systemic toxicity, relative to larger ones [235]. Pulmonary exposure to ceria NPs aggravates vascular toxicity in rats with vascular injury induced by the anticancer cisplatin, through mechanisms that involve oxidative stress, inflammation, and DNA damage [236]. Inhaled ceria NPs can penetrate deeply into the lungs and it was shown that they can inhibit the formation of the pulmonary surfactant lining of alveoli. The effect depends on various physicochemical properties of the NPs, such as size, hydrophobicity, dissolution rate, and aggregation state at the biological interface [237]. However, no genotoxicity was noted in rat blood cells after inhalation of ceria NPs over 6 months [238].

Another potential route for ceria uptake is through the skin. Surprisingly there are scarce studies on this area. One recent work analyzed cerium oxide NPs dispersed in synthetic sweat using excised human skin on Franz cells and revealed very low dermal absorption and transdermal permeation of cerium [239].

Hemocompatibility and anticoagulant, anti-inflammatory, and anti-senescence activity of ceria NPs were explored in vitro, with no significant effects on the coagulation process, hemolysis, or platelet aggregation. In human endothelial cells, ceria did not affect cell viability, although it reduced oxidative stress and inhibited the expression of an inflammatory phenotype. Notably, it reduced telomere shortening, thus demonstrating the potential to counteract premature senescence [240]. However, ceria also impaired neuronal differentiation of neural stem cells [241]. Finally, ceria NPs' interaction with lipid bilayers has been investigated as concerns were raised for possible cell membrane disruption [242].

Overall, it appears that no major toxicity of nanosized ceria was manifest, albeit some evidence was provided on adverse repercussions on specific biological studies, depending on several parameters, also including the route of administration, the dose, the extent, and frequency of exposure, and so on. There are many parameters that can influence ceria nanoparticle biocompatibility including size, surface charge, and crystalline phase [243]. Therefore, more studies in this area are imperative to make progress in laying the roadmap of nanosized ceria used in biological organisms and in the environment. A key aspect will be to employ models and dosages that are relevant to realistic conditions of exposure depending on the intended use for this nanomaterial.

5. Conclusions

In conclusion, the remarkable redox properties and oxygen-binding ability of ceria have been long-exploited by industry, especially relevant to catalytic applications, preservation, and remediation of the environment and the energy sector. In medicine, nanosized ceria has been attracting growing interest in recent years [244], and we have witnessed the development of a wide number of applications, especially in proof-of-concept studies that used its nanozyme activity either in sensing or to mitigate the effects of oxidative stress associated with many existing pathologies, ranging from cancer to inflammation diseases. In particular, a recent area of investigation that warrants scope for further studies is the application of ceria anti-inflammatory activity to address unmet clinical needs, such as systemic inflammatory syndromes [245] or brain diseases [246]. However, such challenging applications will certainly need many more years of preclinical studies to reach patients [247], as opposed to nanoceria use in dental nanocomposites, where the first clinical trials have already appeared, showing excellent promise [248].

The use of greener production methods for nanosized ceria has also been gaining momentum, as well as for other types of nanomaterials [249,250], and the use of biotemplates is becoming one of the most popular approaches. However, thus far, the vast majority of studies either exploited naturally occurring microstructures to impart specific morphology to ceria at the microscale, or natural extracts typically displaying a rather complex mixture of molecular compounds, not often very well defined.

It thus appears that there are unexplored routes for the nanoscale definition of ceria using well-defined biomolecular templates, for instance vastly explored for metal NPs using self-assembling peptides [251]. In line with this principle, it could be possible to use different supramolecular geometries to template ceria into various nanomorphologies, for enhanced properties. For instance, ceria nanorods were recently described with persistent porosity for engineered catalytic sites [252], and nanoengineering methods for the crystal microenvironment are highly sought after to tailor the redox performance of nanosized ceria [253].

Furthermore, using molecular gelators as templates [254] can offer further advantages as it can yield nanostructured gels as functional materials with increased NP stability [255,256]. The combination of different molecular and supramolecular components can lead to a qualitative leap in innovation in areas spanning from pollutants' removal to catalysis, soft robotics, medicine, and agriculture [257–261]. Therefore, the future for nanosized ceria is bright, especially if combined with other innovative nanomaterials and green approaches.

Author Contributions: Writing—original draft preparation, P.R.; writing—review and editing, P.F., M.M., and S.M. All authors have read and agreed to the published version of the manuscript.

Funding: Part of the described research was funded by the University of Trieste (FRA2021 to M.M.).

Conflicts of Interest: The authors declare no conflict of interest.

References

1. Montini, T.; Melchionna, M.; Monai, M.; Fornasiero, P. Fundamentals and catalytic applications of CeO_2-based materials. *Chem. Rev.* **2016**, *116*, 5987–6041. [CrossRef]
2. Melchionna, M.; Fornasiero, P. The role of ceria-based nanostructured materials in energy applications. *Mater. Today* **2014**, *17*, 349–357. [CrossRef]
3. Kang, T.; Kim, Y.G.; Kim, D.; Hyeon, T. Inorganic nanoparticles with enzyme-mimetic activities for biomedical applications. *Coord. Chem. Rev.* **2020**, *403*, 213092. [CrossRef]
4. Kaplin, I.Y.; Lokteva, E.S.; Golubina, E.V.; Lunin, V.V. Template synthesis of porous ceria-based catalysts for environmental application. *Molecules* **2020**, *25*, 4242. [CrossRef]
5. Krishnaveni, P.; Priya, M.L.; Annadurai, G. Biosynthesis of nanoceria from *Bacillus subtilis*: Characterization and antioxidant potential. *Res. J. Life Sci. Bioinf. Pharm. Chem. Sci.* **2019**, *5*, 632–644. [CrossRef]

6. Ishak, N.A.I.M.; Kamarudin, S.K.; Timmiati, S.N. Green synthesis of metal and metal oxide nanoparticles via plant extracts: An overview. *Mater. Res. Express* **2019**, *6*, 112004. [CrossRef]
7. Elahi, B.; Mirzaee, M.; Darroudi, M.; Sadri, K.; Kazemi Oskuee, R. Bio-based synthesis of Nano-Ceria and evaluation of its bio-distribution and biological properties. *Colloids Surf. B* **2019**, *181*, 830–836. [CrossRef]
8. Elahi, B.; Mirzaee, M.; Darroudi, M.; Kazemi Oskuee, R.; Sadri, K.; Amiri, M.S. Preparation of cerium oxide nanoparticles in salvia macrosiphon boiss seeds extract and investigation of their photo-catalytic activities. *Ceram. Int.* **2019**, *45*, 4790–4797. [CrossRef]
9. Antony, D.; Yadav, R. Facile fabrication of green nano pure CeO_2 and Mn-decorated CeO_2 with *Cassia angustifolia* seed extract in water refinement by optimal photodegradation kinetics of malachite green. *Environ. Sci. Pollut. Res.* **2021**, *28*, 18589–18603. [CrossRef]
10. Zamani, A.; Marjani, A.P.; Alimoradlu, K. Walnut shell-templated ceria nanoparticles: Green synthesis, characterization and catalytic application. *Int. J. Nanosci.* **2018**, *17*, 1850008. [CrossRef]
11. Wang, C.; Chen, F.; Tang, Y.; Chen, X.; Qian, J.; Chen, Z. Advanced visible-light photocatalytic property of biologically structured carbon/ceria hybrid multilayer membranes prepared by bamboo leaves. *Ceram. Int.* **2018**, *44*, 5834–5841. [CrossRef]
12. Thirumamagal, R.; Fowziya, S.A.; Mohideen, A.M.U.; Beevi, A.H.; Ayeshamariam, A.; Saleem, A.M.; Jayachandran, M. Evaluation of the cytotoxicity effect on HAp doped with Ce_2O_3 and its assessment with breast cancer cell line of MCF-7. *J. Bionanosci.* **2018**, *12*, 350–356. [CrossRef]
13. Singh, A.; Hussain, I.; Singh, N.B.; Singh, H. Uptake, translocation and impact of green synthesized nanoceria on growth and antioxidant enzymes activity of *Solanum lycopersicum* L. *Ecotoxicol. Environ. Saf.* **2019**, *182*, 109410. [CrossRef]
14. Rajan, A.R.; Vilas, V.; Rajan, A.; John, A.; Philip, D. Synthesis of nanostructured CeO_2 by chemical and biogenic methods: Optical properties and bioactivity. *Ceram. Int.* **2020**, *46*, 14048–14055. [CrossRef]
15. Qian, J.; Zhang, W.; Wang, Y.; Chen, Z.; Chen, F.; Liu, C.; Lu, X.; Li, P.; Wang, K.; Chen, A. Visible-light driven nitrogen-doped petal-morphological ceria nanosheets for water splitting. *Appl. Surf. Sci.* **2018**, *444*, 118–125. [CrossRef]
16. Wu, B.; Shan, C.; Zhang, X.; Zhao, H.; Ma, S.; Shi, Y.; Yang, J.; Bai, H.; Liu, Q. CeO_2/Co_3O_4 porous nanosheet prepared using rose petal as biotemplate for photocatalytic degradation of organic contaminants. *Appl. Surf. Sci.* **2021**, *543*, 148677. [CrossRef]
17. Muthuvel, A.; Jothibas, M.; Mohana, V.; Manoharan, C. Green synthesis of cerium oxide nanoparticles using *Calotropis procera* flower extract and their photocatalytic degradation and antibacterial activity. *Inorg. Chem. Commun.* **2020**, *119*, 108086. [CrossRef]
18. Zhou, M.; Zhang, K.; Chen, F.; Chen, Z. Synthesis of biomimetic cerium oxide by bean sprouts bio-template and its photocatalytic performance. *J. Rare Earths* **2016**, *34*, 683–688. [CrossRef]
19. Patil, S.N.; Paradeshi, J.S.; Chaudhari, P.B.; Mishra, S.J.; Chaudhari, B.L. Bio-therapeutic potential and cytotoxicity assessment of pectin-mediated synthesized nanostructured cerium oxide. *Appl. Biochem. Biotechnol.* **2016**, *180*, 638–654. [CrossRef]
20. Gnanasekaran, L.; Rajendran, S.; Priya, A.K.; Durgalakshmi, D.; Vo, D.-V.N.; Cornejo-Ponce, L.; Gracia, F.; Soto-Moscoso, M. Photocatalytic degradation of 2,4-dichlorophenol using bio-green assisted TiO_2-CeO_2 nanocomposite system. *Environ. Res.* **2021**, *195*, 110852. [CrossRef] [PubMed]
21. Yang, W.; Li, J.; Yang, J.; Liu, Y.; Xu, Z.; Sun, X.; Wang, F.; Ng, D.H.L. Biomass-derived hierarchically porous CoFe-LDH/CeO_2 hybrid with peroxidase-like activity for colorimetric sensing of H_2O_2 and glucose. *J. Alloys Compd.* **2020**, *815*, 152276. [CrossRef]
22. Stegmayer, M.A.; Milt, V.G.; Miro, E.E. Biomorphic synthesis of cobalt oxide and ceria microfibers. Their application in diesel soot oxidation. *Catal. Commun.* **2020**, *139*, 105984. [CrossRef]
23. Charbgoo, F.; Ahmad, M.B.; Darroudi, M. Cerium oxide nanoparticles: Green synthesis and biological applications. *Int. J. Nanomed.* **2017**, *12*, 1401–1413. [CrossRef] [PubMed]
24. Wang, Y.; Su, H.-J.; Hua, Q.-R.; Wang, S.-D. From nanoparticles to nanorods: Insights into the morphology changing mechanism of ceria. *Ceram. Int.* **2018**, *44*, 23232–23238. [CrossRef]
25. Smith, L.R.; Sainna, M.A.; Douthwaite, M.; Davies, T.E.; Dummer, N.F.; Willock, D.J.; Knight, D.W.; Catlow, C.R.A.; Taylor, S.H.; Hutchings, G.J. Gas phase glycerol valorization over ceria nanostructures with well-defined morphologies. *ACS Catal.* **2021**, *11*, 4893–4907. [CrossRef]
26. Yang, D.; Fa, M.; Gao, L.; Zhao, R.; Luo, Y.; Yao, X. The effect of DNA on the oxidase activity of nanoceria with different morphologies. *Nanotechnology* **2018**, *29*, 385101–385110. [CrossRef] [PubMed]
27. Fisher, T.J.; Zhou, Y.; Wu, T.-S.; Wang, M.; Soo, Y.-L.; Cheung, C.L. Structure-activity relationship of nanostructured ceria for the catalytic generation of hydroxyl radicals. *Nanoscale* **2019**, *11*, 4552–4561. [CrossRef] [PubMed]
28. Mehmood, R.; Wang, X.; Koshy, P.; Yang, J.L.; Sorrell, C.C. Engineering oxygen vacancies through construction of morphology maps for bio-responsive nanoceria for osteosarcoma therapy. *CrystEngComm* **2018**, *20*, 1536–1545. [CrossRef]
29. Molinari, M.; Symington, A.R.; Sayle, D.C.; Sakthivel, T.S.; Seal, S.; Parker, S.C. Computer-aided design of nanoceria structures as enzyme mimetic agents: The role of bodily electrolytes on maximizing their activity. *ACS Appl. Bio Mater.* **2019**, *2*, 1098–1106. [CrossRef]
30. Patel, V.; Singh, M.; Mayes, E.L.H.; Martinez, A.; Shutthanandan, V.; Bansal, V.; Singh, S.; Karakoti, A.S. Ligand-mediated reversal of the oxidation state dependent ROS scavenging and enzyme mimicking activity of ceria nanoparticles. *Chem. Commun.* **2018**, *54*, 13973–13976. [CrossRef] [PubMed]
31. Kaygusuz, H.; Erim, F.B. Biopolymer-assisted green synthesis of functional cerium oxide nanoparticles. *Chem. Pap.* **2020**, *74*, 2357–2363. [CrossRef]

32. Zhang, L.-H.; Zhou, J.; Liu, Z.-Q.; Guo, J.-B. Mesoporous CeO_2 catalyst synthesized by using cellulose as template for the ozonation of phenol. *Ozone Sci. Eng.* **2019**, *41*, 166–174. [CrossRef]
33. Hasanzadeh, L.; Kazemi Oskuee, R.; Sadri, K.; Nourmohammadi, E.; Mohajeri, M.; Mardani, Z.; Hashemzadeh, A.; Darroudi, M. Green synthesis of labeled CeO_2 nanoparticles with ^{99m}Tc and its biodistribution evaluation in mice. *Life Sci.* **2018**, *212*, 233–240. [CrossRef] [PubMed]
34. Wu, L.; Liu, G.; Wang, W.; Liu, R.; Liao, L.; Cheng, N.; Li, W.; Zhang, W.; Ding, D. Cyclodextrin-modified CeO_2 nanoparticles as a multifunctional nanozyme for combinational therapy of psoriasis. *Int. J. Nanomed.* **2020**, *15*, 2515–2527. [CrossRef]
35. Xiong, Z.-B.; Li, Z.-Z.; Li, C.-X.; Wang, W.; Lu, W.; Du, Y.-P.; Tian, S.-L. Green synthesis of tungsten-doped CeO_2 catalyst for selective catalytic reduction of NO_x with NH_3 using starch bio-template. *Appl. Surf. Sci.* **2021**, *536*, 147719. [CrossRef]
36. Salimi, K. Self-assembled bio-inspired Au/CeO_2 nano-composites for visible white LED light irradiated photocatalysis. *Colloids Surf. A* **2020**, *599*, 124908. [CrossRef]
37. Wang, Z.; Zhao, H.; Gao, Q.; Chen, K.; Lan, M. Facile synthesis of ultrathin two-dimensional graphene-like CeO_2-TiO_2 mesoporous nanosheet loaded with Ag nanoparticles for non-enzymatic electrochemical detection of superoxide anions in HepG2 cells. *Biosens. Bioelectron.* **2021**, *184*, 113236. [CrossRef] [PubMed]
38. Deepthi, N.H.; Darshan, G.P.; Basavaraj, R.B.; Prasad, B.D.; Nagabhushana, H. Large-scale controlled bio-inspired fabrication of 3D CeO_2:Eu^{3+} hierarchical structures for evaluation of highly sensitive visualization of latent fingerprints. *Sens. Actuators B* **2018**, *255*, 3127–3147. [CrossRef]
39. Reddy Kannapu, H.P.; Kim, M.; Jeong, C.; Suh, Y.-W. An efficient Cu-CeO_2 citrate catalyst for higher aliphatic ketone synthesis via alkali-free alkylation of acetone with butanol. *Mater. Chem. Phys.* **2019**, *229*, 402–411. [CrossRef]
40. Prabha, J.P.S.; Tharayil, N.J. Crystal plane effect on antioxidant efficacy of nanoceria synthesized with assistance of DNA. *J. Phys. Chem. Solids* **2020**, *141*, 109421. [CrossRef]
41. Jyothi, P.S.P.; Anitha, B.; Smitha, S.; Vibitha, B.V.; Krishna, P.G.A.; Tharayil, N.J. DNA-assisted synthesis of nanoceria, its size dependent structural and optical properties for optoelectronic applications. *Bull. Mater. Sci.* **2020**, *43*, 119. [CrossRef]
42. Wang, M.; Wang, M.-F.; Wang, Y.-M.; Shen, J.-W.; Wang, Z.-Y.; Gao, H.; Wang, L.-L.; Ouyang, X. DNA assisted synthesis of CeO_2 nanocrystals with enhanced peroxidase-like activity. *CrystEngComm* **2018**, *20*, 4075–4079. [CrossRef]
43. Gong, K.; Zhou, K.; Yu, B. Superior thermal and fire safety performances of epoxy-based composites with phosphorus-doped cerium oxide nanosheets. *Appl. Surf. Sci.* **2020**, *504*, 144314. [CrossRef]
44. Zou, S.; Zhu, X.; Zhang, L.; Guo, F.; Zhang, M.; Tan, Y.; Gong, A.; Fang, Z.; Ju, H.; Wu, C.; et al. Biomineralization-inspired synthesis of cerium-doped carbonaceous nanoparticles for highly hydroxyl radical scavenging activity. *Nanoscale Res. Lett.* **2018**, *13*, 76. [CrossRef]
45. Yang, Z.; Luo, S.; Zeng, Y.; Shi, C.; Li, R. Albumin-mediated biomineralization of shape-controllable and biocompatible ceria nanomaterials. *ACS Appl. Mater. Interfaces* **2017**, *9*, 6839–6848. [CrossRef] [PubMed]
46. Okuda, M.; Suzumoto, Y.; Yamashita, I. Bioinspired synthesis of homogenous cerium oxide nanoparticles and two- or three-dimensional nanoparticle arrays using protein supramolecules. *Cryst. Growth Des.* **2011**, *11*, 2540–2545. [CrossRef]
47. Lach, M.; Kuenzle, M.; Beck, T. Free-standing metal oxide nanoparticle superlattices constructed with engineered protein containers show in crystallo catalytic activity. *Chem.—Eur. J.* **2017**, *23*, 17482–17486. [CrossRef] [PubMed]
48. Curran, C.D.; Lu, L.; Jia, Y.; Kiely, C.J.; Berger, B.W.; McIntosh, S. Direct single-enzyme biomineralization of catalytically active ceria and ceria-zirconia nanocrystals. *ACS Nano* **2017**, *11*, 3337–3346. [CrossRef] [PubMed]
49. Basavaraj, R.B.; Navami, D.; Deepthi, N.H.; Venkataravanappa, M.; Lokesh, R.; Sudheer Kumar, K.H.; Sreelakshmi, T.K. Novel orange-red emitting Pr^{3+} doped CeO_2 nanopowders for white light emitting diode applications. *Inorg. Chem. Commun.* **2020**, *120*, 108164. [CrossRef]
50. Rahdar, A.; Aliahmad, M.; Hajinezhad, M.R.; Samani, M. Xanthan gum-stabilized nano-ceria: Green chemistry based synthesis, characterization, study of biochemical alterations induced by intraperitoneal doses of nanoparticles in rat. *J. Mol. Struct.* **2018**, *1173*, 166–172. [CrossRef]
51. Villa, S.; Maggioni, D.; Hamza, H.; Di Nica, V.; Magni, S.; Morosetti, B.; Parenti, C.C.; Finizio, A.; Binelli, A.; Della Torre, C. Natural molecule coatings modify the fate of cerium dioxide nanoparticles in water and their ecotoxicity to *Daphnia magna*. *Environ. Pollut.* **2020**, *257*, 113597. [CrossRef]
52. Yang, J.; Cohen Stuart, M.A.; Kamperman, M. Jack of all trades: Versatile catechol crosslinking mechanisms. *Chem. Soc. Rev.* **2014**, *43*, 8271–8298. [CrossRef] [PubMed]
53. D'Ischia, M.; Ruiz-Molina, D. Bioinspired catechol-based systems: Chemistry and applications. *Biomimetics* **2017**, *2*, 25. [CrossRef] [PubMed]
54. Chu, C.; Deng, J.; Man, Y.; Qu, Y. Green tea extracts epigallocatechin-3-gallate for different treatments. *BioMed Res. Int.* **2017**, *2017*, 5615647. [CrossRef] [PubMed]
55. Hauser, D.; Septiadi, D.; Turner, J.; Petri-Fink, A.; Rothen-Rutishauser, B. From bioinspired glue to medicine: Polydopamine as a biomedical material. *Materials* **2020**, *13*, 1730. [CrossRef] [PubMed]
56. Ball, V. Composite materials and films based on melanins, polydopamine, and other catecholamine-based materials. *Biomimetics* **2017**, *2*, 12. [CrossRef]

57. Kralj, S.; Longobardo, F.; Iglesias, D.; Bevilacqua, M.; Tavagnacco, C.; Criado, A.; Delgado Jaen, J.J.; Makovec, D.; Marchesan, S.; Melchionna, M.; et al. Ex-solution synthesis of sub-5-nm FeO_x nanoparticles on mesoporous hollow N,O-doped carbon nanoshells for electrocatalytic oxygen reduction. *ACS Appl. Nano Mater.* **2019**, *2*, 6092–6097. [CrossRef]
58. El Yakhlifi, S.; Ball, V. Polydopamine as a stable and functional nanomaterial. *Colloids Surf. B Biointerfaces* **2020**, *186*, 110719. [CrossRef]
59. Wang, Z.; Zou, Y.; Li, Y.; Cheng, Y. Metal-containing polydopamine nanomaterials: Catalysis, energy, and theranostics. *Small* **2020**, *16*, 1907042. [CrossRef]
60. Uzair, B.; Liaqat, A.; Iqbal, H.; Menaa, B.; Razzaq, A.; Thiripuranathar, G.; Fatima Rana, N.; Menaa, F. Green and cost-effective synthesis of metallic nanoparticles by algae: Safe methods for translational medicine. *Bioengineering* **2020**, *7*, 129. [CrossRef]
61. Shan, D.; Hsieh, J.T.; Bai, X.; Yang, J. Citrate-based fluorescent biomaterials. *Adv. Healthc. Mater.* **2018**, *7*, e1800532. [CrossRef]
62. Pautler, R.; Kelly, E.Y.; Huang, P.-J.J.; Cao, J.; Liu, B.; Liu, J. Attaching DNA to nanoceria: Regulating oxidase activity and fluorescence quenching. *ACS Appl. Mater. Interfaces* **2013**, *5*, 6820–6825. [CrossRef] [PubMed]
63. Xu, C.; Liu, Z.; Wu, L.; Ren, J.; Qu, X. Nucleoside triphosphates as promoters to enhance nanoceria enzyme-like activity and for single-nucleotide polymorphism typing. *Adv. Funct. Mater.* **2014**, *24*, 1624–1630. [CrossRef]
64. Huber, R.; Stoll, S. Protein affinity for titanium oxide and cerium oxide manufactured nanoparticles. From ultra-pure water to biological media. *Colloids Surf. A* **2018**, *553*, 425–431. [CrossRef]
65. Kavok, N.; Grygorova, G.; Klochkov, V.; Yefimova, S. The role of serum proteins in the stabilization of colloidal $LnVO_4:Eu^{3+}$ (Ln = La, Gd, Y) and CeO_2 nanoparticles. *Colloids Surf. A* **2017**, *529*, 594–599. [CrossRef]
66. Mazzolini, J.; Weber, R.J.M.; Khan, A.; Guggenheim, E.; Chipman, J.K.; Viant, M.R.; Chen, H.-S.; Shaw, R.K.; Rappoport, J.Z. Protein Corona modulates uptake and toxicity of nanoceria via clathrin-mediated endocytosis. *Biol. Bull.* **2016**, *231*, 40–60. [CrossRef]
67. Vlasova, N.N. Adsorption of amino acids on a cerium dioxide surface. *Colloid J.* **2016**, *78*, 747–752. [CrossRef]
68. Jiang, X.; Huang, X.; Zeng, W.; Huang, J.; Zheng, Y.; Sun, D.; Li, Q. Facile morphology control of 3D porous CeO_2 for CO oxidation. *RSC Adv.* **2018**, *8*, 21658–21663. [CrossRef]
69. Kuenzle, M.; Eckert, T.; Beck, T. Binary protein crystals for the assembly of inorganic nanoparticle superlattices. *J. Am. Chem. Soc.* **2016**, *138*, 12731–12734. [CrossRef] [PubMed]
70. Ozaki, M.; Sakashita, S.; Hamada, Y.; Usui, K. Peptides for silica precipitation: Amino acid sequences for directing mineralization. *Prot. Pept. Lett.* **2018**, *25*, 15–24. [CrossRef] [PubMed]
71. Wang, Z.; Yu, R. Hollow micro/nanostructured ceria-based materials: Synthetic strategies and versatile applications. *Adv. Mater.* **2019**, *31*, e1800592. [CrossRef]
72. Schweke, D.; Mordehovitz, Y.; Halabi, M.; Shelly, L.; Hayun, S. Defect chemistry of oxides for energy applications. *Adv. Mater.* **2018**, *30*, e1706300. [CrossRef]
73. Zhang, Y.; Liu, J.; Singh, M.; Hu, E.; Jiang, Z.; Raza, R.; Wang, F.; Wang, J.; Yang, F.; Zhu, B. Superionic conductivity in ceria-based heterostructure composites for low-temperature solid oxide fuel cells. *Nano-Micro Lett.* **2020**, *12*, 178. [CrossRef]
74. Kaneda, K.; Mitsudome, T. Metal-support cooperative catalysts for environmentally benign molecular transformations. *Chem. Rec.* **2017**, *17*, 4–26. [CrossRef]
75. Fauzi, A.A.; Jalil, A.A.; Hassan, N.S.; Aziz, F.F.A.; Azami, M.S.; Hussain, I.; Saravanan, R.; Vo, D.V.N. A critical review o relationship of CeO_2-based photocatalyst towards mechanistic degradation of organic pollutant. *Chemosphere* **2021**, *286*, 131651. [CrossRef] [PubMed]
76. Jampaiah, D.; Chalkidis, A.; Sabri, Y.M.; Bhargava, S.K. Role of ceria in the design of composite materials for elemental mercury removal. *Chem. Rec.* **2019**, *19*, 1407–1419. [CrossRef] [PubMed]
77. Marchesan, S.; Prato, M. Nanomaterials for (nano)medicine. *ACS Med. Chem. Lett.* **2013**, *4*, 147–149. [CrossRef]
78. Uppal, S.; Aashima; Kumar, R.; Sareen, S.; Kaur, K.; Mehta, S.K. Biofabrication of cerium oxide nanoparticles using emulsification for an efficient delivery of Benzyl isothiocyanate. *Appl. Surf. Sci.* **2020**, *510*, 145011. [CrossRef]
79. Battaglini, M.; Tapeinos, C.; Cavaliere, I.; Marino, A.; Ancona, A.; Garino, N.; Cauda, V.; Palazon, F.; Debellis, D.; Ciofani, G. Design, fabrication, and in vitro evaluation of nanoceria-loaded nanostructured lipid carriers for the treatment of neurological diseases. *ACS Biomater. Sci. Eng.* **2019**, *5*, 670–682. [CrossRef]
80. Singh, S.; Ly, A.; Das, S.; Sakthivel, T.S.; Barkam, S.; Seal, S. Cerium oxide nanoparticles at the nano-bio interface: Size-dependent cellular uptake. *Artif. Cells Nanomed. Biotechnol.* **2018**, *46*, S956–S963. [CrossRef] [PubMed]
81. Peskova, M.; Heger, Z.; Dostalova, S.; Fojtu, M.; Castkova, K.; Ilkovics, L.; Vykoukal, V.; Pekarik, V. Investigation of detergent-modified enzymomimetic activities of TEMED-templated nanoceria towards fluorescent detection of their cellular uptake. *ChemistrySelect* **2018**, *3*, 10139–10146. [CrossRef]
82. Gao, W.; Zhao, Y.; Li, X.; Sun, Y.; Cai, M.; Cao, W.; Liu, Z.; Tong, L.; Cui, G.; Tang, B. H_2O_2-responsive and plaque-penetrating nanoplatform for mTOR gene silencing with robust anti-atherosclerosis efficacy. *Chem. Sci.* **2018**, *9*, 439–445. [CrossRef]
83. Niemiec, S.M.; Hilton, S.A.; Wallbank, A.; Azeltine, M.; Louiselle, A.E.; Elajaili, H.; Allawzi, A.; Xu, J.; Mattson, C.; Dewberry, L.C.; et al. Cerium oxide nanoparticle delivery of microRNA-146a for local treatment of acute lung injury. *Nanomedicine* **2021**, *34*, 102388. [CrossRef]

84. Jurcik, J.; Sivakova, B.; Cipakova, I.; Selicky, T.; Stupenova, E.; Jurcik, M.; Osadska, M.; Barath, P.; Cipak, L. Phosphoproteomics meets chemical genetics: Approaches for global mapping and deciphering the phosphoproteome. *Int. J. Mol. Sci.* **2020**, *21*, 7637. [CrossRef] [PubMed]
85. Qiu, W.; Evans, C.A.; Landels, A.; Pham, T.K.; Wright, P.C. Phosphopeptide enrichment for phosphoproteomic analysis—A tutorial and review of novel materials. *Anal. Chim. Acta* **2020**, *1129*, 158–180. [CrossRef] [PubMed]
86. Piovesana, S.; Iglesias, D.; Melle-Franco, M.; Kralj, S.; Cavaliere, C.; Melchionna, M.; Laganà, A.; Capriotti, A.L.; Marchesan, S. Carbon nanostructure morphology templates nanocomposites for phosphoproteomics. *Nano Res.* **2020**, *13*, 380–388. [CrossRef]
87. Lv, N.; Wang, Z.; Bi, W.; Li, G.; Zhang, J.; Ni, J. C_8-modified $CeO_2//SiO_2$ Janus fibers for selective capture and individual MS detection of low-abundance peptides and phosphopeptides. *J. Mater. Chem. B* **2016**, *4*, 4402–4409. [CrossRef] [PubMed]
88. Fatima, B.; Najam-ul-Haq, M.; Jabeen, F.; Majeed, S.; Ashiq, M.N.; Musharraf, S.G.; Shad, M.A.; Xu, G. Ceria-based nanocomposites for the enrichment and identification of phosphopeptides. *Analyst* **2013**, *138*, 5059–5067. [CrossRef]
89. Yildirim, D.; Gokcal, B.; Buber, E.; Kip, C.; Demir, M.C.; Tuncel, A. A new nanozyme with peroxidase-like activity for simultaneous phosphoprotein isolation and detection based on metal oxide affinity chromatography: Monodisperse-porous cerium oxide microspheres. *Chem. Eng. J.* **2021**, *403*, 126357. [CrossRef]
90. Xu, H.; Liu, M.; Huang, X.; Min, Q.; Zhu, J.-J. Multiplexed quantitative MALDI MS approach for assessing activity and inhibition of protein kinases based on postenrichment dephosphorylation of phosphopeptides by metal-organic framework-templated porous CeO_2. *Anal. Chem.* **2018**, *90*, 9859–9867. [CrossRef]
91. Walther, R.; Huynh, T.H.; Monge, P.; Fruergaard, A.S.; Mamakhel, A.; Zelikin, A.N. Ceria nanozyme and phosphate prodrugs: Drug synthesis through enzyme mimicry. *ACS Appl. Mater. Interfaces* **2021**, *13*, 25685–25693. [CrossRef] [PubMed]
92. Janos, P.; Henych, J.; Pfeifer, J.; Zemanova, N.; Pilarova, V.; Milde, D.; Opletal, T.; Tolasz, J.; Maly, M.; Stengl, V. Nanocrystalline cerium oxide prepared from a carbonate precursor and its ability to breakdown biologically relevant organophosphates. *Environ. Sci. Nano* **2017**, *4*, 1283–1293. [CrossRef]
93. Jiang, L.; Sun, Y.; Chen, Y.; Nan, P. From DNA to nerve agents—The biomimetic catalysts for the hydrolysis of phosphate esters. *ChemistrySelect* **2020**, *5*, 9492–9516. [CrossRef]
94. Vernekar, A.A.; Das, T.; Mugesh, G. Vacancy-engineered nanoceria: Enzyme mimetic hotspots for the degradation of nerve agents. *Angew. Chem. Int. Ed.* **2016**, *55*, 1412–1416. [CrossRef] [PubMed]
95. Fang, R.; Liu, J. Cleaving DNA by nanozymes. *J. Mater. Chem. B* **2020**, *8*, 7135–7142. [CrossRef] [PubMed]
96. Wang, X.; Lopez, A.; Liu, J. Adsorption of phosphate and polyphosphate on nanoceria probed by DNA oligonucleotides. *Langmuir* **2018**, *34*, 7899–7905. [CrossRef]
97. Wang, X.; Liu, B.; Liu, J. DNA-functionalized nanoceria for probing oxidation of phosphorus compounds. *Langmuir* **2018**, *34*, 15871–15877. [CrossRef]
98. Kato, K.; Lee, S.; Nagata, F. Catalytic performance of ceria fibers with phosphatase-like activity and their application as protein carriers. *Adv. Powder Technol.* **2020**, *31*, 2880–2889. [CrossRef]
99. Singh, S.; Kumar, U.; Gittess, D.; Sakthivel, T.S.; Babu, B.; Seal, S. Cerium oxide nanomaterial with dual antioxidative scavenging potential: Synthesis and characterization. *J. Biomater. Appl.* **2021**, 8853282211013451. [CrossRef]
100. Janos, P.; Ederer, J.; Dosek, M.; Stojdl, J.; Henych, J.; Tolasz, J.; Kormunda, M.; Mazanec, K. Can cerium oxide serve as a phosphodiesterase-mimetic nanozyme. *Environ. Sci. Nano* **2019**, *6*, 3684–3698. [CrossRef]
101. Li, H.; Meng, F.; Gong, J.; Fan, Z.; Qin, R. Template-free hydrothermal synthesis, mechanism, and photocatalytic properties of core-shell CeO_2 nanospheres. *Electron. Mater. Lett.* **2018**, *14*, 474–487. [CrossRef]
102. Huang, X.; Zhu, N.; Mao, F.; Ding, Y.; Zhang, S.; Liu, H.; Li, F.; Wu, P.; Dang, Z.; Ke, Y. Enhanced heterogeneous photo-fenton catalytic degradation of tetracycline over $yCeO_2/Fh$ composites: Performance, degradation pathways, Fe^{2+} regeneration and mechanism. *Chem. Eng. J.* **2020**, *392*, 123636. [CrossRef]
103. Sun, Y.; Zhao, C.; Gao, N.; Ren, J.; Qu, X. Stereoselective nanozyme based on ceria nanoparticles engineered with amino acids. *Chem.—Eur. J.* **2017**, *23*, 18146–18150. [CrossRef]
104. Tian, Z.; Yao, T.; Qu, C.; Zhang, S.; Li, X.; Qu, Y. Photolyase-like catalytic behavior of CeO_2. *Nano Lett.* **2019**, *19*, 8270–8277. [CrossRef] [PubMed]
105. Das, S.; Neal, C.J.; Ortiz, J.; Seal, S. Engineered nanoceria cytoprotection in vivo: Mitigation of reactive oxygen species and double-stranded DNA breakage due to radiation exposure. *Nanoscale* **2018**, *10*, 21069–21075. [CrossRef] [PubMed]
106. Popova, N.R.; Popov, A.L.; Ermakov, A.M.; Reukov, V.V.; Ivanov, V.K. Ceria-containing hybrid multilayered microcapsules for enhanced cellular internalisation with high radioprotection efficiency. *Molecules* **2020**, *25*, 2957. [CrossRef] [PubMed]
107. Omri, M.; Becuwe, M.; Courty, M.; Pourceau, G.; Wadouachi, A. Nitroxide-grafted nanometric metal oxides for the catalytic oxidation of sugar. *ACS Appl. Nano Mater.* **2019**, *2*, 5200–5205. [CrossRef]
108. Qian, J.; Cao, Y.; Chen, Z.; Liu, C.; Lu, X. Biomimetic synthesis of cerium oxide nanosquares on RGO and their enhanced photocatalytic activities. *Dalton Trans.* **2017**, *46*, 547–553. [CrossRef] [PubMed]
109. Meng, J.; Li, H.; Chen, R.; Sun, X.; Sun, X. Enzyme-like catalytic activity of porphyrin-functionalized ceria nanotubes for water oxidation. *ChemPlusChem* **2019**, *84*, 1816–1822. [CrossRef]
110. Marzorati, S.; Cristiani, P.; Longhi, M.; Trasatti, S.P.; Traversa, E. Nanoceria acting as oxygen reservoir for biocathodes in microbial fuel cells. *Electrochim. Acta* **2019**, *325*, 134954. [CrossRef]

111. Wang, Z.; Shen, X.; Gao, X.; Zhao, Y. Simultaneous enzyme mimicking and chemical reduction mechanisms for nanoceria as a bio-antioxidant: A catalytic model bridging computations and experiments for nanozymes. *Nanoscale* **2019**, *11*, 13289–13299. [CrossRef]
112. Naganuma, T. Shape design of cerium oxide nanoparticles for enhancement of enzyme mimetic activity in therapeutic applications. *Nano Res.* **2017**, *10*, 199–217. [CrossRef]
113. Liu, X.; Wu, J.; Liu, Q.; Lin, A.; Li, S.; Zhang, Y.; Wang, Q.; Li, T.; An, X.; Zhou, Z.; et al. Synthesis-temperature-regulated multi-enzyme-mimicking activities of ceria nanozymes. *J. Mater. Chem. B* **2021**. [CrossRef]
114. Guo, W.; Zhang, M.; Lou, Z.; Zhou, M.; Wang, P.; Wei, H. Engineering nanoceria for enhanced peroxidase mimics: A solid solution strategy. *ChemCatChem* **2019**, *11*, 737–743. [CrossRef]
115. Attar, F.; Shahpar, M.G.; Rasti, B.; Sharifi, M.; Saboury, A.A.; Rezayat, S.M.; Falahati, M. Nanozymes with intrinsic peroxidase-like activities. *J. Mol. Liq.* **2019**, *278*, 130–144. [CrossRef]
116. Vinothkumar, G.; Arunkumar, P.; Mahesh, A.; Dhayalan, A.; Suresh Babu, K. Size- and defect-controlled anti-oxidant enzyme mimetic and radical scavenging properties of cerium oxide nanoparticles. *New J. Chem.* **2018**, *42*, 18810–18823. [CrossRef]
117. Gupta, A.; Sakthivel, T.S.; Neal, C.J.; Koul, S.; Singh, S.; Kushima, A.; Seal, S. Antioxidant properties of ALD grown nanoceria films with tunable valency. *Biomater. Sci.* **2019**, *7*, 3051–3061. [CrossRef]
118. Bhagat, S.; Srikanth Vallabani, N.V.; Shutthanandan, V.; Bowden, M.; Karakoti, A.S.; Singh, S. Gold core/ceria shell-based redox active nanozyme mimicking the biological multienzyme complex phenomenon. *J. Colloid Interface Sci.* **2018**, *513*, 831–842. [CrossRef]
119. Weng, Q.; Sun, H.; Fang, C.; Xia, F.; Liao, H.; Lee, J.; Wang, J.; Xie, A.; Ren, J.; Guo, X.; et al. Catalytic activity tunable ceria nanoparticles prevent chemotherapy-induced acute kidney injury without interference with chemotherapeutics. *Nat. Commun.* **2021**, *12*, 1436. [CrossRef]
120. Fernandez-Varo, G.; Perramon, M.; Carvajal, S.; Oro, D.; Casals, E.; Boix, L.; Oller, L.; Macias-Munoz, L.; Marfa, S.; Casals, G.; et al. Bespoken nanoceria: An effective treatment in experimental hepatocellular carcinoma. *Hepatology* **2020**, *72*, 1267–1282. [CrossRef]
121. Dong, S.; Dong, Y.; Jia, T.; Liu, S.; Liu, J.; Yang, D.; He, F.; Gai, S.; Yang, P.; Lin, J. GSH-depleted nanozymes with hyperthermia-enhanced dual enzyme-mimic activities for tumor nanocatalytic therapy. *Adv. Mater.* **2020**, *32*, 2002439. [CrossRef]
122. Panda, S.R.; Singh, R.K.; Priyadarshini, B.; Rath, P.P.; Parhi, P.K.; Sahoo, T.; Mandal, D.; Sahoo, T.R. Nanoceria: A rare-earth nanoparticle as a promising anti-cancer therapeutic agent in colon cancer. *Mater. Sci. Semicond. Process.* **2019**, *104*, 104669. [CrossRef]
123. Wason, M.S.; Lu, H.; Yu, L.; Lahiri, S.K.; Mukherjee, D.; Shen, C.; Das, S.; Seal, S.; Zhao, J. Cerium oxide nanoparticles sensitize pancreatic cancer to radiation therapy through oxidative activation of the JNK apoptotic pathway. *Cancers* **2018**, *10*, 303. [CrossRef]
124. Kwon, H.J.; Kim, D.; Seo, K.; Kim, Y.G.; Han, S.I.; Kang, T.; Soh, M.; Hyeon, T. Ceria nanoparticle systems for selective scavenging of mitochondrial, intracellular, and extracellular reactive oxygen species in Parkinson's disease. *Angew. Chem. Int. Ed.* **2018**, *57*, 9408–9412. [CrossRef] [PubMed]
125. Yu, D.; Ma, M.; Liu, Z.; Pi, Z.; Du, X.; Ren, J.; Qu, X. MOF-encapsulated nanozyme enhanced siRNA combo: Control neural stem cell differentiation and ameliorate cognitive impairments in Alzheimer's disease model. *Biomaterials* **2020**, *255*, 120160. [CrossRef]
126. Guan, Y.; Li, M.; Dong, K.; Gao, N.; Ren, J.; Zheng, Y.; Qu, X. Ceria/POMs hybrid nanoparticles as a mimicking metallopeptidase for treatment of neurotoxicity of amyloid-β peptide. *Biomaterials* **2016**, *98*, 92–102. [CrossRef]
127. Ni, D.; Wei, H.; Chen, W.; Bao, Q.; Rosenkrans, Z.T.; Barnhart, T.E.; Ferreira, C.A.; Wang, Y.; Yao, H.; Sun, T.; et al. Ceria nanoparticles meet hepatic ischemia-reperfusion injury: The perfect imperfection. *Adv. Mater.* **2019**, *31*, 1902956. [CrossRef] [PubMed]
128. Casals, G.; Perramón, M.; Casals, E.; Portolés, I.; Fernández-Varo, G.; Morales-Ruiz, M.; Puntes, V.; Jiménez, W. Cerium oxide nanoparticles: A new therapeutic tool in liver diseases. *Antioxidants* **2021**, *10*, 660. [CrossRef]
129. Kobyliak, N.; Virchenko, O.; Falalyeyeva, T.; Kondro, M.; Beregova, T.; Bodnar, P.; Shcherbakov, O.; Bubnov, R.; Caprnda, M.; Delev, D.; et al. Cerium dioxide nanoparticles possess anti-inflammatory properties in the conditions of the obesity-associated NAFLD in rats. *Biomed. Pharmacother.* **2017**, *90*, 608–614. [CrossRef]
130. Hong, S.-E.; An, J.H.; Yu, S.-L.; Kang, J.; Park, C.G.; Lee, H.Y.; Lee, D.C.; Park, H.-W.; Hwang, W.-M.; Yun, S.-R.; et al. Ceria-zirconia antioxidant nanoparticles attenuate hypoxia-induced acute kidney injury by restoring autophagy flux and alleviating mitochondrial damage. *J. Biomed. Nanotechnol.* **2020**, *16*, 1144–1159. [CrossRef] [PubMed]
131. Pinna, A.; Torki Baghbaderani, M.; Vigil Hernandez, V.; Naruphontjirakul, P.; Li, S.; McFarlane, T.; Hachim, D.; Stevens, M.M.; Porter, A.E.; Jones, J.R. Nanoceria provides antioxidant and osteogenic properties to mesoporous silica nanoparticles for osteoporosis treatment. *Acta Biomater.* **2021**, *122*, 365–376. [CrossRef]
132. Kalashnikova, I.; Chung, S.-J.; Nafiujjaman, M.; Hill, M.L.; Siziba, M.E.; Contag, C.H.; Kim, T. Ceria-based nanotheranostic agent for rheumatoid arthritis. *Theranostics* **2020**, *10*, 11863–11880. [CrossRef]
133. Jansman, M.M.T.; Liu, X.; Kempen, P.; Clergeaud, G.; Andresen, T.L.; Hosta-Rigau, L.; Thulstrup, P.W. Hemoglobin-based oxygen carriers incorporating nanozymes for the depletion of reactive oxygen species. *ACS Appl. Mater. Interfaces* **2020**, *12*, 50275–50286. [CrossRef] [PubMed]
134. Tang, S.; Zhou, L.; Liu, Z.; Zou, L.; Xiao, M.; Huang, C.; Xie, Z.; He, H.; Guo, Y.; Cao, Y.; et al. Ceria nanoparticles promoted the cytotoxic activity of CD8+ T cells by activating NF-KB signaling. *Biomater. Sci.* **2019**, *7*, 2533–2544. [CrossRef]

135. Jia, J.; Huang, Y.; Jia, J.; Sun, J.; Peng, S.; Xie, Q.; Yi, L.; Li, C.; Zhang, T. CeO_2@PAA-LXW7 attenuates LPS-induced inflammation in BV2 microglia. *Cell. Mol. Neurobiol.* **2019**, *39*, 1125–1137. [CrossRef]
136. Choi, B.; Soh, M.; Manandhar, Y.; Kim, D.; Han, S.I.; Baik, S.; Shin, K.; Koo, S.; Kwon, H.J.; Ko, G.; et al. Highly selective microglial uptake of ceria-zirconia nanoparticles for enhanced analgesic treatment of neuropathic pain. *Nanoscale* **2019**, *11*, 19437–19447. [CrossRef]
137. Cha, B.G.; Jeong, H.-G.; Kang, D.-W.; Nam, M.-J.; Kim, C.K.; Kim, D.Y.; Choi, I.-Y.; Ki, S.K.; Kim, S.I.; Han, J.H.; et al. Customized lipid-coated magnetic mesoporous silica nanoparticle doped with ceria nanoparticles for theragnosis of intracerebral hemorrhage. *Nano Res.* **2018**, *11*, 3582–3592. [CrossRef]
138. Wang, Y.; Zhang, P.; Li, M.; Guo, Z.; Ullah, S.; Rui, Y.; Lynch, I. Alleviation of nitrogen stress in rice (*Oryza sativa*) by ceria nanoparticles. *Environ. Sci. Nano* **2020**, *7*, 2930–2940. [CrossRef]
139. Mortimer, M.; Li, D.; Wang, Y.; Holden, P.A. Physical properties of carbon nanomaterials and nanoceria affect pathways important to the nodulation competitiveness of the symbiotic N2-fixing bacterium bradyrhizobium diazoefficiens. *Small* **2020**, *16*, 1906055. [CrossRef] [PubMed]
140. Charbgoo, F.; Ramezani, M.; Darroudi, M. Bio-sensing applications of cerium oxide nanoparticles: Advantages and disadvantages. *Biosens. Bioelectron.* **2017**, *96*, 33–43. [CrossRef]
141. Kim, H.Y.; Park, K.S.; Park, H.G. Glucose oxidase-like activity of cerium oxide nanoparticles: Use for personal glucose meter-based label-free target DNA detection. *Theranostics* **2020**, *10*, 4507–4514. [CrossRef]
142. Gao, Y.; Zhou, Y.; Chandrawati, R. Metal and metal oxide nanoparticles to enhance the performance of enzyme-linked immunosorbent assay (ELISA). *ACS Appl. Nano Mater.* **2020**, *3*, 1–21. [CrossRef]
143. Alizadeh, N.; Salimi, A.; Sham, T.-K.; Bazylewski, P.; Fanchini, G. Intrinsic enzyme-like activities of cerium oxide nanocomposite and its application for extracellular H_2O_2 detection using an electrochemical microfluidic device. *ACS Omega* **2020**, *5*, 11883–11894. [CrossRef]
144. Pratsinis, A.; Kelesidis, G.A.; Zuercher, S.; Krumeich, F.; Bolisetty, S.; Mezzenga, R.; Leroux, J.-C.; Sotiriou, G.A. Enzyme-mimetic antioxidant luminescent nanoparticles for highly sensitive hydrogen peroxide biosensing. *ACS Nano* **2017**, *11*, 12210–12218. [CrossRef]
145. Mu, J.; Zhao, X.; Li, J.; Yang, E.-C.; Zhao, X.-J. Coral-like CeO_2/NiO nanocomposites with efficient enzyme-mimetic activity for biosensing application. *Mater. Sci. Eng. C* **2017**, *74*, 434–442. [CrossRef]
146. Hosseini, M.; Sadat Sabet, F.; Khabbaz, H.; Aghazadeh, M.; Mizani, F.; Ganjali, M.R. Enhancement of the peroxidase-like activity of cerium-doped ferrite nanoparticles for colorimetric detection of H_2O_2 and glucose. *Anal. Methods* **2017**, *9*, 3519–3524. [CrossRef]
147. Gao, W.; Wei, X.; Wang, X.; Cui, G.; Liu, Z.; Tang, B. A competitive coordination-based CeO_2 nanowire-DNA nanosensor: Fast and selective detection of hydrogen peroxide in living cells and in vivo. *Chem. Commun.* **2016**, *52*, 3643–3646. [CrossRef]
148. Xie, J.; Cheng, D.; Zhou, Z.; Pang, X.; Liu, M.; Yin, P.; Zhang, Y.; Li, H.; Liu, X.; Yao, S. Hydrogen peroxide sensing in body fluids and tumor cells via in situ produced redox couples on two-dimensional holey $CuCo_2O_4$ nanosheets. *Microchim. Acta* **2020**, *187*, 469. [CrossRef]
149. Liu, Z.; Cao, Y.; Zhang, X.; Yang, H.; Zhao, Y.; Gao, W.; Tang, B. A dual-targeted CeO_2-DNA nanosensor for real-time imaging of H_2O_2 to assess atherosclerotic plaque vulnerability. *J. Mater. Chem. B* **2020**, *8*, 3502–3505. [CrossRef]
150. Yang, L.; An, B.; Yin, X.; Li, F. A competitive coordination-based immobilization-free electrochemical biosensor for highly sensitive detection of arsenic(V) using a CeO_2-DNA nanoprobe. *Chem. Commun.* **2020**, *56*, 5311–5314. [CrossRef]
151. Lopez, A.; Zhang, Y.; Liu, J. Tuning DNA adsorption affinity and density on metal oxide and phosphate for improved arsenate detection. *J. Colloid Interface Sci.* **2017**, *493*, 249–256. [CrossRef] [PubMed]
152. Anthony, E.T.; Ojemaye, M.O.; Okoh, A.I.; Okoh, O.O. Synthesis of CeO_2 as promising adsorbent for the management of free-DNA harboring antibiotic resistance genes from tap-water. *Chem. Eng. J.* **2020**, *401*, 125562. [CrossRef]
153. Mustafa, F.; Othman, A.; Andreescu, S. Cerium oxide-based hypoxanthine biosensor for Fish spoilage monitoring. *Sens. Actuators B* **2021**, *332*, 129435. [CrossRef]
154. Nguyet, N.T.; Yen, L.T.H.; Doan, V.Y.; Hoang, N.L.; Van Thu, V.; Lan, H.; Trung, T.; Pham, V.-H.; Tam, P.D. A label-free and highly sensitive DNA biosensor based on the core-shell structured CeO_2-NR@Ppy nanocomposite for *Salmonella* detection. *Mater. Sci. Eng. C* **2019**, *96*, 790–797. [CrossRef]
155. Nguyet, N.T.; Hai Yen, L.T.; Van Thu, V.; Lan, H.; Trung, T.; Vuong, P.H.; Tam, P.D. Highly sensitive DNA sensors based on cerium oxide nanorods. *J. Phys. Chem. Solids* **2018**, *115*, 18–25. [CrossRef]
156. Kim, H.Y.; Ahn, J.K.; Kim, M.I.; Park, K.S.; Park, H.G. Rapid and label-free, electrochemical DNA detection utilizing the oxidase-mimicking activity of cerium oxide nanoparticles. *Electrochem. Commun.* **2019**, *99*, 5–10. [CrossRef]
157. Tian, J.; Wei, W.; Wang, J.; Ji, S.; Chen, G.; Lu, J. Fluorescence resonance energy transfer aptasensor between nanoceria and graphene quantum dots for the determination of ochratoxin A. *Anal. Chim. Acta* **2018**, *1000*, 265–272. [CrossRef]
158. Mandal, D.; Biswas, S.; Chowdhury, A.; De, D.; Tiwary, C.S.; Gupta, A.N.; Singh, T.; Chandra, A. Hierarchical cage-frame type nanostructure of CeO_2 for bio sensing applications: From glucose to protein detection. *Nanotechnology* **2021**, *32*, 025504. [CrossRef]
159. Li, H.; Lu, Y.; Pang, J.; Sun, J.; Yang, F.; Wang, Z.; Liu, Y. DNA-scaffold copper nanoclusters integrated into a cerium(III)-triggered Fenton-like reaction for the fluorometric and colorimetric enzymatic determination of glucose. *Microchim. Acta* **2019**, *186*, 862. [CrossRef] [PubMed]

160. Liu, Q.; Yang, Y.; Lv, X.; Ding, Y.; Zhang, Y.; Jing, J.; Xu, C. One-step synthesis of uniform nanoparticles of porphyrin functionalized ceria with promising peroxidase mimetics for H_2O_2 and glucose colorimetric detection. *Sens. Actuators B* **2017**, *240*, 726–734. [CrossRef]
161. Kutova, O.; Dusheiko, M.; Klyui, N.I.; Skryshevsky, V.A. C-reactive protein detection based on ISFET structure with gate dielectric SiO_2-CeO_2. *Microelectron. Eng.* **2019**, *215*, 110993. [CrossRef]
162. Gao, Z.; Li, Y.; Zhang, C.; Zhang, S.; Jia, Y.; Li, F.; Ding, H.; Li, X.; Chen, Z.; Wei, Q. AuCuxO-embedded mesoporous CeO_2 nanocomposites as a signal probe for electrochemical sensitive detection of amyloid-beta protein. *ACS Appl. Mater. Interfaces* **2019**, *11*, 12335–12341. [CrossRef]
163. Wang, J.-X.; Zhuo, Y.; Zhou, Y.; Wang, H.-J.; Yuan, R.; Chai, Y.-Q. Ceria doped zinc oxide nanoflowers enhanced luminol-based electrochemiluminescence immunosensor for amyloid-β detection. *ACS Appl. Mater. Interfaces* **2016**, *8*, 12968–12975. [CrossRef]
164. Liu, Q.; Chen, X.; Kang, Z.-W.; Zheng, C.; Yang, D.-P. Facile synthesis of eggshell membrane-templated Au/CeO_2 3D nanocomposite networks for nonenzymatic electrochemical dopamine sensor. *Nanoscale Res. Lett.* **2020**, *15*, 24. [CrossRef]
165. Ge, C.; Ramachandran, R.; Wang, F. CeO_2-based two-dimensional layered nanocomposites derived from a metal-organic framework for selective electrochemical dopamine sensors. *Sensors* **2020**, *20*, 4880. [CrossRef] [PubMed]
166. Uzunoglu, A.; Stanciu, L.A. Novel CeO_2-CuO-decorated enzymatic lactate biosensors operating in low oxygen environments. *Anal. Chim. Acta* **2016**, *909*, 121–128. [CrossRef]
167. Li, F.; Hu, X.; Wang, F.; Zheng, B.; Du, J.; Xiao, D. A fluorescent "on-off-on" probe for sensitive detection of ATP based on ATP displacing DNA from nanoceria. *Talanta* **2018**, *179*, 285–291. [CrossRef]
168. Li, L.; Zhang, Y.; Yan, Z.; Chen, M.; Zhang, L.; Zhao, P.; Yu, J. Ultrasensitive photoelectrochemical detection of microRNA on paper by combining a cascade nanozyme-engineered biocatalytic precipitation reaction and target-triggerable DNA motor. *ACS Sens.* **2020**, *5*, 1482–1490. [CrossRef]
169. Liang, L.; Lan, F.; Yin, X.; Ge, S.; Yu, J.; Yan, M. Metal-enhanced fluorescence/visual bimodal platform for multiplexed ultrasensitive detection of microRNA with reusable paper analytical devices. *Biosens. Bioelectron.* **2017**, *95*, 181–188. [CrossRef]
170. Zhao, Y.; He, J.; Niu, Y.; Chen, J.; Wu, J.; Yu, C. A new sight for detecting the ADRB1 gene mutation to guide a therapeutic regimen for hypertension based on a CeO_2-doped nanoprobe. *Biosens. Bioelectron.* **2017**, *92*, 402–409. [CrossRef]
171. Li, Y.; Chang, Y.; Yuan, R.; Chai, Y. Highly efficient target recycling-based netlike Y-DNA for regulation of electrocatalysis toward methylene blue for sensitive DNA detection. *ACS Appl. Mater. Interfaces* **2018**, *10*, 25213–25218. [CrossRef] [PubMed]
172. Bulbul, G.; Hayat, A.; Mustafa, F.; Andreescu, S. DNA assay based on nanoceria as fluorescence quenchers (NanoCeracQ DNA assay). *Sci. Rep.* **2018**, *8*, 2426. [CrossRef]
173. Li, L.; Yang, L.; Zhang, S.; Sun, Y.; Li, F.; Qin, T.; Liu, X.; Zhou, Y.; Alwarappan, S. A $NiCo_2S_4$@N/S-CeO_2 composite as an electrocatalytic signal amplification label for aptasensing. *J. Mater. Chem. C* **2020**, *8*, 14723–14731. [CrossRef]
174. Tian, J.; Wang, J.; Li, Y.; Huang, M.; Lu, J. Electrochemically driven omeprazole metabolism via cytochrome P450 assembled on the nanocomposites of ceria nanoparticles and graphene. *J. Electrochem. Soc.* **2017**, *164*, H470–H476. [CrossRef]
175. Hartati, Y.W.; Komala, D.R.; Hendrati, D.; Gaffar, S.; Hardianto, A.; Sofiatin, Y.; Bahti, H.H. An aptasensor using ceria electrodeposited-screen-printed carbon electrode for detection of epithelial sodium channel protein as a hypertension biomarker. *R. Soc. Open Sci.* **2021**, *8*, 202040. [CrossRef]
176. Ding, Y.; Zhang, M.; Li, C.; Xie, B.; Zhao, G.; Sun, Y. A reusable aptasensor based on the dual signal amplification of Ce@AuNRs-PAMAM-Fc and DNA walker for ultrasensitive detection of TNF-α. *J. Solid State Electrochem.* **2021**. [CrossRef]
177. Wang, M.; Hu, M.; Hu, B.; Guo, C.; Song, Y.; Jia, Q.; He, L.; Zhang, Z.; Fang, S. Bimetallic cerium and ferric oxides nanoparticles embedded within mesoporous carbon matrix: Electrochemical immunosensor for sensitive detection of carbohydrate antigen 19-9. *Biosens. Bioelectron.* **2019**, *135*, 22–29. [CrossRef]
178. Wang, S.; Wang, F.; Fu, C.; Sun, Y.; Zhao, J.; Li, N.; Liu, Y.; Ge, S.; Yu, J. AgInSe2-sensitized ZnO nanoflower wide-spectrum response photoelectrochemical/visual sensing platform via Au@nanorod-anchored CeO_2 octahedron regulated signal. *Anal. Chem.* **2020**, *92*, 7604–7611. [CrossRef] [PubMed]
179. Shen, H.; Deng, W.; He, Y.; Li, X.; Song, J.; Liu, R.; Hua, l.; Yang, G.; Li, L. Ultrasensitive aptasensor for isolation and detection of circulating tumor cells based on CeO_2@Ir nanorods and DNA walker. *Biosens. Bioelectron.* **2020**, *168*, 112516. [CrossRef]
180. Ling, J.; Zhao, M.; Chen, F.; Zhou, X.; Li, X.; Ding, S.; Tang, H. An enzyme-free electrochemiluminescence biosensor for ultrasensitive assay of Group B *Streptococci* based on self-enhanced luminol complex functionalized CuMn-CeO_2 nanospheres. *Biosens. Bioelectron.* **2019**, *127*, 167–173. [CrossRef]
181. Pandey, A.; Patel, A.K.; Ariharan, S.; Kumar, V.; Sharma, R.K.; Kanhed, S.; Nigam, V.K.; Keshri, A.; Agarwal, A.; Balani, K. Enhanced tribological and bacterial resistance of carbon nanotube with Ceria- and silver-incorporated hydroxyapatite biocoating. *Nanomaterials* **2018**, *8*, 363. [CrossRef]
182. De Santis, S.; Sotgiu, G.; Porcelli, F.; Marsotto, M.; Iucci, G.; Orsini, M. A simple cerium coating strategy for titanium oxide nanotubes' bioactivity enhancement. *Nanomaterials* **2021**, *11*, 445. [CrossRef]
183. Saito, M.M.; Onuma, K.; Yamamoto, R.; Yamakoshi, Y. New insights into bioactivity of ceria-stabilized zirconia: Direct bonding to bone-like hydroxyapatite at nanoscale. *Mater. Sci. Eng. C* **2021**, *121*, 111665. [CrossRef]
184. Shao, D.; Li, K.; You, M.; Liu, S.; Hu, T.; Huang, L.; Xie, Y.; Zheng, X. Macrophage polarization by plasma sprayed ceria coatings on titanium-based implants: Cerium valence state matters. *Appl. Surf. Sci.* **2020**, *504*, 144070. [CrossRef]

185. You, G.; Wang, C.; Wang, P.; Hou, J.; Xu, Y.; Miao, L.; Feng, T. Insights into spatial effects of ceria nanoparticles on oxygen mass transfer in wastewater biofilms: Interfacial microstructure, in-situ microbial activity and metabolism regulation mechanism. *Water Res.* **2020**, *176*, 115731. [CrossRef]
186. Pandiyan, N.; Murugesan, B.; Sonamuthu, J.; Samayanan, S.; Mahalingam, S. Facile biological synthetic strategy to morphologically aligned CeO_2/ZrO_2 core nanoparticles using Justicia adhatoda extract and ionic liquid: Enhancement of its bio-medical properties. *J. Photochem. Photobiol. B* **2018**, *178*, 481–488. [CrossRef] [PubMed]
187. Frerichs, H.; Puetz, E.; Pfitzner, F.; Reich, T.; Gazanis, A.; Panthoefer, M.; Hartmann, J.; Jegel, O.; Heermann, R.; Tremel, W. Nanocomposite antimicrobials prevent bacterial growth through the enzyme-like activity of Bi-doped cerium dioxide (Ce1-xBixO$_2$-δ). *Nanoscale* **2020**, *12*, 21344–21358. [CrossRef] [PubMed]
188. Bhatt, L.; Chen, L.; Guo, J.; Klie, R.F.; Shi, J.; Pesavento, R.P. Hydrolyzed Ce(IV) salts limit sucrose-dependent biofilm formation by *Streptococcus mutans*. *J. Inorg. Biochem.* **2020**, *206*, 110997. [CrossRef]
189. Xie, C.; Zhang, J.; Ma, Y.; Ding, Y.; Zhang, P.; Zheng, L.; Chai, Z.; Zhao, Y.; Zhang, Z.; He, X. *Bacillus subtilis* causes dissolution of ceria nanoparticles at the nano-bio interface. *Environ. Sci. Nano* **2019**, *6*, 216–223. [CrossRef]
190. Shu, Z.; Zhang, Y.; Ouyang, J.; Yang, H. Characterization and synergetic antibacterial properties of ZnO and CeO_2 supported by halloysite. *Appl. Surf. Sci.* **2017**, *420*, 833–838. [CrossRef]
191. Estes, L.M.; Singha, P.; Singh, S.; Sakthivel, T.S.; Garren, M.; Devine, R.; Brisbois, E.J.; Seal, S.; Handa, H. Characterization of a nitric oxide (NO) donor molecule and cerium oxide nanoparticle (CNP) interactions and their synergistic antimicrobial potential for biomedical applications. *J. Coll. Interface Sci.* **2021**, *586*, 163–177. [CrossRef]
192. Sadidi, H.; Hooshmand, S.; Ahmadabadi, A.; Javad Hosseini, S.; Baino, F.; Vatanpour, M.; Kargozar, S. Cerium oxide nanoparticles (Nanoceria): Hopes in soft tissue engineering. *Molecules* **2020**, *25*, 4559. [CrossRef] [PubMed]
193. Purohit, S.D.; Singh, H.; Bhaskar, R.; Yadav, I.; Chou, C.-F.; Gupta, M.K.; Mishra, N.C. Gelatin-alginate-cerium oxide nanocomposite scaffold for bone regeneration. *Mater. Sci. Eng. C* **2020**, *116*, 111111. [CrossRef]
194. Wei, F.; Neal, C.J.; Sakthivel, T.S.; Kean, T.; Seal, S.; Coathup, M.J. Multi-functional cerium oxide nanoparticles regulate inflammation and enhance osteogenesis. *Mater. Sci. Eng. C* **2021**, *124*, 112041. [CrossRef]
195. Bonciu, A.F.; Orobeti, S.; Sima, L.E.; Icriverzi, M.; Filipescu, M.; Moldovan, A.; Popescu, A.; Dinca, V.; Dinescu, M. Pyramidal shaped ceria nano-biointerfaces for studying the early bone cell response. *Appl. Surf. Sci.* **2020**, *533*, 147464. [CrossRef]
196. Li, J.; Kang, F.; Gong, X.; Bai, Y.; Dai, J.; Zhao, C.; Dou, C.; Cao, Z.; Liang, M.; Dong, R.; et al. Ceria nanoparticles enhance endochondral ossification-based critical-sized bone defect regeneration by promoting the hypertrophic differentiation of BMSCs via DHX15 activation. *FASEB J.* **2019**, *33*, 6378–6389. [CrossRef] [PubMed]
197. Jun, S.-K.; Yoon, J.-Y.; Mahapatra, C.; Park, J.H.; Kim, H.-W.; Kim, H.-R.; Lee, J.-H.; Lee, H.-H. Ceria-incorporated MTA for accelerating odontoblastic differentiation via ROS downregulation. *Dent. Mater.* **2019**, *35*, 1291–1299. [CrossRef]
198. Wu, H.; Li, F.; Shao, W.; Gao, J.; Ling, D. Promoting angiogenesis in oxidative diabetic wound microenvironment using a nanozyme-reinforced self-protecting hydrogel. *ACS Cent. Sci.* **2019**, *5*, 477–485. [CrossRef]
199. Kandasamy, S.; Zhang, B.; He, Z.; Chen, H.; Feng, H.; Wang, Q.; Wang, B.; Ashokkumar, V.; Siva, S.; Bhuvanendran, N.; et al. Effect of low-temperature catalytic hydrothermal liquefaction of *Spirulina platensis*. *Energy* **2020**, *190*, 116236. [CrossRef]
200. Kohansal, K.; Tavasoli, A.; Bozorg, A. Using a hybrid-like supported catalyst to improve green fuel production through hydrothermal liquefaction of *Scenedesmus obliquus* microalgae. *Bioresour. Technol.* **2019**, *277*, 136–147. [CrossRef]
201. Chen, D.; Ma, Q.; Wei, L.; Li, N.; Shen, Q.; Tian, W.; Zhou, J.; Long, J. Catalytic hydroliquefaction of rice straw for bio-oil production using Ni/CeO_2 catalysts. *J. Anal. Appl. Pyrolysis* **2018**, *130*, 169–180. [CrossRef]
202. Deka, K.; Nath, N.; Saikia, B.K.; Deb, P. Kinetic analysis of ceria nanoparticle catalysed efficient biomass pyrolysis for obtaining high-quality bio-oil. *J. Therm. Anal. Calorim.* **2017**, *130*, 1875–1883. [CrossRef]
203. Marin, C.M.; Li, L.; Bhalkikar, A.; Doyle, J.E.; Zeng, X.C.; Cheung, C.L. Kinetic and mechanistic investigations of the direct synthesis of dimethyl carbonate from carbon dioxide over ceria nanorod catalysts. *J. Catal.* **2016**, *340*, 295–301. [CrossRef]
204. Zhao, X.; Suo, H.; Zhang, Z.; Guo, C. Upconverting CeO_2: Yb^{3+}/Tm^{3+} hollow nanospheres for photo-thermal sterilization and deep-tissue imaging in the first biological window. *Ceram. Int.* **2019**, *45*, 21910–21916. [CrossRef]
205. Shehata, N.; Samir, E.; Gaballah, S.; Hamed, A.; Saad, M.; Salah, M. Fluorescent nanocomposite of embedded ceria nanoparticles in electrospun chitosan nanofibers. *J. Fluoresc.* **2017**, *27*, 767–772. [CrossRef]
206. Garcia, A.M.; Melchionna, M.; Bellotto, O.; Kralj, S.; Semeraro, S.; Parisi, E.; Iglesias, D.; D'Andrea, P.; De Zorzi, R.; Vargiu, A.V.; et al. Nanoscale assembly of functional peptides with divergent programming elements. *ACS Nano* **2021**, *15*, 3015–3025. [CrossRef]
207. Rozhin, P.; Charitidis, C.; Marchesan, S. Self-assembling peptides and carbon nanomaterials join forces for innovative biomedical applications. *Molecules* **2021**, *26*, 4084. [CrossRef] [PubMed]
208. Sibhghatulla, S.; Nazia, N.; Syed Mohd Danish, R.; Talib, H.; Aisha, F.; Inho, C. Anti-amyloid aggregating gold nanoparticles: Can they really be translated from bench to bedside for Alzheimer's disease treatment? *Curr. Prot. Pept. Sci.* **2020**, *21*, 1184–1192. [CrossRef]
209. Zand, Z.; Khaki, P.A.; Salihi, A.; Aziz, F.M.; Salihi, A.; Sharifi, M.; Falahati, M.; Sharifi, M.; Qadir, N.N.M.; Qadir, N.N.M.; et al. Cerium oxide NPs mitigate the amyloid formation of α-synuclein and associated cytotoxicity. *Int. J. Nanomed.* **2019**, *14*, 6989–7000. [CrossRef]
210. Marchesan, S.; Ballerini, L.; Prato, M. Nanomaterials for stimulating nerve growth. *Science* **2017**, *356*, 1010. [CrossRef]

211. Kim, D.; Kwon, H.J.; Hyeon, T.; Kim, D.; Kwon, H.J.; Hyeon, T. Magnetite/ceria nanoparticle assemblies for extracorporeal cleansing of amyloid-β in Alzheimer's disease. *Adv. Mater.* **2019**, *31*, e1807965. [CrossRef]
212. Deng, X.-Y.; Cheng, J.; Hu, X.-L.; Wang, L.; Li, D.; Gao, K. Biological effects of TiO_2 and CeO_2 nanoparticles on the growth, photosynthetic activity, and cellular components of a marine diatom *Phaeodactylum tricornutum*. *Sci. Total Environ.* **2017**, *575*, 87–96. [CrossRef]
213. Taylor, N.S.; Merrifield, R.; Williams, T.D.; Chipman, J.K.; Lead, J.R.; Viant, M.R. Molecular toxicity of cerium oxide nanoparticles to the freshwater alga Chlamydomonas reinhardtii is associated with supra-environmental exposure concentrations. *Nanotoxicology* **2016**, *10*, 32–41. [CrossRef]
214. Koehle-Divo, V.; Pain-Devin, S.; Bertrand, C.; Devin, S.; Mouneyrac, C.; Giamberini, L.; Sohm, B. Corbicula fluminea gene expression modulated by CeO_2; nanomaterials and salinity. *Environ. Sci. Pollut. Res.* **2019**, *26*, 15174–15186. [CrossRef]
215. Sendra, M.; Volland, M.; Balbi, T.; Fabbri, R.; Yeste, M.P.; Gatica, J.M.; Canesi, L.; Blasco, J. Cytotoxicity of CeO_2 nanoparticles using in vitro assay with Mytilus galloprovincialis hemocytes: Relevance of zeta potential, shape and biocorona formation. *Aquat. Toxicol.* **2018**, *200*, 13–20. [CrossRef]
216. Noventa, S.; Hacker, C.; Rowe, D.; Elgy, C.; Galloway, T. Dissolution and bandgap paradigms for predicting the toxicity of metal oxide nanoparticles in the marine environment: An in vivo study with oyster embryos. *Nanotoxicology* **2018**, *12*, 63–78. [CrossRef] [PubMed]
217. Sizochenko, N.; Leszczynska, D.; Leszczynski, J. Modeling of interactions between the zebrafish hatching enzyme ZHE1 and A series of metal oxide nanoparticles: Nano-QSAR and causal analysis of inactivation mechanisms. *Nanomaterials* **2017**, *7*, 330. [CrossRef]
218. Milenkovic, I.; Radotic, K.; Despotovic, J.; Nikolic, A.; Loncarevic, B.; Ljesevic, M.; Spasic, S.Z.; Beskoski, V.P. Toxicity investigation of CeO_2 nanoparticles coated with glucose and exopolysaccharides levan and pullulan on the bacterium *Vibrio fischeri* and aquatic organisms *Daphnia magna* and *Danio rerio*. *Aquat. Toxicol.* **2021**, *236*, 105867. [CrossRef]
219. Lawrence, J.R.; Swerhone, G.D.W.; Roy, J.; Paule, A.; Grigoryan, A.A.; Chekabab, S.M.; Korber, D.R.; Dynes, J.J. Microscale and molecular analyses of river biofilm communities treated with microgram levels of cerium oxide nanoparticles indicate limited but significant effects. *Environ. Pollut.* **2020**, *256*, 113515. [CrossRef]
220. Servin, A.D.; De la Torre-Roche, R.; Castillo-Michel, H.; Pagano, L.; Hawthorne, J.; Musante, C.; Pignatello, J.; Uchimiya, M.; White, J.C. Exposure of agricultural crops to nanoparticle CeO_2 in biochar-amended soil. *Plant Physiol. Biochem.* **2017**, *110*, 147–157. [CrossRef]
221. Barrios, A.C.; Medina-Velo, I.A.; Zuverza-Mena, N.; Dominguez, O.E.; Peralta-Videa, J.R.; Gardea-Torresdey, J.L. Nutritional quality assessment of tomato fruits after exposure to uncoated and citric acid coated cerium oxide nanoparticles, bulk cerium oxide, cerium acetate and citric acid. *Plant Physiol. Biochem.* **2017**, *110*, 100–107. [CrossRef]
222. Zhang, P.; Xie, C.; Ma, Y.; He, X.; Zhang, Z.; Ding, Y.; Zheng, L.; Zhang, J. Shape-dependent transformation and translocation of ceria nanoparticles in cucumber plants. *Environ. Sci. Technol. Lett.* **2017**, *4*, 380–385. [CrossRef]
223. Ma, Y.; Xie, C.; He, X.; Zhang, B.; Yang, J.; Sun, M.; Luo, W.; Feng, S.; Zhang, J.; Wang, G.; et al. Effects of ceria nanoparticles and CeCl3 on plant growth, biological and physiological parameters, and nutritional value of soil grown common bean (*Phaseolus vulgaris*). *Small* **2020**, *16*, 1907435. [CrossRef] [PubMed]
224. Maqbool, Q. Green-synthesised cerium oxide nanostructures (CeO_2-NS) show excellent biocompatibility for phyto-cultures as compared to silver nanostructures (Ag-NS). *RSC Adv.* **2017**, *7*, 56575–56585. [CrossRef]
225. Reichman, J.R.; Rygiewicz, P.T.; Johnson, M.G.; Bollman, M.A.; Smith, B.M.; Krantz, Q.T.; King, C.J.; Andersen, C.P. Douglas-Fir (*Pseudotsuga menziesii* (Mirb.) Franco) Transcriptome profile changes induced by diesel emissions generated with CeO_2 nanoparticle fuel borne catalyst. *Environ. Sci. Technol.* **2018**, *52*, 10067–10077. [CrossRef]
226. Tumburu, L.; Andersen, C.P.; Rygiewicz, P.T.; Reichman, J.R. Molecular and physiological responses to titanium dioxide and cerium oxide nanoparticles in Arabidopsis. *Environ. Toxicol. Chem.* **2017**, *36*, 71–82. [CrossRef] [PubMed]
227. You, T.; Liu, D.; Chen, J.; Yang, Z.; Dou, R.; Gao, X.; Wang, L. Effects of metal oxide nanoparticles on soil enzyme activities and bacterial communities in two different soil types. *J. Soils Sediments* **2018**, *18*, 211–221. [CrossRef]
228. Stowers, C.; King, M.; Rossi, L.; Zhang, W.; Arya, A.; Ma, X. Initial sterilization of soil affected interactions of cerium oxide nanoparticles and soybean seedlings (*Glycine max* (L.) Merr.) in a Greenhouse study. *ACS Sustain. Chem. Eng.* **2018**, *6*, 10307–10314. [CrossRef]
229. Xie, C.; Ma, Y.; Yang, J.; Zhang, B.; Luo, W.; Feng, S.; Zhang, J.; Wang, G.; He, X.; Zhang, Z. Effects of foliar applications of ceria nanoparticles and $CeCl_3$ on common bean (*Phaseolus vulgaris*). *Environ. Pollut.* **2019**, *250*, 530–536. [CrossRef]
230. Cotena, M.; Auffan, M.; Tassistro, V.; Resseguier, N.; Rose, J.; Perrin, J. In vitro co-exposure to CeO_2 nanomaterials from diesel engine exhaust and benzo(a)pyrene induces additive DNA damage in sperm and cumulus cells but not in oocytes. *Nanomaterials* **2021**, *11*, 478. [CrossRef] [PubMed]
231. Preaubert, L.; Tassistro, V.; Auffan, M.; Sari-Minodier, I.; Rose, J.; Courbiere, B.; Perrin, J. Very low concentration of cerium dioxide nanoparticles induce DNA damage, but no loss of vitality, in human spermatozoa. *Toxicol. Vitro* **2018**, *50*, 236–241. [CrossRef]
232. Cotena, M.; Auffan, M.; Robert, S.; Tassistro, V.; Resseguier, N.; Rose, J.; Perrin, J. CeO_2 nanomaterials from diesel engine exhaust induce DNA damage and oxidative stress in human and rat sperm in vitro. *Nanomaterials* **2020**, *10*, 2327. [CrossRef]
233. Sundararajan, V.; Dan, P.; Kumar, A.; Venkatasubbu, G.D.; Ichihara, S.; Ichihara, G.; Sheik Mohideen, S. Drosophila melanogaster as an in vivo model to study the potential toxicity of cerium oxide nanoparticles. *Appl. Surf. Sci.* **2019**, *490*, 70–80. [CrossRef]

234. Cappellini, F.; Di Bucchianico, S.; Karri, V.; Latvala, S.; Malmloef, M.; Kippler, M.; Elihn, K.; Hedberg, J.; Wallinder, I.O.; Gerde, P.; et al. Dry generation of CeO_2 nanoparticles and deposition onto a co-culture of A549 and THP-1 cells in air-liquid interface-dosimetry considerations and comparison to submerged exposure. *Nanomaterials* **2020**, *10*, 618. [CrossRef]
235. Wu, J.; Ma, Y.; Ding, Y.; Zhang, P.; He, X.; Zhang, Z. Toxicity of two different size ceria nanoparticles to mice after repeated intranasal instillation. *J. Nanosci. Nanotechnol.* **2019**, *19*, 2474–2482. [CrossRef]
236. Nemmar, A.; Al-Salam, S.; Beegam, S.; Yuvaraju, P.; Ali, B.H. Aortic oxidative stress, inflammation and DNA damage following pulmonary exposure to cerium oxide nanoparticles in a rat model of vascular injury. *Biomolecules* **2019**, *9*, 376. [CrossRef]
237. Yang, Y.; Xu, L.; Dekkers, S.; Zhang, L.G.; Cassee, F.R.; Zuo, Y.Y. Aggregation state of metal-based nanomaterials at the pulmonary surfactant film determines biophysical inhibition. *Environ. Sci. Technol.* **2018**, *52*, 8920–8929. [CrossRef]
238. Cordelli, E.; Keller, J.; Eleuteri, P.; Villani, P.; Ma-Hock, L.; Schulz, M.; Landsiedel, R.; Pacchierotti, F. No genotoxicity in rat blood cells upon 3- or 6-month inhalation exposure to CeO_2 or $BaSO_4$ nanomaterials. *Mutagenesis* **2017**, *32*, 13–22. [CrossRef]
239. Mauro, M.; Crosera, M.; Monai, M.; Montini, T.; Fornasiero, P.; Bovenzi, M.; Adami, G.; Turco, G.; Larese Filon, F. Cerium oxide nanoparticles absorption through intact and damaged human skin. *Molecules* **2019**, *24*, 3759. [CrossRef]
240. Del, T.S.; Caselli, C.; Sabatino, L.; Basta, G.; Ciofani, G.; Ciofani, G.; Cappello, V.; Parlanti, P.; Gemmi, M.; Ragusa, R.; et al. Effects of cerium oxide nanoparticles on hemostasis: Coagulation, platelets, and vascular endothelial cells. *J. Biomed. Mater. Res. A* **2019**, *107*, 1551–1562.
241. Gliga, A.R.; Edoff, K.; Caputo, F.; Kaellman, T.; Blom, H.; Karlsson, H.L.; Ghibelli, L.; Traversa, E.; Ceccatelli, S.; Fadeel, B. Cerium oxide nanoparticles inhibit differentiation of neural stem cells. *Sci. Rep.* **2017**, *7*, 9284. [CrossRef]
242. Acharya, S.; Lu, B.; Edwards, S.; Toh, C.; Petersen, A.; Yong, C.; Lyu, P.; Huang, A.; Schmidt, J. Disruption of artificial lipid bilayers in the presence of transition metal oxide and rare earth metal oxide nanoparticle. *J. Phys. D Appl. Phys.* **2019**, *52*, 044002. [CrossRef]
243. Rivero Arze, A.; Manier, N.; Chatel, A.; Mouneyrac, C. Characterization of the nano-bio interaction between metallic oxide nanomaterials and freshwater microalgae using flow cytometry. *Nanotoxicology* **2020**, *14*, 1082–1095. [CrossRef]
244. Li, H.; Xia, P.; Pan, S.; Qi, Z.; Fu, C.; Yu, Z.; Kong, W.; Chang, Y.; Wang, K.; Wu, D.; et al. The advances of ceria nanoparticles for biomedical applications in orthopaedics. *Int. J. Nanomed.* **2020**, *15*, 7199–7214. [CrossRef]
245. Jeong, H.-G.; Cha, B.G.; Kang, D.-W.; Kim, D.Y.; Yang, W.; Ki, S.-K.; Kim, S.I.; Han, J.; Kim, C.K.; Kim, J.; et al. Ceria nanoparticles fabricated with 6-aminohexanoic acid that overcome systemic inflammatory response syndrome. *Adv. Healthc. Mater.* **2019**, *8*, 1801548. [CrossRef]
246. Choi, S.W.; Kim, J. Recent progress in autocatalytic ceria nanoparticles-based translational research on brain diseases. *ACS Appl. Nano Mater.* **2020**, *3*, 1043–1062. [CrossRef]
247. Banavar, S.; Deshpande, A.; Sur, S.; Andreescu, S. Ceria nanoparticle theranostics: Harnessing antioxidant properties in biomedicine and beyond. *J. Phys. Mater.* **2021**, *4*, 042003. [CrossRef]
248. Hüttig, F.; Keitel, J.P.; Prutscher, A.; Spintzyk, S.; Klink, A. Fixed dental prostheses and single-tooth crowns based on ceria-stabilized tetragonal zirconia/alumina nanocomposite frameworks: Outcome after 2 Years in a clinical trial. *Int. J. Prosthodont.* **2017**, *30*, 461–464. [CrossRef] [PubMed]
249. Adorinni, S.; Cringoli, M.C.; Perathoner, S.; Fornasiero, P.; Marchesan, S. Green approaches to carbon nanostructure-based biomaterials. *Appl. Sci.* **2021**, *11*, 2490. [CrossRef]
250. Castillo-Henríquez, L.; Alfaro-Aguilar, K.; Ugalde-Álvarez, J.; Vega-Fernández, L.; Montes de Oca-Vásquez, G.; Vega-Baudrit, J.R. Green synthesis of gold and silver nanoparticles from plant extracts and their possible applications as antimicrobial agents in the agricultural area. *Nanomaterials* **2020**, *10*, 1763. [CrossRef]
251. Corra, S.; Shoshan, M.S.; Wennemers, H. Peptide mediated formation of noble metal nanoparticles-controlling size and spatial arrangement. *Curr. Opin. Chem. Biol.* **2017**, *40*, 138–144. [CrossRef] [PubMed]
252. Brambila, C.; Sayle, D.C.; Molinari, M.; Nutter, J.; Flitcroft, J.M.; Sayle, T.X.T.; Sakthivel, T.; Seal, S.; Möbus, G. Tomographic study of mesopore formation in ceria nanorods. *J. Phys. Chem. C* **2021**, *125*, 10077–10089. [CrossRef] [PubMed]
253. Seal, S.; Jeyaranjan, A.; Neal, C.J.; Kumar, U.; Sakthivel, T.S.; Sayle, D.C. Engineered defects in cerium oxides: Tuning chemical reactivity for biomedical, environmental, & energy applications. *Nanoscale* **2020**, *12*, 6879–6899. [CrossRef]
254. Bellotto, O.; Cringoli, M.C.; Perathoner, S.; Fornasiero, P.; Marchesan, S. Peptide gelators to template inorganic nanoparticle formation. *Gels* **2021**, *7*, 14. [CrossRef]
255. Cringoli, M.C.; Marchesan, S.; Melchionna, M.; Fornasiero, P. Nanostructured gels for energy and environmental applications. *Molecules* **2020**, *25*, 5620. [CrossRef]
256. Centomo, P.; Zecca, M.; Biffis, A. Cross-linked polymers as scaffolds for the low-temperature preparation of nanostructured metal oxides. *Chem.—Eur. J.* **2020**, *26*, 9243–9260. [CrossRef]
257. Jahović, I.; Zou, Y.-Q.; Adorinni, S.; Nitschke, J.R.; Marchesan, S. Cages meet gels: Smart materials with dual porosity. *Matter* **2021**, *4*, 2123–2140. [CrossRef]
258. Sener, G.; Hilton, S.A.; Osmond, M.J.; Zgheib, C.; Newsom, J.P.; Dewberry, L.; Singh, S.; Sakthivel, T.S.; Seal, S.; Liechty, K.W.; et al. Injectable, self-healable zwitterionic cryogels with sustained microRNA—Cerium oxide nanoparticle release promote accelerated wound healing. *Acta Biomater.* **2020**, *101*, 262–272. [CrossRef]
259. Zheng, N.; Xu, Y.; Zhao, Q.; Xie, T. Dynamic covalent polymer networks: A molecular platform for designing functions beyond chemical recycling and self-healing. *Chem. Rev.* **2021**, *121*, 1716–1745. [CrossRef] [PubMed]

260. Adorinni, S.; Rozhin, P.; Marchesan, S. Smart hydrogels meet carbon nanomaterials for new frontiers in medicine. *Biomedicines* **2021**, *9*, 570. [CrossRef]
261. Soto, F.; Karshalev, E.; Zhang, F.; Esteban Fernandez de Avila, B.; Nourhani, A.; Wang, J. Smart materials for microrobots. *Chem. Rev.* **2021**. [CrossRef] [PubMed]

nanomaterials

MDPI

Review

From Impure to Purified Silver Nanoparticles: Advances and Timeline in Separation Methods

Catarina S. M. Martins , Helena B. A. Sousa and João A. V. Prior *

LAQV, REQUIMTE, Laboratory of Applied Chemistry, Department of Chemical Sciences, Faculty of Pharmacy of University of Porto, Rua de Jorge Viterbo Ferreira, n°. 228, 4050-313 Porto, Portugal; up201304479@edu.ff.up.pt (C.S.M.M.); hbffup@gmail.com (H.B.A.S.)

* Correspondence: joaoavp@ff.up.pt

Abstract: AgNPs have exceptional characteristics that depend on their size and shape. Over the past years, there has been an exponential increase in applications of nanoparticles (NPs), especially the silver ones (AgNPs), in several areas, such as, for example, electronics; environmental, pharmaceutical, and toxicological applications; theragnostics; and medical treatments, among others. This growing use has led to a greater exposure of humans to AgNPs and a higher risk to human health and the environment. This risk becomes more aggravated when the AgNPs are used without purification or separation from the synthesis medium, in which the hazardous synthesis precursors remain unseparated from the NPs and constitute a severe risk for unnecessary environmental contamination. This review examines the situation of the available separation methods of AgNPs from crude suspensions or real samples. Different separation techniques are reviewed, and relevant data are discussed, with a focus on the sustainability and efficiency of AgNPs separation methods.

Keywords: silver nanoparticles; AgNPs; synthesis; separation; purification

Citation: Martins, C.S.M.; Sousa, H.B.A.; Prior, J.A.V. From Impure to Purified Silver Nanoparticles: Advances and Timeline in Separation Methods. *Nanomaterials* **2021**, *11*, 3407. https://doi.org/10.3390/nano11123407

Academic Editors: Ivan Stoikov and Pavel Padnya

Received: 28 October 2021
Accepted: 10 December 2021
Published: 16 December 2021

1. Introduction

In the literature, nanoparticles are defined as particles with a size between 1 and 100 nm and have been widely used because of their unique physical and chemical properties. Over the last years, their applications have been increasing in areas like medicine, the pharmaceutical industry, cosmetics, textiles, the food industry, and others of everyday products [1]. Nanoparticles (NPs) can be divided into two categories: the organics, and the inorganics. Belonging to the class of inorganics, noble metal NPs, especially silver NPs (AgNPs), are the most used [2], since these present intrinsic properties that make them exceptional. Apart from being useful as drug carriers or nanotheranostic sensors, some studies reported bactericidal, antifungal, and antiviral effects [3], which makes it possible to use them in the treatment of certain diseases and medical devices, such as wound and burn dressings, breathing masks, and implantable catheters [4].

AgNPs can be synthesized through two different strategies, the top-down and the bottom-up, with various materials and coatings, sizes, and shapes [5,6]. These strategies include a variety of physical, chemical, and biological methods. Recently, a sustainable method (a greener approach) to synthesize NPs emerged and enabled some reduction in the hazardous impact on the environment. This synthesis approach is based on green-chemistry concepts that rely on a "set of principles that reduces or eliminates the use or generation of hazardous substances in the design, manufacture and application of chemical products" [7]. This eco-friendly approach allows to synthesize NPs using solvents that are non-toxic to the environment, and using greener energy sources like microwave and ultrasound, at low-temperature and pressure conditions. Obviously, the green-based synthesis approaches that rely on relatively innocuous and bio-friendly reagents give raise to lower toxicity and higher biocompatibility of the AgNPs, due to the reagents' chemical source and composition, determining the range of applications for the NPs. Additionally, by manipulating the synthesis parameters, it is possible to obtain AgNPs of different sizes and shapes [6]. This way, the controllable physicochemical properties can also be exploited for different applications of AgNPs.

Despite the reduced use of dangerous reagents or solvents in AgNPs synthesis through green-based synthesis [6], it may still be needed to use some reagents and precursors that have some toxicity, and, consequently, this might limit the possible applications of the nanomaterials in medical, biochemical, analytical, and clinical areas. For example, in bio-related applications, the by-products of the reactions' synthesis might influence, interfere, or interact with natural occurring and complex processes in living cells or other organisms, originating unintentional effects, in comparison if only purified AgNPs were used. The common use of $NaBH_4$ as a reducing agent in many syntheses might be a good example of a toxic reagent that, if not isolated from the raw suspension, will interfere with regular biological processes [6]. To circumvent this limitation, the purification of AgNPs from the crude synthesis media allows to obtain pure suspensions of the NPs, in opposition to using a mixture of the NPs and residues from the synthesis. The use of purified AgNPs in research studies increases by far the efficiency of the assays, allowing to reduce the number of assays necessary to conduct the research. Additionally, the purification of AgNPs allows to properly separate the synthesis' remains, which are considered waste products, reducing their impact on the environment. Therefore, the search for novel separation techniques of AgNPs from the reaction medium is important, especially if it simultaneously allows for the sorting of NPs by size and shape, with this aspect being crucial for the proper usage of AgNPs in research studies.

Moreover, the growing use of AgNPs in consumer products leads to a higher release of these NPs to the environment, creating new hazards to health and the ecosystem, from surface waters to human consumption [8]. The persistent AgNPs exposure can be dangerous for humans because the accumulation of silver in the organism causes serious diseases, like argyria or argyrosis, which promote an irreversible discoloration of the skin or the eyes. The chemical monitoring of AgNPs in environmental, food, textiles, and biomedical samples is of paramount importance, and the analytical methods often rely on initial sample-treatment approaches based on separation or pre-concentration methods.

This review aimed at presenting a brief discussion of the principal methods used to separate AgNPs, between 2004 and 2020, and at providing an overview of the current developments in the separation techniques, pointing out advantages and disadvantages. Due to the wide range of distinct applications of AgNPs, this literature revision was restricted to the works where AgNPs, from crude suspensions or real samples, were used for the development, improvement, and optimization studies of several separation methods. The methods available for separation of AgNPs from the synthesis' media are based on, for example, magnetic analytical schemes, hydrodynamic forces, chromatography, density gradient centrifugation, electrophoresis, selective precipitation, membrane filtration, and liquid extraction techniques, among others.

In 2015, a review about the use of AgNPs as nano-adsorbents for separation and preconcentration of environmental pollutants [9] highlighted the potential use of AgNPs to remove the contaminants of different environmental samples. In that review, the focus was on the use of AgNPs to separate and not its purification. Another work, by Wang et al. (2020), revised AgNPs and Ag ions speciation' methods based exclusively on separation techniques coupled with atomic spectroscopy, but those only included solid-phase extraction, cation-exchange reactions, chromatography, and single-particle detection [10]. So, this revision work constitutes an advance in the general overview of eight different separation techniques, with a focus on the AgNPs separation from crude suspensions and real samples. For this purpose, scientific research was conducted at the global citation database Web of Science^{TM} using an advanced search for each separation method analyzed in this study. Considering the number of obtained articles, the analysis of each separation method was limited to some examples. A discussion about potentialities and limitations of the available methods was also included, while providing, at the same time, an overview of conceptual advances in a timeline approach.

2. Separation Methods of AgNPs

2.1. Magnetic-Based Schemes

Magnetic nanoparticles can be separated by magnetic forces according to their magnetic susceptibilities and/or their sizes [11]. Despite the fact that the theoretical discussion of the physical/magnetic details behind the separations is outside the scope of this review, very briefly, during capture of magnetic NPs there is a competition between magnetic forces and thermal diffusion [12]. The magnetic force (F_M) acting on a nanoparticle can be calculated by the equation, $F_M = \mu_0 \chi V_p H.\nabla H$, in which μ_0 is the permeability of free space, χ is the magnetic susceptibility, V_p is the volume of the particle, and H is an external magnetic field [13,14]. More information can be consulted in the mentioned references, which explain the physical/magnetic details in a more-theoretical perspective. A revision on scientific literature about the separation of AgNPs by magnetic fields resulted in the compilation of the works presented in the Table 1, which is discussed more thoroughly in the following.

Table 1. Overview of AgNPs separation based on magnetic schemes.

Separation Method (Ref.)	Size (nm)	Matrix	Recovery (%)	Optimal Separation Conditions	Year
Surface-modified magnetic capture particles (UMP, GMP, DMP, and Mix D–G) [15]	SEM: 10 and 75	Environmental water	>99%	Add 2 µg/mL AgNPs suspension to 2 mg/mL magnetic particles; shake the mixture for 30 min at 100 rpm to disperse the particles; incubate for 15 min to facilitate absorption of the AgNPs	2014
Magnetic reduced graphene oxide [16]	TEM: 30, 50, 80, and 100 (citrate); 60 and 100 (PVP)	Environmental water	98%	Add 10 mg of adsorbent to 10 mL of the AgNP/Ag suspension; at RT (25 ± 0.5 °C), oscillate at 200 rpm	2017

UMP: unmodified magnetic particles; GMP: glutathione-functionalized magnetic particles; DMP: dopamine-functionalized magnetic particles; Mix D–G: equal mass mixture of DMP and GMP; TEM: transmission electronic microscopy; SEM: scanning electronic microscopy; RT: room temperature.

In 2013, Mwilu and his co-workers [15] developed a separation method using surface-modified magnetic-capture particles. The authors started by synthesizing unmodified magnetic particles (UMP), followed by glutathione-functionalized magnetic particles (GMP) and dopamine-functionalized magnetic particles (DMP). Next, they prepared a mixture of equal masses of DMP and GMP (Mix D–G) to assess their capabilities for separating AgNPs from aqueous media. For that purpose, the magnetic capturing particles were incubated with AgNPs suspensions (of different capping, namely, citrate and polyvinylpyrrolidone (PVP), and different sizes: 10 nm and 75 nm), and then the mixtures were exposed to a neodymium magnet, incorporated into a flow cell. After the application of a magnetic field, there was a visual clearing of the sample suspension, indicating the separation of the AgNPs from the medium flowing out the flow chamber. This methodology enabled separation of the eluate and the captured particles and quantification for total silver by inductively coupled plasma mass spectrometry (ICP-MS) upon acid digestion. The authors verified that the UMP-based magnetic particles allowed a high degree of selectivity for AgNPs over silver ions, with less than 5% of Ag^+ adsorbed. The innovation of this work was the possibility of conducting pre-concentration of the separated AgNPs by magnetic particles, before quantification by ICP-MS, allowing the analysis of trace levels of AgNPs in samples, as well as the selective separation of nanosilver species in mixtures containing silver ions (Ag^+). The selectivity and easiness of execution of the procedure was tested in the pre-concentration and detection of AgNPs content in environmental water samples, which revealed recoveries of spiked AgNPs > 96%. Additionally, the proposed methodology revealed precision and accuracy, free of matrix interference. One main drawback of the methodology was that AgNPs lose their individuality when they get adsorbed to the magnetic particles. After adsorption of AgNPs by the magnetic particles, further characterizations and use of the AgNPs are not possible due to the adsorption phenomena. Moreover, analysis of the particle-size distribution or nanoparticles' quantification by particle counting using single-particle ICP-MS is also unfeasible. Single-particle inductively coupled plasma mass spectrometry (SP-ICP-MS) [17] has emerged as an instrumental method of analysis with proved potential [18,19] when it comes to a more-accurate approach for nanoparticle's size characterization and quantification, which is of paramount importance in environmental and biological toxicological analyses. The SP-ICP-MS is a relatively recent analytical technique that provides information about the composition of each particle present in the sample, their size and size-distribution, their number density, and, also, their concentration.

The literature contains many works involving the separation of AgNPs by magnetic approaches but only when they are incorporated in magnetic nanocomposites, such as magnetic Fe_3O_4-Ag(0) [20]. Hence, the number of works involving the separation of single AgNPs by magnetic schemes is scarce, probably since AgNPs do have not magnetic properties.

2.2. Hydrodynamic Forces

Another method for the separation of AgNPs can be accomplished by exploiting the field-flow fractionation (FFF) technique, which is based on flow concepts [21]. FFF is a method of choice for the size-sorting of NPs based on their hydrodynamic size. The separation of NPs occurs inside a thin channel, where various fields actuate perpendicular to the laminar flow, causing different retention rates of the particles because of their distinct diffusion coefficients, accordingly with their size distribution and physicochemical properties [1,11]. This separation occurs because there is a balance between diffusibility and the external field forces, which leads to the formation of different equilibrium states among the laminar-flow streamlines. When compared with smaller NPs, the larger ones interact more strongly with the external field, being more retained in the accumulation wall, and hence they have longer elution times [11,21]. There are many other variants of FFF, including thermal FFF, sedimentation FFF (SdFFF), electrical FFF, magnetic FFF (MFFF), and flow FFF (F4) [21,22]. Additionally, F4 can be sub-divided into three technical variants: the asymmetrical F4 (AF4), the symmetrical F4, and the hollow fiber F4 (HF5) [23], which will be discussed together with selected literature works for better comprehension (Table 2).

Table 2. Overview of AgNPs separation based on hydrodynamic forces.

Separation Method (Ref.)	Size (nm)	Matrix	Recovery (%)	Optimal Separation Conditions	Year
SdFFF [24]	FE-SEM: 20–100 and 60–150	Environmental water	-	Carrier liquid: water with 0.1% FL-70; injection volume 5~30 μL; vortex for 30 s before the injection; RT	2007
Flow FFF [25]	TEM: 15	-	-	Channel flow: 1 mL/min; cross flow: 0.4–1 mL/min	2009
AF4 coupled with ICP-MS [26]	TEM: 10 nm, 20 ± 5, 40 ± 5, 60 ± 5, and 80 ± 7 nm	Two consumer products, an antiseptic, and a dietary supplement	83 ± 8% and 93 ± 4%	Ultrafiltration membranes, cut-off 1 and 4 kDa; flow rate of 0.8 mL/min; mobile phase: 0.01% SDS, at pH 8	2011
AF4 coupled with ICP-MS [27]	TEM: 42 ± 10 nm	Aqueous medium	<1%	Carrier liquid: 0.5 mM NH_4HCO_3, pH 7.4; PES membrane, cut-off 10 kDa; flow rate: 1.0 mL/min; injection volume: 0.2 mL/min	2013
AF4 coupled with ICP-MS [28]	TEM: 10, 40, and 60 nm	Commercial nutraceutical Products and Korean beer	97 ± 2 and 106 ± 1%	Carrier liquid: ultrapure water and SDS 0.01% at pH 8; regenerated cellulose Membrane, cut-off 10 kDa; injection volume: 200 μL	2014

Table 2. *Cont.*

Separation Method (Ref.)	Size (nm)	Matrix	Recovery (%)	Optimal Separation Conditions	Year
HF5 coupled with ICP-MS [22]	TEM: (tannic acid) 11.4, 5.9, 9.1, 26.5, 8.9; (citrate) 10 and 15.5	-	Similar for both capping agents	Carrier liquid: 30 mM TRIS buffer (pH 8) and 0.02% (*w/v*)—FL-70 and NaN_3 (pH 10); flow rate: pump A—0.5 mL/min, pump B—2.0 mL/min.	2015
HF5 coupled with MALS [23]	TEM: 20 and 140 nm	Aqueous media	>90%	Polymeric membrane, cut-off 100 kDa; mobile phase: water; injection volume: 4 µL	2015
HF5 coupled with multiple detectors (UV-Vis, DLS, and ICP-MS) [29]	TEM: 1.4, 10, 20, 40, and 60 nm	Lake and river waters	70.7−108%	Carrier liquid: 0.1% (*v/v*) FL-70 with 0.02% (*w/v*) NaN_3; inlet flow rate, 1.50 mL/min; radial flow rate, 0.70 mL/min; axial flow, 0.80 mL/min; focusing time: 4 min	2015

FFF: field-flow fractionation; SdFFF: sedimentation field-flow fractionation; AF4: asymmetrical flow field-flow fractionation; HF5: hollow fiber field-flow fractionation; MALS: multi-angle light scattering; ICP-MS: inductive coupled plasma mass spectrometry; DLS: dynamic light scattering; UV–Vis: ultraviolet–visible spectrometry; FL-70TM: commercially available mixture of nonionic and anionic surfactants that included oleic acid, sodium carbonate, tergitol, tetrasodium EDTA, polyethylene glycol, and triethanolamine; TRIS buffer: 2-amino-2-hydroxymethyl-propane-1,3-diol; TEM: transmission electronic microscopy; SEM: scanning electronic microscopy; FE-SEM: field emission scanning electron microscopy; RT: room temperature.

In 2007, SdFFF was employed, by Kim et al. [24], to separate and determine the mean size and the size distributions of AgNPs of about 100 nm in diameter. SdFFF, also known as centrifugal FFF, separates nanoparticles according to their size and density. The separation occurs because there is a centrifugal force, which is generated inside the SdFFF channel when this is spinning at a high rate. The larger and denser nanoparticles are retained more inside the channel than the smallest nanoparticles, because the first ones accumulate on the wall. So, the smallest NPs are eluted first then the largest ones. This work aimed to implement the SdFFF method and optimize the experimental conditions for the AgNPs separation. Some experimental factors, among others, the flow rate, the carrier composition, and the field strength (channel rotation rate), were studied to find the optimal SdFFF conditions for separation of AgNPs. Regarding the proper carrier/eluent, which influences the NPs–NPs and NPs–channel wall interactions, the studies revealed that the separation of AgNPs was not achieved when using pure water as the carrier. Other aqueous-based compositions were tried out, with NaN_3, SDS, or FL-70^{TM} commercial solution (containing a mixture of nonionic and anionic surfactants, including oleic acid, sodium carbonate, tergitol, tetrasodium EDTA, polyethylene glycol, and triethanolamine). However, only when using as carrier a 0.1% or 0.2% FL-70^{TM} commercial solution was the separation of two different populations of AgNPs achieved as intended, with more resolution when using a dilution of 0.1% for the FL-70 commercial solution. In this work, the authors also used mathematical-based deconvolution techniques to determine relative mass contents of AgNPs mixtures. Facing the obtained results, the authors recognized the potential of the methods but concluded that more work was needed for the optimization of experimental conditions, to make SdFFF a useful tool to separate and characterize metal NPs of a broader range of size and chemical nature.

Considering FFF, the most widely used separation technique to isolate commercial NPs is AF4. Figure 1 represents the schematic principle of the separation by asymmetric-flow FFF.

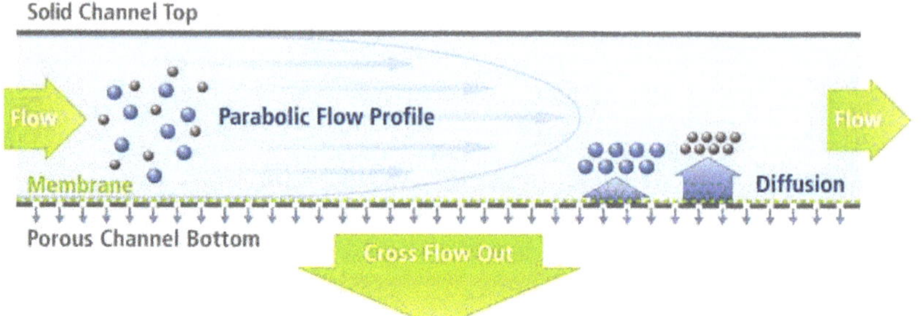

Figure 1. Scheme representative of asymmetric-flow field-flow fractionation. Reproduced from Ref. [30] with permission from Frontiers in Chemistry.

This subtype of 4F bases its separation process on the force that is generated by a cross flow field inside the channel that concentrates the nanoparticles towards a membrane, which is located at the bottom wall of a chamber. This semi-permeable membrane prevents the NPs from passing through its pores but allows the exit of the solvent. Through the flow channel, the sample fractions are separated and eluted towards the detectors, out of the channel. The separation process can be resumed into three steps: the injection, the focusing, and the elution. In the literature, some studies reported the use of AF4 to separate AgNPs from food, nutraceuticals products, and beverages (Table 2). In fact, in 2011, for the first time, Bolea et al. [26] used AF4 coupled with ICP-MS to separate and quantify AgNPs in two consumer products, an antiseptic, and a dietary supplement. The parameters of the separation process that influenced the recovery and resolution were studied, comprising the mobile phase composition, the injection and focusing stages, and the membrane nature. The authors verified that to obtain reproducible results and high recoveries, the most-influencing factors were the mobile phase composition and the membrane nature. The optimal conditions obtained for the highest resolution in the separation of AgNPs were a mobile phase containing an anionic surfactant such as SDS at pH 8 and a polyether sulfone (PES) membrane. Recovery values of 83 $\pm$ 8% for the antiseptic and 93 $\pm$ 4% for the dietary supplement, with respect to the content of AgNPs, were achieved.

Two years later, in 2013, Loeschner et al. [27] developed and optimized an AF4-based method to separate AgNPs stabilized by PVP in aqueous medium. They studied the key factors that had an influence on the separation process, such as the carrier liquid composition, the type of membrane material, the cross flow rate, the spacer height, the focus flow rate, the focus time, and the injected mass. The optimized AF4 parameters, depicted in Table 2, originated relative recoveries > 95% approximately. After the optimization studies, the authors conducted four different approaches (i−iv), based on AF4, to determine the size distribution of AgNPs in suspensions. One of the methods (i) relied on the establishment of a calibration curve using different sizes of polystyrene (PSNP) beads as standards, namely, between 20−100 nm in diameter. After the confirmation that the same separation conditions of AgNPs could be used to properly separate the PSNP beads, a conversion of retention times to diameters was implemented to determine the size distribution of test samples (intensity of absorbance at 400 nm vs. diameter (nm)), assuming the same shape and water–surface interactions of AgNPs and PSNP beads. A second approach (ii) implied the conversion of retention times to hydrodynamic diameters, all dependent on AF4 theory, assuming the behavior of spherical particles at controlled conditions. However, as with most of the theoretical-based deductions, these are dependent on fixing some variables. This work evidenced the influence of the parameters' focus position and channel height on the calculated size distributions. Considering that different nanomaterials for size-distribution calculations imply different separations conditions to achieve the closest to ideal, the previous-mentioned parameters must assume different values for calculations for each experimental separation. The results obtained using this approach were higher and lower than the results obtained by approach (i) when using some extreme values for focus position and channel height obtained by testing the AgNPs and PSNPs, respectively. Yet, overall, the results obtained by (i) and (ii) were similar (Figure 2).

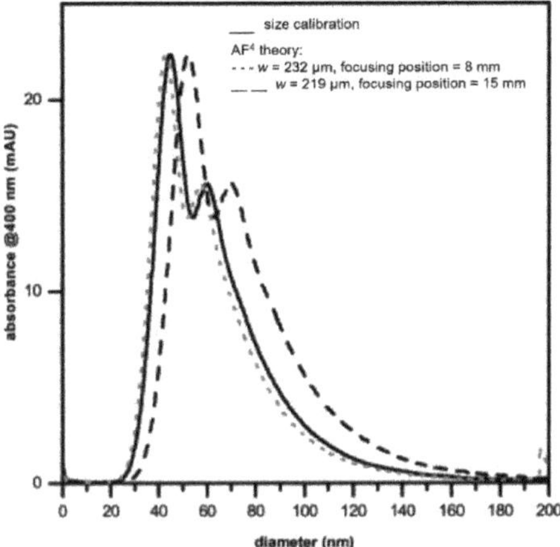

Figure 2. Size distributions of AgNPs by two different approaches: (i) size calibration using polystyrene beads (PSNPs) and (ii) conversion of retention times on AF4 analysis to diameters by AF4 theoretical calculations. Reprinted from [27] copyright (2013), with permission from Elsevier.

The third approach (iii) tested by the authors was based on independent measurements of the AgNPs sizes by TEM. For this purpose, the authors collected AF4-based separated fractions of AgNPs and analyzed each fraction by TEM. The results are depicted in Figure 3.

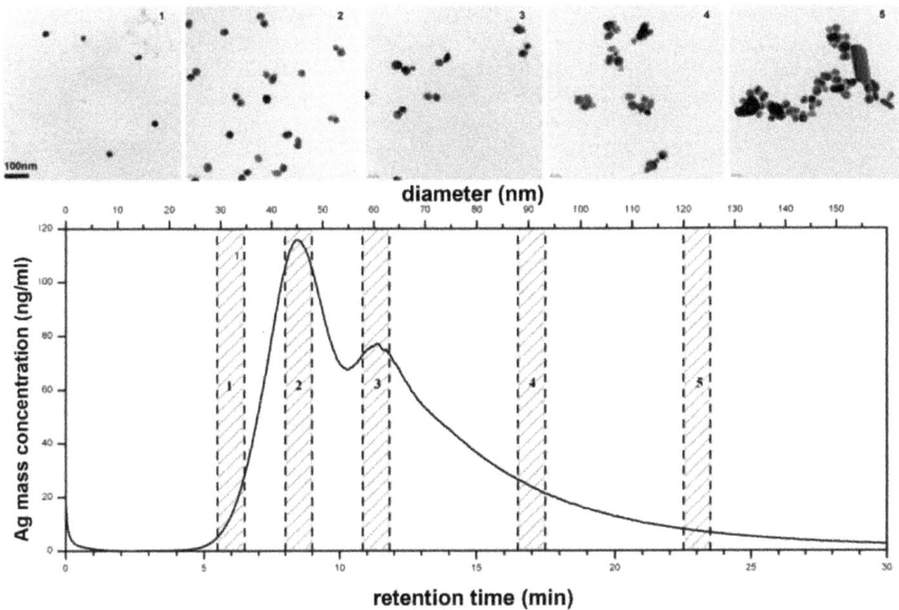

Figure 3. TEM images corresponding to collected fractions of AgNPs separated by AF4 at the conditions: m_{inj} = 10 μg, carrier liquid $(NH_4)_2CO_3$ at pH 9, channel and cross flow rate 1.0 mL/min, spacer height 350 μm, PES membrane. Reprinted from [27] copyright (2013), with permission from Elsevier.

One important aspect about AF4 was that it could not differentiate larger particles from doublets or aggregates of AgNPs, which all appeared as larger diameters in the AF4 fractogram. However, by crossing the AF4 results with the optical spectrum, one could relate some signals apparently corresponding to higher-dimension particles to doublets or aggregates of more particles. The fourth approach (iv) tried by the authors involved in-line measurements with dynamic light scattering (DLS) or multi-angle light scattering (MALS), but these proved unable for the tested suspension of AgNPs. The DLS results revealed a lack of correlation between all ranges of available nanoparticles' diameters, making impossible a complete size determination of a sample. On the other hand, with MALS, the scattered light intensity could not be correlated with the scattering angle due to plasmon resonance effects of AgNPs, for example. From this work, a very important conclusion was that all parameters influencing the AF4-based separation of AgNPs must be carefully studied and optimized or adapted whenever a new type of NPs is analyzed, due to different physical–chemical properties. Another limitation of the proposed method is related with obtaining the quantitative size information, that is, particle mass concentration-based size distributions. Some limitations of AF4 for the separation of AgNPs are due to electrostatic interactions between nanoparticles or nanoparticles and the membrane, caused by the surface charge of the AgNPs, which can hinder their separation behavior. Due to these possible interactions and expected different conditions for the ideal separation of AgNPs of different samples, the authors advised to use to approaches (i) and (ii) together with TEM imaging to successfully determine the size distribution of AgNPs suspensions.

In the next year, Ramos and co-workers [28] studied the feasibility of AF4 combined with ICP-MS for separation, characterization, and quantification of AgNPs in one beverage sample labelled as containing AgNPs ((i) Korean beer) and four commercial nutraceutical products ((ii) antibacterial, antiviral, and antifungal; (iii) antiseptic, disinfectant, and reinforced immune system; (iv) menstrual cycle regulation; and (v) flu prevention and allergies), which claimed to have biocide properties due to containing colloidal silver. In this work the authors emphasized the difficulty of separating, characterizing, and quantifying AgNPs in complex sample matrices like beverages and nutraceuticals, which can also contain other silver species and colloidal forms. In these cases, the sample treatments to extract the AgNPs without causing any kind of aggregation or oxidation processes present difficulties.

To determine the hydrodynamic size of the nanoparticles by AF4, the authors made use of the FFF equation and the Stokes–Einstein equation and compared with the values furnished by the manufacturer. The obtained results with this technique corroborated the TEM results, but with an advantage: the AF4 liquid samples could be analyzed directly, in opposition to what happens with TEM analysis in which the samples must be dried before imaging, with the risk for chemical changes in the sample in consequence of the drying process. After the optimization of the proposed method, the detection limit (LOD) obtained was <28 ng/L. The analytical recovery for total silver, for the nutraceutical product with "antiseptic, disinfectant, reinforce immune system" properties was 97 $\pm$ 2%, when spiked with 12.5 mg/L of ionic silver, and 106 $\pm$ 1%, with AgNPs of 40 nm of dimension. In the authors' opinion, this methodology could become a significant alternative to more-conventional techniques, namely, ultracentrifugation and acid digestion, for silver speciation in complex matrix. The obtained results showed an efficient speciation of the AgNPs from other silver chemical forms, not being affected by the presence of ionic silver.

A miniaturized variant of the F4 technique also exists, which is named hollow-fiber flow field flow fractionation (HF5). In HF5, there is a hollow-fiber made of a porous membrane, where the separation occurs. A forced flow of a mobile phase through the fiber crosses the fiber towards the outlet with a laminar flow profile but also penetrates the pores of the tubular membrane, creating a radial flow (named cross-flow) that is perpendicular to the longitudinal carrier flow. In the process, the smaller nanoparticles (higher diffusion coefficient) of a sample reach faster the fiber outlet, whilst the higher-in-dimension or heavier nanoparticles take more time to complete elution. Thus, the size-fractionation of the nanoparticles is dependent on their diffusion coefficients accordingly with the radial cross-flow, having an influence on the molar mass and hydrodynamic radius of the nanoparticles, for example. In this technique, there are some improvements comparing with AF4, allowing for an increase in the separation efficiency in some cases: a low sample dilution due to the small channel volumes ($\leq$100 μL) and low detector flow rates, allowing to couple to some mass-spectrometry detection methods; the hollow fibers are low-cost materials with a possible disposable usage [23]. The HF5 variant of the F4 technique was also exploited for the separation of AgNPs, with three reported studies presently found in the literature. The first one, described in 2015 from Marassi and co-workers [23], demonstrated the use of HF5 coupled to multi-angle light scattering (MALS) for size-separation and characterization of PVP-stabilized AgNPs in aqueous media. In this pioneering work, the results showed the presence of mainly two different average size populations of AgNPs: one with about 20 nm, and the other one with roughly 140 nm. The analytical recovery higher than 90% confirmed the efficiency of the proposed technique. This novel approach seems to be great to simultaneously separate by size and thoroughly characterize the AgNPs because it provides independent size information. This technique also adds the advantage of separating the Ag^+ ions from AgNPs during the procedure, allowing to overcome potential hazards originated by Ag^+ ions.

Additionally, in the same year, Saenmuangchin et al. [22] successfully developed a homemade HF5 coupled with ICP-MS for the separation of AgNPs, and it was able to circumvent the problem of different retention behaviors of AgNPs when different stabilizing agents were used in the synthesis of AgNPs. Firstly, the authors studied the influence of the carrier solution (FL-70 and TRIS buffer) and the stabilizing agent (tannic acid and citrate) on the retention behavior of AgNPs on the developed HF5 system. Depending on the carrier solution used, different elution profiles for the AgNPs were verified. When using the FL-70 carrier, the citrate- and tannic-acid-capped AgNPs were eluted from the separation system, being the separation depending on the nanoparticles size and on the nature of the capping. This constituted a problem, since AgNPs of similar sizes, but with different capping, had different retention times and were eluted in different fractions. When using the TRIS buffer solution, only the tannic-acid-stabilized AgNPs were eluted from the system. So, to circumvent this problem, tannic acid was added to the TRIS carrier solution to balance the retention behaviors of citrate- and tannic-acid-capped AgNPs. This approach had two purposes: (i) the citrate ligand exchange by tannic acid at the AgNPs surface and (ii) modification of the hollow fiber membrane to become negatively charged. The stabilizer's exchange with tannic acid became necessary because citrate was known to have weak interactions with the nanoparticle's surface, and, during the focusing phase in the HF5 separation procedure, citrate could be unbound from the surface of the nanoparticles, causing these to destabilize. Their destabilization could cause aggregation phenomena of the AgNPs and/or simultaneously increase the nanoparticle–membrane interaction, resulting in higher retention times. The strategy tested by the authors, with the use of 0.1 mM tannic acid in 30 mM TRIS buffer, resulted in similar retention behaviors and good recoveries for both tannic-acid and citrate-stabilized AgNPs. Yet, the proposed separation process must be optimized whenever other types of NPs with other surface coatings are involved, as already recognized by Loeschner et al. [27] when exploiting the AF4 separation-based concept. Overall, the work of Saenmuangchin et al. represents an interesting and valuable contribution for the separation of AgNPs based on the HF5 concept, by proposing a homemade HF5 system capable of separating citrate- and tannic-acid-capped AgNPs, reinforcing the potential of the FFF concept in the separation of nanoparticles, while at the same time providing a low-cost alternative to high-end FFF commercial systems.

Still in 2015, Tan et al. [29] employed, for the first time, the HF5 method for separation and fractionation of AgNPs (>2 nm) and various ionic silver species Ag(I), coupled with multiple detectors (namely, UV−Vis spectrometry, dynamic light scattering—DLS, and ICP-MS) for a full-spectrum speciation analysis and characterization of different sized

AgNPs and Ag(I) species. To discriminate the sizes of AgNPs and Ag(I) species (<2 nm), a minicolumn packed with Amberlite IR120 resin was coupled to the HF5 system (Figure 4). The authors proceeded to the optimization of the HF5 system, and the optimal conditions obtained for the separation are described in Table 2. In the end, by resorting to multiple detectors, the separation, identification, quantification, and characterization of AgNPs and Ag(I) species were successfully achieved. By testing the developed analytical system with spiked lake and river water samples, the authors obtained recoveries between 70.7–108% for seven Ag species: five AgNPs (1.4 nm, 10 nm, 20 nm, 40 nm, and 60 nm), Ag(I), the adduct of Ag(I), and cysteine. The addition of a minicolumn packed with Amberlite IR120 resin to an on-line coupled HF5/MCC-UV/DLS/ICPMS analytical system for Ag(I) speciation analysis and characterization of AgNPs successfully enabled the multi-detection and characterization of different related silver species, including AgNPs of different sizes, reinforcing the potentialities of this novel NPs' separation methodology.

Overall, the separation of AgNPs based on hydrodynamic forces provides researchers with good methods for that purpose, but the posterior application of the separated nanoparticles for other assays, in a purified format, was not verified in the reviewed works.

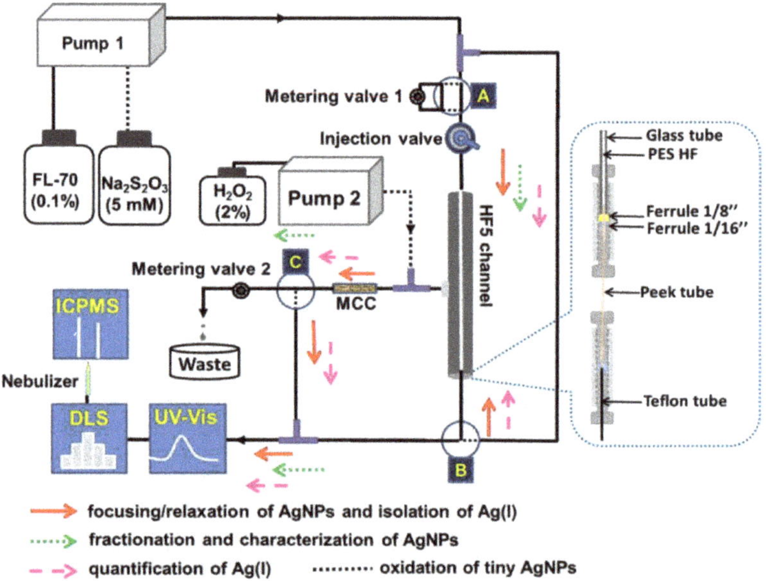

Figure 4. Schematic representation of the on-line coupled HF5/MCC-UV/DLS/ICPMS analytical system. Reprinted from [29] copyright (2015), with permission from ACS (further permission should be directed to the ACS).

2.3. *Chromatography*

Chromatography is a separation technique based on the distribution of the components of a mixture between a fluid (mobile or eluent phase) and an adsorbent (stationary phase). The stationary phase can be a solid or liquid deposited in an inert solid, packed in a column or spread over a surface forming a thin layer [13]. Several scientific reports can be found in the literature about the use of different variants of chromatography to separate AgNPs from the crude suspension or other matrices (Table 3).

Table 3. Overview of AgNPs separation based on chromatography.

Separation Method	Size (nm)	Matrix	Recovery (%)	LOD Value	Optimal Separation Conditions	Year
Reversed-HPLC coupled with ICP-MS [31]	TEM: 10, 20, and 40 nm	Fetal bovine serum and textile products	>80%	0.08 and 0.4 ng/L	Column: Nucleosil, 7 μm particle size, C18, 1000 Å pore size, 250 mm × 4.6 mm; flow rate: 0.5 mL/min; injection volume: 10 μL; mobile phase: 10 mmol/L ammonium acetate at pH 6.8 and 10 mmol/L SDS	2013
Hydrodynamic chromatography coupled with ICP-MS [32]	TEM: <100	Sewage sludge supernatant	-	2.3 ng/mL	Mobile phase: 0.002 M Na_2HPO_4; 0.2% non-ionic surfactant; 0.05% SDS; 0.2% formaldehyde; pH~7.5; injection volume: 20 μL; flow rate: 1.7 mL/min	2009
Reversed-HPLC coupled with ICP-MS in combination with isotope dilution analysis [33]	20, 30, and 40 nm	-	-	1000 Å column: 0.09–3.73 μg/L	Column: Nucleosil, 7 μm particle size, C18, 1000 Å pore size, 250 mm × 4.6 mm; mobile phase: 10 mmol/L SDS, 10 mmol/L ammonium acetate, penicillamine at pH 6.7; flow rate: 0.5 mL/min	2016
SEC coupled with ICP-MS [34]	TEM: 10, 20, 40, 60, and 100 nm	Antibacterial products and environmental waters	84.7–96.4% for Ag(I) and 81.3−106.3% for NAg	0.019 μg/L	Column: 500 Å pore-size; mobile phase: water containing 0.1% (*v/v*) FL-70 and 2 mM $Na_2S_2O_3$; flow rate: of 0.7 mL/min	2014
SEC coupled with ICP-MS [35]	HR-TEM: 10, 20, and 30 nm	Biological tissues (rat liver)	73.7–113% in swine liver; 84.0–104% in rat liver	0.1 μg/g	Column: 5 μm particle size, 1000 Å pore size, 4.6 mm × 250 mm; mobile phase: 2% (*v/v*) FL-70 and 2 mmol/L sodium thiosulfate; flow rate: 0.5 mL/min	2018
Counter-current chromatography [36]	SEM: 13.7 ± 1.9, 14.1 ± 3.5, 19.2 ± 4.3, and 22.2 ± 4.9 nm	Phosphate buffer (20 mM, pH 11)	-	-	Mobile phase: hexane/toluene (1:1, *v/v*), 0.02 mM TOAB; injection volume: 5 mL, flow rate: 1 mL/min; oven temperature 20 °C; rotation of the chromatograph: 700 rpm	2009

ICP-MS: inductive coupled plasma mass spectrometry: HPLC: high-performance liquid chromatography; SEC: size exclusion chromatography; FL-70TM: commercially available mixture of nonionic and anionic surfactants that included oleic acid, sodium carbonate, tergitol, tetrasodium EDTA, polyethylene glycol, and triethanolamine; TEM: transmission electronic microscopy; HR-TEM: high-resolution transmission electronic microscopy; SEM: scanning electronic microscopy; FE-SEM: field emission scanning electron microscopy.

Soto-Alvaredo et al. (2013) [31] developed a method of coupling reversed-phase high-performance liquid chromatography (HPLC) with ICP-MS for the separation/detection of AgNPs and Ag(I) species. The direct coupling of HPLC to ICP-MS was accomplished by using only PEEK tubing connectors, a dual piston pump, and a six-way injection valve, between the column outlet and the nebulizer of ICP-MS. The studied AgNPs were stabilized with citrate and had sizes ranging between 10, 20, and 40 nm. In a single chromatographic run, all silver species (AgNPs and Ag(I)) were detected. For this purpose, thiosulfate was added into the mobile phase to elute Ag(I) species at the end of the separation, without influencing the stability of the AgNPs. The authors tested as a real sample sports' socks, considering that these are examples of textiles known to contain AgNPs and other closely related additives. The proposed methodology allowed to obtain three different chromatogram peaks: the first two peaks were related to the presence of nanoparticles of different dimensions (the first peak in the range of 20−40 nm and the second one ~7 nm); the third peak was related to the presence of Ag(I) species. Interestingly, the authors could only conclude about the presence of silver species, as part of silver nanoparticles or bound to particles of other nature, as well the presence of silver ionic species. The origin of the detected Ag(I) species remained uncertain, as the AgNPs used in the study did not reveal to disintegrate and release silver ions, even during the extraction procedure. So, the authors believe that those species were already present in the textile sample and not formed during the extraction process. However, the assays carried out could not determine with certainty the origin of the ionic silver species. The results obtained from the extracts of sports socks showed that the quantitative data essentially depended on the extraction conditions, namely, the part of the textile sampled for extraction. So, despite the fact that more than 80% of recovery the of both the AgNPs and Ag(I) species was verified—and the calculated LOD was between 0.08 and 0.4 ng/L—it was proved that the extraction method had a marked influence in the developed methodology to be used as a routine laboratory assay in real samples.

Later, in 2016, C.A. Sötebier et al. [33] applied a combination of isotope dilution analysis (IDA) with the HPLC procedure reported by Soto-Alvaredo [31], mentioned above, and coupled to ICP-MS. The aim of the work was the separation and simultaneous quantification of AgNPs and Ag ions. The authors investigated the separation mechanism of AgNPs of different origins and dimensions, and Ag ions, through a comparative study of two different pore-size chromatographic columns: 1000 Å and 4000 Å. For the quantification of total Ag, AgNPs, and Ag ions concentrations, the authors resorted to a post-column IDA approach, using two different silver isotopes ($^{107}Ag/^{109}Ag$). The study showed a decrease in the recovery rates with the increase in AgNPs size when a 1000 Å column was used, whereas, with a 4000 Å column, the recovery rate could be considered independent of the AgNPs with sizes between 18 nm to 55 nm (<60 nm). The obtained results for a 1000 Å column could be due to interactions of the nanoparticles with the stationary phase, despite the fact that these often do not occur, considering that larger-sized nanoparticles can only interact with the larger pores of the column material. Regarding the LOD values, these were verified to increase with the particles' sizes, when using the 1000 Å column, while, with the 4000 Å column, this effect only was verified for larger AgNPs. In fact, larger AgNPs were not eluted from the column of the 1000 Å column leading to conclude about the interaction more strongly with the smaller pores of the stationary phase, due to the higher negative charge and steric hindrance of larger AgNPs, which made it impossible to determine the exclusion limit for each of the columns. To around the problem of the detection of the exclusion value to the 1000 Å column, the authors performed HPLC-ICP-MS experiments in the single-particle mode (HPLC-spICP-MS) combination with IDA. They verified that a specific exclusion limit does not exist, which supports the argument that there are interactions of nanoparticles with the stationary phase. The authors pointed out that the proposed separation method could be applied to different nanoparticle systems like gold and polystyrene, but further investigation was required to eliminate the effects of interaction with the stationary phase, by choosing another eluent or column materials with even larger pore sizes.

Among chromatographic-based techniques to fractionate NPs, size exclusion chromatography (SEC) is probably the most popular [13]. Over the years, this technique has been known by various names, namely, liquid-exclusion chromatography, gel-filtration chromatography, and gel-permeation chromatography [37]. In SEC, the separation is not based on the interaction of NPs with the stationary phase but on the differences in their hydrodynamic volumes [13]. The chromatography column is filled with a porous matrix, which creates flow channels. The smaller particles can permeate deep inside the column, because they have a minor diameter compared with the pore size of the matrix, whereas the larger ones are immediately excluded or conditioned to permeate between larger pores. The larger the particles, the shorter the retention time [1]. When compared with other separation methods, SEC presents several advantages such as being easy to scale-up, with a low chance of sample loss and less time for separation being needed. However, it may originate separations with low resolution, since it depends on certain factors such as the flow rate, the column dimensions, and the packing material, limiting to some extent its application [13,37]. Figure 5 shows a schematic representation of the separation principle of SEC.

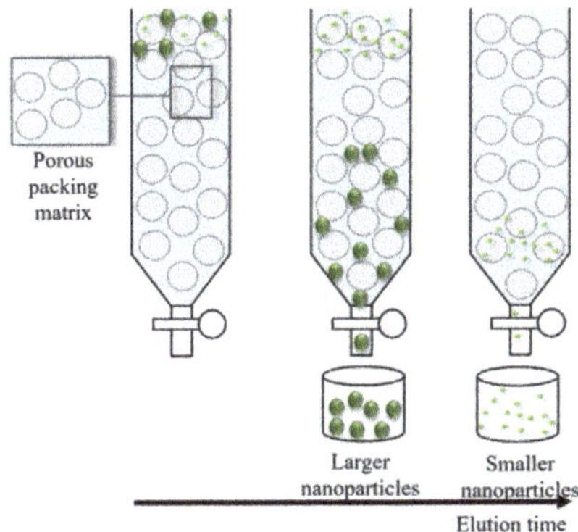

Figure 5. Representative scheme of SEC.

In 2014, Zhou et al. [34] developed a novel SEC method for a rapid and high-resolution separation of dissoluble Ag(I) species from AgNPs, in five antibacterial products and three environmental water samples. The optimal conditions for obtaining the best results were found to be based on the use of a 500 Å pore-size amino column, an aqueous mobile phase containing 0.1% (*v/v*) FL-70^{TM}, and 2 mM $Na_2S_2O_3$, at a flow rate of 0.7 mL/min. As a result, AgNPs and Ag_2S were eluted in one fraction, whereas dissoluble Ag(I) was eluted as a baseline separated peak. This efficient approach allowed to separate a whole range of AgNPs' sizes between 1 to 100 nm, in only 5 min. The authors obtained excellent analytical recoveries, for environmental water samples, ranging from 84.7−96.4% for Ag(I) and 81.3−106.3% for nanoparticulate Ag. For the studied antibacterial products, the analytical recoveries obtained for Ag(I) ranged from 94.8−102.7%. For the first time, a successful separation between Ag(I) from ~1 nm Ag nanoclusters was reported. With these results the authors concluded that the proposed SEC-based methodology was a valuable alternative to use as a rapid and facile separation of dissoluble ions from the AgNPs bulk suspensions.

For the first time, in 2018, Dong et al. [35] developed a method by SEC coupled to ICP-MS, for accurate size characterization of AgNPs with biomolecule corona (AgNP@BCs) and for mass quantification of different Ag(I) species in biological tissues (rat and swine liver). The information provided by these analyses is of paramount importance for the understanding of AgNPs activity in vivo, that is, in which processes where the NPs are involved, interfere, or initiate in vivo, and which physical–chemical transformations they are subject to. The separation via SEC was based on the previous report of Zhou et al. [34]. The application of the proposed method to a rat liver after 24 h of exposure in vivo to the nanoparticles was verified as an excellent separation between Ag(I)-biomolecule complex and AgNP@BCs, which made possible the quantification of free Ag. The obtained concentrations were 36.4 ± 5.6 μg/g for the Ag(I)-biomolecule complex and 25.9 ± 2.8 μg/g for AgNP@BCs. To demonstrate the high accuracy of the proposed method to quantify the amount of Ag in vivo, the authors compared the total Ag mass from the two species (the Ag(I)-biomolecule complex and AgNP@BCs) derived from SEC-ICP-MS (62.2 ± 6.4 μg/g) with the total Ag content determined by microwave digestion followed by ICP-MS (64.3 ± 5.5 μg/g). Due to the similarity of the obtained values, it was possible to demonstrate the high precision of this method. This quantification allowed also to verify that after 24 h exposure to AgNPs, more than half (56.6%) of the total Ag accumulated in the liver was present as Ag(I).

Separation of AgNPs by size, through a counter-current chromatography (CCC) approach, a type of a support-free liquid chromatography, was studied in 2009 by Shen et al. [36]. This chromatography technique is composed by two immiscible solvents, which, after settling down, form two layers that act as the stationary phase and as the mobile phase. Usually, one layer is hydrophilic, while the other is hydrophobic. The separation of different chemical species or compounds by CCC occurs due to the different solubilities of the components between the stationary and the mobile phase. Because of the lack of a solid support to run the separation, there is no chance of irreversible adsorption of materials, which represents an advantage over other techniques [36,38]. The work of Shen et al. involved the synthesis of AgNPs modified with 11-mercaptoundecanoic acid (MUA), to study the potentialities of CCC in separating aqueous-dispersible nanoparticles. The solvent system was constituted by hexane/toluene 1:1, *v/v*, containing tetraoctylammonium bromide (TOAB) to act as a phase-transfer catalyst. The separation was achieved through ion-pair formation between tetraoctylammonium cations (TOA^+) and the carboxyl group (anions) present on the AgNPs surfaces. To achieve the best separation and recovery conditions, several concentrations of TOAB were tested following a continuous extraction procedure. The assay revealed that the optimal concentration of TOAB for the continuous extraction was 0.02 mM. Following this, the authors also studied the influence of opting for continuous or stepwise extraction procedures, concluding that, after continuous extractions, a successful size discrimination was accomplished, with four fractions of AgNPs collected: 13.7 ± 1.9, 14.1 ± 3.5, 19.2 ± 4.3, and 22.2 ± 4.9 nm. On the other hand, after a stepwise extraction, the synthesized AgNPs sample only originated one fraction of 15.8 ± 5.3 nm. The obtained results demonstrated that the batch step-gradient extraction approach provided better size discrimination than the stepwise extraction. However, the application of the described separation methodology with real samples was not studied.

2.4. Density Gradient Centrifugation

Centrifugation is a technique used to separate particles from a solid–liquid mixture, according to their size, shape, and density. However, to achieve the separation of extreme small particles, like NPs, a centrifugal force is required to compensate or overcome the balance between gravitational forces, Brownian motions, and thermal diffusion, which allow the maintenance of the NPs in suspension [13]. The centrifugal forces originate sufficient energy to move the nanoparticles radially away from the axis of rotation, at pre-determined speeds, separating the NPs by shape and size. When the NPs have similar size and/or shape, the separation process by centrifugation is very difficult, and so a more-powerful technique like the density gradient centrifugation is required. This type of centrifugation is based on the creation of a density gradient [13]. The density gradient can be prepared, typically, with sucrose, glycerol, or another aqueous solution, and it is created inside of a centrifuge tube [1]. The solution that fills the centrifuge tube originates a decreasing density gradient from the bottom to the top of the tube [39].

Isopycnic centrifugation and rate zonal centrifugation are the two variants of density gradient centrifugation. In the isopycnic centrifugation, the process continues until most of the particles reach their isopycnic position in the centrifuge tube with the density gradient, that is, a position where their density equals the density of the medium. This type of centrifugation separates different particles based only on their different densities [39]. This way, one disadvantage of the isopycnic method is its incompatibility with metallic NPs separation, since these are denser than the highest densities attainable in aqueous media gradients (<1.7 g/cm^3), making it not possible to separate metallic NPs with the isopycnic centrifugation method. For this purpose, rate zonal centrifugation is chosen [13]. Rate zonal centrifugation, also known as rate zonal ultracentrifugation, or typically as sucrose density gradient ultracentrifugation, is a classical separation and purification technique used in the laboratories to purify bulk suspensions, namely, nanoparticles [40]. In this technique the fractionation of nanoparticles occurs by size, shape, and density, sedimenting through the stationary gradient at different rates. This rate of sedimentation will depend on all the influencing factors, namely: the size, shape, and density of the nanoparticles; the viscosity and density of the gradient; and finally, the centrifugal force. For example, the larger nanoparticles will sediment closer to the bottom, and smaller ones will be retained closer to the top of the gradient [41]. Briefly, different solutions of sucrose concentrations (hence, densities) are prepared and layered on top of each other, from the bottom to the top of a centrifuge tube, originating a density gradient. Then, the sample containing AgNPs is dropped on top of the prepared sucrose gradient and centrifuged. Later, the separated layers are collected, washed repeatedly, and centrifuged again for further studies [42].

In general, centrifugation is considered a low-cost, straightforward, and appropriate method to isolate and purify AgNPs, especially to remove the residues from newly prepared suspensions [1,42]. Table 4 depicts the most-important aspects derived from the analysis of selected works in the literature where AgNPs were separated by centrifugation techniques. The selected works were reduced to the ones in which the authors included a detailed description of the experimental parameters used, to enable a proper analysis in this revision. Most of other works in the literature mentioning AgNPs separation by centrifugation do not include the valuable experimental data used for separation (parameters and conditions used in the centrifugations), making impossible its inclusion in the present review, which intends to be a comparison analysis.

A sucrose-density-gradient-centrifugation method was studied and optimized by Y. Asnaashari Kahnouji et al. [42] in 2019 for the separation of AgNPs produced through a chemical precipitation method and coated with chitosan, with sizes ranging between 15 and 235 nm. The best separation conditions were obtained with 10%, 20%, 30%, and 40% sucrose gradients, during 2 h at 6000 rpm and 5 °C. At the end of the separation process, the fractionated layers with the AgNPs were collected by syringe and washed thrice with deionized water. Then, the collect fractions were centrifuged to remove remaining residues, and, finally, they were dispersed in deionized water for characterization studies by FTIR, DLS, and UV–Vis analysis. The synthesized AgNPs characterized by DLS revealed a range of sizes between 15–235 nm, but most of the particles were in the range of 2.7–6.3 nm. Additionally, with the DLS analysis, the authors identified the size distribution of the separated AgNPs in the four sucrose layers. For the first layer, the nanoparticles ranged from 4.9 nm to 6.3 nm; in the second layer, the NPs were in the range of 3.9–4.9 nm; and, for the third layer, the nanoparticles were in the range of 2.7–3.4 nm. In the fourth layer, the NPs ranged from 98.3 to 235 nm. This study revealed that the AgNPs suffered from some type of capping disintegration, leading to the reduction in their hydrodynamic sizes after the separation. The authors related this observation with the washing step necessary after the separation, which lead to the loss of some of the chitosan stabilizing agent at the NPs surface.

Table 4. Overview of AgNPs separation based on centrifugation.

Separation Method	Size [nm]	Shape	Matrix	Optimal Separation Conditions	Year
Sucrose density gradient centrifugation method [42]	FE-SEM and TEM: 15–235 nm	-	Chitosan-coated AgNPs	10%, 20%, 30%, and 40% sucrose gradient, during 2 h at 6000 rpm	2019
Sucrose density gradient centrifugation method [43]	HR-TEM: 52–117 nm for AuNPs, and from 38–61 nm for AgNPs	Spherical, pentagonal, triangular, and hexagonal	*Magnolia kobus* leaf extract AgNPs	40 min at 3500 rpm for AuNPs, and 90 min at 3500 rpm for AgNPs	2014
Centrifuging process [44]	-	Quasi-spherical	PVP-coated AgNPs	1st centrifugation: 8000 rpm; 2nd centrifugation: 16,000 rpm; 3rd centrifugation: 24,000 rpm	2015

HR-TEM: high-resolution transmission electronic microscopy; FE-SEM: field emission scanning electronic microscopy.

In 2014, Lee and co-workers [43] synthesized Au and AgNPs of different sizes and shapes using *Magnolia kobus* leaf extract. To separate the Au and AgNPs synthesized from a plant-mediated process by size, the authors used sucrose density gradient centrifugation. The mother suspension of both Au and AgNPs was composed of NPs of different shapes such as pentagons, triangles, cubes, spheres, and hexagons of different sizes. A TEM analysis of the crude suspensions revealed nanoparticles sizes between 5 and 300 nm for AuNPs and 15 to 500 nm for the AgNPs. After some studies, the best separation results were obtained when using as conditions of the method, 40 min at 3500 rpm for AuNPs and 90 min at 3500 rpm for AgNPs. The smaller NPs were observed at the lower densities of the sucrose gradient, while the larger ones were observed at the higher-density layers. After the separation, for the AuNPs, the TEM analysis showed a particle size ranging from 52 to 117 nm and from 38 to 61 nm for AgNPs, from the lower- to the higher-density gradient. Up to date, all studies found in the literature have referred to the use of high values of centrifugal forces (5000–6000 rpm) for the successful separation of nanoparticles of different sizes. However, the described work allowed the use of 3500 rpm for the separation of nanoparticles synthesized using *Magnolia kobus*. Additionally, it was concluded in the work that the separation of the different shapes of the nanoparticles occurred in some way related to the density of the nanoparticles rather the shape; that is, there was the separation of NPs of different shapes along the different sucrose densities, but the analyzed TEM of the separated fractions revealed a mix of shapes per fraction, instead of a single shape type. For example, using a 30% sucrose concentration, it was possible to separate most of the small nanoparticles of spherical shape and, to a lesser extent, some triangle, and hexagon-shaped NPs.

Hyun et al. [44] proposed in 2015 the separation of AgNPs based on their surface-plasmon-resonance (SPR) bands. In the work, the authors synthesized AgNPs through a polyol reaction, with stabilization by PVP, to control the target SPR band of polydisperse AgNPs. The AgNPs obtained were quasi-sphere, and their size depended on the PVP concentration: when the PVP concentration decreased, the size of AgNPs increased. The quasi-sphere AgNPs can have a surface-plasmon-resonance band ranging from 320 to 450 nm. Afterwards, the synthesized raw AgNPs suspensions were subjected to separation by a centrifuging process. For that, the authors studied the influence of three different speed conditions (8000, 16,000, and 24,000 rpm) for separating AgNPs accordingly with the SPR bands. The monitoring of the SPR bands was conducted by UV–Vis spectrometry. The results revealed a successful separation based on SPR peaks. In fact, the AgNPs samples were separated accordingly with the SPR bands varying between 406 and 435 nm. Thus, this work positively revealed that the centrifugation-based separation technique was a promising approach to separate the AgNPs accordingly with the surface-plasmon-resonance band, and, thus, it allowed a more-thorough characterization of the optical properties of AgNPs.

The use of centrifugation-based methodologies to separate AgNPs allowed the recovery of the separated NPs fractions and, in some cases, the near preservation of their physical–chemical properties, enabling the posterior use of the purified AgNPs in specific chemical, biological, or biochemical assays. The discussed results place the centrifugation-based methodologies as one separation method that successfully separates NPs by size and shape, whilst efficient synthesis methodologies in order to obtain monodispersed, shape-segregated nanoparticle dispersions are still required to be improved.

2.5. Electrophoresis

Charged molecules or particles in a uniform electrical field can be separated by electrophoretic techniques. Among these techniques, gel electrophoresis is the most popular. In this technique, the charged particles are forced to migrate in a gel matrix, by an electric field, toward the electrode of the opposite polarity, being in the course separated into distinct bands depending on their charge, size, or shape [13,45]. The separations of NPs by gel electrophoresis with agarose gel or polyacrylamide gel are the most popular used, while tris-borate-EDTA (TBE) is the most common electrolyte [11,45]. In Table 5, one can find the collected information from the scientific literature regarding the separation of AgNPs by electrophoresis.

Table 5. Overview of the AgNPs separation by electrophoresis.

Separation Method	Analyte	Size [nm]	Matrix	Optimal Separation Conditions	Year
Agarose gel electrophoresis [46]	AgNPs with PEG	-	-	0.2% agarose gel; 30 min at 150 V; 0.5× TBE buffer (pH ≈ 9)	2007
CE with diode-array detection [47]	AgNPs	SEM: 36.3 ± 5.9 nm	-	Background electrolyte: 20 mM SDS,10 mM Tris, pH 8.5; voltage: 20 kV	2004
CE coupled with ICP-MS [48]	AgNPs with citrate acid, lipoic acid, PVP, and bovine serum albumin	TEM: 10–110 nm	Consumer products (six dietary supplements)	Background electrolyte: CHES 10 mM, TX-100 30 mM, pH 9.5; voltage: 25 kV	2015
CE [49]	AgNPs with honey or glucose	TEM: 12 and 18 nm	-	Background electrolyte: 20 mM sodium borate, 20 mM SDS, pH 8.5; voltage: 20 kV	2017
MEKC [50]	Wound dressings: Atrauman® Ag, Aquacel® Ag, and FKDP-AgNPs	PSD: 284.5 nm	Wound dressings	Background electrolyte: 0.02 M borate buffer solution, 0.03 M SDS; pH 9; voltage: 20 kV	2019

ICP-MS: inductive coupled plasma mass spectrometry; PSD: particle-size distribution; CE: capillary electrophoresis; MEKC: micellar electrokinetic chromatography; TEM: transmission electronic microscopy; SEM: scanning electronic microscopy; PVP: polyvinylpyrrolidone; PEG: polyethylene glycol.

The successful separation of AgNPs, coated with a charged polymer layer of polyethylene glycol (PEG), by agarose gel electrophoresis, according to their size and shape, was demonstrated by Hanauer et al. [46] in 2007. In fact, in this work, the authors studied the potentiality of electrophoretic-based separation for AuNPs and AgNPs of very different shapes and sizes. The TEM analysis of the sample before the separation revealed the presence of 13% rods, 34% spheres (including hexagons), 44% triangles, and 9% other shapes (Figure 6a,b). The separation results were monitored spectrophotometrically of the surface-plasmon-resonance bands (and recorded by photographs of the gel) and confirmed by TEM. With a true-color photograph of the gel, it was possible to verify that the authors achieved the separation of the NPs with a 0.2% agarose gel run, for 30 min at 150 V, in 0.5× TBE buffer (pH ≈ 9). The photos of the resultant gel showed four different colors, which were related to the AgNPs separation according to their morphology and size. The colors that appeared in the gel were due to the size- and shape-dependent optical properties of AgNPs, but it was not possible to obtain a distinct separation of these band colors (Figure 6c).

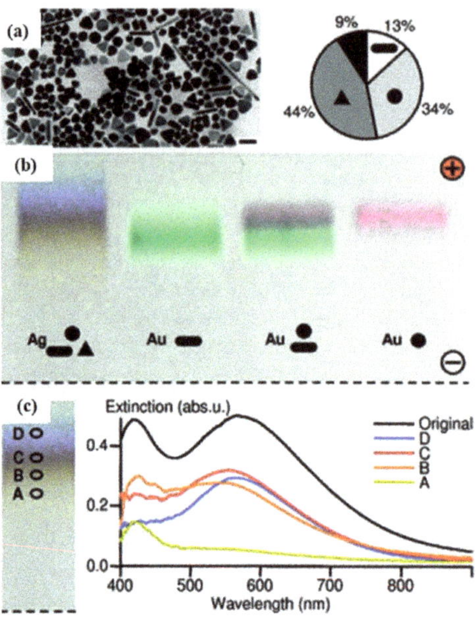

Figure 6. (**a**) TEM image of AgNPs sample and graphical distribution of the different shapes of the nanoparticles; (**b**) photograph of an agarose gel run for separation of nanoparticles (0.2% agarose, 30 min run, 150 V, 0.5× TBE buffer); (**c**) separated fractions of silver nanoparticles in agarose gel and their extinction spectra. Reprinted and adapted with permission from [46], copyright (2007) American Chemical Society.

After the separation by gel electrophoresis, the TEM results revealed the presence of rods in the fraction with the particles with the lowest mobility, while there were spheres and triangles in the faster fractions. The average particle size was lower in the slowest fraction (41 ± 2 nm) and higher in the faster fraction (65 ± 2 nm). The authors demonstrated the capacity to separate AgNPs through a gel electrophoresis method, according with their size and shape.

Another potential technique to separate AgNPs is capillary electrophoresis (CE). In CE, the separation process is based on the difference in the electrophoretic mobility of the NPs [48,49], and thus separation occurs accordingly with size and/or surface charge density. In the literature, there are some examples of CE applications, such as the work of Liu and co-workers [47], in 2005, which combined CE with a diode-array detection (DAD) system, allowing to achieve simultaneously the separation and characterization of AgNPs. To obtain a fully resolved separation, the authors found that the addition of an anionic surfactant, like sodium dodecyl sulphate (SDS), in the running electrolyte, enhanced the resolution of the separation, because it prevented the coagulation of AgNPs during the separation process. The optimal concentration of SDS was found to be 20 mM. This work, without question, constituted an important landmark in the field, because it showed that the combination of CE with DAD system was a powerful method to simultaneous separate and characterize AgNPs.

In 2016, Qu and co-workers [48] combined CE with ICP-MS for a rapid separation and quantification of AgNPs and Ag ions, in consumer products (six dietary supplements). The AgNPs analyzed were of different capping nature and were coated with citric acid, lipoic acid, PVP, and bovine serum albumin (BSA). The nanoparticles coated with BSA are more likely to adsorb free ionic Ag^+ on their surface, resulting in a significant underestimation of the amount of Ag^+ and overestimation of the amount of AgNPs. To facilitate the separation, and to keep the oxidation state of silver, preventing interaction of BSA with Ag^+, the compound tiopronin was added to the background electrolyte, in the concentration of 1 mM. Even with an excess amount of BSA, the authors obtained a recovery > 93%, for both ionic silver and AgNPs. With further studies in the commercially available products, they achieved the speciation of AgNPs and Ag^{+} in six minutes, under optimized conditions, detailed in Table 5. The robustness of the method was evaluated by the analysis of six dietary supplements by the proposed system CE-ICP-MS. The obtained results constituted by the amounts of nanoparticulate Ag and free Ag ions were compared with ICP-MS analysis of total Ag after the acid digestion of the same samples. A good accordance between the sum of the amounts of ionic and AgNPs by CE-ICP-MS and the obtained total silver quantities by ICP-MS after acid digestion was obtained.

More recently, in 2017, Fa et al. [49], with the aim of simultaneously obtaining information about zeta potential, size distribution, and the colloidal stability of AgNPs, used CE to separate AgNPs synthesized at pH 5.0 and 10.0 with honey or glucose as reducing agents (green-chemistry approach). In the separation, an electrolyte solution composed of 20 mM sodium borate and 20 mM SDS at pH 8.5 was used. The developed method allowed the separation of AgNPs within a short run time (<12 min). The obtained characterization results with the proposed method were compared with the DLS and TEM analysis. The electrophoretic mobility and zeta potential values were calculated through Smoluchowski's equation and Ohshima's equation, respectively. The calculated electrophoretic-mobility values were in accordance with the ones obtained by DLS. Additionally, Ohshima's equation provided similar results for the zeta potential, when compared to those obtained by conventional characterization techniques. The proposed method required minimal volumes of samples and reagents, produced small amounts of residues, was easily handled and low-cost, constituting a good alternative for the separation of green-synthetized AgNPs while maintaining the same sustainability standards imposed by green chemistry.

One other variant of CE is micellar electrokinetic chromatography (MEKC). Nanoparticles are known to be positively or negatively charged, accordingly with the chemical nature of their capping or stabilizing agent used during the synthesis process. These capping agents promote the stabilization of the NPs via electrostatic or steric repulsion. The importance of the charge relies on preventing the NPs aggregation over the time and also gives stability to the nanoparticles while in suspension [6]. MEKC is a hybrid technique that combines an electrophoresis process with chromatography, and it can be used for the separation of AgNPs, because they are charged.

In MEKC, both ionic and neutral substances can be separated using surfactant micelles, instead of what happens in CE, which only separates ionic compounds. Briefly, the separation principle in MEKC is based on the addition of a surfactant, above its critical micellar concentration (CMC) to the buffer solution, acting like a micellar phase. This micellar phase will interact with the analytes present in the sample, by a partitioning mechanism, and then the analytes that have greater affinity for the micelles are trapped inside the micelle. The migration time of the analytes will depend on the affinity to the micelles. The analytes that have more affinity will have a slow velocity of migration, while the other analytes will migrate faster, and the retention time will be lower [51]. One of the first studies exploiting MEKC to separate AgNPs from wound dressing, for example, was developed in 2019 by M. Konop et al. [50]. The wound dressings used were two commercially available (Atrauman® Ag (Sydney, NSW, Australia), Aquacel® Ag (ConvaTec, Princeton, NJ, USA)) and one experimental (FKDP-AgNPs) dressing. Each sample was previously prepared before MECK analysis, by mixing at 300 rpm with fetal bovine serum (FBS) for a total of 72 h. Several aliquots were analyzed at different time intervals: 1, 24, 48, and 72 h. The study allowed to conclude that the best separation conditions for AgNPs were attained using 20 mM borate buffer solution at pH 9 and with 20 mM SDS addition. As a result, Atrauman Ag was the dressing where the quickest release of silver from the matrix was observed, whereas, in Aquacel Ag and FKDP-AgNPs, a slower silver release was observed. The observed results were very important, allowing to conclude that the dressing matrix influences the silver release. The proposed MEKC method required only a very small volume of sample extract (100 μL), and the execution of the procedure was fast and simple. However, due to undesired interactions between the synthetized AgNPs, used to produce in the lab a wound dressing, with the electrolyte solution, some instability of the NPs resulting in some aggregation phenomena was confirmed. In fact, the measured zeta potential value was about −42.8 ± 6.65 mV. Additionally, the method presented low reproducibility of the analytical signals on the electropherograms because AgNPs tended to sediment. The present work revealed that more studies of MEKC must be done to optimize the conditions of separation and quantification of AgNPs released from dressing materials, and other samples, involving more studies of interaction between the AgNPs and FBS used to extract the analytes from the samples.

2.6. Selective Precipitation

Polydisperse NPs can be separated into fractions through a size-selective precipitation technique, according to size-dependent physical and chemical properties, reactivity, and/or stability [13,52]. To achieve a selective precipitation of the NPs, by modifying their solubility in the medium, a miscible substance (gas, liquid, or salt) where the NPs do not aggregate is added into the suspension [11,52]. Following this addition and mixture, the nanoparticles gradually start precipitating, from the largest- to the smallest-sized, originating several fractions with size-separated NPs, which can be collected as they are formed. To attain this separation, there are four possible methods: (i) using a supercritical fluid that acts like a solvent with density-tunable dissolving power; (ii) adding a non-solvent where NPs do not solubilize; (iii) adding salts; and (iv) using a gas-expanded liquid [11,52]. Not all these approaches were found in the literature for separation of AgNPs, which was the aim of the present revision. Table 6 compiles the works properly identified in the literature containing a detailed description of selective size-selective precipitation of AgNPs. The analysis of the table leads to conclude that these separation processes are not much applied for the separation of AgNPs.

Table 6. Overview of the AgNPs separation by selective precipitation.

Separation Method	Size (nm)	Optimal Separation Conditions	Year
CO_2-expanded liquid approach [53]	TEM: 2 and 10 nm, having a mean size of 5.5 nm	Pressurization series: 500, 550, 600, 625, and 650 psi	2005
Surfactant-assisted shape [54]	-	1st separation: 0.4 mL of 0.4–0.5 M CTAB; 40–80 °C; 12 h aged 2nd separation: centrifugation at 400× *g*, 16 min; 0.2 mL of 0.2 M CTAB	2018

TEM: transmission electronic microscopy.

In 2005, McLeod et al. [53] developed a method to separate AgNPs using a CO_2-expanded liquid approach. By adding a compressed gas (CO_2) to an organic solvent (hexane), the liquid volume is expanded several times, which is why the resultant liquid CO_2/hexane is described as a gas-expanded liquid or abbreviated by GEL. Briefly, the separation process was developed inside of a spiral glass tube. Here, 200 μL of a hexane solution containing the AgNPs were inserted, and the pressure was slowly elevated to 500 psi, which allowed to reach the equilibrium state over 20 min

When the GEL equilibrium was reached, the fraction with the largest AgNPs, which were no longer soluble in the expanded liquid, precipitated. Next, the spiral glass tube was rotated 180° to achieve other AgNPs populations that were attached in the tube walls. Increasing the gas pressure, a second fraction of AgNPs with small dimensions precipitated as well. The process continued until all AgNPs became precipitated. The gradual changes of gas pressure, from 500 psi to 650 psi, selectively precipitated AgNPs, from the largest to the smallest. In the end, six fractions of AgNPs were collected and analyzed by TEM. This technique allowed a precise size-separation of AgNPs in a single step, taking no longer than 1.5 h and requiring a reduced volume of organic solvent (200 μL); it was possible to select the nanoparticles to be separated, by size, by choosing the necessary pressure value. By testing a sample constituted by AgNPs of sizes ranging between 2 and 10 nm, with a mean size of 5.5 nm, the separation into six different mean diameter-sized fractions of AgNPs: 6.7 ± 1.4 nm, 6.6 ± 1.0 nm, 5.8 ± 1.1 nm, 5.3 ± 0.5 nm, 4.8 ± 0.5 nm, and 4.1 ± 0.6 nm was achieved. This new utilization of CO_2-gas-expanded liquids provides the area of nanotechnology with a promising method for NPs purification and separation by size.

To purify AgNPs of different shapes and sizes, prepared by a seed-mediated technique, in 2018, Hu et al. [54] applied a surfactant-assisted shape-separation method. The authors took a previously reported method used to purify AuNPs [55] and adapted it to purify AgNPs. The sample suspension was centrifuged at 10,000 rpm for 10 min, and then different volumes of different concentrations of CTAB solution (0.2–0.5 M) were added into the precipitate and mixed at different temperatures (25–80 °C). The authors studied the influence on the separation of AgNPs (triangular shape) of different CTAB volumes, CTAB concentrations, CTAB temperatures, and aging times. Following this, the study proceeded with secondary separations controlled by different CTAB concentrations and volumes. Finally, the authors tested the optimized procedure to separate silver nanotriangles and nanospheres. The bulk suspension was a mixture of silver nanotriangles, nanospheres, and nanorods. The initial UV–Vis spectrum related to the as-synthesized AgNPs presented three different absorption peaks (~342 nm, ~422 nm, and ~584 nm), while, after the first separation step, the supernatant showed a UV–Vis spectrum with a single absorbance peak at 416 nm, which is related with the presence of silver nanospheres with different sizes. After the second separation step, the supernatant had three absorption peaks (~342, ~416, and ~584 nm), which means that, in this phase, the AgNPs were mainly silver nanotriangles. With this approach, a successful separation was achieved, monitored by UV–Vis spectra (Figure 7). The relative ratio between CTAB and AgNPs seems to be determinant, together with the temperature, to separate the triangular-shaped AgNPs. Probably, the procedure requires the optimization of the separation conditions for each type of AgNPs under analysis. The authors concluded that this separation methodology could be used to separate not only AuNPs but also silver or other metal NPs.

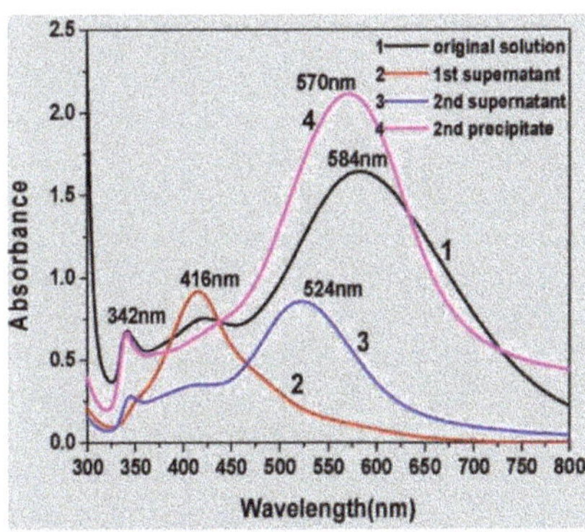

Figure 7. Separation of AgNPs by shape, monitored by UV–Vis spectrophotometry, after two sequences of selective precipitation. Reprinted from [54], copyright (2018), with permission from Elsevier.

2.7. Membrane Filtration

An alternative to the purification and size-separation of NPs is filtration through a porous membrane. The distribution of sizes of the membrane pores influences the time of retention and the elution of the NPs. For example, for nanoparticles with sizes ranging from 2–50 nm, an ultrafiltration process is used, whereas, for particles with a size ranging between 20–500 nm, the process is known as microfiltration [11,13]. In this method, the separation process is based on the migration of the particles through a porous membrane, because the concentration on one side is different from the other side. These concentrations' differences make the nanoparticles diffuse from the higher-concentration to the lower-concentration area. The advantages of this method are the minimal equipment requirements and small solvent volumes, the recyclability of the feed solution, the high resolution, and the facility to be scaled up [11,56]. Nowadays, the most commonly used membranes are made of polymer or ceramic [56]. The size of the membrane pore is a key factor for an effective separation. So, if the membrane has a pore size smaller than the size of the nanoparticles to be purified, no separation will occur, because the NPs cannot simply cross through the pores by diffusion. Additionally, adsorption and aggregation of the nanoparticles can occur in these situations, causing the blockage of the membrane surface. [11]. The better uniformity of membrane's porosity, the better the performance of nanoparticle separation [13].

To our knowledge, this technique was not used to separate AgNPs, but it was successfully used in the separation of AuNPs by Krieg et al. [56], in 2011. As with other separation techniques previously described, its use in noble-metal nanoparticles based on gold indicates the possible use in silver-based nanoparticles as well. Considering the recognized potentiality of the separation methods based on porous membranes, some discussion about related practical works found in the scientific literature was included here. In the work of Krieg et al., some supramolecular membranes were prepared from PP2b (5,5′-bis(1-ethynyl-7-polyethylene glycol-N,N′-bis(ethylpropyl) perylene-3,4,9,10-tetracarboxylic diimide)-2,2′-bipyridine) in water. To study the applicability of the fabricated membrane for ultrafiltration, AuNPs (red suspension) of various sizes were filtered over a 12-mm-thick PP2b layer. As a result, a pale-yellow filtrate was obtained, and characterization by UV–Vis spectrophotometry suggested that NPs larger than 5 nm were removed, by the absence of a surface plasmon band in the filtrate. To confirm these results, a TEM analysis was performed. The TEM images confirmed that NPs with a particle size > 5 nm were effectively removed and remained in the retentate, and, also, allowed to confirm that the membrane cut-off was 5 nm. To evaluate the membrane performance, the authors made an additional ten experiments, where five of them were performed with freshly prepared PP2b for membrane fabrication, and the other five were performed with the recycled membrane. As a result, and after TEM analysis, the authors obtained a particle size in the filtrates of about 2.3 ± 0.2 nm for the first five experiments (freshly membrane) and, for the recycled one, a particle size of about 2.3 ± 0.1 nm. These supramolecular membranes have the advantage of simple fabrication, versatility, and the capacity to be used multiple times, after being cleaned and recovered. The authors of the reviewed work concluded that these membranes allow to recover the separated nanoparticles, obtaining purified suspensions of AuNPs to further use in biological, analytical, and biochemical applications, which is an important feature to highlight, and that the method could be extrapolated to purify other nanoparticles with a cutoff of roughly 5 nm. Additionally, since the membranes are robust, and the supramolecular structure retains its adaptivity, they are susceptible to recycling.

2.8. Liquid Extraction

Extraction is a separation process where the analytes present on the sample (a liquid or a solid mixture) are selectively separated with a liquid immiscible solvent. Usually, the two immiscible phases are water and an organic solvent, and the separation is dependent on the relative solubility of the target analytes. Widely used, this is a method for organic and inorganic compounds [11,13].

In the scope of the extraction methods available, the concept of cloud-point refers to the minimum temperature in which a clear solution undergoes a liquid–liquid phase separation to form an emulsion. Thus, cloud-point extraction (CPE) is based on the solubilization ability and on the cloud points of non-ionic surfactants. It is considered an approach for the separation, extraction, and preconcentration of trace elements, before their determination [57], such as for beryllium and chromium in water samples [58], silver in environmental waters [59], cobalt in water and food samples [60], and cadmium, copper, and lead in biological fluids [61], among others. The extraction protocol for trace elements is based on three steps. In the first one, a non-ionic surfactant, for example, Triton X-114, is added into the sample solution with a concentration that exceeds the critical micelle concentration. Temperature, pressure, or pH can be changed to form micelles, in which trace elements are entrapped. At the cloud-point, the solution with the surfactant becomes turbid, and it is separated into two distinct phases (the surfactant-rich phase and a dilute aqueous phase). The analytes separated in the surfactant-rich phase can then be extracted and concentrated, due to the analyte–micelle interaction [1,57]. In the case of separation of AgNPs, the addition of some salts, such as $NaNO_3$ or $Na_2S_2O_3$, improves the separation of the phases and increases the efficiency of the extraction. In this situation, the separation is facilitated because the salt has the ability to reduce the Coulomb repulsion between charged AgNPs [62]. Additionally, the increase in the extraction's efficiency occurs because $S_2O_3^{2-}$ anion acts like a chelating reagent with silver, binding to the silver ions forming a complex, which prevents the transfer of Ag^+ to the surfactant-rich phase, thus eliminating the Ag^+ interference completely, allowing only the extraction of AgNPs. Nonetheless, $S_2O_3^{2-}$ may cause some interference in the characterization of AgNPs by UV–Vis spectrophotometry. Thus, the authors recommend the use of $NaNO_3$ when the objective is a UV–Vis characterization, but in the case of other analyses, the use of $Na_2S_2O_3$ is preferable [59]. Figure 8 shows the schematic representation of the CPE protocol steps.

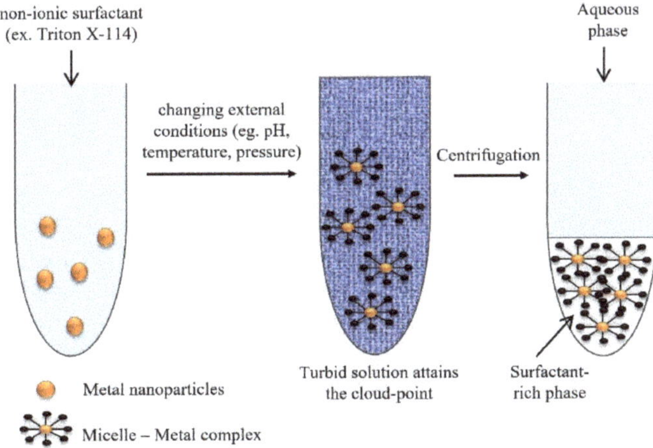

Figure 8. Representative scheme of CPE protocol.

CPE is also a method of choice for extracting pollutants (e.g., metal NPs) from environmental and biological samples, without modifying their sizes and shapes. It is a simple, low-cost, and environment-friendly technique, which has a high capacity to concentrate trace elements [57]. Nevertheless, a limitation of this method is the low extraction efficacy when the analytes are protein-coated NPs, like AgNPs functionalized with BSA. As an alternative, Qu et al. [48] proposed a capillary electrophoresis approach for separating those AgNPs, allowing to achieve excellent recovery values in consumer products tested (higher than 93%).

Table 7 contains the analysis of some works found in the scientific literature, involving the separation of AgNPs based on extraction procedures.

From the analysis of the Table 7, it is evident the prevalence of CPE techniques to separate the AgNPs. For example, in the work of Wu et al. [63] in 2011, a methodology to separate AgNPs and Ag^+ in environmental waters was developed, while at the same time, the quantification of AgNPs was developed. For this purpose, the authors combined CPE and a colorimetric assay using Tween 20-stabilized AuNPs to allow the indirect quantification of AgNPs. The extracting agent Triton-X 114 was used in the preconcentration step of CPE, and Tween 20-stabilized AuNPs were used as a colorimetric probe, for Ag^+ obtained after, by oxidation with H_2O_2 of the separated fraction of AgNPs. Figure 9 illustrates the procedure used by the authors for the identification of AgNPs through the AuNPs-based colorimetric probe. For the detection of Ag^+ through colorimetric assay after CPE, the supernatant was discarded, and an amount of H_2O_2 was added to the concentrated AgNPs present in the Triton-X-114-rich phase. H_2O_2-oxidized AgNPs were added to Ag^+ under acidic conditions. The Tween 20-AuNPs act as sensing probes to the silver ions. The presence of the remains of citrate ions on the AuNPs surface reduces Ag^+ to Ag on the surface of AuNPs, and, as consequence, the stabilizer Tween 20 is removed from the AuNPs. As a result, the AuNPs became unstable and aggregate. The developed method proposed by the authors showed high selectivity for AgNPs and allowed to separate the AgNPs from Ag^+.

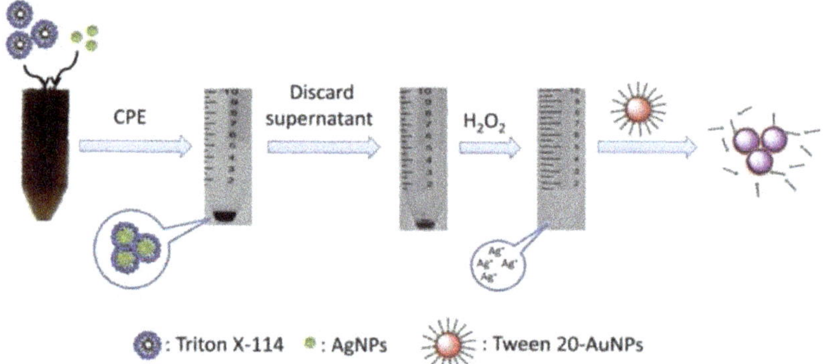

Figure 9. Separation of AgNPs by Triton X-114-based CPE. Reproduced from Ref. [63] with permission from the Royal Society of Chemistry.

With the developed work, the authors were able to propose for the first time a new methodology, combining CPE with AuNPs-based sensors, to separate and detect AgNPs. Furthermore, even in the presence of low AgNPs concentrations in the samples, this approach can be used without requiring equipment, since the changes in color of the AuNPs can be observed with the naked eye. So, in the future, this methodology can hopefully be widely applied to the in situ AgNPs screening in environmental waters. This potentiality of the method is worthy of much consideration for its importance in the real-time control of AgNPs pollution in the environment.

Table 7. Overview of the AgNPs separation by liquid extraction procedures.

Separation Method	Detection Method	Analyte	Size (nm)	Matrix	Recovery (%)	LOD Value	Optimal Separation Conditions	Year
Selective extraction with *n* alkanoic acids (*n*-decanoic acid) [64]	-	Au and Ag dendrimer encapsulated NPs (DENs)	HR-TEM: 1.4 ± 0.4 for G6-OH (Au_{147}) and 1.7 ± 0.4 for G6-OH (Ag_{110})	Aqueous medium	-	-	4.0 mL of 0.25 M *n*-decanoic acid/hexane solution; 6.0 mL of an aqueous mixture containing 0.18 mM G6-OH(Ag110) and 0.20 mM G6-OH(Au147); vortex for 30 s	2004
Triton X-114-based CPE [65]	-	Au NPs, Ag NPs, C60 fullerene, TiO_2, Fe_3O_4 NPs, CdSe/ZnS, and SWCNTs	TEM: 5–100 nm	Aqueous medium	92–97%	-	Triton X-114 surfactant (3.6 mM); NaCl (3.4 mM); heat the suspension above the CPT (23–25 °C)	2009
Triton X-114-based CPE [63]	Spectrophotometry	AgNPs	TEM: 10 and 54 nm	Environmental water	Tap water: 102 ± 3%; seawater: 98 ± 5%	Tap water: 4.3 ng/mL; seawater: 43 ng/mL	0.01 M $Na_2S_2O_3$; 0.2% Triton X-114; incubate at 40 °C, 30 min; centrifuge at 750× *g* at room temperature, 5 min	2011
Triton X-114-based CPE [59]	ICP-MS	AgNPs	TEM and SEM: 9–94 nm	Environmental water	57–116%	0.006 μg/L	1 M $Na_2S_2O_3$ or 3.5 M $NaNO_3$; 5% (*w/v*) TX-114; pH 3; incubate at 40 °C, 30 min; centrifuge at 2000 rpm, at room temperature, 5 min	2009
Triton X-114-based CPE [62]	ICP-MS	AgNPs	TEM: 12.4 ± 0.3 nm	HepG2 Cells	Approx. 92%.	2.94 μg/L for AgNPs and 2.40 μg/L for Ag^+	1 mol/L $Na_2S_2O_3$; 10% (*w/v*) TX-114; pH 3,5; incubate at 40 °C, 30 min; centrifuge at 3000 rpm, at room temperature, 5 min	2013
Triton X-114-based CPE [66]	ICP-MS	AgNPs	-	Environmental waters and antibacterial products	1.2–10% for Ag^+ and 71.7–103% for AgNPs	0.4 μg/kg for AgNPs and 0.2 μg/kg for Ag^+	1 M $Na_2S_2O_3$, and 10% (*w/v*) TX-114; pH 3; incubate at 40 °C, 30 min; centrifuge at 2000 rpm, at room temperature, 5 min	2011
Triton X-114-based CPE [67]	ETAAS	AgNPs	-	Environmental water	>88%	0.7 ng/L	1.0 mL of saturated EDTA solution; 400 mL of 1 M sodium acetate; 100 mL 1.25 M acetic acid; 1 mL of 10% (*w/w*) TX-114; incubate at 40 °C, 2 h	2013
Triton X-114-based CPE [68]	ETAAS	AgNPs	20, 40, and 60 nm	Waste water	110 ± 6% and 101 ± 10%	0.04 μg/L	8.6% (*v/v*) Triton X-114; saturated EDTA; pH 7; incubate 60 °C for 20 min; Centrifuge at 7000 rpm, at 4 °C, 20 min	2018
CPE [69]	TXRF	AgNPs	SEM: 40–100 nm	Soil extracts and consumer products water extracts	-	0.7–0.8 μg/L	1 M $Na_2S_2O_3$; 5% Triton X-114; pH 3,7; incubate at 40 °C, 30 min; centrifuge at 2000 rpm, at room temperature, 5 min; cool in a freezer for 15 min	2018

CPE: cloud-point extraction; ICP-MS: inductive coupled plasma mass spectrometry: ETAAS: electrothermal atomic absorption spectrometry; TXRF: total reflection X-ray fluorescence spectrometry; TEM: transmission electronic microscopy; SEM: scanning electronic microscopy.

In 2009, Liu et al. [59] developed a CPE method with Triton-X 114 to preconcentrate AgNPs from environmental water, without disturbing their sizes and shapes. The AgNPs were concentrated into the Triton-X 114-rich phase, and then they were characterized by TEM, SEM, and UV–Vis and quantified by ICP-MS after microwave digestion. Furthermore, the authors studied the optimization of the CPE parameters, studying, for example, the addition of some salts, like $NaNO_3$ and $Na_2S_2O_3$, and they proved that the addition improved the phase separation and the extraction of AgNPs, as well as the preservation of Ag^+ in the upper aqueous phase, as previously explained in the introduction to the CPE concept. To evaluate the applicability of the optimized method (the optimized conditions can be consulted in the Table 7), the authors analyzed four types of real environmental water samples (influents, effluents, lake water, and river water). As a result, recoveries of 57–116% and a LOD of 0.006 μg/L were obtained. In conclusion, the proposed methodology showed to be efficient for the selective extraction of AgNPs and concentrations of trace AgNPs from environmental-water samples. Taking into account the previously described method in 2009 [59], in 2011, Chao et al. [66] had as the main goal the study of the applicability of the Triton-X-114-based CPE for the speciation analysis of AgNPs and Ag^+ in environmental waters and antibacterial products. Six antibacterial products were tested, but AgNPs were detected in only three of them. The authors justified the results with the hypothesis that the actual values of AgNPs might be below the method's LOD. Extraction efficiencies of Ag^+ ranging from 1.2–10% were obtained, while the recoveries of AgNPs were between 71.7–103%, in antibacterial products, and 74.5–108% in environmental waters. The recoveries obtained for water samples were in agreement with those obtained in the previous study. The authors described this technique as fast (the maximum of the extraction efficiency was reached in 10 min), simple, and low-cost.

In 2013, Yu et al. [62] proposed to separate AgNPs and Ag^+ in the cell lysates of exposed HepG2 cells using Triton-X-114-based CPE. After the exposure of HepG2 cells to AgNPs, the cells were centrifuged to cause cell lysis, and cell lysates were obtained. Then, $Na_2S_2O_3$ was added to the cell lysates, to be subjected to CPE. As already mentioned, the addition of the salt allowed the preservation of Ag^+ in the upper aqueous phase and facilitated the transference of AgNPs into the Triton-X-114-rich phase. The AgNPs and Ag^+ contents were determined by ICP-MS, after microwave digestion. The LOD of this method was calculated as 2.94 μg/L for AgNPs and 2.40 μg/L for Ag^+. The results showed an average total recovery of ≈92%. The proposed methodology allowed the quantification of AgNPs and Ag^+ contents in a complex matrix, like cell lysates. For the first time, a method that allowed this determination in biological samples was developed.

A quantification procedure for AgNPs based on electrothermal atomic absorption spectrometry (ETAAS) was developed by Hartmann et al. [67] (2013). The authors had to first extract the analytes through a CPE method, from environmental samples. The authors re-evaluated the optimized conditions for the AgNPs separation by CPE, since, in the proposed work, the detection system was based on ETAAS. Following this, it was concluded that a pH value ≥ 5 and ethylenediamine tetraacetic acid (EDTA) as the complex agent were better conditions for the CPE enrichment factor. In Table 7, one can find the optimized conditions for the assay, including the ETAAS measurement parameters. Under the optimized conditions, a LOD of 0.7 ng/L was obtained, and the environmental analysis of the samples (river water and treated and untreated municipal wastewater) revealed a mean recovery > 88% in all cases. The described methodology was fast; allowed for an easy sample preparation, by avoiding the microwave digestion before the quantification by ETAAS; and allowed to achieve a low LOD value. Despite the encouraging results, the authors concluded that more studies must be done to evaluate this separation technique, such as using AgNPs of different syntheses with several stabilizing agents. In fact, the authors had to re-evaluate the CPE conditions for the application of the extraction method coupled to ETAAS.

More recently, in 2018, Lopez-Mayan et al. [68] developed a reliable and simple CPE method to purify AgNPs from wastewater samples from three water-treatment plants from

Galicia, Spain, followed by quantification by ETAAS. Through multivariate analysis (and experimental-design approaches), the parameters related to the CPE method (surfactant concentration, type of complexing agent (EDTA or $Na_2S_2O_3$), pH value, incubation temperature, incubation time, and centrifugation time) were optimized. The selected optimal conditions were 8.6% (*v*/*v*) Triton X-114, 750 μL saturated EDTA, and pH 7. To evaluate the sensitivity of the procedure, the LOD and the limit of quantification (LOQ) were calculated as 0.04 μg/L and 0.13 μg/L, respectively. The recovery assays revealed results of 110 ± 6% and 101 ± 10% for the sample spiked with 1.0 and 2.0 μg/L of AgNPs, respectively. The authors concluded that the combination of CPE and ETAAS was a good approach to analyze AgNPs in wastewater samples and that it was a valid alternative to CPE coupled to ICP-MS.

The combination of the CPE method with total reflection X-ray fluorescence spectrometry (TXRF) was made for the first time, in 2018, by Torrent et al. [69]. TXRF allows materials characterization and trace-element determination based on the principle of total reflection of X-rays. Briefly, when an incident X-ray beam, with an angle below the critical angle of the substrate, falls upon a flat polished surface, a fluorescence signal characteristic of the present element is emitted from the sample [70]. This combination procedure was used to separate and quantify AgNPs, with different coatings (citrate, PVP, and PEG), in complex aqueous samples, including soil and consumer-product water extracts. In all cases, the LOD obtained was 0.7–0.8 μg/L. The authors also studied the influence of the presence of Ag ions in the extraction and recovery of AgNPs, by the developed CPE-TXRF methodology. It was concluded that for Ag^+/AgNPs ratios higher than three, the procedure was not adequate since it was verified that, in those conditions, the separation between Ag^+ and AgNPs was not efficient. Yet, considering the usual ratios present in most real samples, this was not an issue. Following this, the CPE-TXRF methodology was applied for the quantification of AgNPs in the samples of Hansaplast Universal Antibacterial Plasters (Beiersdorf AG, Hamburg, Germany) and spiked soils. The samples were submitted to an aqueous extraction procedure, followed by filtration using an acetate cellulose membrane filter of 0.45 μm pore size. The CPE procedure followed by the authors was based on the work of Liu et al. [59], with the proper adaptions derived from optimizations studied in the work. The obtained results by the proposed methodology were confirmed by single-particle inductively coupled plasma mass spectrometry (SP-ICP-MS) analysis.

A good agreement was obtained by comparing both analytical techniques, CPE-TXRF and SP-ICP-MS, with respect to AgNPs quantifications in two soil samples spiked with AgNPs. In relation to the extract sample from the Hansaplast band aid, the estimated concentrations were 60 ± 5 μg/L with CPE-TXRF and 91% of AgNPs by SP-ICP-MS. In this case, the concentrations estimated by CPE-TXRF were much lower than the actual total Ag concentration (76% lower comparing with the obtained result by ICP-OES, which estimated the total silver content). Since the SI-ICP-MS technique originated results that indicated the presence of a large distribution in the sizes population of AgNPs, it could be that the CPE extraction efficiency was affected by the presence of larger AgNPs, which is also supported by some reports found in the literature [71] regarding this same problem. The authors concluded that more studies must be performed to improve the combination techniques for screening AgNPs in mixtures with different nanoparticles, namely, of different coatings and particle sizes.

3. Timeline Conceptual Evolution

A scientific research at the global citation database Web of Science^{TM} was conducted, using an advanced search for each separation method analyzed in this study. By restricting the research to silver nanoparticles and the separation method, this survey initially revealed the following number of studies for each method: 318 for the magnetic-based method, 49 for hydrodynamic forces, 203 for chromatography, 55 for centrifugation, 99 for electrophoresis, 14 for selective precipitation, 72 for membrane filtration, and 120 for liquid extraction. Yet, after consulting the studies, it was found that most of them were not in fact focusing on silver nanoparticles as the target of separation from crude suspensions or other samples. Additionally, some studies were also excluded due to other aspects, namely, the inclusion of AgNPs in the separating materials (like in magnetic nanocomposites or as antiseptic additives in filtration membranes) or their use as probes. Consequently, after a thorough and refined analysis of the previous studies, the numbers of the survey were reduced to: 5 for the magnetic-based method, 21 for hydrodynamic forces, 16 for chromatography, 11 for density gradient centrifugation, 22 for electrophoresis, 2 for selective precipitation, 0 for membrane filtration, and 20 for liquid extraction.

Using the numbers mentioned before, a timeline conceptual evolution study was conducted to allow the perception of any possible tendency in the use of a specific separation method and if there is evidence of more investment in research to a separation procedure, during the time between 2004 and 2020.

A graphic representation of the distribution of the reviewed works by years, accounting for the number of each separation technique, is depicted in Figure 10. A relatively higher number of published studies in the years 2014 and 2016 was identified, whilst no studies in 2006 were identified. As already stated, the most-exploited methods of separation of silver nanoparticles identified for analytical purposes were electrophoresis, hydrodynamic, and liquid-extraction-based methods.

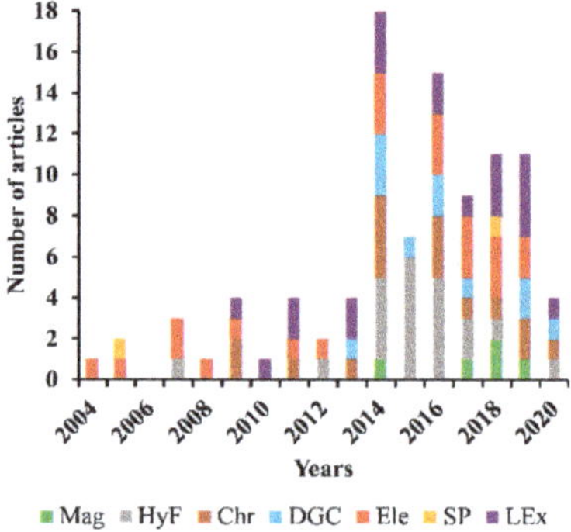

Figure 10. Timeline of separation methods: graphic representation of the number of published articles discriminated by separation method (separation methods: Mag—magnetic-based schemes; HyF—hydrodynamic forces; Chr—chromatography; DGC—density gradient centrifugation; Ele—electrophoresis; SP—selective precipitation; Lex—liquid extraction).

Unfortunately, the identified number of articles found in the literature in the scope of the present review was low, revealing that if one considers the large number of studies per month or year that are published about silver nanoparticles, most of them do not detail or reveal any separation or purification methods for the nanoparticles. In our opinion, this should change, because the influence of the surrounding medium resultant from the synthesis procedures contains several remains of compounds that influence the assays where the AgNPs are intervenient. This is important because it leads to biased conclusions about the real influence or effect of silver nanoparticles in, for example, cellular assays, electronic applications, drug-delivery effects, decontamination assays of the environment, etc. For example, in cellular assays, it is of utmost importance to purify the nanoparticles to be tested with cells, to achieve true and real conclusions about the nanoparticles' effects, reflected, for example, in calculated IC50 values. In the work of Morais et al. (2020) [72], the authors discussed the problem of hundreds of works dealing with AgNPs synthesized by green chemistry, with natural products (as plant extracts) in which the authors did not purify or separate the nanoparticles but nonetheless conducted cell assays with cancer-cell lines, calculating IC50 values. However, in the literature, one can also find works using the same plant extracts tested in tumor cells, with the respective IC50 values also determined. So, there is no way to be sure if the IC50 values were only due to the action of the AgNPs. This doubt has a significant impact on the nanotechnology-based health-treatment evolution. Fahad A. Alharbi et al. (2020) [73] synthesized AgNPs using the leaves extract of a xerophytic plant (*Neurada procumbens*). After the synthesis, the authors tested the antibacterial properties of the produced AgNPs against a group of multi-drugs-resistant (MDR) bacteria (*Klebsiella pneumoniae, Acinetobacter baumannii,* and *Escherichia coli*). As a result, the AgNPs exhibited a high level of antimicrobial activity against MDR bacteria. However, after synthesis, the suspension of AgNPs did not undergo any purification process, so, probably the effects observed against the MDR bacteria might in fact be due to the action of the silver nanoparticles capped with the leaves-extract compounds, but they could also be due to the residues of the natural product that resulted from the entire synthesis process.

So, one can conclude that more research should be performed focusing on the theme of this review since there is a noticeable lack of innovation or application of separation methods in original works involving silver nanoparticles and their use in a purified format.

4. Conclusions

The separation/purification process of AgNPs is of utmost importance since the applicability of the NPs depend on its purity. In all possible applications for AgNPs, one wants to take advantage only of the physical–chemical properties of the NPs and of from the precursors and reagents remaining from the syntheses. A variety of analytical procedures have been developed to separate AgNPs from the crude synthesis material or other matrices, allowing its purification and size sorting. In this review, we thoroughly described the current options for the separation/purification methods of NPs, mainly the silver ones from diverse sample matrices.

Over the last few years, significant progress has been made, derived from technology advances. We reviewed successful cases of separation of AgNPs. Despite considering the advantages and disadvantages of these processes, it is not possible to conclude which, among all, is the best method to apply. The range of existing separation and purification methods is diverse, and the most-suitable method will depend on the type of NPs to be separated and on the purpose that will be given to them. Depending on that purpose, it might be imperative to maintain the integrity of the AgNPs for further applications. In these cases, one must be aware that some separation methods end up disintegrating the AgNPs making unfeasible their further use, restricting the choice of the method. However, traditional methods still must undergo some optimizations to be more efficient and to follow environmental sustainability demands. There is a greater need to develop novel methods or to innovate others already implemented, allowing a more-efficient AgNPs separation, while, at the same time, ensuring a low-cost, versatile, and eco-friendly method.

Author Contributions: Conceptualization, C.S.M.M. and J.A.V.P.; methodology, C.S.M.M. and J.A.V.P.; formal analysis, C.S.M.M. and J.A.V.P.; investigation, C.S.M.M. and H.B.A.S.; writing—original draft preparation, C.S.M.M.; writing—review and editing, H.B.A.S., C.S.M.M. and J.A.V.P.; supervision, J.A.V.P.; project administration, J.A.V.P.; funding acquisition, J.A.V.P. All authors have read and agreed to the published version of the manuscript.

Funding: This work received financial support from National Funds (FCT, Fundação para a Ciência e Tecnologia) through project PTDC/MED-QUI/29800/2017.

Data Availability Statement: Not applicable.

Acknowledgments: We gratefully acknowledge the support of LAQV-REQUIMTE (UIDB/50006/2020 and UIDP/50006/2020).

Conflicts of Interest: The authors declare no conflict of interest.

References

1. Liu, J.F.; Yu, S.J.; Yin, Y.G.; Chao, J.B. Methods for separation, identification, characterization and quantification of silver nanoparticles. *Trac Trends Anal. Chem.* **2012**, *33*, 95–106. [CrossRef]
2. Sajid, M.; Plotka-Wasylka, J. Nanoparticles: Synthesis, characteristics, and applications in analytical and other sciences. *Microchem. J.* **2020**, *154*, 104623. [CrossRef]
3. Alarcon, E.I.; Griffith, M.; Udekwu, K.I. (Eds.) *Silver Nanoparticle Applications*; Springer: Berlin/Heidelberg, Germany; New York, NY, USA, 2015.
4. Burdușel, A.-C.; Gherasim, O.; Grumezescu, A.M.; Mogoantă, L.; Ficai, A.; Andronescu, E. Biomedical Applications of Silver Nanoparticles: An Up-to-Date Overview. *Nanomaterials* **2018**, *8*, 681. [CrossRef]
5. Moeinzadeh, S.; Jabbari, E. Nanoparticles and Their Applications. In *Springer Handbook of Nanotechnology*; Bhushan, B., Ed.; Springer: Berlin/Heidelberg, Germany, 2017; pp. 335–361.
6. Mukherji, S.; Bharti, S.; Shukla, G.; Mukherji, S. 1. Synthesis and characterization of size- and shape-controlled silver nanoparticles. In *Volume 1 B Metallic Nanomaterials (Part B)*; Kumar, C., Ed.; De Gruyter: Berlin, Germany, 2018. [CrossRef]
7. Anastas, P.; Eghbali, N. Green chemistry: Principles and practice. *Chem. Soc. Rev.* **2010**, *39*, 301–312. [CrossRef] [PubMed]
8. Yu, S.J.; Yin, Y.G.; Liu, J.F. Silver nanoparticles in the environment. *Environ. Sci. Process. Impacts* **2013**, *15*, 78–92. [CrossRef] [PubMed]
9. Dastafkan, K.; Khajeh, M.; Ghaffari-Moghaddam, M.; Bohlooli, M. Silver nanoparticles for separation and preconcentration processes. *Trac Trends Anal. Chem.* **2015**, *64*, 118–126. [CrossRef]
10. Wang, X.; Yang, H.Y.; Li, K.J.; Xiang, Y.; Sha, Y.; Zhang, M.; Yuan, X.; Huang, K. Recent developments of the speciation analysis methods for silver nanoparticles and silver ions based on atomic spectrometry. *Appl. Spectrosc. Rev.* **2020**, *55*, 509–524. [CrossRef]
11. Mori, Y. Size-Selective Separation Techniques for Nanoparticles in Liquid. *Kona Powder Part J.* **2015**, *32*, 102–114. [CrossRef]
12. Kelland, D.R. Magnetic separation of nanoparticles. *IEEE Trans. Magn.* **1998**, *34*, 2123–2125. [CrossRef]
13. Kowalczyk, B.; Lagzi, I.; Grzybowski, B.A. Nanoseparations: Strategies for size and/or shape-selective purification of nanoparticles. *Curr. Opin. Colloid Interface Sci.* **2011**, *16*, 135–148. [CrossRef]
14. Moeser, G.D.; Roach, K.A.; Green, W.H.; Hatton, T.A.; Laibinis, P.E. High-gradient magnetic separation of coated magnetic nanoparticles. *AIChE J.* **2004**, *50*, 2835–2848. [CrossRef]
15. Mwilu, S.K.; Siska, E.; Baig, R.B.; Varma, R.S.; Heithmar, E.; Rogers, K.R. Separation and measurement of silver nanoparticles and silver ions using magnetic particles. *Sci. Total Environ.* **2014**, *472*, 316–323. [CrossRef] [PubMed]
16. Luo, L.; Yang, Y.; Li, H.; Ding, R.; Wang, Q.; Yang, Z. Size characterization of silver nanoparticles after separation from silver ions in environmental water using magnetic reduced graphene oxide. *Sci. Total Environ.* **2018**, *612*, 1215–1222. [CrossRef]
17. Degueldre, C.; Favarger, P.Y.; Bitea, C. Zirconia colloid analysis by single particle inductively coupled plasma–mass spectrometry. *Anal. Chim. Acta* **2004**, *518*, 137–142. [CrossRef]
18. Ramos, K.; Gomez-Gomez, M.M.; Camara, C.; Ramos, L. Silver speciation and characterization of nanoparticles released from plastic food containers by single particle ICPMS. *Talanta* **2016**, *151*, 83–90. [CrossRef] [PubMed]
19. Yang, Y.; Long, C.-L.; Li, H.-P.; Wang, Q.; Yang, Z.-G. Analysis of silver and gold nanoparticles in environmental water using single particle-inductively coupled plasma-mass spectrometry. *Sci. Total Environ.* **2016**, *563–564*, 996–1007. [CrossRef]
20. Inglezakis, V.J.; Kurbanova, A.; Molkenova, A.; Zorpas, A.A.; Atabaev, T.S. Magnetic Fe_3O_4-Ag(0)Nanocomposites for Effective Mercury Removal from Water. *Sustainability* **2020**, *12*, 5489. [CrossRef]
21. Kato, H. Chapter 3.3.2—Field-flow fractionation (FFF) with various detection systems. In *Characterization of Nanoparticles*; Hodoroaba, V.-D., Unger, W.E.S., Shard, A.G., Eds.; Elsevier: Amsterdam, The Netherlands, 2020; pp. 249–264.
22. Saenmuangchin, R.; Mettakoonpitak, J.; Shiowatana, J.; Siripinyanond, A. Separation of silver nanoparticles by hollow fiber flow field-flow fractionation: Addition of tannic acid into carrier liquid as a modifier. *J. Chromatogr. A* **2015**, *1415*, 115–122. [CrossRef] [PubMed]
23. Marassi, V.; Casolari, S.; Roda, B.; Zattoni, A.; Reschiglian, P.; Panzavolta, S.; Tofail, S.A.; Ortelli, S.; Delpivo, C.; Blosi, M.; et al. Hollow-fiber flow field-flow fractionation and multi-angle light scattering investigation of the size, shape and metal-release of silver nanoparticles in aqueous medium for nano-risk assessment. *J. Pharm. Biomed. Anal.* **2015**, *106*, 92–99. [CrossRef]
24. Kim, S.T.; Kang, D.Y.; Lee, S.H.; Kim, W.S.; Lee, J.T.; Cho, H.S.; Kim, S.H. Separation and quantitation of silver nanoparticles using sedimentation field-flow Fractionation. *J. Liq. Chromatogr. Relat. Technol.* **2007**, *30*, 2533–2544. [CrossRef]
25. Cumberland, S.A.; Lead, J.R. Particle size distributions of silver nanoparticles at environmentally relevant conditions. *J. Chromatogr. A* **2009**, *1216*, 9099–9105. [CrossRef] [PubMed]
26. Bolea, E.; Jimenez-Lamana, J.; Laborda, F.; Castillo, J.R. Size characterization and quantification of silver nanoparticles by asymmetric flow field-flow fractionation coupled with inductively coupled plasma mass spectrometry. *Anal. Bioanal. Chem.* **2011**, *401*, 2723–2732. [CrossRef] [PubMed]
27. Loeschner, K.; Navratilova, J.; Legros, S.; Wagner, S.; Grombe, R.; Snell, J.; von der Kammer, F.; Larsen, E.H. Optimization and evaluation of asymmetric flow field-flow fractionation of silver nanoparticles. *J. Chromatogr. A* **2013**, *1272*, 116–125. [CrossRef]
28. Ramos, K.; Ramos, L.; Camara, C.; Gomez-Gomez, M.M. Characterization and quantification of silver nanoparticles in nutraceuticals and beverages by asymmetric flow field flow fractionation coupled with inductively coupled plasma mass spectrometry. *J. Chromatogr. A* **2014**, *1371*, 227–236. [CrossRef]

29. Tan, Z.Q.; Liu, J.F.; Guo, X.R.; Yin, Y.G.; Byeon, S.K.; Moon, M.H.; Jiang, G.B. Toward full spectrum speciation of silver nanoparticles and ionic silver by on-line coupling of hollow fiber flow field-flow fractionation and minicolumn concentration with multiple detectors. *Anal. Chem.* **2015**, *87*, 8441–8447. [CrossRef] [PubMed]
30. Müller, D.; Cattaneo, S.; Meier, F.; Welz, R.; de Mello, A.J. Nanoparticle separation with a miniaturized asymmetrical flow field-Reflow fractionation cartridge. *Front. Chem.* **2015**, *3*, 45. [CrossRef]
31. Soto-Alvaredo, J.; Montes-Bayon, M.; Bettmer, J. Speciation of silver nanoparticles and silver(I) by reversed-phase liquid chromatography coupled to ICPMS. *Anal. Chem.* **2013**, *85*, 1316–1321. [CrossRef]
32. Tiede, K.; Boxall, A.B.A.; Tiede, D.; Tear, S.P.; David, H.; Lewis, J. A robust size-characterisation methodology for studying nanoparticle behaviour in 'real' environmental samples, using hydrodynamic chromatography coupled to ICP-MS. *J. Anal. At. Spectrom.* **2009**, *24*, 964–972. [CrossRef]
33. Sotebier, C.A.; Weidner, S.M.; Jakubowski, N.; Panne, U.; Bettmer, J. Separation and quantification of silver nanoparticles and silver ions using reversed phase high performance liquid chromatography coupled to inductively coupled plasma mass spectrometry in combination with isotope dilution analysis. *J. Chromatogr. A* **2016**, *1468*, 102–108. [CrossRef] [PubMed]
34. Zhou, X.X.; Liu, R.; Liu, J.F. Rapid chromatographic separation of dissoluble Ag(I) and silver-containing nanoparticles of 1–100 nanometer in antibacterial products and environmental waters. *Environ. Sci. Technol.* **2014**, *48*, 14516–14524. [CrossRef]
35. Dong, L.; Zhou, X.; Hu, L.; Yin, Y.; Liu, J. Simultaneous size characterization and mass quantification of the in vivo core-biocorona structure and dissolved species of silver nanoparticles. *J. Environ. Sci.* **2018**, *63*, 227–235. [CrossRef] [PubMed]
36. Shen, C.W.; Yu, T. Size-fractionation of silver nanoparticles using ion-pair extraction in a counter-current chromatograph. *J. Chromatogr. A* **2009**, *1216*, 5962–5967. [CrossRef] [PubMed]
37. Hong, P.; Koza, S.; Bouvier, E.S.P. Size-Exclusion Chromatography for the Analysis of Protein Biotherapeutics and their Aggregates. *J. Liq. Chromatogr. Relat. Technol.* **2012**, *35*, 2923–2950. [CrossRef]
38. Rahman, M. Chapter 4—Application of Computational Methods in Isolation of Plant Secondary Metabolites. In *Computational Phytochemistry*; Sarker, S.D., Nahar, L., Eds.; Elsevier: Amsterdam, The Netherlands, 2018; pp. 107–139.
39. Hull, R. Chapter 13—Assay, Detection, and Diagnosis of Plant Viruses. In *Plant Virology*, 5th ed.; Hull, R., Ed.; Academic Press: Boston, MA, USA, 2014; pp. 755–808.
40. Nagarajan, T.; Ertl, H.C.J. Chapter 14—Human and animal vaccines. In *Rabies*, 4th ed.; Fooks, A.R., Jackson, A.C., Eds.; Academic Press: Boston, MA, USA, 2020; pp. 481–508.
41. Serwer, P. *Encyclopedia of Separation Science*; Wilson, I.D., Ed.; Academic Press: Oxford, UK, 2000; pp. 2102–2109.
42. Asnaashari Kahnouji, Y.; Mosaddegh, E.; Bolorizadeh, M.A. Detailed analysis of size-separation of silver nanoparticles by density gradient centrifugation method. *Mater. Sci. Eng. C Mater. Biol. Appl.* **2019**, *103*, 109817. [CrossRef]
43. Lee, S.H.; Salunke, B.K.; Kim, B.S. Sucrose density gradient centrifugation separation of gold and silver nanoparticles synthesized using Magnolia kobus plant leaf extracts. *Biotechnol. Bioprocess Eng.* **2014**, *19*, 169–174. [CrossRef]
44. Hyun, J.Y.; Yun, C.; Kim, K.H.; Kim, W.H.; Jeon, S.W.; Bin, I.W.; Kim, J.P. Control of the plasmon resonance from poly-dispersed silver nanoparticles. *Jpn. J. Appl. Phys.* **2015**, *54*, 02BD02. [CrossRef]
45. Surugau, N.; Urban, P.L. Electrophoretic methods for separation of nanoparticles. *J. Sep. Sci.* **2009**, *32*, 1889–1906. [CrossRef] [PubMed]
46. Hanauer, M.; Pierrat, S.; Zins, I.; Lotz, A.; Sonnichsen, C. Separation of nanoparticles by gel electrophoresis according to size and shape. *Nano Lett.* **2007**, *7*, 2881–2885. [CrossRef] [PubMed]
47. Liu, F.K.; Ko, F.H.; Huang, P.W.; Wu, C.H.; Chu, T.C. Studying the size/shape separation and optical properties of silver nanoparticles by capillary electrophoresis. *J. Chromatogr. A* **2005**, *1062*, 139–145. [CrossRef] [PubMed]
48. Qu, H.; Mudalige, T.K.; Linder, S.W. Capillary electrophoresis coupled with inductively coupled mass spectrometry as an alternative to cloud point extraction based methods for rapid quantification of silver ions and surface coated silver nanoparticles. *J. Chromatogr. A* **2016**, *1429*, 348–353. [CrossRef] [PubMed]
49. Fa, A.J.G.; Cerutti, I.; Springer, V.; Girotti, S.; Centurion, M.E.; Di Nezio, M.S.; Pistonesi, M.F. Simple Characterization of Green-Synthesized Silver Nanoparticles by Capillary Electrophoresis. *Chromatographia* **2017**, *80*, 1459–1466. [CrossRef]
50. Konop, M.; Klodzinska, E.; Borowiec, J.; Laskowska, A.K.; Czuwara, J.; Konieczka, P.; Cieslik, B.; Waraksa, E.; Rudnicka, L. Application of micellar electrokinetic chromatography for detection of silver nanoparticles released from wound dressing. *Electrophoresis* **2019**, *40*, 1565–1572. [CrossRef] [PubMed]
51. Hancu, G.; Simon, B.; Rusu, A.; Mircia, E.; Gyéresi, A. Principles of micellar electrokinetic capillary chromatography applied in pharmaceutical analysis. *Adv. Pharm. Bull.* **2013**, *3*, 1–8. [CrossRef] [PubMed]
52. Tovstun, S.A.; Razumov, V.F. Theory of size-selective precipitation. *J. Nanopart. Res.* **2016**, *19*, 8. [CrossRef]
53. McLeod, M.C.; Anand, M.; Kitchens, C.L.; Roberts, C.B. Precise and rapid size selection and targeted deposition of nanoparticle populations using CO_2 gas expanded liquids. *Nano Lett.* **2005**, *5*, 461–465. [CrossRef] [PubMed]
54. Hu, G.S.; Jin, W.X.; Zhang, W.; Wu, K.; He, J.H.; Zhang, Y.; Chen, Q.Y.; Zhang, W.Z. Surfactant-assisted shape separation from silver nanoparticles prepared by a seed-mediated method. *Colloids Surf. A Physicochem. Eng. Asp.* **2018**, *540*, 136–142. [CrossRef]
55. Jana, N.R. Nanorod shape separation using surfactant assisted self-assembly. *Chem. Commun.* **2003**, 1950–1951. [CrossRef] [PubMed]
56. Krieg, E.; Weissman, H.; Shirman, E.; Shimoni, E.; Rybtchinski, B. A recyclable supramolecular membrane for size-selective separation of nanoparticles. *Nat. Nanotechnol.* **2011**, *6*, 141–146. [CrossRef] [PubMed]

57. Hussain, C.M.; Keçili, R. Sustainable Development and Environmental Analysis. In *Modern Environmental Analysis Techniques for Pollutants*; Hussain, C.M., Keçili, R., Eds.; Elsevier: Amsterdam, The Netherlands, 2020; pp. 343–379.
58. Macháčková, L.; Žemberyová, M. Cloud point extraction for preconcentration of trace beryllium and chromium in water samples prior to electrothermal atomic absorption spectrometry. *Anal. Methods* **2012**, *4*, 4042–4048. [CrossRef]
59. Liu, J.F.; Chao, J.B.; Liu, R.; Tan, Z.Q.; Yin, Y.G.; Wu, Y.; Jiang, G.B. Cloud Point Extraction as an Advantageous Preconcentration Approach for Analysis of Trace Silver Nanoparticles in Environmental Waters. *Anal. Chem.* **2009**, *81*, 6496–6502. [CrossRef]
60. Wang, S.; Meng, S.; Guo, Y. Cloud Point Extraction for the Determination of Trace Amounts of Cobalt in Water and Food Samples by Flame Atomic Absorption Spectrometry. *J. Spectrosc.* **2013**, *2013*, 735702. [CrossRef]
61. Giné, M.F.; Patreze, A.F.; Silva, E.L.; Sarkis, J.E.S.; Kakazu, M.H. Sequential cloud point extraction of trace elements from biological samples and determination by inductively coupled plasma mass spectrometry. *J. Braz. Chem. Soc.* **2008**, *19*, 471–477. [CrossRef]
62. Yu, S.J.; Chao, J.B.; Sun, J.; Yin, Y.G.; Liu, J.F.; Jiang, G.B. Quantification of the uptake of silver nanoparticles and ions to HepG2 cells. *Environ. Sci. Technol.* **2013**, *47*, 3268–3274. [CrossRef] [PubMed]
63. Wu, Z.H.; Tseng, W.L. Combined cloud point extraction and Tween 20-stabilized gold nanoparticles for colorimetric assay of silver nanoparticles in environmental water. *Anal. Methods* **2011**, *3*, 2915–2920. [CrossRef]
64. Wilson, O.M.; Scott, R.W.J.; Garcia-Martinez, J.C.; Crooks, R.M. Separation of dendrimer-encapsulated Au and Ag nanoparticles by selective extraction. *Chem. Mater.* **2004**, *16*, 4202–4204. [CrossRef]
65. Liu, J.F.; Liu, R.; Yin, Y.G.; Jiang, G.B. Triton X-114 based cloud point extraction: A thermoreversible approach for separation/concentration and dispersion of nanomaterials in the aqueous phase. *Chem. Commun.* **2009**, *12*, 1514–1516. [CrossRef] [PubMed]
66. Chao, J.B.; Liu, J.F.; Yu, S.J.; Feng, Y.D.; Tan, Z.Q.; Liu, R.; Yin, Y.G. Speciation analysis of silver nanoparticles and silver ions in antibacterial products and environmental waters via cloud point extraction-based separation. *Anal. Chem.* **2011**, *83*, 6875–6882. [CrossRef] [PubMed]
67. Hartmann, G.; Hutterer, C.; Schuster, M. Ultra-trace determination of silver nanoparticles in water samples using cloud point extraction and ETAAS. *J. Anal. At. Spectrom.* **2013**, *28*, 567–572. [CrossRef]
68. Lopez-Mayan, J.J.; Cerneira-Temperan, B.; Pena-Vazquez, E.; Barciela-Alonso, M.C.; Dominguez-Gonzalez, M.R.; Bermejo-Barrera, P. Evaluation of a cloud point extraction method for the preconcentration and quantification of silver nanoparticles in water samples by ETAAS. *Int. J. Environ. Anal. Chem.* **2018**, *98*, 1434–1447. [CrossRef]
69. Torrent, L.; Iglesias, M.; Hidalgo, M.; Margui, E. Determination of silver nanoparticles in complex aqueous matrices by total reflection X-ray fluorescence spectrometry combined with cloud point extraction. *J. Anal. At. Spectrom.* **2018**, *33*, 383–394. [CrossRef]
70. Wobrauschek, P. Total reflection X-ray fluorescence analysis—A review. *X-ray Spectrom.* **2007**, *36*, 289–300. [CrossRef]
71. Duester, L.; Fabricius, A.L.; Jakobtorweihen, S.; Philippe, A.; Weigl, F.; Wimmer, A.; Schuster, M.; Nazar, M.F. Can cloud point-based enrichment, preservation, and detection methods help to bridge gaps in aquatic nanometrology? *Anal. Bioanal. Chem.* **2016**, *408*, 7551–7557. [CrossRef] [PubMed]
72. Morais, M.; Teixeira, A.L.; Dias, F.; Machado, V.; Medeiros, R.; Prior, J.A.V. Cytotoxic Effect of Silver Nanoparticles Synthesized by Green Methods in Cancer. *J. Med. Chem.* **2020**, *63*, 14308–14335. [CrossRef] [PubMed]
73. Alharbi, F.A.; Alarfaj, A.A. Green synthesis of silver nanoparticles from Neurada procumbens and its antibacterial activity against multi-drug resistant microbial pathogens. *J. King Saud Univ. Sci.* **2020**, *32*, 1346–1352. [CrossRef]

 nanomaterials

MDPI

Article

Physical–Chemical Exfoliation of *n*-Alkylamine Derivatives of Layered Perovskite-like Oxide $H_2K_{0.5}Bi_{2.5}Ti_4O_{13}$ into Nanosheets

Iana A. Minich [1], Oleg I. Silyukov [1,*], Sergei A. Kurnosenko [1], Veronika V. Gak [1], Vladimir D. Kalganov [2], Petr D. Kolonitskiy [3] and Irina A. Zvereva [1]

1 Department of Chemical Thermodynamics and Kinetics, Institute of Chemistry, Saint Petersburg State University, 199034 Saint Petersburg, Russia; st048953@student.spbu.ru (I.A.M.); st040572@student.spbu.ru (S.A.K.); st062004@student.spbu.ru (V.V.G.); irina.zvereva@spbu.ru (I.A.Z.)
2 Interdisciplinary Resource Centre for Nanotechnology, Saint Petersburg State University, 199034 Saint Petersburg, Russia; v.kalganov@spbu.ru
3 Centre for Innovative Technologies of Composite Nanomaterials, Saint Petersburg State University, 199034 Saint Petersburg, Russia; petr.kolonitckii@spbu.ru
* Correspondence: oleg.silyukov@spbu.ru

Citation: Minich, I.A.; Silyukov, O.I.; Kurnosenko, S.A.; Gak, V.V.; Kalganov, V.D.; Kolonitskiy, P.D.; Zvereva, I.A. Physical–Chemical Exfoliation of *n*-Alkylamine Derivatives of Layered Perovskite-like Oxide $H_2K_{0.5}Bi_{2.5}Ti_4O_{13}$ into Nanosheets. *Nanomaterials* **2021**, *11*, 2708. https://doi.org/10.3390/nano11102708

Academic Editors: Juan Francisco Sánchez Royo and Arthur P. Baddorf

Received: 15 August 2021
Accepted: 5 October 2021
Published: 14 October 2021

Abstract: In the present work, we report the results on exfoliation and coating formation of inorganic–organic hybrids based on the layered perovskite-like bismuth titanate $H_2K_{0.5}Bi_{2.5}Ti_4O_{13}\cdot H_2O$ that could be prepared by a simple ion exchange reaction from a Ruddlesden–Popper phase $K_{2.5}Bi_{2.5}Ti_4O_{13}$. The inorganic–organic hybrids were synthesized by intercalation reactions. Exfoliation into nanosheets was performed for the starting hydrated protonated titanate and for the derivatives intercalated by *n*-alkylamines to study the influence of preliminary intercalation on exfoliation efficiency. The selected precursors were exfoliated in aqueous solutions of tetrabutylammonium hydroxide using facile stirring and ultrasonication. The suspensions of nanosheets obtained were characterized using UV–vis spectrophotometry, dynamic light scattering, inductively coupled plasma spectroscopy, and gravimetry. Nanosheets were coated on preliminarily polyethyleneimine-covered Si substrates using a self-assembly procedure and studied using atomic force and scanning electron microscopy.

Keywords: layered oxides; perovskites; bismuth titanates; exfoliation; nanosheets; coating

1. Introduction

Two-dimensional materials obtained from bulk materials have become an important research topic since the inception of graphene technologies [1]. One of the most promising directions in this field is the production of nanolayers via topdown approaches such as the exfoliation of bulk layered oxides. In particular, layered compounds with a perovskite structure are of great interest.

Layered perovskite-like oxides consist of negatively charged perovskite-like layers that fold into a two-dimensional layered structure with cations occupying the interlayer space to compensate for the negative charge of the layers [2,3]. In recent years, layered perovskite-like oxides have attracted great attention due to their structural features and associated anisotropic physical effects. Such compounds exhibit various interesting functional properties such as superconductivity [4,5], dielectric, ferroelectric and piezoelectric properties [6–10], photoluminescence [11–13], and photocatalytic activity [14–17]. In addition, as a rule, they demonstrate high thermal and mechanical stability and can be relatively easily and economically obtained via high-temperature solid-phase synthesis methods [18]. It is important to note that these materials allow useful characteristics to be fine-tuned, since the A sites of layered perovskites can be easily modified with metal ions using a conventional solid-state method [11,19–23].

Such compounds can replace interlayer cations of alkali metals by ion exchange with other cations, in particular, cations of other metals, complex cationic units, or protons, which, in the latter case, allows one to obtain so-called protonated forms [24–27]. One of the significant features of the protonated forms of perovskite-like oxides is their acidic properties, which allow the intercalation of organic bases to be conducted into their interlayer space, forming organic–inorganic hybrids [27–30]. Some perovskite-like oxides have been successfully exfoliated into two-dimensional perovskite nanosheets using chemical exfoliation in solutions of bulky quaternary ammonium ions [13,31–34]. Oxide nanosheets (about 1 nm thick) with the perovskite structure have completely different physical properties than their bulk counterparts. They can serve as multipurpose precursors for creating multilayer films, porous composites, and layered nanocomposites [33,35–38]. The resulting nanolayers already show intriguing properties with potential uses in photoluminescence [39,40], catalysis [16,41–43], and energy storage [44,45].

Even though for some layered perovskite-like oxides, the processes of protonation and subsequent splitting into nanolayers have been studied in great detail, the number of known structures amenable to exfoliation remains rather limited. In this regard, the task of finding new similar compounds is important.

Recently Liu et al. [46] synthesized a new layered Ruddlesden–Popper phase $K_{2.5}Bi_{2.5}Ti_4O_{13}$ (KBT_4) with mixed K/Bi co-occupancy on the perovskite A site, showing stoichiometric hydration and being amenable to form protonated compounds $H_2K_{0.5}Bi_{2.5}Ti_4O_{13}\cdot yH_2O$ [47,48] via substitution of K^+ by H^+. The $H_2K_{0.5}Bi_{2.5}Ti_4O_{13}\cdot H_2O$ ($HKBT_4\cdot H_2O$) structure can be described as an alternation of perovskite slabs ($K_{0.5}Bi_{2.5}Ti_4O_{13}$) with interlayer protons and water molecules. It was shown that this protonated derivative is capable of intercalation of *n*-alkylamines [49] and grafting of *n*-alcohols in the interlayer space [50]. This work reports on exfoliating $HKBT_4\cdot H_2O$ and its *n*-alkylamine derivatives in a tetrabutylammonium hydroxide (TBAOH) aqueous solution to obtain stable suspensions of perovskite nanolayers and their coatings.

2. Materials and Methods

2.1. Sample Preparation and Characterization

2.1.1. Initial Substances

KNO_3 (Vekton, St Petersburg, Russia, 99.9%), Bi_2O_3 (Vekton, St Petersburg, Russia, 99.9%), and TiO_2 (Vekton, St Petersburg, Russia, 99.9%) were dried at 200, 600, and 1000 °C, respectively. Methylamine ($MeNH_2$, 38% solution in water, Chemical line, St Petersburg, Russia,), ethylamine ($EtNH_2$, 70% solution in water, Merck, Darmstadt, Germany), *n*-propylamine ($PrNH_2$, Sigma-Aldrich, St Louis, MI, USA, 98%), *n*-butylamine ($BuNH_2$, 99.9%, Chemical line, St Petersburg, Russia,), *n*-hexylamine ($HxNH_2$, Sigma-Aldrich, St Louis, MI, USA, 99.9%), *n*-heptane (ECOS, Moscow, Russia, 99.9%), TBAOH (40 wt. % solution in water, Acros Organics, NJ, USA); polyethyleneimine (PEI, approx. M.N. 60,000, branched, 50 wt. % solution in water, Acros Organics, NJ, USA) were used as received.

Preparation of Inorganic Hosts

The starting inorganic matrix, KBT_4, and its protonated form were prepared by previously reported methods [46,48]. KBT_4 was synthesized using a solid-state reaction from Bi_2O_3, TiO_2, and KNO_3. The oxides were taken in stoichiometric ratios; the 20% excess of KNO_3 was used to compensate for its loss during the calcination due to the volatilization. $HKBT_4\cdot H_2O$ was obtained via the ion exchange reaction, which was carried out for 1 week in 1M HNO_3 at 20 °C.

Preparation of Inorganic-Organic Hybrids

Inorganic–organic hybrids with *n*-amines $HKBT_4 \times RNH_2$ were obtained by acid–base intercalation in solutions of corresponding *n*-amines. The reactions were performed in optimized conditions reported earlier. [49] Namely, hybrids with methylamine, ethylamine, *n*-propylamine, and *n*-butylamine were obtained via the 24 h reaction at 60 °C using 38% (for methylamine) and 50% aqueous solutions. The reaction with *n*-hexylamine was performed at 80 °C in 50% solution in *n*-heptane. After the reactions, solid products were washed with acetone or *n*-hexane (for $HxNH_2$ derivative).

2.1.2. Precursors Characterization

The fact of formation and phase purity of the precursors prepared were controlled using XRD analysis. The detailed qualitative and quantitative characterization of hydrated and dehydrated protonated forms obtained and *n*-alkylamine intercalated hybrids were performed via XRD, Raman, FTIR, and ^{13}C NMR spectroscopy, and TG, STA, and CNH analysis, details of which we presented earlier [50]. Starting from the ethylamine derivative, the intercalation of *n*-alkylamines led to a linear increase in the interlayer distance, *d*, compared to that of the hydrated protonated sample, which confirms the fact of the intercalation and the resulting weakening of direct binding of perovskite layers to each other.

2.1.3. Exfoliation

The protonated and amine-intercalated forms obtained were exfoliated using a physical–chemical method that implies the ultrasound treatment of the precursors in an aqueous solution of TBAOH. Specifically, the solid samples were taken in amounts according to a $TBA^+:H^+$ ratio of 1:1 and suspended in 30 mL of 0.004M TBAOH. The suspensions were sonicated using an ultrasonic homogenizer (UP200St) equipped with a 7 mm sonotrode (Hielscher, Teltow, Germany) at a half ultrasound amplitude, A = 50%, for 5 min (total energy ~8000 Ws) twice with intermediate stirring of the suspension at room temperature (~20 °C) for 24 h and 1 week. After the second sonication step, the fraction of large particles containing an unexfoliated residue was separated using centrifugation at 1000 RCF for 60 min.

2.1.4. Concentration Determination

The concentration of suspension was studied firstly by two direct approaches—inductively coupled plasma spectrometry (ICP) analysis and the gravimetry method for the suspension obtained from the *n*-butylamine precursor (for higher concentrations, A = 100% sonication was used). It should be noted that the concentration of obtained suspensions for other layered compounds presented in the literature is often disregarded [31,34,51], and, in several works, it is determined by one of these direct methods [13,39,52]. However, the concentration of the suspensions is of high importance for their future applications and specifically for composite preparations and nanosheets' deposition.

Gravimetry was performed as follows: the liquid phase was evaporated at 300 °C in a platinum crucible, preliminarily calcined at 900 °C to constant mass. Then, the sediment was calcined at 800 °C for 10 min several times until constant mass was reached. The temperature was chosen according to the complete decomposition of the protonated form into $K_{0.5}Bi_{2.5}Ti_4O_{13}$. For ICP analysis, the suspension was mineralized under microwave-assisted heating in a concentrated HNO_3/HF mixture. It was found that, in general, the gravimetrical method shows slightly higher concentrations compared to ICP analysis. Higher values calculated for K^+ content compared to Bi^{3+} and Ti^{4+} may be explained by possible incomplete mineralization of the sample. Thus, for further express determination of the concentration, we prepared spectrophotometric calibrations using the nanosheets' concentration determined by the gravimetrical method.

To control and compare the efficiency of exfoliation for different precursors, spectrophotometry calibrations based on those obtained by direct methods concentrations were prepared. For this, the suspension was diluted 5, 10, 15, 20, 25, 35, and 50 times using a 0.004 M TBAOH solution. The UV–vis spectra and calibrations prepared are shown in Figure A1. As may be seen, the shape of the bands upon dilutions is preserved, and two clear shoulders with maximums at 215 and 248 nm are presented. Two spectral maximums were compared, and the maximum at 248 nm was used for the calibration equation.

Density calibration was also considered as an alternative method that is less dependent on the particles' size distribution and their morphology. It was found that, in this case, the dependence is also close to linear in the concentration range studied and, therefore, densimetry may be used as a versatile approach to the suspensions' analysis. The densimetry calibrations obtained are presented in Figure A2.

2.1.5. Sample Preparation for SEM and AFM Measurements

For further characterization of the nanosheets using SEM and AFM methods, their deposition and flocculation were performed. Nanoparticle reassembly for SEM characterization was carried out following the previously reported method via drop-casting of 1M KCl solution [53]. The precipitate obtained was separated by centrifugation, washed with distilled H_2O, and dried under CaO.

Nanosheets' deposition on Si wafers was performed using the self-assembly method shown earlier for other layered niobates and titanates [37,53]. The deposition of the nanosheets was optimized by varying the pH and concentration of the suspensions. The tested concentrations for the original pH (11.9) were 25, 50, 100, 250, 500 and 750 mg/L. Additionally, deposition for suspensions with pH = 9 and concentrations 25, 50, and 100 mg/L was performed. Silicon substrates were washed via treatment in 50% HCl/methanol solution at 60 °C for 30 min and the surface was hydroxylated using concentrated H_2SO_4 for 30 min at 60 °C. After that, for better adhesion of negatively charged nanosheets, the surface of the substrate was coated with positively charged polymer, PEI, via treatment with 2.5 g/L of its aqueous solution at pH = 9. The coating time was 20 min, according to methods outlined in the literature [37,53]. After coating, the wafers were washed with water and calcined at 250 °C for 30 min to remove surface-adsorbed residual TBAOH. Additionally, the reliability of the coating technique was confirmed with the quartz crystal microbalance–dissipation (QCM-D) experiments. The experiments were carried out on SiO_2-coated quartz resonators, with a fundamental vibration frequency of 5 MHz. A detector was fixed in the cell with deionized water. The baseline drift was less than 0.1 Hz/h (which corresponds to 1.77 ng/sm^2). The experiment procedure was as follows: firstly, deionized water was replaced by PEI solution, and the cell was left for 20 min to form the coating; then, the stability of the PEI coating obtained was tested in a flow of water, after which the cell was filled with TBAOH solution. The results of QCM-D experiments showed that PEI coating is already formed after two minutes of treatment. The water flow leads to the partial diffusion of the PEI from the surface; after that, the coating is stabilized. Treatment of the sensor with the TBAOH solution leads to a similar effect caused by the adsorption of TBAOH molecules.

2.2. Instrumentation

Powder X-ray diffraction (XRD) analysis of the samples was performed on a Rigaku Miniflex II benchtop diffractometer (Tokyo, Japan) with CuKα radiation, angle range, $2\theta = 3 - 60°$, scanning rate, 10°/min, step, 0.02°. The lattice parameters were calculated on the basis of all the reflections observed using DiffracPlus Topas software (Bruker, Billerica, MA, USA). Concentrations of perovskite nanosheets in colloidal solutions used for building spectrophotometric calibration plots were determined by inductively coupled plasma atomic emission spectroscopy (ICP) on a Shimadzu ICPE-9000 spectrometer (Shimadzu, Kyoto, Japan) after preliminary acid digestion. Spectrophotometric analysis was performed on a Thermo Scientific GENESYS 10S UV–Vis spectrophotometer (Waltham, MA, USA). The pH values of nanosheets suspensions' media were measured using a laboratory pH-meter Toledo SevenCompact S220 (Mettler Toledo, Greifensee, Switzerland) equipped with an InLab Expert Pro-ISM electrode. Particle size distribution in colloidal suspensions, as well as their ζ-potentials, were determined using the method of dynamic light scattering (DLS) on a Photocor Compact-Z analyzer (Photocor, Moscow, Russia) using DynaLS software (version, Photocor, Moscow, Russia). Morphology of the coated particles was investigated on a Zeiss Merlin (Oberkochen, Germany) scanning electron microscope (SEM). Particle thickness and surface topography were studied using an NT-MDT Integra-Aura (manufacturer, Zelenograd, Russia) atomic-force microscope (AFM) and Gwyddion software (version,, Czech Metrology Institute, Brno, Czech Republic). Quartz crystal microbalance-dissipation experiments were performed on a BiolinScientific Q-Sense E-4 (BiolinScientific, Gothenburg, Sweden) device.

3. Results and Discussion

3.1. Optimization of Suspensions Preparations

In the present work, physical–chemical exfoliation was used. This approach implies two driving forces for exfoliation: the intercalation of the bulky exfoliation agent into the interlayer space, which facilitates the intercalation of solvent molecules and leads to the swelling of the sample, and subsequent sonication, which generates cavitation bubbles, resulting in high temperature and pressure local zones, which break apart the original samples' blocks into nanosheets. To determine the impact of these factors on the exfoliation efficiency and to optimize the conditions to prepare highly concentrated suspensions, we considered a row of samples with varied interlayer molecules, namely an as-prepared protonated form, which only includes water molecules in the interlayer space and organically modified samples with preliminarily intercalated *n*-amines: methylamine, ethylamine, *n*-propylamine, *n*-butylamine, and *n*-hexylamine. In addition, the duration of the intermediate stirring between sonication steps was varied for some of the samples.

Examples of UV–vis spectra, obtained for the suspensions of protonated and *n*-amine-intercalated samples with 24 h holding, are presented in Figure 1. As it may be seen in the case of organically modified precursors Me-Bu, two wide shoulders with maximums at 215 and 248 nm are presented on spectral bands, which may be the result of the size distribution of particles in the suspension. In the case of the protonated and *n*-hexylamine samples, a sharp peak at 210 nm appears. It should also be noted that visible agglomeration was detected for these suspensions, which was later additionally confirmed by DLS measurements for the dispersed protonated form. So, we associate the appearance of this band with the formation of bigger spherical agglomerates providing different scattering effects. In the case of the *n*-hexylamine sample, the floating and poor mixing of the powder in the TBAOH water solution were observed, which is most possibly caused by the strong hydrophobicity of the surface-adsorbed *n*-hexylamine. Although, as it may be seen from the spectra, sonication still made its partial exfoliation possible, implied by the presence of the wide shoulder, as in the case of other *n*-amine intercalates. The sample obtained was strongly unstable and the residue of floating agglomerates could not be separated by centrifugation, so this sample is not considered further in this work.

The concentrations of the suspensions prepared from the row of amine-intercalated samples with intermediate stirring for 24 h and 1 week are presented in Table 1. In the case of 24 h experiments, exfoliation was performed three times with different precursors, prepared under the same conditions, and the average concentration is presented. As it may be seen, the concentration insignificantly varies for suspensions, prepared using stirring for 24 h and 1 week, so it may be concluded that the insertion of exfoliant and solvents molecules is not time-dependent for these samples. In contrast, the choice of precursor for exfoliation strongly influences the exfoliation efficiency. In all of the cases, the most concentrated suspensions could be prepared using the *n*-propylamine and *n*-butylamine intercalates. The highest nanosheet concentration achieved using the *n*-propylamine derivative was about 780 mg/L. Earlier, we tested other Ruddlesden–Popper titanates, $HLnTiO_4$ and $H_2Ln_2Ti_3O_{10}$ (Ln = La, Nd) [54], the exfoliation of which was more efficient for the methylamine derivatives in comparison with *n*-butylamine ones, so we expected similar results for the present bismuth titanate. This was mainly explained by the stronger methylamine affinity to water, compared to less hydrophilic longer-chain *n*-amines, which leads to the more effective intercalation of water molecules during the stirring period, promoting subsequent exfoliation. Another possible reason may be associated with stronger alkyl–alkyl interactions between alkyl chains in the interlayer space for larger *n*-amines, preventing the lamination of the layered structure. The present work uses similar approaches to compare the exfoliation process for two different matrices and, as a result, it may be concluded that the choice of precursor for each matrix should be found experimentally rather than theoretically.

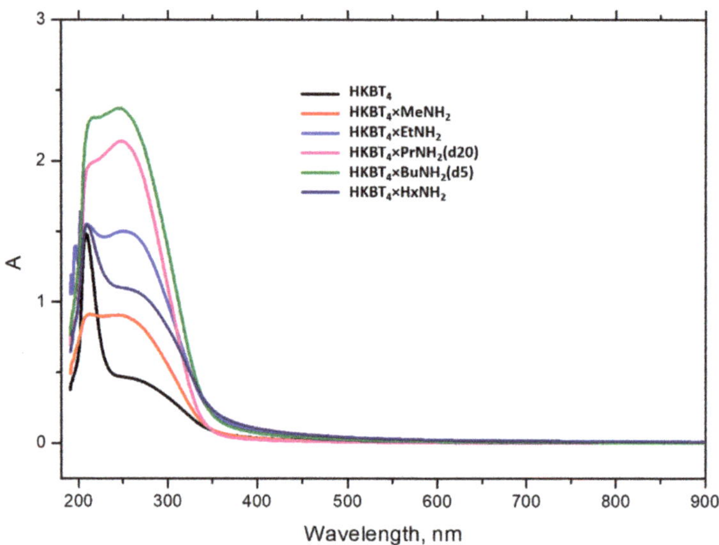

Figure 1. UV–vis spectra of suspensions obtained from protonated titanate $HKBT_4{\cdot}H_2O$ and its inorganic–organic hybrids.

Table 1. Concentrations of suspensions obtained from various inorganic–organic hybrids with 24 h and 1 week intermediate stirring time calculated from UV–vis spectra.

Sample	Concentration (24 h), mg/L	Concentration (1 w), mg/L
$HKBT_4 \times MeNH_2$	18	13
$HKBT_4 \times EtNH_2$	26	23
$HKBT_4 \times PrNH_2$	730	770
$HKBT_4 \times BuNH_2$	156	245

3.2. DLS Measurements

Average sizes of particles and size distribution in the suspensions prepared from various precursors, determined by DLS measurements, are shown in Table 2 and Figure 2. DLS measurements were performed for all of the suspensions the next day after preparation, and suspensions were shaken before the measurements. The presented results for samples with the narrowest size distribution (for *n*-propylamine) and wider size distribution (for *n*-butylamine) are refracted in the calculated average size for each sample. As it may be seen, in all of the cases, the average particle sizes were about 100 nm, except for slightly decreased sizes for $HKBT_4 \times PrNH_2$ suspension. The poor stability of the suspension prepared from protonated sample led to the aggregation of the particles into ~2 μm agglomerates after 24 h. However, the resonication right before the measurement shows that the original sizes of the particles correlate with the ones for amine-containing samples.

Table 2. Average sizes of prepared nanoparticles in suspensions determined by DLS.

Sample	Average Size, nm
$HKBT_4$ (aggregated)	~2000
$HKBT_4$ (resonicated)	96
$HKBT_4 \times MeNH_2$	77
$HKBT_4 \times EtNH_2$	83
$HKBT_4 \times PrNH_2$	65
$HKBT_4 \times BuNH_2$	88

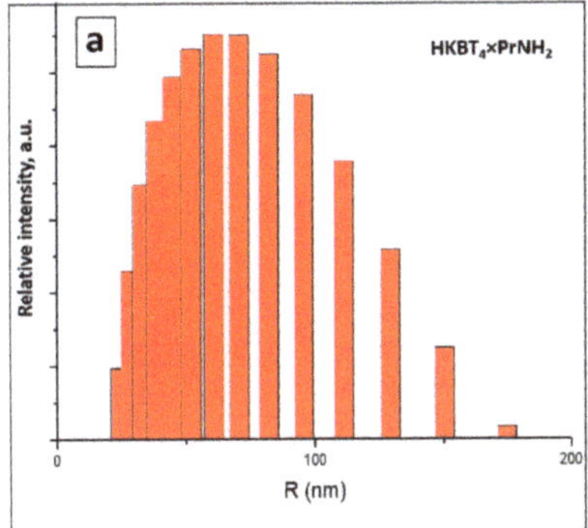

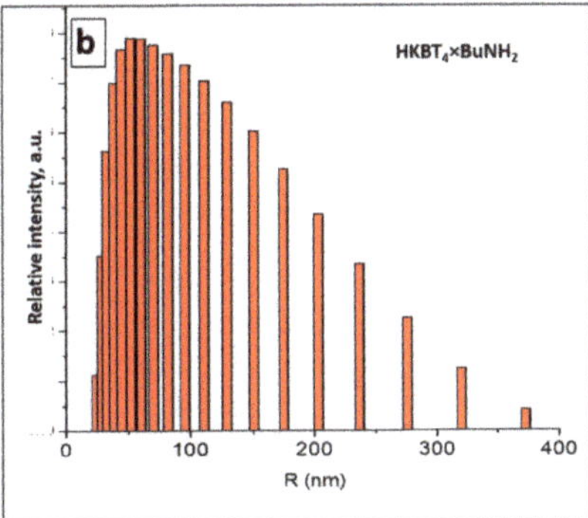

Figure 2. Particles size distribution for selected (**a**) *n*-propylamine and (**b**) *n*-butylamine non-aggregated samples.

3.3. Stability of Nanoparticles Suspensions

The stability of the suspensions obtained is an important characteristic defining the possibility of their storage and practical use. Moreover, stability should be considered for the further deposition of particles. In addition to visual and spectrometric control of suspension stability over time and during centrifugation (as we use centrifugation to separate unexfoliated residue, it also serves as a factor of the detected "exfoliation efficiency"), the aggregative stability of suspensions was also evaluated by measuring zeta potentials. ζ-potentials were measured for as-prepared suspensions from various precursors and for suspensions, prepared from protonated and *n*-propylamine intercalated forms at different pHs. ζ-potentials for suspensions prepared from various amine-intercalated precursors are presented in Figure 3a. As it may be seen, the ζ values obtained vary in the range from −17 to −27 mV, which corresponds to the average stability of metal oxide suspensions. However, the concentrated suspensions obtained did not show any visible agglomeration after 2 months of undisturbed storage (in the case of Me-Bu samples).

The main factors affecting ζ-potential considered are the pH of the suspensions, surface conductivity [55], and the concentration of suspensions [56]. It was shown earlier that strong dilution may lead to the lower stability of suspensions, which was observed in the case of the exfoliated protonated form. It should also be noted that the most concentrated suspension obtained using the *n*-propylamine sample showed a similar value for zeta potential. Nevertheless, practically, this suspension showed high stability.

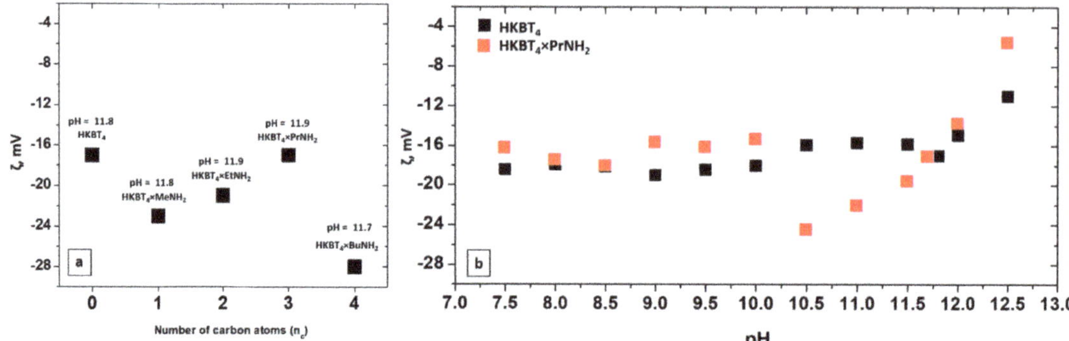

Figure 3. Potentials for (**a**) suspensions, prepared with various inorganic–organic precursors, (**b**) $HKBT_4$ (black) and $HKBT_4 \times PrNH_2$ (red) suspensions with varied pH.

The dependence of ζ-potentials on pH for the suspensions obtained from protonated and *n*-propylamine forms is presented in Figure 3b. In both cases, the original pH of the suspensions was ~11.8–11.9 and it was lowered by the dropwise addition of HCl and raised using a KOH solution. The step was 0.5 and the maximum studied pH was 12.5 due to the visible strong agglomeration of the particles at this pH. Both of the suspensions show similar values of ζ-potentials at their original pH (~17 mV). The values for the protonated sample at pH = 7.5–11.8 vary in the range of ~16–19 mV and the lowering is observed at higher pH with the minimum by module at pH = 12.5 ($\zeta = -11$ mV). A similar plot is obtained for the *n*-propylamine sample in the pH range of 7.5–10 with the local maximum (by module) at pH = 10.5 ($\zeta = -24$ mV). A more crucial decrease in ζ-potential is observed at a high pH with minimal value $\zeta = -5.5$ mV).

The pH of the obtained suspensions was also considered to verify that the difference in stability and obtained concentration does not arise from the difference in pH. In all of the cases, the pH of the suspensions prepared with 24 h stirring was in the range of ~11.7–11.9, which corresponds to the original pH of the TBAOH solution (11.8), while the pH for suspensions prepared with 1 w stirring was found to be in the range of ~9.5–9.9, which implies that pH does not depend on the precursor chosen for exfoliation and the concentration of suspensions obtained. Thus, there is no evident correlation between the pH of suspensions and exfoliation efficiency in the pH range studied.

3.4. Particles Deposition on Silicon Wafers

Deposition of the samples was performed from suspensions prepared using the *n*-propylamine precursor, as they had the highest original concentration. In the self-assembly approach used before for niobates and titanates [37,53], relatively diluted suspensions were used and the pH was usually adjusted to pH = 9, as it favors the presence of a higher percentage of cationic groups in the PEI coating [57]. In this work, we varied the concentration, and in some cases, the pH of suspensions to optimize particles' deposition, as such experiments have not been performed before with bismuth titanates. The efficiency of deposition and the lateral size and thickness of the obtained particles were studied using SEM and AFM. SEM images for selected samples are shown in Figure 4.

Figure 4. SEM images for initial $HKBT_4$ $PrNH_2$ sample (**a**); reassembled after exfoliation sample (**b**); Si substrate coating with 50 mg/L suspension at pH = 11.9 (**c**) and Si substrate coating with 750 mg/L suspension at pH = 11.9 (**d**).

The bulk powder precursors obtained had a plate-like morphology with relatively big lateral sizes. After exfoliation and reassembly, the agglomerated nanosheets were formed. SEM of the wafers obtained showed that in the case of low concentrations, only some single particles could be detected on the surface and using high concentrations leads to the formation of a more dense coating.

The successful deposition of particles was further proven using AFM. The corresponding images for selected samples are shown in Figure 5.

The deposition of low concentrated samples (25 mg/L) was ineffective and almost no particles were detected on the wafers in both of the cases (at original pH = 11.9 and pH = 9). An increase in suspensions' concentrations generally was shown to result in more dense coatings. Single deposited particles may be observed for the samples, prepared using 50, 100, and 250 mg/L suspensions. The thickness of the particles obtained was found to be in the range of 2.5–5 nm, which corresponds to the monolayers and bilayers of the starting inorganic matrix. (Figures 6 and A3). Using the highly concentrated suspensions (500 and 750 mg/L) resulted in the formation of a relatively dense coating of the silicon surface.

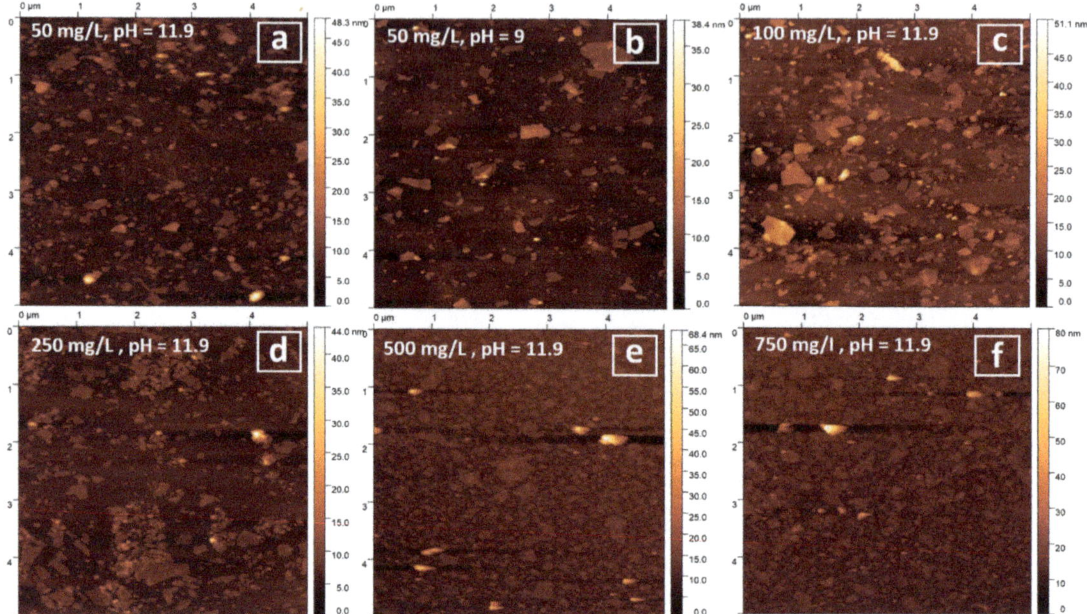

Figure 5. AFM images for coatings obtained in various conditions: 50 mg/L, pH 11.9 (**a**); 50 mg/L, pH 9 (**b**); 100 mg/L, pH 11.9 (**c**); 250 mg/L, pH 11.9 (**d**); 500 mg/L, pH 11.9 (**e**); 750 mg/L, pH 11.9 (**f**).

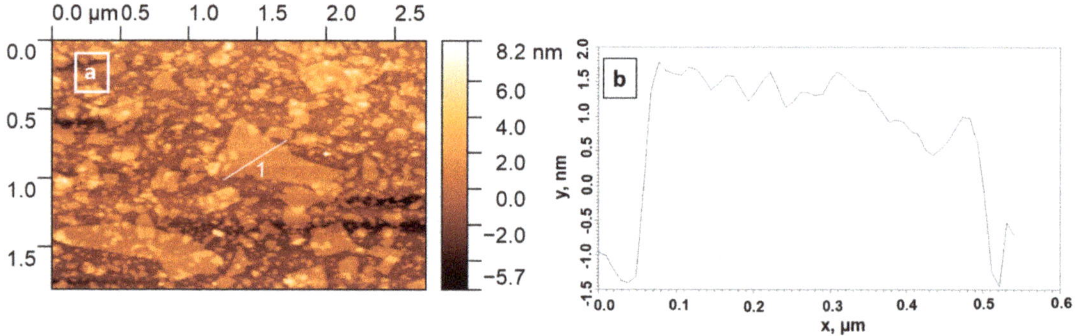

Figure 6. AFM image area (**a**) and height profile for a selected particle (**b**) for 500 mg/L, pH 11.9 suspension.

4. Conclusions

The physical–chemical exfoliation of a new Ruddlesden–Popper bismuth titanate, $H_2K_{0.5}Bi_{2.5}Ti_4O_{13}$, and its inorganic–organic hybrids with *n*-alkylamines was studied for the first time in an aqueous solution of TBAOH. Although it was possible to carry out the process of splitting and obtain a suspension of nanolayers for all samples, it was shown that the choice of precursor has a strong impact on the efficiency of exfoliation. To determine the concentration of the obtained suspensions, the two direct approaches—ICP analysis and the gravimetry method—were used. For express analysis, indirect UV–vis spectroscopy and the densimetry method were proposed. The most concentrated suspensions (up to 770 mg/L) after the removal of the unexfoliated part of oxide via centrifugation were obtained using the *n*-propylamine derivative. At the same time, the initially protonated derivative without intercalated amine was shown to have the worst exfoliation ability and stability of the resulting suspension. No correlation was found between ζ-potentials of the suspensions obtained and their concentrations for various samples, which implies that the stability of the resultant suspension does not influence the exfoliation procedure, considering the possible precipitation of unstable exfoliated particles during the centrifugation. According to the DLS data for the suspension obtained from the n-propylamine derivative, the average particle size lies at around 65 nm, which is in agreement with SEM and AFM data for the deposited particles. The negatively charged nanosheets' deposition on the surface of the Si substrate coated with positively charged polymer, PEI, was studied, considering the concentration, and in some cases, the pH of the suspension. As a result, the dense coatings were obtained using suspensions with high concentrations for the first time for Ruddlesden–Popper-type titanates. The thickness of the major particles deposited was found to be around 2.5–3 nm, which corresponds to the single-layer thickness of the starting inorganic matrix; this indicates that under the conditions used, it is possible to obtain suspensions of monolayers of the corresponding oxide.

Thus, the results presented in this work expand the range of objects for the creation of oxide nanolayers. The resulting nanolayers and coatings can be used as functional materials for catalysis and electronics.

Author Contributions: Conceptualization, I.A.Z. and O.I.S.; methodology, I.A.M. and O.I.S.; investigation, I.A.M., S.A.K., V.V.G., V.D.K. and P.D.K.; writing—original draft preparation, I.A.M.; writing—review and editing, I.A.Z., S.A.K. and O.I.S.; supervision, I.A.Z.; project administration, I.A.Z.; funding acquisition, I.A.Z. and O.I.S. All authors have read and agreed to the published version of the manuscript.

Funding: This research was funded by the Russian Science Foundation, grant No. 20-73-00027 (the QCM-D experiments, surface coatings formation, and characterization methodology development) and by the Russian Foundation for Basic Research, grant No. 19-33-90050 (initial compound preparation, exfoliation, and sample characterization experiments).

Data Availability Statement: The data presented in this study are available in the article.

Acknowledgments: The study was conducted using the equipment of the Saint Petersburg State University Research Park: Centre for X-ray Diffraction Studies, Centre for Optical and Laser Research, Centre for Chemical Analysis and Materials Research, Centre for Thermal Analysis and Calorimetry, Interdisciplinary Centre for Nanotechnology, Centre for Innovative Technologies of Composite Nanomaterials, Centre for Diagnostics of Functional Materials for Medicine, Pharmacology and Nanoelectronics.

Conflicts of Interest: The authors declare no conflict of interest.

Appendix A

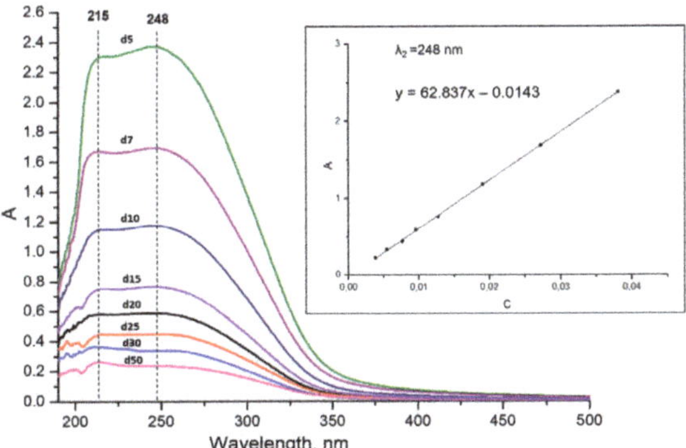

Figure A1. UV–vis spectra of $HKBT_4 \times BuNH_2$ suspensions in TBAOH with varied dilution and calibration obtained (inset). The initial concentration of undiluted suspension was 0.195 g/L.

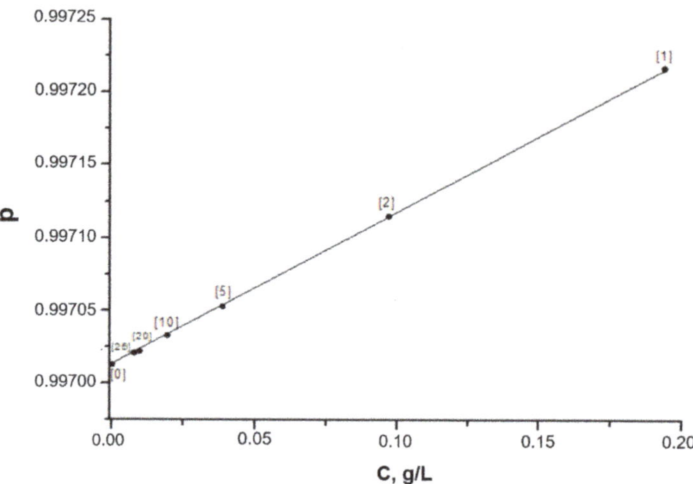

Figure A2. Density calibration obtained for $HKBT_4 \times BuNH_2$ suspension diluted to various concentrations (dilution degree presented in [] commas).

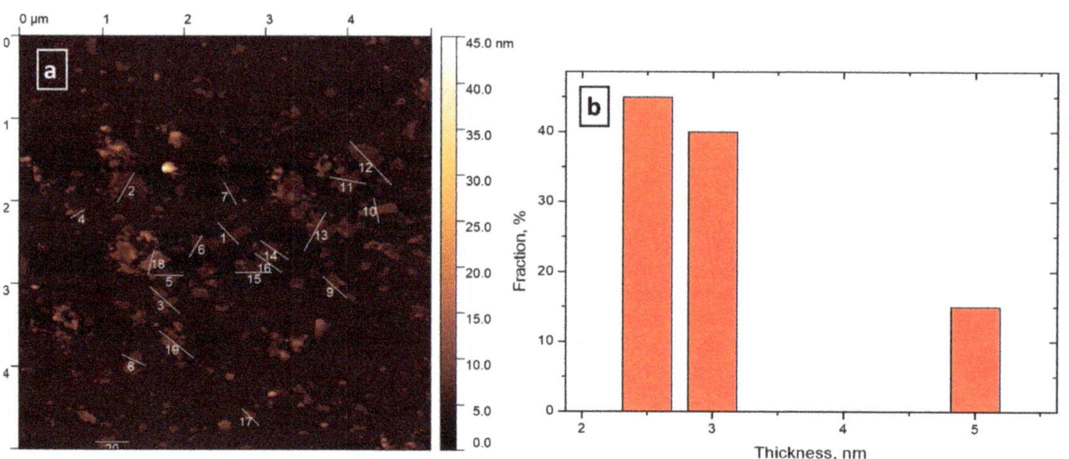

Figure A3. (**a**) AFM image for particles deposed from 250 mg/L suspension with marked particles used for thickness distribution calculations and (**b**) obtained thickness distribution.

References

1. Geim, A.K.; Novoselov, K.S. The rise of graphene. *Nat. Mater.* **2007**, *6*, 183–191. [CrossRef] [PubMed]
2. Dion, M.; Ganne, M.; Tournoux, M. Nouvelles familles de phases MIMII2Nb3O10 a feuillets "perovskites". *Mater. Res. Bull.* **1981**, *16*, 1429–1435. [CrossRef]
3. Jacobson, A.J.; Johnson, J.W.; Lewandowski, J.T. Interlayer Chemistry between Thick Transition-Metal Oxide Layers: Synthesis and Intercalation Reactions of $K[Ca_2Na_{n-3}Nb_nO3_{n+1}]$ ($3 \leq n \leq 7$). *Inorg. Chem.* **1985**, *24*, 3727–3729. [CrossRef]
4. Kato, M.; Kajita, T.; Hanakago, R.; Koike, Y. Search for new superconductors by the Li-intercalation into layered perovskites. *Phys. C Supercond.* **2006**, *445–448*, 26–30. [CrossRef]
5. Toda, K.; Teranishi, T.; Sato, M. Possibility of superconductivity in new reduced tantalate and titanate with the layered perovskite structure. *J. Eur. Ceram. Soc.* **1999**, *19*, 1525–1529. [CrossRef]
6. Chen, C.; Ning, H.; Lepadatu, S.; Cain, M.; Yan, H.; Reece, M.J. Ferroelectricity in Dion–Jacobson $ABiNb_2O_7$ (A = Rb, Cs) compounds. *J. Mater. Chem. C* **2015**, *3*, 19–22. [CrossRef]
7. Benedek, N.A. Origin of ferroelectricity in a family of polar oxides: The dion-jacobson phases. *Inorg. Chem.* **2014**, *53*, 3769–3777. [CrossRef]

8. Fennie, C.J.; Rabe, K.M. Ferroelectricity in the Dion-Jacobson CsBiNb2O7 from first principles. *Appl. Phys. Lett.* **2006**, *88*, 262902. [CrossRef]
9. Li, B.-W.; Osada, M.; Ebina, Y.; Ozawa, T.C.; Ma, R.; Sasaki, T. Impact of perovskite layer stacking on dielectric responses in $KCa_2Na_{n-3}Nb_nO_{3n+1}$ (n = 3–6) Dion–Jacobson homologous series. *Appl. Phys. Lett.* **2010**, *96*, 182903. [CrossRef]
10. Gou, G.; Shi, J. Piezoelectricity enhancement in Dion-Jacobson $RbBiNb_2O_7$ via negative pressure. *EPL (Europhys. Lett.)* **2014**, *108*, 67006. [CrossRef]
11. Zhang, N.; Guo, C.; Jing, H. Photoluminescence and cathode-luminescence of Eu^{3+}-doped $NaLnTiO_4$ (Ln = Gd and Y) phosphors. *RSC Adv.* **2013**, *3*, 7495. [CrossRef]
12. Gu, H.; Hu, Z.; Hu, Y.; Yuan, Y.; You, J.; Zou, W. The structure and photoluminescence of Bi4Ti3O12 nanoplates synthesized by hydrothermal method. *Colloids Surf. A Physicochem. Eng. Asp.* **2008**, *315*, 294–298. [CrossRef]
13. Ida, S.; Ogata, C.; Eguchi, M.; Youngblood, W.J.; Mallouk, T.E.; Matsumoto, Y. Photoluminescence of perovskite nanosheets prepared by exfoliation of layered oxides, $K_2Ln_2Ti_3O_{10}$, $KLnNb_2O_7$, and $RbLnTa_2O_7$ (Ln: Lanthanide ion). *J. Am. Chem. Soc.* **2008**, *130*, 7052–7059. [CrossRef]
14. Rodionov, I.A.; Silyukov, O.I.; Zvereva, I.A. Study of photocatalytic activity of layered oxides: $NaNdTiO_4$, $LiNdTiO_4$, and $HNdTiO_4$ titanates. *Russ. J. Gen. Chem.* **2012**, *82*, 635–638. [CrossRef]
15. Domen, K.; Yoshimura, J.; Sekine, T.; Kondo, J.; Tanaka, A.; Maruya, K.; Onishi, T. A Novel Series of Photocatalysts with an Ion-Exchangeable Layered Structure of Niobate. *Stud. Surf. Sci. Catal.* **1993**, 2159–2162. [CrossRef]
16. Hu, Y.; Mao, L.; Guan, X.; Tucker, K.A.; Xie, H.; Wu, X.; Shi, J. Layered perovskite oxides and their derivative nanosheets adopting different modification strategies towards better photocatalytic performance of water splitting. *Renew. Sustain. Energy Rev.* **2020**, *119*, 109527. [CrossRef]
17. Rodionov, I.A.; Silyukov, O.I.; Utkina, T.D.; Chislov, M.V.; Sokolova, Y.P.; Zvereva, I.A. Photocatalytic properties and hydration of perovskite-type layered titanates $A_2Ln_2Ti_3O_{10}$ (A = Li, Na, K.; Ln = La, Nd). *Russ. J. Gen. Chem.* **2012**, *82*, 1191–1196. [CrossRef]
18. Lichtenberg, F.; Herrnberger, A.; Wiedenmann, K. Synthesis, structural, magnetic and transport properties of layered perovskite-related titanates, niobates and tantalates of the type $A_nB_nO3_{n+2}$, $A'A_{k-1}B_kO_{3k+1}$ and $A_mB_{m-1}O_{3m}$. *Prog. Solid State Chem.* **2008**, *36*, 253–387. [CrossRef]
19. Yang, Y.; Chen, Q.; Yin, Z.; Li, J. Study on the photocatalytic activity of $K_2La_2Ti_3O_{10}$ doped with vanadium (V). *J. Alloys Compd.* **2009**, *488*, 364–369. [CrossRef]
20. Ozawa, T.C.; Fukuda, K.; Akatsuka, K.; Ebina, Y.; Sasaki, T. Preparation and Characterization of the Eu^{3+} Doped Perovskite Nanosheet Phosphor: $La_{0.90}Eu_{0.05}Nb_2O_7$. *Chem. Mater.* **2007**, *19*, 6575–6580. [CrossRef]
21. Wei, T.; Li, C.P.; Zhou, Q.J.; Zou, Y.L.; Zhang, L.S. Upconversion luminescence and ferroelectric properties of Er^{3+} doped $Bi_4Ti_3O_{12}$–$SrBi_4Ti_4O_{15}$. *Mater. Lett.* **2014**, *118*, 92–95. [CrossRef]
22. Ya-Hui, Y.; Qi-Yuan, C.; Zhou-Lan, Y.; Jie, L. Study on the photocatalytic activity of $K_2La_2Ti_3O_{10}$ doped with zinc(Zn). *Appl. Surf. Sci.* **2009**, *255*, 8419–8424. [CrossRef]
23. Huang, Y.; Xie, Y.; Fan, L.; Li, Y.; Wei, Y.; Lin, J.; Wu, J. Synthesis and photochemical properties of La-doped $HCa_2Nb_3O_{10}$. *Int. J. Hydrog. Energy* **2008**, *33*, 6432–6438. [CrossRef]
24. Gopalakrishnan, J. Chimie Douce Approaches to the Synthesis of Metastable Oxide Materials. *Am. Chem. Soc.* **1995**, *7*, 1265–1275. [CrossRef]
25. Schaak, R.E.; Mallouk, T.E. Perovskites by Design: A Toolbox of Solid-State Reactions. *Chem. Mater.* **2002**, *14*, 1455–1471. [CrossRef]
26. Yafarova, L.V.; Silyukov, O.I.; Myshkovskaya, T.D.; Minich, I.A.; Zvereva, I.A. New data on protonation and hydration of perovskite-type layered oxide $KCa_2Nb_3O_{10}$. *J. Therm. Anal. Calorim.* **2021**, *143*, 87–93. [CrossRef]
27. Jacobson, A.J.; Johnson, J.W.; Lewandowski, J. Intercalation of the layered solid acid $HCa_2Nb_3O_{10}$ by organic amines. *Mater. Res. Bull.* **1987**, *22*, 45–51. [CrossRef]
28. Hong, Y.; Kim, S.-J. Intercalation of Primary Diamines in the Layered Perovskite Oxides, $HSr_2Nb_3O_{10}$. *Bull. Korean Chem. Soc.* **1996**, *17*, 730–735.
29. Kurnosenko, S.A.; Silyukov, O.I.; Mazur, A.S.; Zvereva, I.A. Synthesis and thermal stability of new inorganic-organic perovskite-like hybrids based on layered titanates $HLnTiO_4$ (Ln = La, Nd). *Ceram. Int.* **2020**, *46*, 5058–5068. [CrossRef]
30. Silyukov, O.I.; Khramova, A.D.; Zvereva, I.A. Synthesis of Organic-Inorganic Derivatives of Perovskite-Like Layered $HCa_2Nb_3O_{10}$ Oxide with Monoethanolamine and Glycine. *Glas. Phys. Chem.* **2020**, *46*, 256–259. [CrossRef]
31. Wang, T.H.; Henderson, C.N.; Draskovic, T.I.; Mallouk, T.E. Synthesis, exfoliation, and electronic/protonic conductivity of the dion-jacobson phase layer perovskite $HLa_2TiTa_2O_{10}$. *Chem. Mater.* **2014**, *26*, 898–906. [CrossRef]
32. Nicolosi, V.; Chhowalla, M.; Kanatzidis, M.G.; Strano, M.S.; Coleman, J.N. Liquid Exfoliation of Layered Materials. *Science* **2013**, *340*, 1226419. [CrossRef]
33. Osada, M.; Sasaki, T. Exfoliated oxide nanosheets: New solution to nanoelectronics. *J. Mater. Chem.* **2009**, *19*, 2503–2511. [CrossRef]
34. Lee, W.-J.; Yeo, H.J.; Kim, D.-Y.; Paek, S.-M.; Kim, Y.-I. Exfoliation of Dion-Jacobson Layered Perovskite into Macromolecular Nanoplatelet. *Bull. Korean Chem. Soc.* **2013**, *34*, 2041–2043. [CrossRef]
35. Li, B.-W.; Osada, M.; Akatsuka, K.; Ebina, Y.; Ozawa, T.C.; Sasaki, T. Solution-Based Fabrication of Perovskite Multilayers and Superlattices Using Nanosheet Process. *Jpn. J. Appl. Phys.* **2011**, *50*, 09NA10. [CrossRef]

36. Pulyalina, A.; Rostovtseva, V.; Minich, I.; Silyukov, O.; Toikka, M.; Saprykina, N.; Polotskaya, G. Specific Structure and Properties of Composite Membranes Based on the Torlon® (Polyamide-imide)/Layered Perovskite Oxide. *Symmetry* **2020**, *12*, 1142. [CrossRef]
37. Sasaki, T.; Ebina, Y.; Tanaka, T.; Harada, M.; Watanabe, M.; Decher, G. Layer-by-layer assembly of titania nanosheet/polycation composite films. *Chem. Mater.* **2001**, *13*, 4661–4667. [CrossRef]
38. Hong, Y.-S.; Kim, S.-J. Synthesis and Characterization of Molecular Composite Prepared from Layered Perovskite Oxide, $HLa_2Ti_2NbO_{10}$. *Bull. Korean Chem. Soc.* **1997**, *18*, 623–628. [CrossRef]
39. Ida, S.; Ogata, C.; Unal, U.; Izawa, K.; Inoue, T.; Altuntasoglu, O.; Matsumoto, Y. Preparation of a blue luminescent nanosheet derived from layered perovskite $Bi_2SrTa_2O_9$. *J. Am. Chem. Soc.* **2007**, *129*, 8956–8957. [CrossRef]
40. Tetsuka, H.; Takashima, H.; Ikegami, K.; Nanjo, H.; Ebina, T.; Mizukami, F. Nanosheet seed-layer assists oriented growth of highly luminescent perovskite films. *Chem. Mater.* **2009**, *21*, 21–26. [CrossRef]
41. Takagaki, A.; Sugisawa, M.; Lu, D.; Kondo, J.N.; Hara, M.; Domen, K.; Hayashi, S. Exfoliated nanosheets as a new strong solid acid catalyst. *J. Am. Chem. Soc.* **2003**, *125*, 5479–5485. [CrossRef]
42. Zhou, Y.; Wen, T.; Zhang, X.; Chang, B.; Kong, W.; Guo, Y.; Yang, B.; Wang, Y. A Multiple Structure-Design Strategy towards Ultra-thin Niobate Perovskite Nanosheets with Thickness-Dependent Photocatalytic Hydrogen-Evolution Performance. *Chem.-Asian J.* **2017**, *12*, 2727–2733. [CrossRef]
43. Hu, S.; Chi, B.; Pu, J.; Jian, L. Novel heterojunction photocatalysts based on lanthanum titanate nanosheets and indium oxide nanoparticles with enhanced photocatalytic hydrogen production activity. *J. Mater. Chem. A* **2014**, *2*, 19260–19267. [CrossRef]
44. Osada, M.; Akatsuka, K.; Ebina, Y.; Kotani, Y.; Ono, K.; Funakubo, H.; Ueda, S.; Kobayashi, K.; Takada, K.; Sasaki, T. Langmuir–Blodgett Fabrication of Nanosheet-Based Dielectric Films without an Interfacial Dead Layer. *Jpn. J. Appl. Phys.* **2008**, *47*, 7556–7560. [CrossRef]
45. Khan, M.S.; Kim, H.; Kim, Y.; Ebina, Y.; Sugimoto, W.; Sasaki, T.; Osada, M. Scalable Design of Two-Dimensional Oxide Nanosheets for Construction of Ultrathin Multilayer Nanocapacitor. *Small* **2020**, *16*, 2003485. [CrossRef] [PubMed]
46. Liu, S.; Avdeev, M.; Liu, Y.; Johnson, M.R.; Ling, C.D. A New n = 4 Layered Ruddlesden–Popper Phase $K_{2.5}Bi_{2.5}Ti_4O_{13}$ Showing Stoichiometric Hydration. *Inorg. Chem.* **2016**, *55*, 1403–1411. [CrossRef] [PubMed]
47. Silyukov, O.I.; Minich, I.A.; Zvereva, I.A. Synthesis of protonated derivatives of layered perovskite-like bismuth titanates. *Glas. Phys. Chem.* **2018**, *44*, 115–119. [CrossRef]
48. Minich, I.A.; Silyukov, O.I.; Kulish, L.D.; Zvereva, I.A. Study on thermolysis process of a new hydrated and protonated perovskite-like oxides $H_2K_{0.5}Bi_{2.5}Ti_4O_{13}{\cdot}yH_2O$. *Ceram. Int.* **2019**, *45*, 2704–2709. [CrossRef]
49. Minich, I.A.; Silyukov, O.I.; Gak, V.V.; Borisov, E.V.; Zvereva, I.A. Synthesis of Organic-Inorganic Hybrids Based on Perovskite-like Bismuth Titanate H2K0.5Bi2.5Ti4O13·H2O and n-Alkylamines. *ACS Omega* **2020**, *5*, 8158–8168. [CrossRef]
50. Minich, I.A.; Silyukov, O.I.; Mazur, A.S.; Zvereva, I.A. Grafting reactions of perovskite-like bismuth titanate $H_2K_{0.5}Bi_{2.5}Ti_4O_{13}{\cdot}H_2O$ with n-alcohols. *Ceram. Int.* **2020**, *46*, 29373–29381. [CrossRef]
51. Ebina, Y.; Sasaki, T.; Watanabe, M. Study on exfoliation of layered perovskite-type niobates. *Solid State Ion.* **2002**, *151*, 177–182. [CrossRef]
52. Gao, H.; Shori, S.; Chen, X.; zur Loye, H.C.; Ploehn, H.J. Quantitative analysis of exfoliation and aspect ratio of calcium niobate platelets. *J. Colloid Interface Sci.* **2013**, *392*, 226–236. [CrossRef] [PubMed]
53. Ebina, Y.; Akatsuka, K.; Fukuda, K.; Sasaki, T. Synthesis and in situ X-ray diffraction characterization of two-dimensional perovskite-type oxide colloids with a controlled molecular thickness. *Chem. Mater.* **2012**, *24*, 4201–4208. [CrossRef]
54. Kurnosenko, S.A.; Silyukov, O.I.; Minich, I.A.Z. Exfoliation of Methylamine and n-Butylamine Derivatives of Layered Perovskite-Like Oxides $HLnTiO_4$ and $H_2Ln_2Ti_3O_{10}$ (Ln = La, Nd) into Nanolayers. *Glas. Phys. Chem.* **2021**, *47*, 372–381. [CrossRef]
55. Li, S.; Leroy, P.; Heberling, F.; Devau, N.; Jougnot, D.; Chiaberge, C. Influence of surface conductivity on the apparent zeta potential of calcite. *J. Colloid Interface Sci.* **2016**, *468*, 262–275. [CrossRef]
56. Tantra, R.; Schulze, P.; Quincey, P. Effect of nanoparticle concentration on zeta-potential measurement results and reproducibility. *Particuology* **2010**, *8*, 279–285. [CrossRef]
57. Suh, J.; Hwang, B.K. Ionization of PEI and PAA at various pH's. *Bioorg. Chem.* **1994**, *22*, 318–327. [CrossRef]

Article

2D Monomolecular Nanosheets Based on Thiacalixarene Derivatives: Synthesis, Solid State Self-Assembly and Crystal Polymorphism

Alena A. Vavilova [1], Pavel L. Padnya [1], Timur A. Mukhametzyanov [1], Aleksey V. Buzyurov [1], Konstantin S. Usachev [2], Daut R. Islamov [3], Marat A. Ziganshin [1], Artur E. Boldyrev [1] and Ivan I. Stoikov [1,*]

1 A.M. Butlerov Chemical Institute, Kazan Federal University, Kremlevskaya Street, 18, 420008 Kazan, Russia; anelia_86@mail.ru (A.A.V.); padnya.ksu@gmail.com (P.L.P.); timur.mukhametzyanov@kpfu.ru (T.A.M.); abuzurov95@gmail.com (A.V.B.); marat.ziganshin@kpfu.ru (M.A.Z.); boldyrev25@gmail.com (A.E.B.)

2 Institute of Fundamental Medicine and Biology, Kazan Federal University, Kremlevskaya Street, 18, 420008 Kazan, Russia; k.usachev@kpfu.ru

3 A.E. Arbuzov Institute of Organic and Physical Chemistry, Russian Academy of Sciences, Arbuzov Street, 8, 420088 Kazan, Russia; daut1989@mail.ru

* Correspondence: ivan.stoikov@mail.ru; Tel.: +7-843-2337241

Received: 25 November 2020; Accepted: 8 December 2020; Published: 14 December 2020

Abstract: Synthetic organic 2D materials are attracting careful attention of researchers due to their excellent functionality in various applications, including storage batteries, catalysis, thermoelectricity, advanced electronics, superconductors, optoelectronics, etc. In this work, thiacalix[4]arene derivatives functionalized by geranyl fragments at the lower rim in *cone* and *1,3-alternate* conformations, that are capable of controlled self-assembly in a 2D nanostructures were synthesized. X-ray diffraction analysis showed the formation of 2D monomolecular-layer nanosheets from synthesized thiacalix[4]arenes, the distance between which depends on the stereoisomer used. It was established by DSC, FSC, and PXRD methods that the obtained macrocycles are capable of forming different crystalline polymorphs, moreover dimethyl sulphoxide (DMSO) is contributing to the formation of a more stable polymorph for *cone* stereoisomer. The obtained crystalline 2D materials based on synthesized thiacalix[4]arenes can find application in material science and medicine for the development of modern pharmaceuticals and new generation materials.

Keywords: nanomaterials; 2D nanostructures; thiacalix[4]arene; terpenoids; geraniol; X-ray crystal analysis; 2D monomolecular-layer nanosheets; polymorphism

1. Introduction

In recent years, two-dimensional (2D) materials have attracted a lot of attention from researchers due to their superior functionality, such as mechanical, electrical, magnetic and optical properties [1–4]. In accordance to their structure, composition and electronic properties 2D materials may have to categorized as thin materials (graphene, silicene, germanene, and their saturated forms; hexagonal boron nitride; silicon carbide), rare earth, semimetals, transition metal chalcogenides and halides, and finally synthetic organic 2D materials, exemplified by 2D covalent organic frameworks [5]. Among the large family of these materials, synthetic organic 2D materials occupy a special place, which are used to create three-dimensional (3D) hybrid superlattices from alternating layers of organic molecules [1]. Creation of 3D superlattices provides new opportunities in various applications, including batteries, catalysis, thermoelectricity, advanced electronics, superconductors, optoelectronics, etc. According to the literature [1,5], the search for organic structures for creating 2D materials is still a difficult problem, since in many cases the orientation of organic molecules can be disordered. As a consequence, 2D materials based on them tend to repackage or aggregate, which leads to a weakening or even disappearance of 2D characteristics. Therefore, obtaining organic molecules capable of forming a monolayer or bilayer nanostructures with ordered packing is an urgent task for creating 3D superlattices with a fixed distance between layers, which plays a decisive role in the characteristics of the materials obtained.

In addition, recently there has been risen sharply the interest to polymorphic systems as an essentially interesting phenomenon and as an increasingly important component in the development and marketing of different materials based on organic molecules. The search of possible polymorphs is important for pharmaceuticals [6], high-energy materials, dyes and pigments [7] because different polymorphs usually have different properties [8]. Polymorphism significantly affects the physicochemical properties of materials, such as stability, density, melting point, solubility, bioavailability, etc. [9]. Hence, characterization of all possible polymorphs, identification of a stable polymorph and the development of a reliable process for sequential production are crucial significance in modern design of pharmaceuticals and materials.

In this work, we propose to use molecules with conformationally flexible and restricted units for creating 2D material. The controlled self-assembly of 2D structures is implemented using the example of well-known phospholipids, which are capable of forming monomolecular and bimolecular layers [10–13]. To obtain such structures, terpenoid fragments, which are found among phospholipids, are of particular interest. It is known that terpenoids such as verbenol [14] and menthol [15] are capable of polymorphism and crystalline modifications. Therefore, geraniol, an available non-toxic natural monoterpene, was chosen as such fragment. In addition, its structure is conformationally limited due to the presence of a double bond. As a result, compounds containing geraniol fragments are capable of packing and organizing into mono- and bilayers [16,17]. Amphiphilic compounds containing a polar fragment are usually used to obtain these layers, similar to surfactants, both ionic and non-ionic. Nonionic surfactants, along with anionic and cationic ones, were used to design 2D nanocrystals [18]. In this work, we propose to use a macrocyclic fragment of thiacalix[4]arene to create structures capable of forming mono- and bimolecular layers. The advantage of this platform is the presence of various stereoisomeric forms [19–25] allowing geranyl fragments to be strictly oriented in the space. Thiacalix[4]arenes are conformationally flexible molecular platforms that allow varying the size of the internal cavity, the number, nature and the spatial arrangement of binding sites. Calixarene derivatives are widely used for construction of different functional materials. Among such examples are known hybrid organic-inorganic silica nanoparticles for separation of model oligonucleotides and proteins [26], biodegradable, biocompatible polylactide polymers with cyclophanes as a core [27], antibacterial and catalytic systems based on calixarenes [28,29]. There are examples of thiacalixarene derivatives for the development of synthetic receptors for biologically important anions [22] and dopamine [23], which can be used in biomimetic materials, diagnostic agents and in the treatment of neurodegenerative diseases. Mention should also be made about polyplexes based on thiacalixarene derivatives, which are used for DNA packaging and have promising applications in biomolecular delivery systems and long-term storage of human genetic material [25]. Among other things, the controlled (smart) formation of polymorphs based on thiacalixarene derivative are also known [30]. Thus, according to previous contributions and applications of thiacalix[n]arenes it may be concluded that these macrocycles are promising candidates for constructing 2D nanostructures, especially different polymorphs that can be used in medicine for the development of modern pharmaceuticals, diagnostic agents and new generation materials.

This work is devoted to the synthesis of thiacalix[4]arene derivatives functionalized at the lower rim with geranyl fragments in *cone* and *1,3-alternate* conformations, the development of approaches to the formation of 2D monomolecular nanosheets based on them in a crystal, and the study of possible polymorphic transitions in the solid phase (Figure 1).

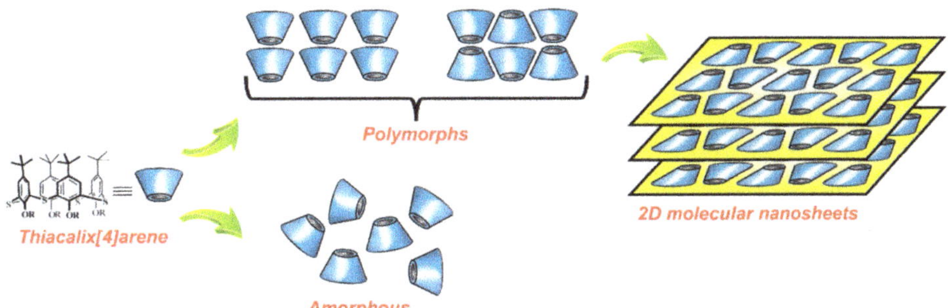

Figure 1. Supposed scheme for the formation of 2D monomolecular nanosheets based on *p-tert*-butylthiacalix[4]arene derivatives in the solid phase.

2. Materials and Methods

2.1. General

Melting points were determined using the Boetius Block apparatus. All chemicals were purchased from Aldrich and most of them used as received without additional purification. Organic solvents were purified by standard procedures. The ^1H and ^{13}C NMR spectra were recorded on the Bruker Avance 400 spectrometer (400.0 MHz for H-atoms and 100.0 MHz for C-atoms), chemical shifts are reported in ppm (See Supplementary Materials, Figures S1–S6). The residual solvent peaks were used as an internal standard ($CDCl_3$). The concentration of the analyzed solutions was 10 mM. The ATR FTIR spectra were recorded on the Spectrum 400 (Perkin Elmer, Seer Green, Lantrisant, UK) IR spectrometer: resolution 1 cm^{-1}, accumulation of 64 scans, recording time 16 s; in the range of wave numbers 400–4000 cm^{-1} (See Supplementary Materials, Figures S11–S13). Electrospray ionization mass spectra (ESI) were obtained on an AmazonX mass spectrometer (Bruker Daltonik GmbH, Bremen, Germany) (See Supplementary Materials, Figures S9 and S10). All experiments were conducted in negative ion polarity mode, in the range of m/z from 100 to 2800. The voltage on the capillary was −4500 V. Nitrogen was used as the gas-drier with a temperature of 300 °C and a flow rate of 10 L min^{-1}, and nebulizer pressure of 55.16 kPa. Data was processed using DataAnalysis 4.0 (Bruker Daltonik GmbH, Bremen, Germany). The crystallinity of the samples were additionally controlled using BXFM polarizing optical microscope with LC30 camera (Olympus, Tokyo, Japan) (See Supplementary Materials, Figures S15 and S16).

2.2. Synthesis

2.2.1. Geranyl Bromoacetate (**2**)

A solution of 0.02 mol of geraniol and 0.021 mol of DIPEA in chloroform (120 mL) was prepared and cooled to −5 °C. A thermometer was installed to control the temperature. A solution of 0.021 mol of bromoacetic acid bromide in 10 mL of chloroform was added dropwise at such a rate that the temperature did not rise above −2 °C. After the addition, the reaction mixture was left for 1 h at room temperature. The resulting reaction mixture was then washed with 5% aqueous Na_2CO_3 solution (2 × 50 mL) and 50 mL of water. The organic solvent was removed on a rotary evaporator under reduced pressure. The resulting crude product was purified by column chromatography on silica gel (eluent hexane-propanol-2 at ratio 20:1). The yield of pale yellow oil of **2** was 4.29 g (78%). ^1H NMR ($CDCl_3$, δ, ppm, J/Hz): 1.60 (3H, s, CH_3), 1.68 (3H, s, CH_3), 1.71 (3H, s, CH_3), 2.04–2.11 (4H, m, CH_2–CH_2), 3.83 (2H, s, O=C–CH_2Br), 4.68 (2H, d, =CH–$\underline{CH_2}$–O, $^3J_{HH}$ = 7.2 Hz), 5.07 (1H, m, =CH), 5.35 (1H, m, =CH). ^{13}C NMR ($CDCl_3$, δ, ppm): 16.54, 17.71, 25.70, 26.11, 26.22, 39.52, 63.13, 117.36, 123.61, 131.93, 143.58, 167.23. IR spectrum (ν/cm^{-1}): 2967 (–CH_3), 2915 (–CH_2–), 2856 (–CH_3), 1735 (C=O),

1444 (=CH–), 1424 (=CH–), 1408 (=CH–), 1376 (–CH_3), 1276 (C–O–C), 1204 (C–O–C), 1153 (C–O–C), 953 (=CH–), 888 (=CH–), 830 (=CH–), 554 (–CH_2Br). Calcd for $C_{12}H_{19}BrO_2$ (%): C, 52.38; H, 6.96; Br, 29.04. Found (%): C, 52.13; H, 6.77; Br, 28.90.

2.2.2. 5,11,17,23-Tetra-*Tert*-Butyl-25,26,27,28-Tetra[(Geranyloxycarbonyl)-Methoxy]-2,8,14,20-Tetrathiacalix[4]arene (*Cone*-**4**)

A mixture of 1.00 g (1.38 mmol) of 5,11,17,23-tetra-*tert*-butyl-25,26,27,28-tetrahydroxy-2,8,14,20-tetrathiacalix[4]arene (**3**), 3.06 g (11.11 mmol) of geranyl bromoacetate (**2**), 1.18 g (11.11 mmol) of anhydrous sodium carbonate in 70 mL of dry acetone was refluxed for 100 h. After cooling the precipitate from the reaction mixture was filtered off and washed with 2 × 10 mL acetone. The filtrate was concentrated under reduced pressure to give an oil that was recrystallized from methanol and cooled to −20 °C. The precipitate formed was filtered off and dried in vacuo over P_2O_5. The yield of thiacalix[4]arene *cone*-**4** (white powder) was 0.83 g (40%). Mp: 53–56 °C. ^{1}H NMR (CDC $_3$, δ, ppm, J/Hz): 1.08 (s, 36H, $(CH_3)_3$C); 1.59 (s, 12H, –CH_3); 1.67 (s, 24H, –CH_3); 2.04 (m, 8H, –$\underline{CH_2}$–CH_2–); 2.08 (m, 8H, –CH_2–$\underline{CH_2}$–); 4.65 (d, 8H, –O–$\underline{CH_2}$–CH=, $^3J_{HH}$ = 7.0 Hz); 5.08 (br.t, 4H, =C$\underline{H}$–CH_2–CH_2–, $^3J_{HH}$ = 7.0 Hz); 5.16 (s, 8H, –O–CH_2–); 5.34 (br.t, 4H, =C$\underline{H}$–CH_2–O–, $^3J_{HH}$ = 7.0 Hz); 7.27 (s, 8H, Ar–H). ^{13}C NMR ($CDCl_3$, δ, ppm): 16.64, 25.85, 26.50, 31.29, 39.72, 61.70, 70.85, 118.44, 123.98, 129.50, 134.21. ^{1}H–^{1}H NOESY NMR spectrum (most important cross-peaks are presented): H^{4b}/H^3, H^9/H^{10}, H^9/H^{12}, H^{10}/H^{13}, H^{12}/H^{13}, H^{13}/H^{15}, H^{14}/H^{17}, H^{15}/H^{17}. IR spectrum (ν/cm^{-1}): 2962 (–CH_3), 2911 (–C_{Ph}–H), 1760 (C=O), 1730 (–C(O)OR), 1669 (C=C), 1477 (–CH_2–), 1443 (–CH_3), 1432 (–CH_2–CH=C–), 1382 (–CH_3), 1266 (C=O), 1177 (–C(O)OR), 1098 (–C(O)OR), 1054 (C_{Ph}–O–CH_2–). MS (ESI): calculated [M^+] *m/z* = 1496.8, found $[M + Na]^+$ *m/z* = 1519.9. El. Anal. Calcd for $C_{88}H_{120}O_{12}S_4$ (%): C, 70.55; H, 8.07; S, 8.56. Found (%): C, 70.43; H, 7.94; S, 8.79.

2.2.3. 5,11,17,23-Tetra-*Tert*-Butyl-25,26,27,28-Tetra[(Geranyloxycarbonyl)-Methoxy]-2,8,14,20-Tetrathiacalix[4]arene (*1,3-Alternate*-**5**)

A mixture of 0.50 g (0.69 mmol) of 5,11,17,23-tetra-*tert*-butyl-25,26,27,28-tetrahydroxy-2,8,14,20-tetrathiacalix[4]arene (**3**), 1.53 g (5.55 mmol) of geranyl bromoacetate (**2**), 1.76 g (5.55 mmol) of anhydrous cesium carbonate in 50 mL of dry acetone was refluxed for 30 h. After cooling, the precipitate from the reaction mixture was filtered off and washed with 2 × 10 mL acetone. The filtrate was concentrated under reduced pressure to give an oil that was recrystallized from methanol and cooled to −20 °C. The precipitate formed was filtered off and dried in vacuo over P_2O_5. The yield of thiacalix[4]arene *1,3-alternate*-**5** (white powder) was 0.56 g (54%). Mp: 96–103 °C. ^{1}H NMR ($CDCl_3$, δ, ppm, J/Hz): 1.24 (s, 36H, $(CH_3)_3$C); 1.60 (s, 12H, -CH_3); 1.69 (s, 12H, –CH_3); 1.72 (s, 12 H, –CH_3); 2.06 (m, 8H, –$\underline{CH_2}$–CH_2–); 2.10 (m, 8H, –CH_2–$\underline{CH_2}$–); 4.63 (s, 8H, –O–CH_2–); 4.70 (d, 8H, –O–$\underline{CH_2}$–CH=, $^3J_{HH}$ = 7.0 Hz); 5.09 (br.t, 4H, =C$\underline{H}$–CH_2–CH_2–, $^3J_{HH}$ = 6.4 Hz); 5.37 (br.t, 4H, =C$\underline{H}$–CH_2–O–, $^3J_{HH}$ = 7.0 Hz); 7.51 (s, 8H, Ar–H). ^{13}C NMR ($CDCl_3$, δ, ppm): 16.69, 17.87; 25.86, 26.45, 31.22, 34.31; 39.75, 61.64, 68.44; 118.39, 123.90, 127.78, 132.05; 133.80; 146.31; 157.44; 168.04. ^{1}H–^{1}H NOESY NMR spectrum (most important cross-peaks are presented): $H^3/H^{9'}$, H^{4b}/H^3, $H^{4b}/H^{7'}$, $H^{4b}/H^{9'}$, $H^{4b}/H^{10'}$, $H^{4b}/H^{12'}$, H^9/H^{12}, H^{10}/H^9, H^{10}/H^{13}, H^{10}/H^{14}, H^{12}/H^{13}, H^{14}/H^{17}, H^{15}/H^{13}, H^{15}/H^{14}, H^{15}/H^{17}. IR spectrum (ν/cm^{-1}): 2964 (–CH_3), 2909 (–C_{Ph}–H), 1764 (–C(O)OR), 1676 (C=C), 1479 (–CH_2–), 1443 (–CH_3), 1424 (–CH_2–CH=C–), 1383 (–CH_3), 1269 (C=O), 1184 (–C(O)OR), 1086 (–C(O)OR), 1062 (C_{Ph}–O–CH_2–). MS (ESI): calculated [M^+] *m/z* = 1496.8, found $[M + K]^+$ *m/z* = 1536.8. El. Anal. Calcd for $C_{88}H_{120}O_{12}S_4$ (%): C, 70.55; H, 8.07; S, 8.56. Found (%): C, 70.32; H, 8.05; S, 8.17.

2.3. *Differential Scanning Calorimetry (DSC)*

The DSC experiments were performed using the DSC204 F1 Phoenix differential scanning calorimeter (Netzsch, Selb, Germany) in an argon atmosphere (flow rate 150 mL/min) with the heating/cooling rate of 10 K/min. The cooling–heating cycle was repeated twice. DSC204 F1 Phoenix was calibrated according to the manufacturer's recommendations by measuring six standard compounds

(Hg, In, Sn, Bi, Zn, and CsCl) as described previously [31]. The error in the temperature and enthalpy determination by this technique were 0.1 K and 3%, respectively. The onset temperatures of processes were determined using the DTG curve. Samples of calixarenes with a mass of 1.73 mg (**5**) and 3.36 mg (**4**) were placed in a 40 μL aluminum crucible with a lid containing 0.5 mm diameter hole.

2.4. Fast Scanning Calorimetry (FSC)

FSC experiments were carried out on a FlashDSC 1 instrument (Mettler-Toledo, Greifensee, Switzerland) using one and the same UFS1 calorimetric chip-sensor for all measurements [32]. Prior to the experiments, the sensor was conditioned and corrected according to the manufacturer's recommendations and calibrated using biphenyl and benzoic acid as standards. The measurements were performed at 27 °C sensor-support temperature and in argon dynamic atmosphere at 80 mL/min. The state of the samples was monitored by an optical microscope BXFM (Olympus, Tokyo, Japan), under crossed polarizers.

2.5. Powder X-ray Diffraction (PXRD) Experiment

X-ray powder diffractograms were determined using a MiniFlex 600 diffractometer (Rigaku, Tokyo, Japan) equipped with a D/teX Ultra detector. In this experiment, CuK_α (λ = 1.54178 Å) radiation (30 kV, 10 mA) was used, and K_β radiation was eliminated with a Ni filter. The diffractograms were determined at RT in the reflection mode, with a scanning speed of 5°/min. The clathrate samples were loaded into a glass holder. The patterns were recorded without sample rotation.

2.6. X-ray Diffraction (XRD)

Data sets for single crystals **4** and **5** were collected on a Rigaku XtaLab Synergy S instrument (Rigaku, Tokyo, Japan) with a HyPix detector and a PhotonJet microfocus X-ray tube using Cu K_α (1.54184 Å) radiation at low temperature. Images were indexed and integrated using the CrysAlisPro data reduction package. Data were corrected for systematic errors and absorption using the ABSPACK module: numerical absorption correction based on Gaussian integration over a multifaceted crystal model and empirical absorption correction based on spherical harmonics according to the point group symmetry using equivalent reflections. The GSPF reference application line (GRAL) module was used for analysis of systematic absences and space group determination. The structures were solved by direct methods using SHELXT (University of Göttingen, Göttingen, Germany) [33] and refined by the full-matrix least-squares on F^2 using SHELXL [34]. Non-hydrogen atoms were refined anisotropically. The hydrogen atoms were inserted at the calculated positions and refined as riding atoms. The positions of the hydrogen atoms of methyl groups were found using rotating group refinement with idealized tetrahedral angles. The figures were generated using Mercury 4.1 program. The *t*-Bu fragment in crystals **5** was located with the aid of rigid fragment [35] and subsequently refined with constraints and restraints.

Crystal Data for **4.** $C_{88}H_{120}O_{12}S_4$, M_r = 1498.07, triclinic, *P*-1 (No. 2), a = 13.9848(3) Å, b = 17.1054(4) Å, c = 19.9615(4) Å, α = 107.312(2)°, β = 105.052(2)°, γ = 98.286(2)°, V = 4273.55(18) Å^3, T = 99.97(15) *K*, *Z* = 2, *Z*′ = 1, μ(Cu K_α) = 1.475, 62,647 reflections measured, 17386 unique (R_{int} = 0.0551) which were used in all calculations. The final wR_2 was 0.3254 (all data) and R_1 was 0.0999 ($I > 2(I)$). CCDC number: 2045930.

Crystal Data for **5.** $C_{88}H_{120}O_{12}S_4$, M_r = 1498.07, orthorhombic, *Ccce* (No. 68), a = 20.2109(11) Å, b = 32.012(2) Å, c = 13.3175(7) Å, $\alpha = \beta = \gamma$ = 90°, V = 8616.4(9) Å^3, T = 100.00(10) K, Z = 4, Z′ = 0.25, μ(Cu K_α) = 1.463, 14,969 reflections measured, 4359 unique (R_{int} = 0.0401) which were used in all calculations. The final wR_2 was 0.4798 (all data) and R_1 was 0.1451 ($I > 2(I)$). CCDC number: 2045929.

3. Results

3.1. Synthesis of Thiacalix[4]arene Derivatives

The synthesis of targeted compounds capable of self-organization in monolayers is an important and current challenge of modern organic chemistry, and also has practical application in materials science and nanotechnology. A suitable platform for creating such structures is *p-tert*-butylthiacalix[4]arene [36,37]. The functionalization of this macrocycle by terpenoid fragments allows one to obtain structures with a rigid framework and conformationally flexible substituents, potentially capable of self-organization.

Initially, an alkylating agent geranyl bromoacetate was obtained according to the original method. The desired product was obtained by the reaction of geraniol **1** with bromoacetic acid bromide in chloroform in the presence of diisopropylethylamine (DIPEA) (Scheme 1). Geranyl bromoacetate **2** was synthesized in good 78% yield and purified by column chromatography on silica gel using hexane/propanol-2 at 20:1 ratio as eluent. The structure of geranyl bromoacetate **2** was confirmed by a complex of physical methods, including ^{1}H, ^{13}C NMR, IR spectroscopy and elemental analysis.

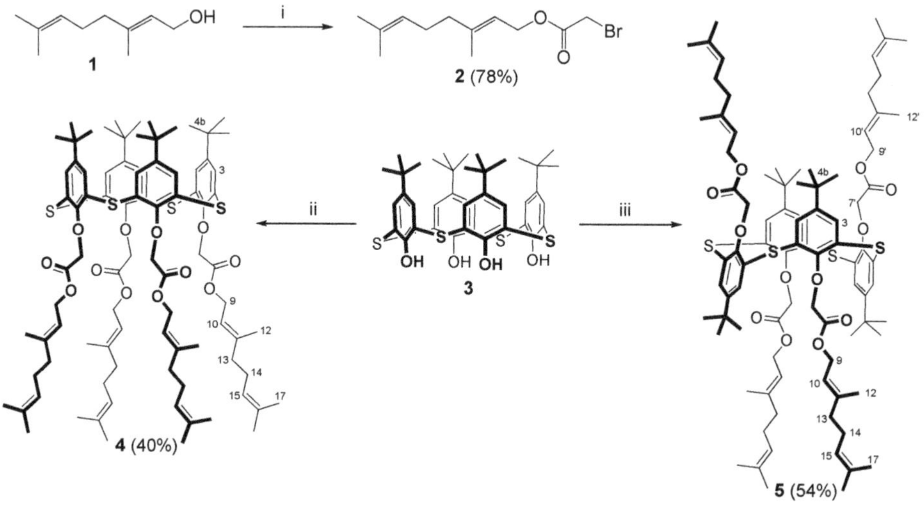

Scheme 1. Synthesis of geranyl bromoacetate **2**, *p-tert*-butylthiacalix[4]arenes **4** and **5**. Reagents and conditions: (i) bromoacetic acid bromide, DIPEA, $CHCl_3$, −5 °C; (ii) compound **2**, Na_2CO_3, acetone; (iii) compound **2**, Cs_2CO_3, acetone.

In order to obtain tetrasubstituted thiacalix[4]arene derivatives containing terpenoid fragments, the interaction of *p-tert*-butylthiacalix[4]arene **3** with geranyl bromoacetate in acetone in the presence of alkali metal (sodium, cesium) carbonates was studied (Scheme 1). The base and solvent were specified in accordance with their efficiency in alkylation of *p-tert*-butylthiacalix[4]arene at the lower rim [38].

As expected, the targeted tetrasubstituted *p-tert*-butylthiacalix[4]arenes in *cone* **4** and *1,3-alternate* **5** conformations were obtained using sodium and cesium carbonates as base, respectively. Low yields (40 and 54%) of the compounds **4** and **5** are concerned with recrystallization and purification steps.

The structure and composition of the new thiacalix[4]arene derivatives **4** and **5** were characterized by ^{1}H, ^{13}C NMR, IR spectroscopy, mass spectrometry (ESI) and elemental analysis. The spatial structure of the synthesized macrocycles was established by 2D NOESY NMR spectroscopy.

In the ^{1}H NMR spectra of the obtained compounds the signals of *tert*-butyl and aromatic protons appear as singlets at 1.08 ppm (macrocycle **4**), 1.24 ppm (macrocycle **5**) and 7.27 ppm (macrocycle **4**), 7.51 ppm (macrocycle **5**), respectively, which indicates the formation of symmetrical products (See Supplementary Materials, Figures S2 and S3). The signals of oxymethylene protons are also observed as singlets in the area of strong fields (4.63 ppm) for *1,3-alternate* **5** stereoisomer compared to *cone* **4** stereoisomer, the signals of which are observed at weaker fields (5.16 ppm). Apparently, this is due to the shielding of oxymethylene protons by aromatic fragments of the macrocycle in *1,3-alternate* conformation. The signals of the methyl protons of the substituent are observed as singlets, the signals of the methylene protons of the substituent are observed as multiplets, and the signals of the methine protons at the double bond are observed as a broadened triplet. Chemical shifts, multiplicity, and integrated intensity of proton signals in the ^{1}H NMR spectrum are in good agreement with the proposed structures of *p-tert*-butylthiacalix[4]arenes.

The spatial structure of the compounds **4** and **5** was studied by two-dimensional NMR ^{1}H–^{1}H NOESY spectroscopy (See Supplementary Materials, Figures S7 and S8). The absence in the 2D NMR NOESY spectrum of the compound **4** of cross peaks due to the dipole-dipole interaction of the *tert*-butyl group protons with oxymethylene protons, as well as cross peaks between the aromatic protons of the macrocycle and oxymethylene protons, confirms that macrocycle **4** is in *cone* conformation (See Supplementary Materials, Figure S7). The presence of cross peaks in the 2D NMR NOESY spectrum of compound **5** due to the dipole-dipole interaction of the protons of the *tert*-butyl group with methyl, oxymethylene and methine protons at the double bond, as well as cross peaks between the aromatic protons of the macrocycle and oxymethylene protons, confirms that the macrocycle **5** is in *1,3-alternate* conformation (See Supplementary Materials, Figure S8).

In the IR spectra of the obtained compounds **4** and **5** (See Supplementary Materials, Figures S12 and S13), an absorption band of stretching vibrations of the ether group (ν, 1750–1790 cm^{-1}), the carbonyl group (ν, 1260–1269 cm^{-1}) and double bonds (ν, 1669–1676 cm^{-1}) were observed. They were absent in the IR spectrum of the initial thiacalix[4]arene **3**. In the IR spectra of thiacalix[4]arenes **4** and **5** the absorption bands of valence (ν, 2962–2964 cm^{-1}) and deformation vibrations of the methyl group (δ_{as}, 1443; δ_s, 1382–1383 cm^{-1}) were observed. The absence of the absorption band of stretching vibrations of the hydroxyl group in the IR spectra of compounds **4** and **5** indicates the complete substitution of the initial *p-tert*-butylthiacalix[4]arene **3**.

The spatial structure of the obtained products **4** and **5** was fully confirmed using X-ray (Figures 2 and 3). In both cases, suitable single crystals were obtained from a DMSO solution. The symmetry group of product **4** is *P*-1 (Figure 2). As can be seen from Figure 2B, in the case of compound **4**, the formation of a zigzag close packing is observed, and in one layer the orientation of the molecules in one direction "head to tail", and in the next layer the orientation is opposite to "tail to head". The distance between these layers is 8 Å. Due to this, the layers are superimposed on each other in a zigzag manner, forming a close packing. The distance between the layers has a subnanometer size.

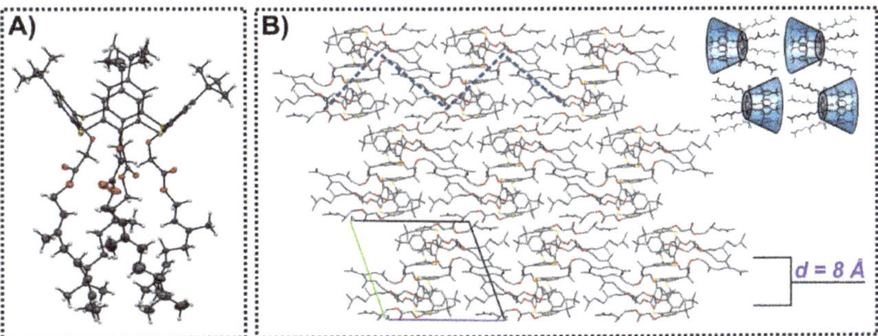

Figure 2. X-ray structure of **4** (**A**) showing *cone* conformation (gray—carbon atoms, white—hydrogen atoms, yellow—sulfur atoms, red—oxygen atoms) and crystal packing (the parallelogram is the elementary cell, and the dashed line represents the zigzag packing) (**B**). H atoms and disorder are not presented for clarity in (**B**).

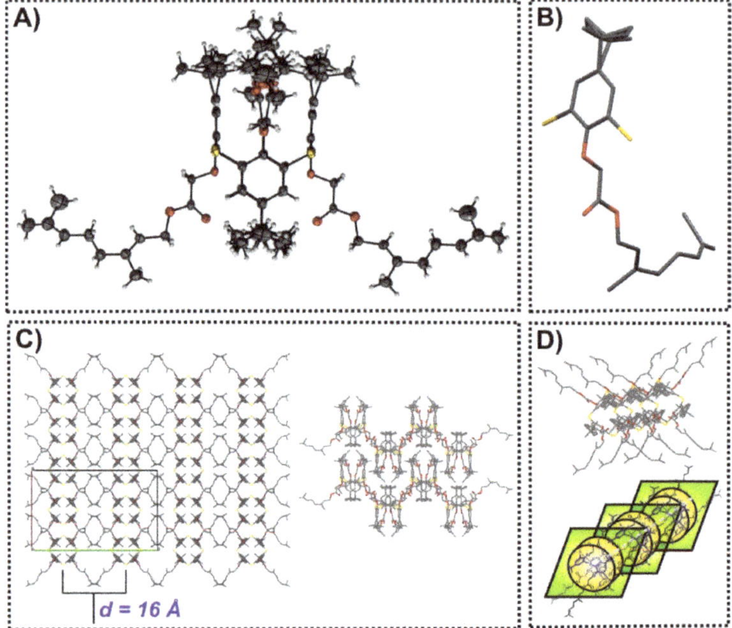

Figure 3. X-ray structure of **5** (**A**) showing the *1,3-alternate* conformation (gray—carbon atoms, white—hydrogen atoms, yellow—sulfur atoms, red—oxygen atoms), asymmetric unit (**B**), crystal packing (the rectangle is the elementary cell) (**C**), stack of molecules in crystals **5** (**D**). H atoms and disorder are not presented for clarity in (**B–D**).

The symmetry group of product **5** is *Ccce* (Figure 3). As can be seen from Figure 3C, in the case of compound **5**, the formation of strictly ordered nanostructured layers is observed, the distance between which is equal to half of the axis *b*, which is approximately 16 Å. The elementary cell is a rectangle. Based on these X-ray analysis, the structure of thiacalix[4]arene **5** in the *1,3-alternate* conformation is characterized by high-order symmetry. Moreover, the fragment shown in Figure 3B is independent, all the others are symmetrical to it.

3.2. DSC Analysis

Since macromolecular structures stabilized by the interaction of many weak forces undergo conformational or phase transitions upon heating, significant information about these structures can be obtained using DSC analysis. The thermal properties of thiacalix[4]arenes **4** and **5** obtained were studied by DSC, Figure 4. The results obtained indicate different thermal properties of the studied thiacalixarenes. Macrocycle **4** does not show any polymorphic transitions during the first heating up to melting point at 52.6 °C, Figure 4A. The enthalpy of melting is 27.3 J/g.

The DSC curve (first heating) obtained for the thiacalixarene **5** has a more complex shape. Above 91 °C, the first endothermic process begins, on which the second endothermic process is overlapped. The total enthalpy of the two processes is 39 J/g. Such behavior is characteristic of calixarenes capable of polymorphism [30,39], and may be due to the presence of two polymorphic forms of **5**. The studied calixarenes do not crystallize on cooling the melts. On reheating, thiacalixarene **4** retains a supercooled liquid state, Figure 4A, while thiacalixarene **5** undergoes several transformations, Figure 4B. The DSC curve at 56 °C exhibits an exothermic effect with $\Delta H = 21.9$ J/g, which corresponds to the phase transition of calixarene from the amorphous state to the crystalline state (cold crystallization). The resulting crystalline phase melts at 95 °C with $\Delta H = 24.9$ J/g. The temperatures of the peaks T_p

and the end T_f of melting in the first and second heating runs are almost identical: T_{1p} = 105.9 °C, T_{2p} = 106.4 °C and T_{1f} = 109.1 °C, T_{2f} = 108.7 °C. Thus, last endothermic effect corresponds to the melting of the same polymorphic modification of thiacalixarene. The lower value of the enthalpy of melting on the second heating can be associated with the partial melting of calixarene at a lower temperature due to the heat of cold crystallization. The overlap of two opposite processes also explains the lower value of the enthalpy of cold crystallization in comparison with the enthalpy of melting during the first heating.

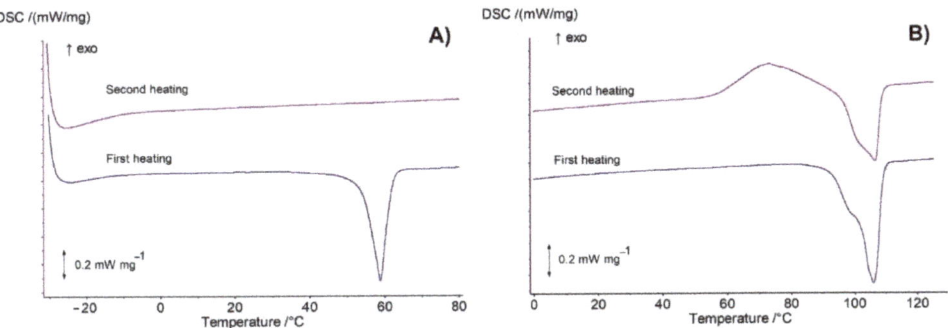

Figure 4. Data of DSC analysis for the thiacalix[4]arenes **4** (**A**) and **5** (**B**).

3.3. PXRD Data

The complex shape of the DSC curve of thiacalix[4]arene **5** indicates the presence of several polymorphic modifications of the macrocycle. To confirm the presence of polymorphs, the method of powder X-ray diffraction analysis was applied (Figure 5).

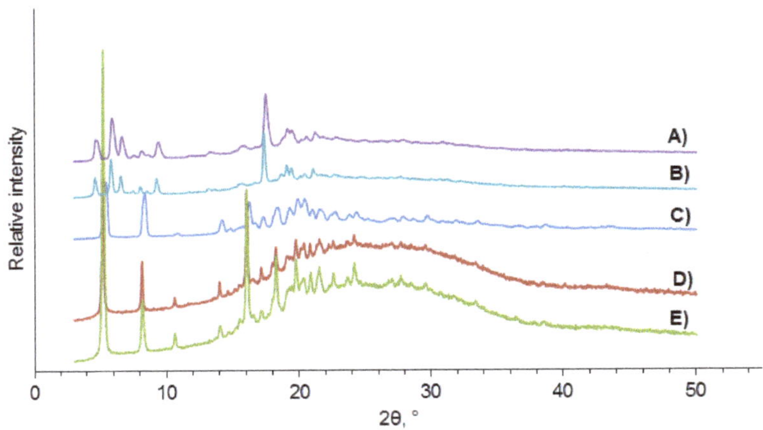

Figure 5. PXRD data for the (**A**) initial form of **4**, (**B**) thiacalix[4]arene **4** melted at 60 °C, (**C**) initial form of **5**, (**D**) thiacalix[4]arene **5** after melting and cooling, (**E**) thiacalix[4]arene **5** reheated to 80 °C.

According to PXRD data (Figure 5A,B), macrocycle **4** does not undergo any conformational transformations after melting of the sample (at 60 °C). This is in good agreement with the DSC data, where no polymorphic forms of thiacalix[4]arene **4** were observed. In the case of thiacalix[4]arene **5** to control the temperature and phase state the sample was heated and cooled using DSC, then the same sample was reheated to 80 °C. The resulting samples were analyzed by PXRD (Figure 5D,E).

Diffraction patterns 5c and 5d differ from each other at an angle of $2\theta = 20°$, which confirms the presence of a polymorphic transition of macrocycle **5** according to DSC data. In this case, after cooling and reheating to 80 °C (Figure 5E), the diffraction pattern remains identical. Apparently, this is due to the formation of a stable polymorph of macrocycle **5** after the melting of the sample.

3.4. Fast Scanning Calorimetry

Compared to the conventional DSC, FSC provides faster temperature scanning rates both on cooling and heating and uses much smaller samples (typically 10–1000 ng). The latter trait is of particular interest in the investigation of the compounds with polymorphism [39].

The quality of the calorimetric curve in FSC greatly depends on the thermal contact between the sample and the chip. Typically, the studied compound will be "premelted," i.e., melted and crystallized so that the resulting crystal has a good and uniform contact surface with the chip [40]. This approach may not be suitable for the compounds with polymorph transitions because the melting may destroy the polymorph of interest, and another, more stable of faster forming polymorph may be produced upon cooling. Taking the above considerations into account, the experiments were performed as follows.

In the first step of the investigation, the scans of the powders after synthesis were performed for both studied compounds (Figure 6A,B).

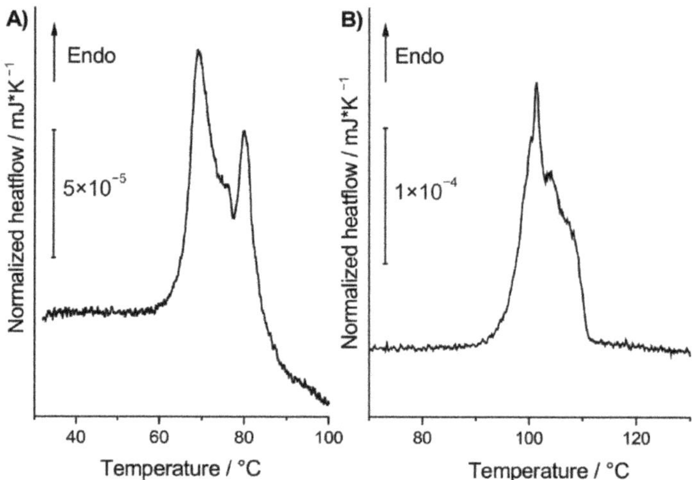

Figure 6. Heating scans of compound **4** (**A**) and compound **5** (**B**) in powder form without premelting. 100 K/s heating rate.

The endotherm of melting is visible in scans of both compounds. The irregular shape of the endotherm can be the result of the poor thermal contact between the sample and the chip, as well as the presence of several polymorph forms at the same time. No thermal effects were visible on the cooling scans and on the second heating scans in both cases.

Next, the samples were crystallized by adding a speckle of the crystals of the respective compound to the amorphous material produced in the previous step of research. The growth of the crystalline phase was monitored by the polarized optical microscopy (POM). The FSC scans of the developed crystals are shown in Figure 7A,B.

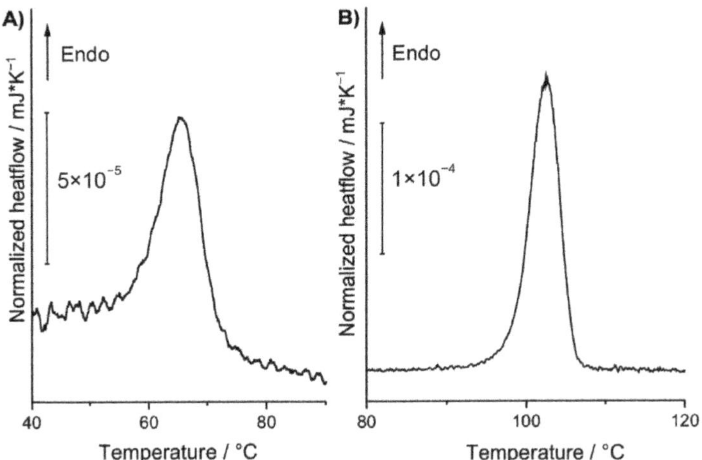

Figure 7. Heating scans of compound **4** (**A**) and compound **5** (**B**) crystallized on the chip-sensor. 100 K/s heating rate.

As seen on the graphs, the melting endotherms have a regular shape with the onset temperature, which is in agreement with the DSC results. However, in the case of compound **5**, a complex melting effect was seen in some scans (See Supplementary Materials, Figure S14). This appears to be stochastic and can be a result of the presence of different polymorphs in the recrystallized sample. Thus, the added seeds may promote growth of different polymorphs simultaneously.

We have also performed FSC scans of the crystals grown from DMSO solutions. Prior to scanning, the solvent was evaporated at 45 °C, which is sufficient due to the high surface to volume ratio of the samples used for FSC. Full evaporation of the solvent was confirmed by monitoring the repeatability of the heat flow in **5** heating and cooling scans between 30 to 45 °C. The final FSC melting scans of the crystals grown from the DMSO solution are shown in Figure 8A,B.

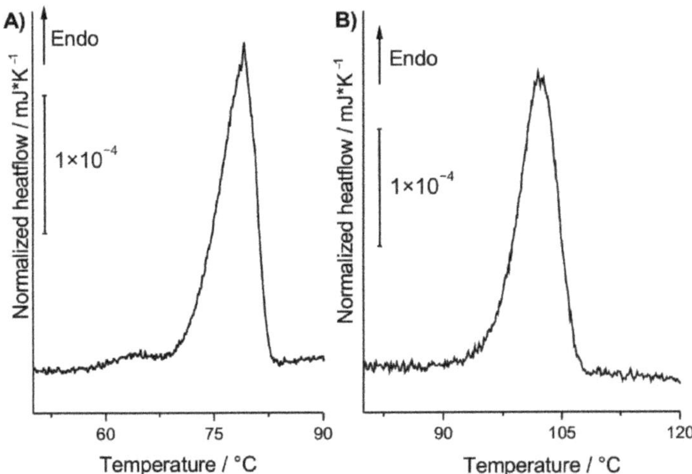

Figure 8. Heating scans of compound **4** (**A**) and compound **5** (**B**) crystallized from DMSO. 100 K/s heating rate.

The melting effect of compound **5** (Figure 8B) is in agreement with the result of the experiments with the recrystallized samples, while in the case of the compound 4, the melting effect is visible at a much higher temperature (Figure 8A). A small peak at lower temperature, which is close to the position of the peak of the recrystallized sample (Figure 7A), is also visible. Thus, compound 4 is also able to form different crystal polymorphs. The formation of the more stable polymorph is promoted by DMSO.

The geranyl substituents are located on the one side of the macrocyclic ring in *cone* stereoisomer **4** and on the opposite sides of the macrocycle in *1,3-alternate* stereoisomer **5**. Like that, in the case of *cone*-**4**, four geranyl substituents are spatially separated from *tert*-butyl fragments, and in the case of *1,3-alternate*-**5**, geranyl fragments alternate with bulky *tert*-butyl groups. As a result, the compounds **4** and **5** have different symmetry of the molecules and different crystalline packaging that provide the formation of different crystalline polymorphs. The above description is in good agreement with X-Ray data. In crystalline form of *cone*-**4** the orientation of the molecules of one layer in one direction "head to tail" alternates with the orientation of next layer "tail to head". Contrary, the structure of *1,3-alternate*-**5** is characterized by high-order symmetry and the molecules overlap each other like "stacks" to form a channel. Since the structures of the synthesized compounds contain geranyl fragments capable of conformational rotations and, consequently, changes in their packing, it is obvious that the obtained thiacalixarenes are capable of the existence different polymorphic forms. It was shown by the results of FSC scans that both compounds can form different crystal polymorphs. The bulky nature of the substituents results in a slow crystallization process, which prevents monotropic polymorph transitions or melting-crystallization-melting effects as seen in faster crystallizing materials [41].

4. Conclusions

Thus, for the first time, an approach for the preparation of 2D nanostructures in crystal based on novel thiacalix[4]arene derivatives functionalized at the lower rim with geranyl fragments in *cone* and *1,3-alternate* conformations was developed. X-ray diffraction analysis showed the formation of 2D monomolecular-layer nanosheets from the thiacalix[4]arenes synthesized. It was found that the distance between the layers in the case of *1,3-alternate* stereoisomer has a nanometer size (16 Å), and in the case of *cone* stereoisomer subnanometer size (8 Å). It was established by DSC, PXRD and FSC methods that the obtained macrocycles are capable of forming different crystalline polymorphs. It was shown that DMSO promotes the formation of a more stable polymorph in the case of thiacalix[4]arene with geranyl fragments in *cone* conformation. However, in the case of *1,3-alternate* stereoisomer, a similar stable polymorph is formed in a single crystal (recrystallization from DMSO) and in crystals obtained by melt crystallization. Finding possible polymorphs is important for pharmaceuticals, high-energy materials, dyes and pigments. Therefore, the obtained stable polymorphs of thiacalix[4]arenes synthesized, which form an ordered packing of monomolecular 2D nanosheets, can find application in nanomaterials science and medicine for the modern development of pharmaceuticals and new generation materials.

Supplementary Materials: The following are available online at http://www.mdpi.com/2079-4991/10/12/2505/s1. Figure S1: ^{1}H NMR spectrum of geranyl bromoacetate **2** ($CDCl_3$, 298 K, 400 MHz), Figure S2: ^{1}H NMR spectrum of thiacalix[4]arene **4** in *cone* conformation ($CDCl_3$, 298 K, 400 MHz), Figure S3: ^{1}H NMR spectrum of thiacalix[4]arene **5** in *1,3-alternate* conformation ($CDCl_3$, 298 K, 400 MHz), Figure S4: ^{13}C NMR spectrum of geranyl bromoacetate **2** ($CDCl_3$, 298 K, 100 MHz), Figure S5: ^{13}C NMR spectrum of thiacalix[4]arene **4** in *cone* conformation ($CDCl_3$, 298 K, 100 MHz), Figure S6: ^{13}C NMR spectrum of thiacalix[4]arene **5** in *1,3-alternate* conformation ($CDCl_3$, 298 K, 100 MHz), Figure S7: 2D NMR NOESY ^{1}H-^{1}H spectrum of thiacalix[4]arene **4** in *cone* conformation ($CDCl_3$, 298 K, 400 MHz), Figure S8: 2D NMR NOESY ^{1}H-^{1}H spectrum of thiacalix[4]arene **5** in *1,3-alternate* conformation ($CDCl_3$, 298 K, 400 MHz), Figure S9: Mass spectrum ESI of thiacalix[4]arene **4** in *cone* conformation, Figure S10: Mass spectrum ESI of thiacalix[4]arene **5** in *1,3-alternate* conformation, Figure S11: ATR FTIR spectrum of geranyl bromoacetate **2**, Figure S12: ATR FTIR spectrum of thiacalix[4]arene **4** in *cone* conformation, Figure S13: ATR FTIR spectrum of thiacalix[4]arene **5** in *1,3-alternate* conformation, Figure S14: Data of FSC analysis: Heating scans of different samples of compound **5** crystallized on the chip-sensor. 100 K/s heating rate, Figure S15: The images of microcrystalline sample **4** (*cone*) under polarized light at heating to 53 °C (A), 60 °C (B) and 63 °C (C), Figure S16:

The images of microcrystalline sample **5** (*1,3-alternate*) under polarized light at heating to 95 °C (A), 106 °C (B) and 112 °C (C).

Author Contributions: Conceptualization, I.I.S. and A.A.V.; methodology, P.L.P.; software, P.L.P., D.R.I., K.S.U.; formal analysis, P.L.P., M.A.Z., T.A.M.; investigation, M.A.Z., T.A.M., A.V.B., A.E.B.; resources, A.E.B. and A.V.B.; writing—original draft preparation, A.A.V.; writing—review and editing, A.A.V., P.L.P., I.I.S.; visualization, A.A.V.; supervision, A.A.V. and I.I.S.; project administration, A.A.V.; funding acquisition, A.A.V., P.L.P., I.I.S. All authors have read and agreed to the published version of the manuscript.

Funding: The work was supported by the grants of the President of the Russian Federation for state support of young Russian scientists—candidates of sciences (MK-12.2020.3, MK-1279.2020.3). The investigation of spatial structure of the compounds by NMR spectroscopy was carried out within the framework of the grant of the President of the Russian Federation for state support of leading scientific schools of the Russian Federation (NSh-2499.2020.3).

Acknowledgments: The authors are grateful to the C. Schick in discussing the DSC and FSC results.

Conflicts of Interest: The authors declare no conflict of interest.

References

1. Huang, Y.; Liang, J.; Wang, C.; Yin, S.; Fu, W.; Zhu, H.; Wan, C. Hybrid superlattices of two-dimensional materials and organics. *Chem. Soc. Rev.* **2020**, *49*, 6866–6883. [CrossRef] [PubMed]
2. Xiong, P.; Ma, R.; Sakai, N.; Nurdiwijayanto, L.; Sasaki, T. Unilamellar metallic MoS_2/graphene superlattice for efficient sodium storage and hydrogen evolution. *ACS Energy Lett.* **2018**, *3*, 997–1005. [CrossRef]
3. Zhao, C.; Zhang, P.; Zhou, J.; Qi, S.; Yamauchi, Y.; Shi, R.; Fang, R.; Ishida, Y.; Wang, S.; Tomsia, A.P.; et al. Layered nanocomposites by shear-flow-induced alignment of nanosheets. *Nature* **2020**, *580*, 210–215. [CrossRef] [PubMed]
4. Kempt, R.; Kuc, A.; Han, J.H.; Cheon, J.; Heine, T. 2D Crystals in Three Dimensions: Electronic Decoupling of Single-Layered Platelets in Colloidal Nanoparticles. *Small* **2018**, *14*, 1803910. [CrossRef] [PubMed]
5. Miró, P.; Audiffred, M.; Heine, T. An atlas of two-dimensional materials. *Chem. Soc. Rev.* **2014**, *43*, 6537–6554. [CrossRef] [PubMed]
6. Lee, A.Y.; Erdemir, D.; Myerson, A.S. Crystal polymorphism in chemical process development. *Annu. Rev. Chem. Biomol. Eng.* **2011**, *2*, 259–280. [CrossRef]
7. Bernstein, J. *Polymorphism in Molecular Crystals*, 2nd ed.; Oxford University Press: Oxford, UK, 2020; Volume 30, p. 608.
8. Lee, E.H. A practical guide to pharmaceutical polymorph screening & selection. *Asian J. Pharm. Sci.* **2014**, *9*, 163–175. [CrossRef]
9. Aitipamula, S. Polymorphism in molecular crystals and cocrystals. In *Advances in Organic Crystal Chemistry*, 1st ed.; Miyata, M., Tamura, R., Eds.; Springer: Tokyo, Japan, 2015; pp. 265–298.
10. Möhwald, H. Phospholipid Monolayers. In *Handbook of Biological Physics—Structure and Dynamics of Membranes*, 1st ed.; Lipowsky, R., Sackman, E., Eds.; Elsevier: Amsterdam, Netherlands, 1995; Volume 1A, pp. 161–211.
11. Santos, H.A.; García-Morales, V.; Pereira, C.M. Electrochemical properties of phospholipid monolayers at liquid–liquid interfaces. *Chem. Phys. Chem.* **2010**, *11*, 28–41. [CrossRef]
12. Pichot, R.; Watson, R.L.; Norton, I.T. Phospholipids at the interface: Current trends and challenges. *Int. J. Mol. Sci.* **2013**, *14*, 11767–11794. [CrossRef]
13. Ariga, K.; Yuki, H.; Kikuchi, J.I.; Dannemuller, O.; Albrecht-Gary, A.M.; Nakatani, Y.; Ourisson, G. Monolayer studies of single-chain polyprenyl phosphates. *Langmuir* **2005**, *21*, 4578–4583. [CrossRef]
14. Harper, J.K.; Grant, D.M. Solid-state ^{13}C chemical shift tensors in terpenes. 3. Structural characterization of polymorphous verbenol. *J. Am. Chem. Soc.* **2000**, *122*, 3708–3714. [CrossRef]
15. Corvis, Y.; Négrier, P.; Massip, S.; Leger, J.M.; Espeau, P. Insights into the crystal structure, polymorphism and thermal behavior of menthol optical isomers and racemates. *CrystEngComm* **2012**, *14*, 7055–7064. [CrossRef]
16. Pozzi, G.; Birault, V.; Werner, B.; Dannenmuller, O.; Nakatani, Y.; Ourisson, G.; Terakawa, S. Single-Chain Polyprenyl Phosphates Form "Primitive" Membranes. *Angew. Chem. Int. Ed.* **1996**, *35*, 177–180. [CrossRef]

17. Streiff, S.; Ribeiro, N.; Wu, Z.; Gumienna-Kontecka, E.; Elhabiri, M.; Albrecht-Gary, A.M.; Ourisson, G.; Nakatani, Y. "Primitive" membrane from polyprenyl phosphates and polyprenyl alcohols. *Chem. Biol.* **2007**, *14*, 313–319. [CrossRef] [PubMed]
18. Aslanov, L.A.; Fetisov, G.V.; Paseshnichenko, K.A.; Dunaev, S.F. Liquid phase methods for design and engineering of two-dimensional nanocrystals. *Coord. Chem. Rev.* **2017**, *352*, 220–248. [CrossRef]
19. Stoikov, I.I.; Yantemirova, A.A.; Nosov, R.V.; Rizvanov, I.K.; Julmetov, A.R.; Klochkov, V.V.; Antipin, I.S.; Konovalov, A.I.; Zharov, I. *p-tert*-Butyl thiacalix[4]arenes functionalized at the lower rim by amide, hydroxyl and ester groups as anion receptors. *Org. Biomol. Chem.* **2011**, *9*, 3225–3234. [CrossRef] [PubMed]
20. Stoikov, I.I.; Yantemirova, A.A.; Nosov, R.V.; Julmetov, A.R.; Klochkov, V.V.; Antipin, I.S.; Konovalov, A.I. Chemo-and stereocontrolled alkylation of 1, 2-disubstituted at the lower rim 1,2-alternate *p-tert*-butylthiacalix[4]arene. *Mendeleev Commun.* **2011**, *21*, 41–43. [CrossRef]
21. Stoikov, I.I.; Mostovaya, O.A.; Yantemirova, A.A.; Antipin, I.S.; Konovalov, A.I. Phosphorylation of *p-tert*-butyl (thia)calixarenes by ethylene chlorophosphite. *Mendeleev Commun.* **2012**, *22*, 21–22. [CrossRef]
22. Vavilova, A.A.; Stoikov, I.I. *p-tert*-Butylthiacalix[4]arenes functionalized by *N*-(4'-nitrophenyl) acetamide and *N,N*-diethylacetamide fragments: Synthesis and binding of anionic guests. *Beilstein J. Org. Chem.* **2017**, *13*, 1940–1949. [CrossRef]
23. Mostovaya, O.A.; Padnya, P.L.; Vavilova, A.A.; Shurpik, D.N.; Khairutdinov, B.I.; Mukhametzyanov, T.A.; Khannanov, A.A.; Kutyreva, M.P.; Stoikov, I.I. Tetracarboxylic acids on a thiacalixarene scaffold: Synthesis and binding of dopamine hydrochloride. *New J. Chem.* **2018**, *42*, 177–183. [CrossRef]
24. Vavilova, A.A.; Gorbachuk, V.V.; Shurpik, D.N.; Gerasimov, A.V.; Yakimova, L.S.; Padnya, P.L.; Stoikov, I.I. Synthesis, self-assembly and the effect of the macrocyclic platform on thermal properties of lactic acid oligomer modified by *p-tert*-butylthiacalix[4]arene. *J. Mol. Liq.* **2019**, *281*, 243–251. [CrossRef]
25. Yakimova, L.S.; Nugmanova, A.R.; Mostovaya, O.A.; Vavilova, A.A.; Shurpik, D.N.; Mukhametzyanov, T.A.; Stoikov, I.I. Nanostructured Polyelectrolyte Complexes Based on Water-Soluble Thiacalix[4]Arene and Pillar[5]Arene: Self-Assembly in Micelleplexes and Polyplexes at Packaging DNA. *Nanomaterials* **2020**, *10*, 777. [CrossRef] [PubMed]
26. Yuskova, E.A.; Ignacio-de Leon, P.A.A.; Khabibullin, A.; Stoikov, I.I.; Zharov, I. Silica nanoparticles surface-modified with thiacalixarenes selectively adsorb oligonucleotides and proteins. *J. Nanopart. Res.* **2013**, *15*, 2012. [CrossRef]
27. Mostovaya, O.A.; Gorbachuk, V.V.; Padnya, P.L.; Vavilova, A.A.; Evtugyn, G.A.; Stoikov, I.I. Modification of Oligo- and Polylactides With Macrocyclic Fragments: Synthesis and Properties. *Front. Chem.* **2019**, *7*, 554. [CrossRef]
28. Rodik, R.V.; Boyko, V.I.; Kalchenko, V.I. Calixarenes in bio-medical researches. *Curr. Med. Chem.* **2009**, *16*, 1630–1655. [CrossRef]
29. Schühle, D.T.; Peters, J.A.; Schatz, J. Metal binding calixarenes with potential biomimetic and biomedical applications. *Coord. Chem. Rev.* **2011**, *255*, 2727–2745. [CrossRef]
30. Gataullina, K.V.; Ziganshin, M.A.; Stoikov, I.I.; Klimovitskii, A.E.; Gubaidullin, A.T.; Suwińska, K.; Gorbatchuk, V.V. Smart polymorphism of thiacalix[4]arene with long-chain amide containing substituents. *Cryst. Growth Des.* **2017**, *17*, 3512–3527. [CrossRef]
31. Frolov, I.N.; Okhotnikova, E.S.; Ziganshin, M.A.; Firsin, A.A. Interpretation of Double-Peak Endotherm on DSC Heating Curves of Bitumen. *Energ. Fuel.* **2020**, *34*, 3960–3968. [CrossRef]
32. Mathot, V.; Pyda, M.; Pijpers, T.; Poel, G.V.; Van de Kerkhof, E.; Van Herwaarden, S.; Van Herwaarden, F.; Leenaers, A. The Flash DSC 1, a power compensation twin-type, chip-based fast scanning calorimeter (FSC): First findings on polymers. *Thermochim. Acta* **2011**, *522*, 36–45. [CrossRef]
33. Sheldrick, G.M. SHELXT: Integrating space group determination and structure solution. *Acta Crystallogr.* **2015**, *71*, 3–8. [CrossRef]
34. Sheldrick, G.M. A Short History of SHELX. *Acta Crystallogr.* **2007**, *64*, 112–122. [CrossRef] [PubMed]
35. Guzei, I.A. An idealized molecular geometry library for refinement of poorly behaved molecular fragments with constraints. *J. Appl. Crystallogr.* **2014**, *47*, 806–809. [CrossRef]
36. Lhotak, P. Chemistry of thiacalixarenes. *Eur. J. Org. Chem.* **2004**, *8*, 1675–1692. [CrossRef]
37. Morohashi, N.; Narumi, F.; Iki, N.; Hattori, T.; Miyano, S. Thiacalixarenes. *Chem. Rev.* **2006**, *106*, 5291–5316. [CrossRef] [PubMed]

38. Iki, N.; Narumi, F.; Fujimoto, T.; Morohashi, N.; Miyano, S. Selective synthesis of three conformational isomers of tetrakis[(ethoxycarbonyl)methoxy]thiacalix[4]arene and their complexation properties towards alkali metal ions. *J. Chem. Soc. Perkin Trans.* **1998**, *12*, 2745–2750. [CrossRef]
39. Gataullina, K.V.; Buzyurov, A.V.; Ziganshin, M.A.; Padnya, P.L.; Stoikov, I.I.; Schick, C.; Gorbatchuk, V.V. Using fast scanning calorimetry to detect guest-induced polymorphism by irreversible phase transitions in the nanogram scale. *CrystEngComm* **2019**, *21*, 1034–1041. [CrossRef]
40. Abdelaziz, A.; Zaitsau, D.H.; Mukhametzyanov, T.A.; Solomonov, B.N.; Cebe, P.; Verevkin, S.P.; Schick, C. Melting temperature and heat of fusion of cytosine revealed from fast scanning calorimetry. *Thermochim. Acta* **2017**, *657*, 47–55. [CrossRef]
41. Perrenot, B.; Widmann, G. Polymorphism by differential scanning calorimetry. *Thermochim. Acta* **1994**, *234*, 31–39. [CrossRef]

Publisher's Note: MDPI stays neutral with regard to jurisdictional claims in published maps and institutional affiliations.

MDPI
St. Alban-Anlage 66
4052 Basel
Switzerland
Tel. +41 61 683 77 34
Fax +41 61 302 89 18
www.mdpi.com

Nanomaterials Editorial Office
E-mail: nanomaterials@mdpi.com
www.mdpi.com/journal/nanomaterials

www.ingramcontent.com/pod-product-compliance
Lightning Source LLC
LaVergne TN
LVHW070201170726
843515LV00007B/1976

* 9 7 8 3 0 3 6 5 4 7 6 4 0 *